Hermann Haken

Laser Theory

With 72 Figures

Springer-Verlag Berlin Heidelberg New York 1983

Professor Dr. Dr. h.c. HERMANN HAKEN
Universität Stuttgart, Institut für Theoretische Physik, Pfaffenwaldring 57/IV
D-7000 Stuttgart 80

This book originally appeared in hardcover as Volume XXV/2c of
Encyclopedia of Physics
© by Springer-Verlag Berlin, Heidelberg 1970
ISBN 3-540-04856-1 Springer-Verlag Berlin Heidelberg New York
ISBN 0-387-04856-1 Springer-Verlag New York Heidelberg Berlin

ISBN 3-540-12188-9 Springer-Verlag Berlin Heidelberg New York
ISBN 0-387-12188-9 Springer-Verlag New York Heidelberg Berlin

Library of Congress Cataloging in Publication Data. Haken, H. Laser theory. "Originally appeared in hardcover as volume XXV/2c of Encyclopedia of physics"-Verso of t.p. Includes index. 1. Lasers. I. Title.
QC688.H34 1983 535.5'8 83-458

2153/3130-543210

Dedicated to the memory of my parents.

H. HAKEN.

Foreword.

This book, written by one of the pioneers of laser theory, is now considered a classic by many laser physicists. Originally published in the prestigious *Encyclopedia of Physics* series, it is now being republished in paperback to make it available not only to professors and scientists, but also to students. It presents a thorough treatment of the theory of laser resonators, the quantum theory of coherence, and the quantization of electromagnetic fields. Especial emphasis is placed on the quantum-mechanical treatment of laser light by means of quantum-mechanical Langevin equations, the density matrix equation, and the Fokker-Planck equation. The semiclassical approach and the rate equation approach are also presented. The principles underlying these approaches are used to derive the relevant equations, from which, in turn, the various properties of laser light are derived.

Preface.

The concept of the laser came into existence more than a decade ago when SCHAWLOW and TOWNES showed that the maser principle could be extended to the optical region. Since then this field has developed at an incredible pace which hardly anybody could have foreseen. The laser turned out to be a meeting place for such different disciplines as optics (e.g. spectroscopy), optical pumping, radio engineering, solid state physics, gas discharge physics and many other fields.

The underlying structure of the laser theory is rather simple. The main questions are: what are the light intensities (a), what are the frequencies (b), what fluctuations occur (c), or, in other words, what are the coherence properties. Roughly speaking these questions are treated by means of the rate equations (a), the semiclassical equations (b), and the fully quantum mechanical equations (c), respectively. The corresponding chapters are written in such a way that they can be read independently from each other. For more details about how to proceed, the reader is advised to consult Chap. I.4.

When a theoretical physicist tries to answer the above questions in detail and in a satisfactory way he will find that the laser is a fascinating subject from whatever viewpoint it is treated. Indeed, mathematical methods from such different fields as resonator theory, nonlinear circuit theory or nonlinear wave theory, quantum theory including quantum electrodynamics, spin resonance theory and quantum statistics had to be applied or were even newly developed for the laser, e.g. several methods in quantum statistics applicable to systems far from thermal equilibrium. A number of these concepts and methods can certainly be used in other branches of physics, such as nonlinear optics, nonlinear spin wave theory, tunnel diodes, Josephson junctions, phase transitions etc. Thus it is hoped that physicists working in those fields, too, will find the present article useful.

I became acquainted with the theoretical problems of the laser during a stay at the Bell Telephone Laboratories in spring and summer 1960, shortly before the first laser was made to work.

I am grateful to Prof. WOLFGANG KAISER who drew my attention to this problem and with whom I had the first discussions on this subject.

The main part of the present article had been completed in 1966, when I became ill. I have used the delay to include a number of topics which have developed in the meantime, e.g. the Fokker-Planck equation referring to quantum systems and the theory of ultrashort pulses.

I am indebted to my colleagues, co-workers and students for many stimulating discussions, in particular to my friend and colleague, W. WEIDLICH. The manuscript has been read critically and checked by several of them, and I owe thanks besides to H. GEFFERS, U. GNUTZMANN, R. GRAHAM, F. HAAKE, Mrs. HÜBNER-PELIKAN, K. KAUFMANN, P. REINEKER, H. RISKEN, H. SAUERMANN, C. SCHMID, H. D. VOLLMER and K. ZEILE. In addition, several of them made a series of valuable suggestions for improving the manuscript, in particular H. RISKEN and H. D. VOLLMER.

The manuscript would never have been completed, however, without the tireless assistance of my secretary, Mrs. U. FUNKE, who not only typed several versions of it with great patience, but also prepared the final form in a perfect way.

Stuttgart, February, 1969. H. HAKEN.

Contents.

Laser Theory.

By

H. HAKEN.

With 72 Figures.

I. Introduction.

I.1. The maser principle. Coherent electromagnetic waves can be generated in the radio frequency range both by classical sources (moving charges) and by the maser* principle, but their generation in the optical range is only possible by the laser** principle. The laser represents an extension (proposed by SCHAWLOW and TOWNES[1] in 1958) of the microwave maser. The latter device had been proposed by BASOV, PROKHOROV[2], TOWNES[3] and WEBER[4]. In the present article*** we mainly treat the laser, although the theoretical results can be readily applied to the maser, and many of the results can be exemplified with it.

The maser[5,6] is composed essentially of two subsystems: The electromagnetic field within a cavity, and a set of maser-active atoms (or molecules). In a cavity

* Maser: *M*icrowave *A*mplification by *S*timulated *E*mission of *R*adiation.

** Laser: *L*ight *A*mplification by *S*timulated *E*mission of *R*adiation. In some of the early literature, the laser was often called "Optical Maser", instead. In the Soviet literature the name "Optical Quantum Generator (OKG)" or "Quantum Generator of Light" is sometimes used.

*** As more than 10,000 papers have been published on lasers and related topics, we refer in the present article only to those papers which were used in its preparation or which are closely related to the special topics treated. For more details the reader is referred to reference books like TOMIYASU: The Laser Literature; ASHBURN: Laser Literature (Western Periodicals) or the Laser Abstracts (European Abstracts Service, Göteborg, Sweden).

[1] A. L. SCHAWLOW and C. H. TOWNES: Phys. Rev. **112**, 1940 (1958).

[2] N. G. BASOV and A. M. PROKHOROV: J. Exptl. Theoret. Phys. USSR **27**, 431 (1954); **28**, 249 (1955).

[3] J. P. GORDON, H. J. ZEIGER, and C. H. TOWNES: Phys. Rev. **95**, 282 (1954); **99**, 1264 (1954).

[4] J. WEBER: Trans. IRE, PGED-3, June 1953; — The three-level maser principle was introduced by BASOV and PROKHOROV, 1955 (see [2]), and its use in the solid state was first suggested by N. BLOEMBERGEN: Phys. Rev. **104**, 324 (1956).

[5] Review articles and books about the maser: J. WEBER: Rev. Mod. Phys. **31**, 681 (1959). J. R. SINGER: Advan. Electron Phys. **15**, 73 (1961). — J. R. SINGER: Masers. New York: John Wiley & Sons 1959. — A. E. SIEGMAN: Microwave Solid-state Masers. New York: McGraw-Hill Book Co. 1964. — A. A. VUYLSTEKE: Elements of Maser Theory. Princeton, New Jersey: D. van Nostrand Co. 1960. — W. LOUISELL: Radiation and Noise in Quantum Electronics. New York: McGraw-Hill Book Co. 1964.

[6] Review articles and books about masers and (mainly) lasers: Note that the word "optical maser" is used in the same sense as "laser". — W. KAISER: Physica Status Solidii **2**, 1117 (1962). — G. J. TROUP: Masers and Lasers. Methuen's Monographs. New York: John Wiley & Sons 1963. — O. S. HEAVENS: Appl. Opt. Suppl. on Optical Masers, p. 1 (1962); — Optical Masers. Methuen's Monographs. London 1964. — W. R. BENNETT JR.: Appl. Opt. Suppl. on Optical Masers, p. 24 (1962). — B. A. LENGYEL: Lasers. New York: John Wiley & Sons 1962. — A. YARIV and J. P. GORDON: Proc. IEEE **51**, 4 (1963). — G. BIRNBAUM: Optical Maser. New York: Academic Press 1964. — C. H. TOWNES, ed.: Quantum Electronics. New York: Columbia University Press 1960. — J. R. SINGER, ed.: Advances in Quantum Electronics. New York:

consisting of highly conducting walls, only a specific set of standing electromagnetic waves, the modes, can exist. These modes are connected with a discrete sequence of frequencies ("eigenfrequencies" or "characteristic frequencies"). The maser-active atoms must possess two "optical" levels whose spacing is in resonance with one of the eigenfrequencies of the cavity (in practical application the cavity has to be tuned, of course, because the atomic transition energy is fixed). After the atoms have been excited to the upper energy level by a "pump" mechanism, a resonant electromagnetic mode causes them to make a transition into their lower energy level, and during this process of "stimulated emission" they transfer their excitation energy to the electromagnetic field. Thus all atomic energy is converted into the energy of a single mode. This device *amplifies* electromagnetic waves. If there are enough excited atoms to compensate for the cavity losses, the system acts as a generator of electromagnetic energy within a specific mode and eigenfrequency or, in other words, as an oscillator. The feasibility of such a device was first demonstrated with ammonia molecules[3].

I.2. The laser condition. The difficulty about extending the maser principle into the optical region[6], is this: the wavelength of light is short compared with a "cavity" of any reasonable dimensions, therefore the spacing between different

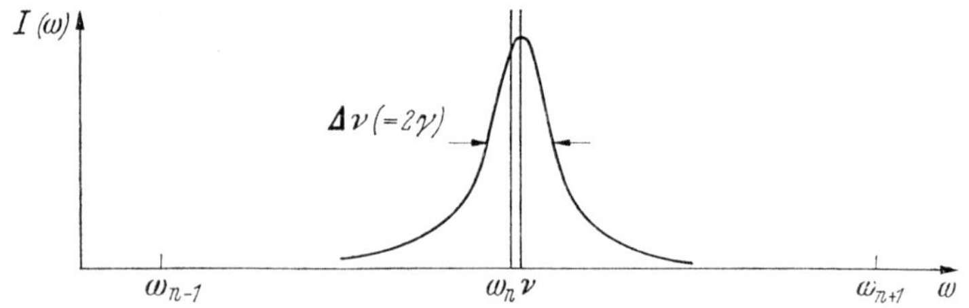

Fig. 1. The intensity distribution of the spontaneously emitted light. ν: atomic line center ω_m: resonances in the cavity; in the *maser* only one (or a few) cavity resonances ly within the atomic width, 2γ. ω_n is tuned close to the atomic resonance, ν.

"eigenfrequencies" is very small, so that there are a great many modes in the frequency range of the width of the atomic transition. Thus a further selection of modes is highly desirable. This can be achieved by omitting the side walls of a rectangular resonator and keeping only the two opposite mirrors. The Fabry-

Columbia University Press 1961. — P. GRIVET and N. BLOEMBERGEN, eds.: Quantum Electronics. Proceedings of the Third Internat. Congr., Paris, 1963. New York: Columbia University Press 1964. — W. S. C. CHANG, ed.: Lasers and Applications. Engineering Station, Ohio State University, Columbus, Ohio, 1961. — J. FOX, ed.: Proceedings of the Symposium on Optical Masers. Microwave Research Inst. Symp. Ser. Vol. XIII. New York: Polytechnic Press of the Polytechnic Inst. of Brooklyn 1963. — P. A. MILES, ed.: Quantum Electronics and Coherent Light. Proc. of the Internat. School of Physics "Enrico Fermi" Course XXXI, 1963 (Director: C. H. TOWNES). New York: Academic Press 1964. — C. DEWITT, A. BLANDIN, and C. COHEN-TANNOUDJI, eds.: Quantum Optics and Electronics. Lectures delivered at Les Houches, during the 1964 session of the Summer School of Theoretical Physics. New York-London-Paris: Gordon & Breach 1965; — Laser Parameter Measurements Handbook (ed. H. G. HEARD). New York-London-Sidney: J. Wiley & Sons 1968. — WILLIAM V. SMITH and PETER P. SOROKIN: The Laser. New York: McGraw-Hill Book Co. 1966. — C. G. B. GARRETT: Gas Lasers. New York: McGraw-Hill Book Co. 1967. — JU. L. KLIMONTOVICH: Quantum Generators of Light and Nonlinear Optics. Moscow 1966. — D. RÖSS: Lichtverstärker und Oszillatoren. Akademische Verlagsgesellschaft Frankfurt a. Main: 1966. — A. YARIV: Quantum Electronics. New York-London-Sideny: John Wiley & Sons 1967. — W. KLEEN u. R. MÜLLER: Laser. Berlin-Heidelberg-New York: Springer 1969.

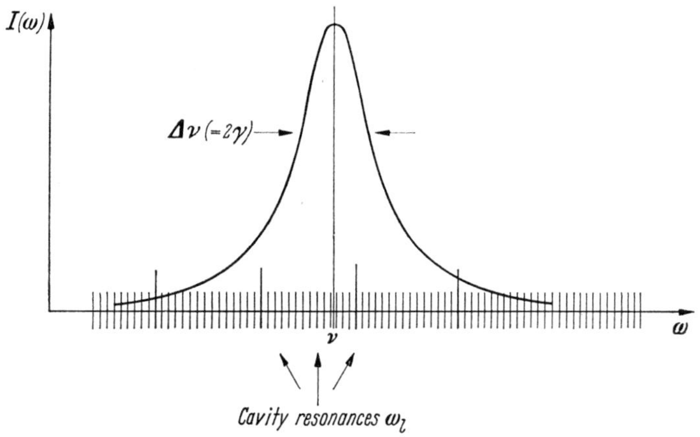

Fig. 2. The same as Fig. 1, but for the laser: usually, a great number of mode frequencies is lying within the atomic line. The longer bars refer to axial modes.

Perot-type resonator, which has been proposed by SCHAWLOW and TOWNES[7], PROKHOROV[8] and DICKE[9], has the following property:

Consider Figs. 3 and 4. Before the laser process starts, the excited atoms emit light spontaneously in all directions. Due to the arrangement of mirrors, only light emitted very close to the axis can remain in the "cavity" long enough to bring about stimulated emission of the atoms, whereas all other "modes" cannot be

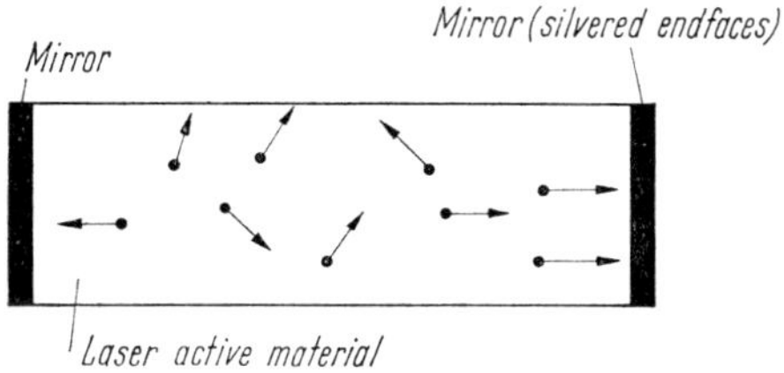

Fig. 3. The spontaneous emission of photons occurs in all directions.

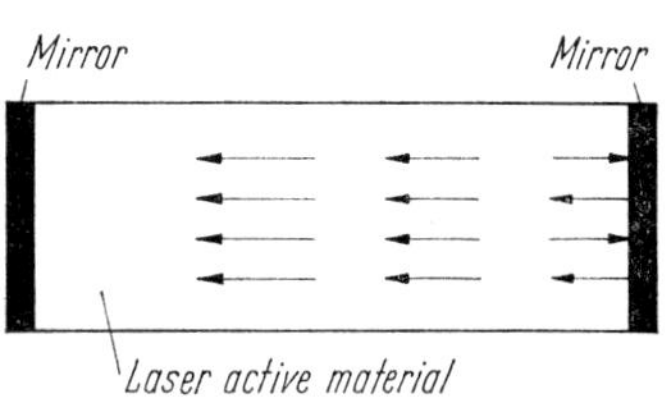

Fig. 4. On account of the mirrors, only axial or near-axial modes stay long enough in the cavity to be enhanced by stimulated emission.

amplified. Thus, the Fabry-Perot interferometer leads to a pronounced discrimination of modes with respect to their lifetimes. Furthermore due to the usual interference, it supports only those modes whose wavelength λ is given by $L = n\lambda/2$ where n is an integer and L the distance between the mirrors (see Fig. 5). FOX and LI[10] have shown in detail by a computer calculation that, even in such a degenerate "cavity" as that represented by the Fabry-Perot interferometer, a well-defined set of decaying modes indeed exists.

For laser action to take place, a laser condition must be fulfilled. The important parameters involved are revealed by the following argument (we follow in essence SCHAWLOW and TOWNES[7]). In order to sustain laser action, the rate of generation of photons must be faster than the rate of loss. Let the radiative life-

[7] A. L. SCHAWLOW and C. H. TOWNES: Phys. Rev. **112**, 1940 (1958).
[8] A. M. PROKHOROV: J. Exptl. Theoret. Phys. USSR **34**, 1658 (1958).
[9] R. H. DICKE: U.S. Patent 2851652 (Sept. 9, 1958).
[10] A. G. FOX and T. LI: Bell System Tech. J. **40**, 489 (1961).

time of an atom be τ, so that it emits $1/\tau$ photons per second spontaneously. We are, however, interested only in one special "mode". We thus have to divide by the number p of modes lying in the linewidth $\Delta\nu$ of the spontaneous emission line of the atom. The number of spontaneously emitted photons per atom per second in a definite mode is thus $\frac{1}{\tau p}$. According to EINSTEIN[11], who first considered stimulated emission, the stimulated emission rate is then $\frac{n}{p\tau}$, where n is the number of photons already present. If there are N_2 atoms in the upper and N_1 in the lower state, the total generation rate is

$$(N_2 - N_1)\frac{n}{p\tau}. \tag{I.2.1}$$

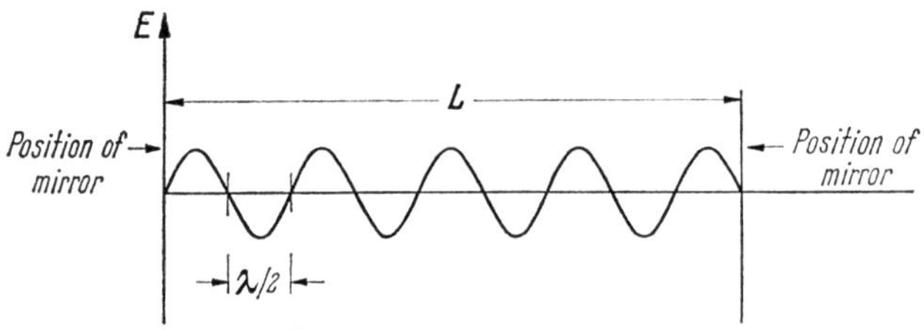

Fig. 5. Only modes with $L = n \cdot \frac{\lambda}{2}$ (n: integer) fit into the "cavity", E is the electric field strength.

The difference $N_2 - N_1$ occurs because the atoms in the lower state *absorb* photons at the rate of $\frac{n}{p\tau}$ per atom, which just equals the stimulated emission rate per excited atom. If the lifetime of the mode in the cavity is t_1, the rate of loss is

$$\frac{n}{t_1}. \tag{I.2.2}$$

Laser action starts, when (I.2.1) is bigger than (I.2.2) or when

$$\frac{(N_2 - N_1)}{p\tau} > \frac{1}{t_1}. \tag{I.2.3}$$

For a Lorentzian line shape, p is given by $\omega^2 V \Delta\nu / c^3 \pi$ so that one obtains as laser condition:

$$\frac{(N_2 - N_1)\, c^3 \pi}{\tau \omega^2 V \Delta\nu} > \frac{1}{t_1}. \tag{I.2.4}$$

V is the volume of the cavity and c the velocity of light. ω is the *circular* frequency of the emitted light, and agrees with that of the atomic transition. In order to fulfill (I.2.4):

a) The lifetime must be long enough. This requires a detailed study of the losses of decaying modes in Fabry-Perot resonators as well as in other resonator types. Also other loss mechanisms must be considered, for example, impurity scattering, or coupling between modes.

b) The atomic linewidth $\Delta\nu$ must be small. This is mainly a question of selecting materials but nevertheless requires a theoretical investigation of broadening mechanisms. Note that $\Delta\nu$ can be much bigger than $1/\tau$, because $\Delta\nu$ includes all kinds of broadening, such as lattice-vibrations etc.

[11] A. EINSTEIN: Physik. Z. **18**, 121 (1917).

c) The inversion N_2-N_1 must be high enough. This requires a detailed knowledge of excitation mechanisms like optical pumping[12] etc.

Beyond the question, under what conditions can laser action be achieved, the following are of practical importance:

What modes are excited?

What are the conditions of single or multiple mode action?

At what frequencies does laser action take place?

Is there a mutual interaction between modes?

Laser action was detected first by MAIMAN[13] in 1960; he excited a ruby rod with silvered plane parallel ends by means of an intense light pulse. His observation of line-narrowing and lifetime-shortening was closely followed by a detailed study by COLLINS, NELSON, SCHAWLOW, BOND, GARRETT and KAISER[14] in 1960 of the output characteristics such as coherence, directionality and spiking. Laser action connected with impurity atoms or even F-centers in crystals and glasses is nowadays observed in a huge variety of materials. It also takes place in gases, being first observed in 1961 in a He-Ne mixture by JAVAN, BENNETT and HERRIOT[15], who also developed the first continuously working laser. The occurrence of laser action in semiconductor p-n-junctions, the first made of GaAs, was found by HALL, FENNER, KINGSLEY, SOLTYS and CARLSON[16], NATHAN, DUMKE, BURNS, DILL and LASHER[17] and QUIST, REDIKER, KEYES, KRAG, LAX, MCWHORTER and ZEIGER[18]. In our present article we do not attempt to survey these materials and their laser properties, partly because their exploration is still developing very fast, and partly because these data should be sought in compilations like "Landoldt Börnstein" etc.

I.3. Properties of laser light.

a) Spatial coherence. As we have stated above, it is mainly axial light which is enhanced by stimulated emission, as shown in Fig. 6. Speaking in terms of a wave picture, the end-mirrors are hit by a plane wave front. Accordingly, when a mask with two slits is put on an end mirror, the far-field pattern is exactly that

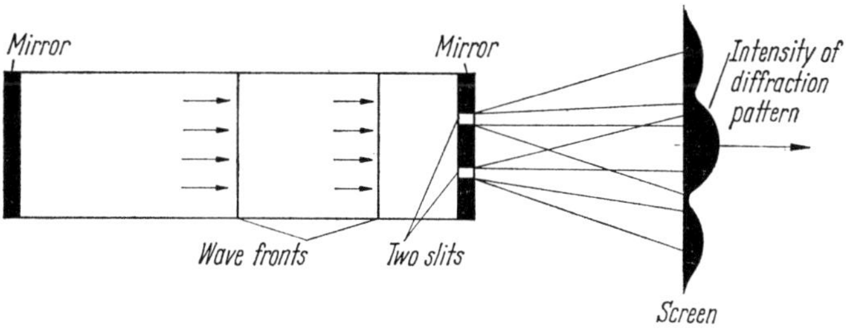

Fig. 6. Spatial coherence of laser light.

[12] For a recent representation of "optical pumping" [A. KASTLER: J. Phys. Radium **11**, 225 (1950)] see J. BROSSEL, in: Quantum Optics and Electronics, p. 189. New York: Gordon & Breach 1965.

[13] T. H. MAIMAN: Brit. Commun. Electron. **7**, 674 (1960); — Nature **187**, 493 (1960).

[14] R. J. COLLINS, D. F. NELSON, A. L. SCHAWLOW, W. BOND, C. G. B. GARRETT, and W. KAISER: Phys. Rev. Letters **5**, 303 (1960).

[15] A. JAVAN, W. R. BENNETT, and D. R. HERRIOTT: Phys. Rev. Letters **6**, 106 (1961).

[16] R. N. HALL, G. E. FENNER, J. D. KINGSLEY, T. D. SOLTYS, and R. O. CARLSON: Phys. Rev. Letters **9**, 366 (1962).

[17] M. I. NATHAN, W. P. DUMKE, G. BURNS, F. H. DILL, and G. LASHER: Appl. Phys. Letters **1**, 62 (1962).

[18] T. M. QUIST, F. H. REDIKER, K. J. KEYES, W. E. KRAG, B. LAX, A. L. MCWHORTER, and H. J. ZEIGER: Appl. Phys. Letters **1**, 21 (1962).

of the Fraunhofer interference of a plane wave passing through two slits. The beam is highly directional, its aperture coming close to the theoretical limit imposed by diffraction (finite extension of the end-faces). The spatial coherence, as proved by the two-slit experiment, is *not to be confused* with the

b) Temporal coherence. When two radio waves with frequencies ω_1, ω_2 are superimposed and analyzed by a square-law detector, a well-defined beat note is found for an extended period. Light from natural sources (thermal light), on the other hand, consists of very many statistically independent wavetracks each of about 10^{-8} sec duration, so that beat notes could be observed only on such a short time scale. Furthermore, because the light amplitude consists of many statistically independent elements, it possesses a Gaussian distribution[19]. Let us now turn to the laser (compare Fig. 7). Because the axial mode which lies

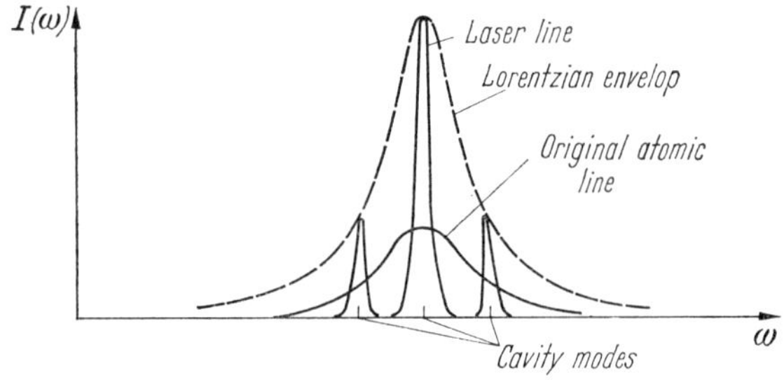

Fig. 7. Laser light below threshold: Strong enhancement of the near resonance mode: A pronounced line narrowing resulting. (Qualitative representation.)

closest to the atomic resonance has the highest gain, laser light concentrates its linewidth around that resonance, so that a *line-narrowing* occurs. Laser light can thus achieve an enormous spectral purity, which comes close to that of the Mössbauer effect. After the discovery of laser action it was for a time believed that laser light is composed of statistically independent wave tracks so that its amplitude possesses a Gaussian distribution analogous to that of natural light, the only difference being that the decay time of the single wave tracks is considerably decreased. This view was backed by theoretical treatments of laser noise. However, in 1964[20] we derived from first principles by establishing and solving a quantum mechanical, nonlinear laser equation that laser light is narrow band Gaussian only below the laser threshold, but that above this threshold the laser acts as a self-sustained oscillator with a highly stabilized (classical) amplitude. On this amplitude, small quantum fluctuations are superimposed. Furthermore, quantum fluctuations cause the phase to diffuse slowly. If we disregard the fluctuations, the corresponding quantum state of the field may be characterized by the so-called coherent state[21]. The predicted decrease of the amplitude fluctuations with increasing laser output[20] was fully substantiated by intensity correlation experiments by ARMSTRONG and SMITH[22], FREED and HAUS[23], and others.

---

[19] For recent representations see: R. J. GLAUBER, in: Quantum Optics and Electronics. New York: Gordon & Breach 1965, and L. MANDEL and E. WOLF: Rev. Mod. Phys. **37**, 251 (1965). This article contains a rather complete list of further references on the quantum and statistical aspects of light.

[20] H. HAKEN: Z. Physik **181**, 96 (1964).

[21] See Footnote[19].

[22] J. A. ARMSTRONG and A. W. SMITH: Phys. Rev. Letters **14**, 68 (1965); — Phys. Letters **19**, 650 (1965); — Phys. Rev. **140**, A 155 (1965).

[23] C. FREED and H. A. HAUS: Appl. Phys. Letters **6**, 85 (1965).

c) Photon statistics. The characteristic difference of laser light below and above the threshold shows up most dramatically in the change of photon statistics. Light from thermal sources and laser light below threshold obey Bose-Einstein statistics and are *chaotic*, so that the photon number mean square deviation is

$$\overline{\Delta n^2} = \bar{n}(\bar{n}+1),$$

whereas laser light above threshold possesses approximately a Poisson distribution with

$$\overline{\Delta n^2} = \bar{n}.$$

The smooth transition between the two regions, quantitatively predicted by RISKEN[24] in 1965, was completely confirmed by detailed experiments of ARMSTRONG and SMITH[25], by ARECCHI[26] and coworkers as well as by other groups.

d) High intensity. Because an essential part of the energy of the excited atoms goes into a very narrow frequency range, the energy density per unit frequency interval is extremely high and exceeds any thermal source by many orders of magnitude. But the total intensity can also be very high, especially in the "Q-switched" laser[27] proposed by HELLWARTH. By this device energy is stored in the excited atoms and then released very quickly by changing the cavity loss or by changing the gain.

Yet higher intensity can be reached in ultrashort pulses.

e) Ultrashort pulses. We finally mention, as another example of the extreme properties of laser light which may have important applications, the possibility of ultrashort pulses of a duration of 10^{-12} sec or less. Here many modes with a frequency spread of the full linewidth or even more are coupled together with fixed phases, so that a wave packet with a very small extension in the time coordinate arises[28].

I.4. Plan of the article. While this article is being written, laser research is still progressing and there are a great many publications appearing every month. On the other hand, the more fundamental aspects of the nature of laser light and many details of laser action now seem to be well understood. For these reasons we decided to confine ourselves to a thorough treatment of questions which are basic from the physical point of view. We are fully aware that in so doing we are omitting many important contributions to this field, and also that another author would have chosen perhaps other aspects as being "basic".

The following outline of our approach may be helpful:

The logical connection between the different aspects of laser theory is presented in Tables I and II. They also represent essentially the structure of the present article:

In Chap. III the Schrödinger equation for the coupling between the quantized electromagnetic field and the electrons of matter is derived and other represen-

[24] H. RISKEN: Z. Physik **186**, 85 (1965). Interpreting the quantum mechanical equation of ref. [20] as a classical one, RISKEN established its Fokker-Planck equation and determined from it the photon distribution function. In the mean time a great number of theoretical and experimental papers appeared dealing with these problems.

[25] A. W. SMITH and J. A. ARMSTRONG: Phys. Letters **19**, 650 (1966); — Phys. Rev. Letters **16**, 1169 (1966).

[26] F. T. ARECCHI, G. S. RODARI, and A. SONA: Phys. Letters **25** A, 59 (1967).

[27] R. W. HELLWARTH, in: Advances in Quantum Electronics (J. SINGER, ed., p. 334). New York: Columbia Univ. Press 1961.

[28] For the first experimental observation see: L. E. HARGROVE, R. L. FORK, and M. A. POLLACK: Appl. Phys. Letters **5**, 4 (1964), and for the first theoretical treatment: M. DI DOMENICO JR.: J. Appl. Phys. **35**, 2870 (1964). — A. YARIV: J. Appl. Phys. **36**, 388 (1965).

Survey I

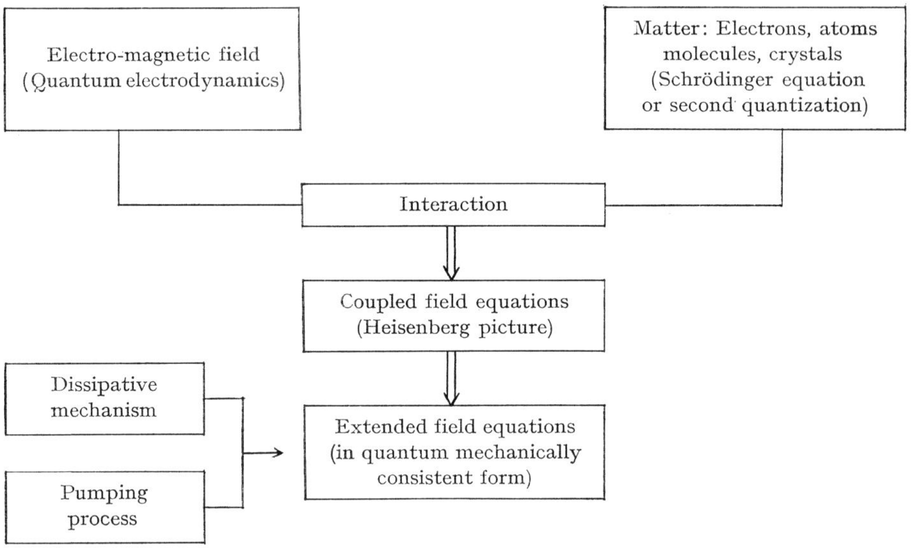

Survey II

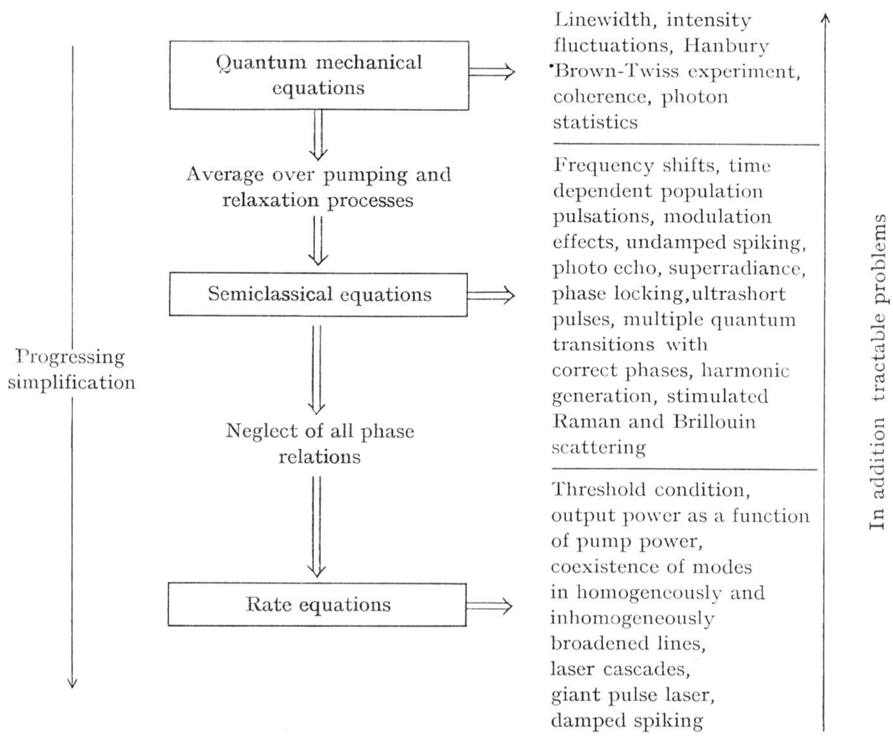

tations (Heisenberg picture, interaction representation) are also given. Because this Schrödinger equation is derived in detail in standard text books, we give only a brief sketch. On the other hand, the laser process can be fully understood only if the coupling of the light field and the laser atoms to the surroundings i.e. the "cavity", the pumping process, lattice vibrations etc., is taken into account. Methods for dealing with this problem in a fully quantum mechanical manner have been developed only recently, so that we treat this problem in more detail. We will show that there now exist several different but essentially equivalent ways which allow all laser phenomena to be treated quantitatively (Chap. IV).

Chap. V is not included in the survey of Tables I and II. It treats quantum and statistical aspects of the light field in general (not necessarily of the laser field) and deals especially with the concept of coherence.

Chaps. VI, VII, VIII are arranged according to Table II.

As an example of the application of the extended field equations we treat in detail a system with a homogeneously broadened line. This allows us to explain the characteristics of the "subthreshold" as well as of the genuine "laser" region in a rather simple way, and then to derive formulae for the linewidth and intensity fluctuations in detail.

After averaging over the fluctuations caused by the pumping and losses, we obtain the "semiclassical equations" (Chap. VII), which are more or less classical. They fail for a treatment of the linewidth, and other fluctuation phenomena, but can well be applied to the calculation of frequency shifts, population pulsations, undamped spiking, the photo-echo, ultrashort pulses etc. As an example we treat here mainly the inhomogeneously broadened, atomic line. The theory of the gas laser is included in this chapter.

Finally when we neglect all phase relations between light-field and atoms, we end up with the rate equations (Chap. VIII) which have been playing an important role in the understanding of several gross features of the laser, such as the threshold condition, ouptut power, coexistence of modes, the giant pulse laser and similar problems.

We have tried to write each of the chapters in a self-contained way, so that each of them should be understandable without a knowledge of the preceding ones. Thus a reader, who is interested in the threshold condition, for instance, can start immediately with Chap. VIII. Similarly, for an understanding of Chap. V the reading of Chap. III is not required.

II. Optical resonators.

II.1. Introduction. While cavities with closed walls are adequate for mode and frequency selection in the microwave region, resonators with open sidewalls give a more pronounced selection in the optical range. The reasons were explained in the introduction on p. 2—3. Because the wavelength is in general much smaller than the dimensions of the resonator (including the radii of curvature of the mirrors), the application of geometrical optics is often permissible and useful. In this way one can obtain a quick survey of the properties of mirror arrangements giving rise to high quality modes. For their achievement there must be a family of rays which are reasonably often reflected between the mirrors before leaving the system. The plane-parallel Fabry-Perot interferometer, arrangements with several plane mirrors, or with two spherical mirrors, or the whispering gallery modes of a sphere are examples of suitable resonators (compare Figs. 8 and 15). In order to obtain a high Q[1], a second criteria, which follows from physical optics (theory of diffraction), must, however, be fulfilled[2]:

In the case of two mirrors with apertures $2A_1$ and $2A_2$ respectively, separated by a distance D, the inequality

$$\frac{A_1 A_2}{\lambda D} > 1 \qquad \text{(II.1.1)}$$

must hold.

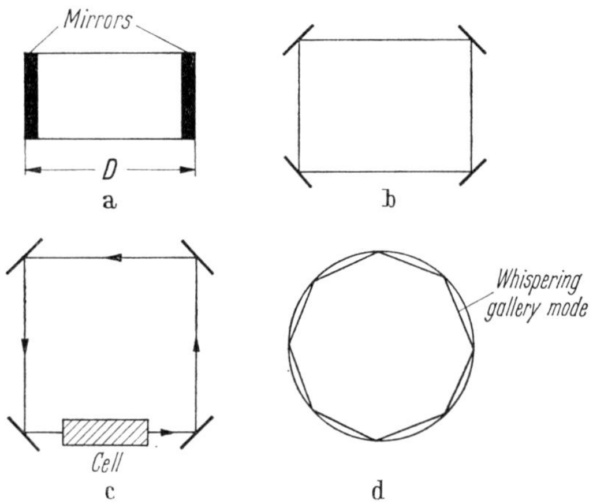

Fig. 8a—d. a) Fabry-Perot interferometer. b) Arrangement with 4 mirrors. c) Inserting of a cell causing Faraday rotation allows to select modes running in one direction. d) A sphere allows for whispering gallery modes.

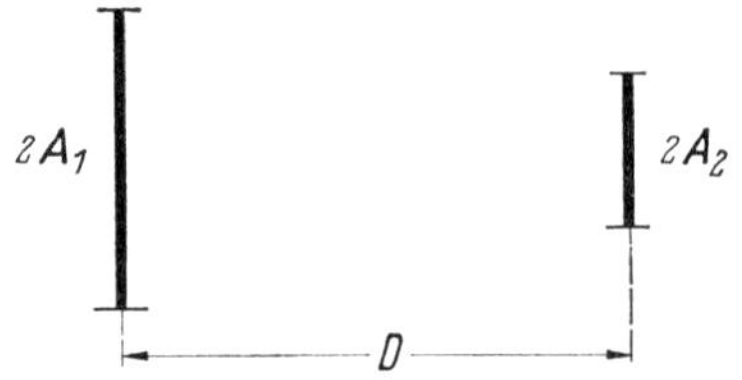

Fig. 9. Arrangement of two plane parallel mirrors.

The parameter $N = \frac{A^2}{\lambda D}$ obtained for $A_1 = A_2 = A$ is called the Fresnel number. It is approximately equal to the number of Fresnel zones seen in one mirror from the center of the other mirror.

A resonator theory should explain the following points:

1. The mode pattern on the mirrors, and, more generally,
2. the mode distribution in the interior of the resonator,
3. the losses due to diffraction, reflection, mirrors, misalignment and aberration,
4. the far-field pattern.

We will be mainly concerned with 1., 2., 3.; 2. and 3. in particular are of fundamental importance for the further theoretical treatment of laser action.

[1] The value is defined as $Q = \omega t_0$, where ω is the mode frequency and t_0 its lifetime in the unloaded cavity. t_0 is here understood as that time, in which the mode intensity drops down to the e'th part of its initial value. In this article we will use the decay constant $\varkappa = \frac{1}{2t_0}$.

[2] A. Yariv and J. P. Gordon: Proc. IEEE **51**, 4 (1963).

In the following we treat two important cases in detail: The Fabry-Perot resonator with plane parallel plates and the confocal resonator. In order to demonstrate how the resonator problem can be treated, two different methods are described:

a) a mode analysis, in which the wave function of the electromagnetic field is determined by solving the time-independent wave equation with a suitable boundary condition.

b) a wave analysis, in which a field distribution is determined that reproduces itself in spatial distribution and phase as the wave bounces back and forth between the two reflectors. At each transit the intensity is lowered by reflection and diffraction losses, therefore "reproduction" means "besides a constant factor". If, however, the resonator is filled with active material, the gain constant can be adjusted in such a way that the absolute value of this constant factor equals one.

As we will show below, the dynamics of the laser process is usually treated in two steps. In the first step the decaying modes with the properties 2) and 3) are determined in the unloaded resonator, i.e. in a resonator where non-laser-active atoms are present. Then in the second step the amplification (or oscillation) and their discrimination by means of the active atoms are calculated. It must be noted that this procedure is only permissible for high Q-modes but may fail for small Q-modes. This is illustrated by the following example. Consider a rod consisting of active materials with no mirrors at the end faces. In this rod there may exist a continuum of decaying modes having all the same decay time. If now laser action takes place, one mode or only a few discrete modes occur.

II.2. The Fabry-Perot resonator with plane parallel reflectors.

a) Spatial distribution of modes. The electric field configuration of the dominant mode and a number of higher-order modes for square and circular mirrors are shown[3] in Fig. 10. The cartesian mode classification ("transverse electric and magnetic fields") applies to plane as well as confocal spherical mirrors. The assumption that the waves are almost transverse is justified, because the curvature of the wavefront away from the transverse is exceedingly small, as demonstrated by Fox and Li. Only for very high-order modes does this assumption begin to fail.

The circular patterns in Fig. 10 have the mode designation TEM_{pl} where p gives the number of nodes in radial and l that of nodes in azimuthal direction. Sometimes, a third index denotes the number of nodes in the axial direction.

We shall now represent an analytical treatment (we closely follow the paper by Risken[4]). The essential features can be seen from a two-dimensional resonator which consists of two plane strip metal mirrors at the ends of the Fabry-Perot resonator. It is assumed that the space between the mirrors is filled with active material, being described by a complex susceptibility, $\chi=\chi'+i\chi''$ (for a derivation see IV.6). Due to the symmetry of the problem and taking into account Maxwell's equation we may put either[5]

$$A)\qquad \left.\begin{aligned} &E_x=U(y,z)\,,\\ &H_y=-\frac{i}{\mu\omega}\frac{\partial U}{\partial z}\,,\qquad H_z=\frac{i}{\mu\omega}\frac{\partial U}{\partial y}\end{aligned}\right\}\tag{II.2.1}$$

or

$$B)\qquad \left.\begin{aligned} &H_x=U(y,z)\,,\\ &E_y=\frac{i}{(\varepsilon+\chi)\omega}\frac{\partial U}{\partial z}\,;\qquad E_z=-\frac{i}{(\varepsilon+\chi)\omega}\frac{\partial U}{\partial y}\end{aligned}\right\}\tag{II.2.2}$$

---

[3] A. G. Fox and T. Li: Bell System Tech. J. **40**, 489 (1961).
[4] H. Risken: Z. Physik **180**, 150 (1964).
[5] M. Born and E. Wolf: Principles of Optics, p. 558. London: Pergamon Press 1959.

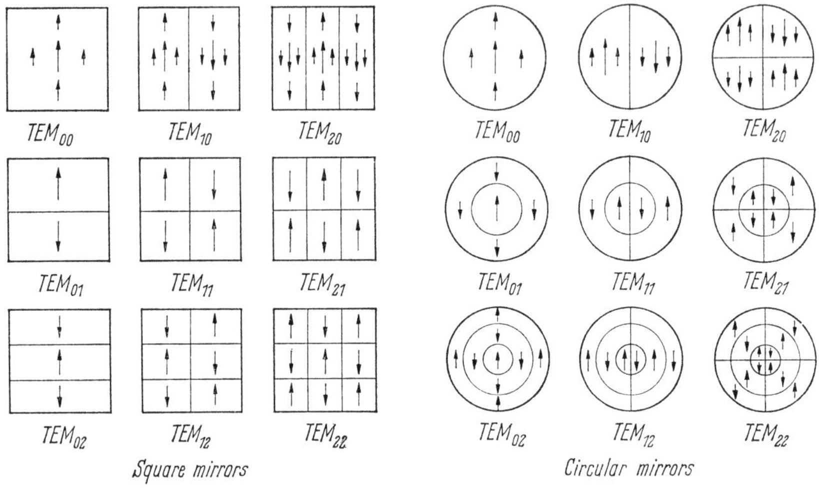

Fig. 10. Field configuration of normal modes for square and and circular mirrors (after Fox and Li, l.c. [3]. Copyright 1961, The American Telephone and Telegraph Co, reprinted by permission.) In the meantime a new mode designation is given in Fig. 17 in the paper of H. Kogelnik and T. Li, l.c. [3a]. According to this new designation the indices lm of TEM_{lm} are now interchanged. We wish to thank Dr. Li for drawing our attention to this change.

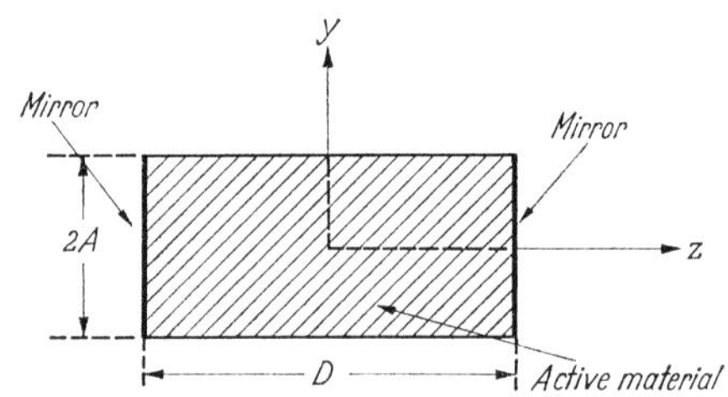

Fig. 11. Two-dimensional Perot-Fabry-Laser.

where U obeys the wave-equation

$$\Delta U + k_M^2 U = 0, \tag{II.2.3}$$

$$k_M^2 = \frac{\omega^2}{c^2}\left(1 + \frac{\chi}{\varepsilon}\right) \tag{II.2.3a}$$

within the active material[6] and

$$\Delta U + k^2 U = 0, \tag{II.2.4}$$

$$k^2 = \frac{\omega^2}{c^2} \tag{II.2.4a}$$

outside it. ε is the dielectric constant.

The metal mirrors are assumed to have a reflectivity, r, close to one, so that the tangential components of $\boldsymbol{E}$ and $\boldsymbol{H}$ must satisfy the so-called Leontovich[7]

[3a] H. Kogelnik and T. Li: Appl. Optics **5**, 1550 (1966). This review paper contains a great number of further references on laser beams and resonators.

[6] It is assumed that k_M is constant over the space of the resonator.

[7] M. A. Leontovich: Bull. Acad. Sci. USSR. Phys. Ser. **8**, 1 (1944). See also: Landau-Lifschitz: Electrodynamics of Continuous Media, p. 280. London: Pergamon Press 1960.

condition

$$\boldsymbol{E}_{tang} = \sqrt{\frac{\mu}{\varepsilon}} \frac{1-i}{4} (1-r) \, \boldsymbol{n} \times \boldsymbol{H} \tag{II.2.5}$$

where $\boldsymbol{n}$ is the normal vector of the mirror surface.

The essential results can be summarized as follows: The electric field or the vector potential has in the lowest approximation (which is shown to be very good for not too small Fresnel numbers) the following spatial dependence:

$$\sin\left[(y+A)\frac{m\pi}{2A}\right] \sin\left[\left(z+\frac{D}{2}\right)\frac{n\pi}{D}\right] \tag{II.2.6a}$$

in two dimensions (with strip mirrors), which can be readily generalized to three dimensions

$$\sin\left[(x+A_1)\frac{l\pi}{2A_1}\right] \sin\left[(y+A_2)\frac{m\pi}{2A_2}\right] \sin\left[\left(z+\frac{D}{2}\right)\frac{n\pi}{D}\right] \tag{II.2.6b}$$

where $2A_1$, $2A_2$ are the lengths of rectangular end mirrors in x- and y-direction, respectively. l, m, n are integer numbers. An essential result of RISKEN's analysis is that these functions vanish approximately at the boundary of the active material. While this is rather obvious for the silvered ends of the rod, it is a remarkable result for the unsilvered boundary between the active material and the vacuum (or air). It should be noted that this result is derived for axial or nearly axial modes, so that n is a big number ($\lambda n \approx D$), whereas l, m are small integer numbers of order unity.

The *resonance condition* reads approximately

$$\frac{\omega^2}{c^2} \approx \left(\frac{l\pi}{2A_1}\right)^2 + \left(\frac{m\pi}{2A_2}\right)^2 + \left(\frac{n\pi}{D}\right)^2, \quad l, m, n = 0, 1, \ldots. \tag{II.2.7}$$

Besides these gross features of the mode-pattern, RISKEN has also treated its fine structure, which is described both by complex l's, m's and n's as well as by additional terms in (II.2.6a).

We now give a derivation of these results:

Because the high-Q-modes have wavefronts nearly perpendicular to the resonator axis, we expect a radiation field outside the cavity for which one can assume that $\left.\frac{\partial U}{\partial y}\right|_{y=\pm A} \approx 0$ outside the cavity. RISKEN's approximation, which turns out to be a very good one, consists in requiring that the field components in the lines $y = \pm A$, i.e. $E_z(\pm A, z)$ and $H_z(\pm A, z)$ vanish outside the cavity, i.e. for $|z| > D/2$.

The field in the outer space for $y > A$ can be represented by

$$U(y, z) = \int_{-\frac{D}{2}}^{\frac{D}{2}} \left.\frac{\partial U(y', z')}{\partial y'}\right|_{y'=A} G(y, z; A, z')\, dz' \tag{II.2.8}$$

with

$$G(y, z; A, z') = -\frac{i}{2} H_0^{(1)}\left(k\sqrt{(y-A)^2 + (z-z')^2}\right), \tag{II.2.9}$$

($H_0^{(1)}$ = Hankel function of the first kind)

where it is assumed that the normal derivative of G taken at $y = A$ vanishes.

Because we make use of the mirror symmetry of the problem with respect to the $x-y$-plane, we do not need the corresponding equation for the lower part of the outer space.

Within the active material we expand the wave amplitude $U(y, z)$ into a complete set of trigonometric functions.

$$U(y,z)=\sum_m \alpha_m \begin{Bmatrix}\cos l_m y\\ \sin l_m y\end{Bmatrix} \begin{Bmatrix}\cos q_m z\\ \sin q_m z\end{Bmatrix}. \tag{II.2.10}$$

Because of the symmetry of the problem, U is either even in y or z (expansion in a cos-series) or odd (expansion in a sin-series). The curly brackets indicate the various possibilities, where all *four* combinations can occur. Taking into account the boundary conditions (II.2.5), we get

$$q_m=\tilde{q}-\frac{2\pi m}{D}-\frac{1+i}{2D}(1-r) \tag{II.2.11}$$

where the m's are integer numbers. We have chosen the index m in such a way, that the value l_0 related to q_0 is of minimal magnitude for all other l_m. The choice of $\tilde{q}$ depends on whether A) or B) (see p. 11) applies and whether $\sin q_m z$ or $\cos(q_m z)$ is taken:

$$\tilde{q}=\frac{\pi}{L}\tilde{n} \quad \text{where} \quad \left.\begin{array}{l} \left.\begin{array}{lll}\tilde{n}: & \text{odd} & (\cos q_m z)\\ \tilde{n}: & \text{even} & (\sin q_m z)\end{array}\right\} \text{case A},\\ \\ \left.\begin{array}{lll}\tilde{n}: & \text{odd} & \cos q_m z\\ \tilde{n}: & \text{even} & \sin q_m z\end{array}\right\} \text{case B}.\end{array}\right\} \tag{II.2.11a}$$

Since u has to obey the wave Eq. (II.2.3), the relation

$$l_m^2+q_m^2=l_0^2+q_0^2=k_M^2 \tag{II.2.12}$$

must be fulfilled. If $l_0 A$ is of the order one and $kD \gg 1$ we find

$$l_m=2\sqrt{\pi}\,\sqrt{k/D}\,\sqrt{m} \qquad (m\neq 0). \tag{II.2.13}$$

Thus all l_m are determined for $m \neq 0$. They are either real for $m \geqq 1$ or imaginary for $m \leqq -1$.

While the continuity of the normal derivative of the u's at the boundary $(y=\pm A)$ is guaranteed by the formula (II.2.8), the condition that the wave amplitude is continuous, must be fulfilled explicitly

$$U(A,z)=-\frac{i}{2}\int_{-\frac{D}{2}}^{\frac{D}{2}} \left.\frac{\partial U(y',z')}{\partial y'}\right|_{y'=A} H_0^{(1)}(k|z-z'|)\,dz' \tag{II.2.14}$$

Inserting (II.2.10) into (II.2.14) multiplying it with $\begin{Bmatrix}\cos q_n z\\ \sin q_n z\end{Bmatrix}$ and integrating over z we obtain the following set of equations for the coefficients α:

$$\alpha_0 \sum_m \frac{\alpha_m}{\alpha_0}\{D l_m g'(l_m A) I_{nm}-g(l_m A)\,\delta_{nm}\}=0 \tag{II.2.15}$$

where

$$g(l_m A)=\begin{Bmatrix}\cos l_m A\\ \sin l_m A\end{Bmatrix}$$

and

$$I_{nm} = -\frac{i}{D^2} \int_{-\frac{D}{2}}^{\frac{D}{2}} \int_{-\frac{D}{2}}^{\frac{D}{2}} \begin{Bmatrix} \cos q_n z \\ \sin q_n z \end{Bmatrix} \begin{Bmatrix} \cos q_m z' \\ \sin q_m z' \end{Bmatrix} H_0^{(1)}(k|z-z'|)\, dz\, dz' \qquad \text{(II.2.16)}$$

δ_{nm} is the Kronecker symbol.

The prime at g in Eq. (II.2.15) means derivation with respect to the argument. In (II.2.15) we have neglected terms of the order $(1-r)/(kD)$.

The I_{nm}'s were calculated under the assumption $|(k-q_0)D| \ll 1$:

$$I_{nm} = -\frac{1}{2} \frac{1+i}{\sqrt{\pi k D}} B_{nm} \qquad \text{(II.2.17)}$$

with

$$B_{00} = \frac{4}{3}$$

and

$$\left.\begin{aligned} B_{nn} &\approx \frac{1+i}{2\sqrt{n}} \qquad n \neq 0, \\ B_{0n} &\approx -\frac{i}{\pi n}(-1)^n \end{aligned}\right\} \qquad \text{(II.2.18)}$$

where already for $n \geqq 1$ the asymptotic formulas of the Fresnel integrals have been used (for more details see Risken, l. c.[4] on p. 11).

Because the off-diagonal elements of I_{nm} are smaller than the diagonal ones we use in a first approximation only the diagonal elements of (II.2.15) leading to

$$\left.\begin{aligned} D\, l_0\, g'(l_0 A)\, I_{00} &= g(l_0 A) \\ \frac{\alpha_m}{\alpha_0} &= 0 \quad \text{for} \quad m \neq 0 \end{aligned}\right\} \qquad \text{(II.2.19)}$$

Since I_{00} is very small compared to one, the solution of (II.2.19) reads

$$l_0 A = x_\mu \left(1 + \frac{D}{A} I_{00}\right) \qquad \text{(II.2.20)}$$

where x_μ is one of the zeros of g.

As will be shown below, this solution is already capable of reproducing the approximate features of the mode pattern and also the losses as derived by Fox and Li (l.c.[3] on p. 11).

It is, however, a simple matter to calculate higher approximations. Inserting the α's obtained in second order from (II.2.15), we find for the field distribution in the interior of the resonator:

$$\left.\begin{aligned} \frac{u}{\alpha_0} &= \begin{Bmatrix} \cos q z \\ \sin q z \end{Bmatrix} g(l_0 A \eta) - \\ &- \frac{x_\mu}{2\pi} g'(x_\mu) \frac{1-i}{\sqrt{2N}} \frac{1}{\pi} \sum_{n \neq 0} \begin{Bmatrix} \cos q_n z \\ \sin q_n z \end{Bmatrix} \frac{(-1)^n}{n} \frac{g(\zeta \sqrt{n}\, \eta)}{i g'(\zeta \sqrt{n}) + g(\zeta \sqrt{n})} \end{aligned}\right\} \qquad \text{(II.2.21)}$$

and for l_0:

$$l_0 A = x_\mu - \frac{1+i}{\sqrt{2N}} \frac{x_\mu}{2\pi} \left\{ \frac{4}{3} + \frac{1+i}{\pi^2} \sum_{n \neq 0} \frac{1}{n^2} \frac{\sqrt{n}\, g'(\zeta \sqrt{n})}{i g'(\zeta \sqrt{n}) + g(\zeta \sqrt{n})} \right\} \qquad \text{(II.2.22)}$$

The parameter ζ is defined by $\zeta = 2\pi\sqrt{2N}$, where $N = A^2/(\lambda D)$ is the Fresnel number;

$$\eta = \frac{y}{A}; \quad g(v) = \begin{Bmatrix} \cos v \\ \sin v \end{Bmatrix}; \quad x_\mu = \begin{Bmatrix} \frac{\pi}{2}, \frac{3}{2}\pi, \ldots \\ \pi, 2\pi, \ldots \end{Bmatrix}. \tag{II.2.23}$$

Obviously the solution depends only on the Fresnel-number as mentioned first by FOX and LI[3].

If we want to use only the first approximation of the secular Eq. (II.2.15), we have to cancel the additional sum terms in (II.2.21) and (II.2.22). This is

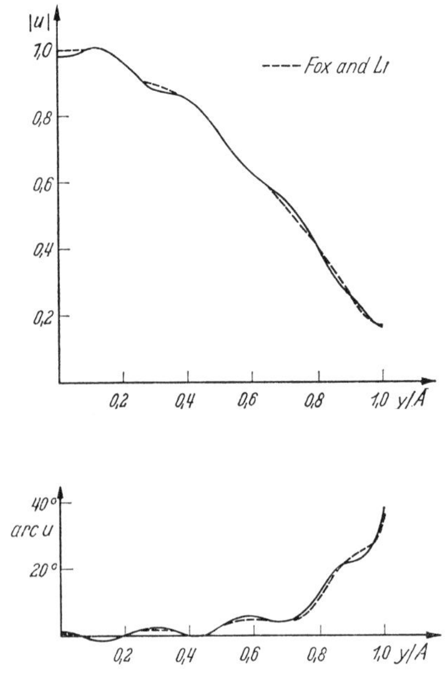

Fig. 12. Amplitude and phase distribution for the lowest mode of even symmetry for the two-dimensional resonator ($N = 6, 25$). (From RISKEN, l.c. [4] on p. 11.)

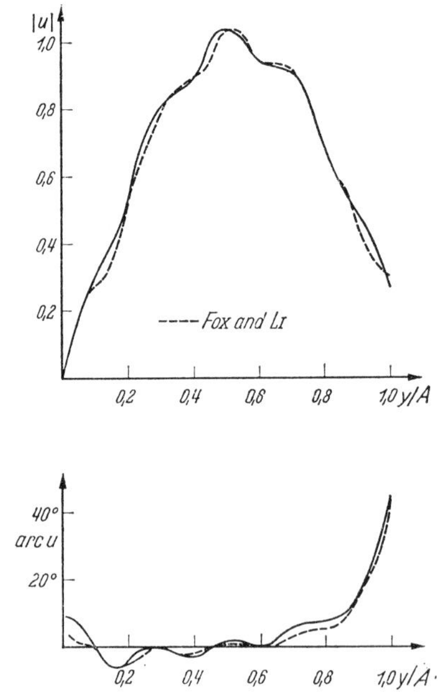

Fig. 13. Amplitude and phase distribution for the lowest mode of odd symmetry for the two-dimensional resonator ($N = 10$). (From RISKEN, l.c. [4] on p. 11.)

exactly the solution obtained by BARONNE[8] and VAINSHTEIN[9] by different methods. Looking at the solution, it may be seen that only the additional sum terms with $n \geqq 1$ in (3,23) are responsible for the fine structure of the mode pattern, i.e. the small ripples, whereas the additional sum terms with $n \leqq -1$ are different from zero only in the neighbourhood of $y = \pm A$. In Figs. 12 and 13 the field distribution on the mirrors taking into account only the sum terms $n = 1$ and $n = -1$ is compared with those of FOX and LI. RISKEN has chosen the parameter N as large as possible, because his treatment is valid only for $N \gg 1$.

Obviously his curves fit very well with those of FOX and LI. By looking at Figs. 12, 13 one has to bear in mind that the field distribution as determined by FOX and LI is defined besides a constant complex factor. Therefore we have multiplied in Figs. 12, 13 the curves of FOX and LI by a proper constant factor and we have added a proper constant phase.

[8] S. R. BARONNE: J. Appl. Phys. **34**, 831 (1963).
[9] L. A. VAINSHTEIN: Soviet Phys. JETP **17**, 709 (1963).

b) Diffraction losses. The diffraction losses are calculated in the following way: Using MAXWELL's equations for the steady state one easily derives

$$\operatorname{div}(\boldsymbol{E}\times\boldsymbol{H}^* + \boldsymbol{E}^*\times\boldsymbol{H}) = -2\omega\,\boldsymbol{E}\boldsymbol{E}^*\,\chi_{(i)} \tag{II.2.24}$$

where $\chi_{(i)}$ is the imaginary part of the susceptibility (permeability and dielectric constant are assumed to be real). Therefore the relative losses, i.e. the ratio between the energy that is dissipated from the active material and the electric energy of the material itself, is given by

$$\gamma_{res} = -\omega\chi_{(i)} = -2l_{(r)}\,l_{(i)}\frac{c^2}{\omega} - 2q_{(r)}\,q_{(i)}\frac{c^2}{\omega}. \tag{II.2.25}$$

The losses in the time D/c thus have the form

$$V = \gamma_{res}\frac{D}{c} = -\frac{2D\,l_{(r)}\,l_{(i)}}{k} + (1-r) = V_d + V_r. \tag{II.2.26}$$

The first term describes the diffraction losses and the second term the reflection losses. Using Eq. (II.2.22) we find

$$\left.\begin{aligned} &g(v) = \cos v; \qquad x_\mu = \frac{\pi}{2}(1+2\mu),\\ &V_d = \frac{(1+2\mu)^2}{4(2N)^{\frac{3}{2}}}\left\{\frac{4}{3} + \frac{1}{\pi^2}\sum_{n=1}^{\infty}\frac{1-\sin(2\zeta\sqrt{n})}{n^{\frac{3}{2}}}\right\}. \end{aligned}\right\} \tag{II.2.27a}$$

$$\left.\begin{aligned} &g(v) = \sin v; \qquad x_\mu = \pi\cdot\mu,\\ &V_d = \frac{\mu^2}{4(2N)^{\frac{3}{2}}}\left\{\frac{4}{3} + \frac{1}{\pi^2}\sum_{n=1}^{\infty}\frac{1+\sin(2\zeta\sqrt{n})}{n^{\frac{3}{2}}}\right\}. \end{aligned}\right\} \tag{II.2.27b}$$

Comparing these results with those of FOX and LI, RISKEN gets an excellent agreement for $N > 6$. Beyond the paper of FOX and LI he obtains a very weak fine structure of the diffraction losses due to the sum terms in (II.2.27a) and (II.2.27b).

RISKEN has also treated a resonator embedded in a material with a different dielectric constant for the case

$$\frac{(\varepsilon-\varepsilon_M)}{\varepsilon} \gg \frac{1}{kD}$$

ε in outer space, ε_M in the material of the resonator. In this case

$$I_{nm} = -i\frac{\delta_{nm}}{D\sqrt{k^2-q_n^2}} \tag{II.2.28}$$

where δ_{nm} is the Kronecker symbol.

His paper also contains the case, in which the mirrors at the endfaces are infinitely extended.

We compare the order of magnitude of the diffraction losses V_d and of the field amplitude, U_b, at the border between the active material and the free space, for the various cases discussed by RISKEN. We see, that in cases I, II and III the amplitude U_b is approximately zero whereas in case IV U_b is of the order one. It is interesting to remark, that the diffraction losses in case IV are of the same order as the diffraction losses used by WAGNER and BIRNBAUM[10].

[10] W. G. WAGNER and G. BIRNBAUM: J. Appl. Phys. 32, 1185 (1961).

	I and II	III	IV
$\frac{U_b}{U_0}$	$\frac{\lambda}{A}\,\frac{\sqrt{\varepsilon}}{\sqrt{\varepsilon-\varepsilon_M}}$	$\frac{\sqrt{D\lambda}}{A}$	1
V_d	$\frac{D}{A}\left(\frac{\lambda}{A}\right)^2 n^2 \sqrt{\frac{\varepsilon^2}{\varepsilon_M(\varepsilon-\varepsilon_M)}}$	$\frac{\sqrt{D\lambda}}{A}\,\frac{D}{A}\,\frac{\lambda}{A}\,n^2$	$\frac{D}{A}\,\frac{\lambda}{A}\,n$

U_0 is an average amplitude. I: finite mirrors $(\varepsilon-\varepsilon_M)/\varepsilon \gg 1/(kD)$. II: infinite mirrors $(\varepsilon-\varepsilon_M)/\varepsilon \gg 1/(kD)$. III: finite mirrors $\varepsilon = \varepsilon_M$. IV: infinite mirrors $\varepsilon = \varepsilon_M$ and the expression $2(kA)\frac{A}{D}(1-r)$ of the order one.

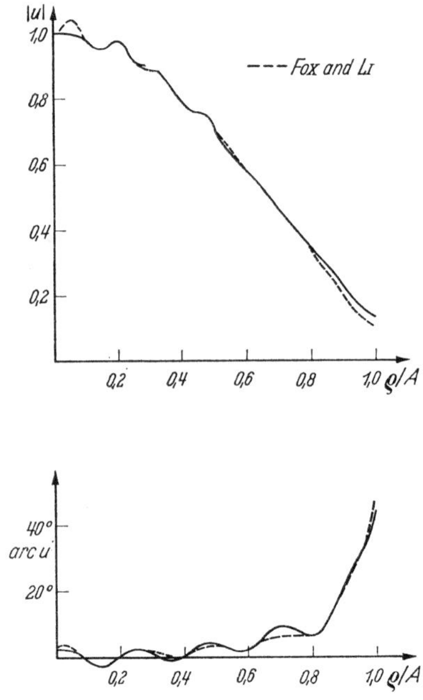

Fig. 14. Amplitude and phase distribution for the lowest mode of circular symmetry for the three-dimensional resonator ($N = 10$). (From RISKEN, l.c. [4] on p. 11.)

c) Three-dimensional resonator. According to RISKEN the problem of a cylindrical resonator can be treated in exactly the same manner as the two-dimensional resonator. The *only difference is,* that the g's occurring in the formulas (II.2.21), (II.2.22) are now defined by

$$g(v) = J_\nu(v)\, e^{i\nu\varphi} \tag{II.2.29}$$

where J_ν are Bessel functions and

$$x_\mu = j_{\nu\mu} \tag{II.2.30}$$

where $j_{\nu\mu}$ are the zeros of J_ν.

A graphical plot for this case is given in Fig. 14, while the losses read:

$$V_d = \left(\frac{j_{\nu\mu}}{\pi}\right)^2 \frac{1}{(2N)^{\frac{3}{2}}} \left[\frac{4}{3} + \frac{1}{\pi^2}\sum_{n=1}^{\infty} \frac{1+(-1)^n \cos(2\zeta\sqrt{n})}{n^{\frac{3}{2}}}\right]. \tag{II.2.31}$$

II.3. Confocal resonator[11]. This resonator is formed by two spherical mirrors of equal curvature separated by their common radius of curvature. The focal length of a spherical mirror is one-half of its radius of curvature, so that the focal points of the reflectors coincide. The reflectors are assumed to be square with a dimension $2A$ (compare Fig. 15) which is small compared to the spacing $D=R$ (R=radius). A and R are large compared to the wavelength. From the symmetry of the problem we can choose the electric field vector either in x- or in y-direction. In

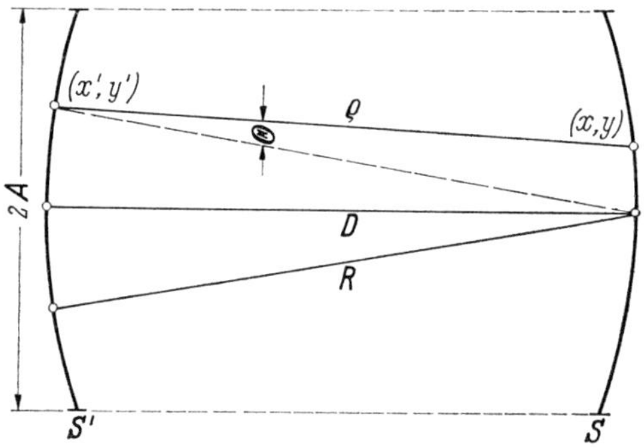

Fig. 15. Arrangement of confocal mirrors. (After BOYD and GORDON[11]. The American Telephone and Telegraph Co, 1961, reprinted by permission.)

the following we drop the corresponding index. According to HUYGEN's principles, the field at a point (x, y) belonging to surface S is determined by that of the field at all points of the other surface, S', by

$$E_S(x, y) = \int\limits_{S'} \frac{ik(1+\cos\Theta)}{4\pi\varrho} e^{-ik\varrho} E_{S'}(x', y')\, dS'. \tag{II.3.1}$$

The distance ϱ between (x, y) and (x', y') is approximately given (due to $2A<R$ and if $A^2/R\lambda \ll R^2/A^2$)

$$\frac{\varrho}{R} \approx 1 - \frac{xx'+yy'}{R^2}. \tag{II.3.2}$$

As is also usual in diffraction theory, ϱ can be made equal to R in the denominator of (II.3.1). The angle Θ (compare Fig. 15) is ≈ 0.

The *modes* are found by requiring that the field over S' reproduce itself over S, besides a constant factor, which is written as

$$\sigma = \sigma_m \cdot \sigma_n \tag{II.3.3}$$

so that

$$E_S(x, y) = \sigma E_{S'}(x, y). \tag{II.3.4}$$

Because the "kernel" $\frac{e^{-ik\varrho}}{\varrho}$ factorizes on account of (II.3.2) it is possible to represent $E(x, y)$ as a product:

$$E_{S'} = E_0 f_m(x)\, g_n(y). \tag{II.3.5}$$

[11] For more details see G. D. BOYD and J. P. GORDON: Bell System Tech. J. **40**, 489 (1961), whom we will follow in this Chap. 3.

Inserting (II.3.2)—(II.3.5) into (II.3.1) yields

$$\sigma_m f_m(x) \cdot \sigma_n g_n(y) = \frac{i k e^{-ikR}}{2\pi R} \int_{-A}^{A} f_m(x')\, e^{ik\frac{x'x}{R}}\, dx' \int_{-A}^{A} g_n(y')\, e^{ik\frac{y'y}{R}}\, dy'. \qquad \text{(II.3.6)}$$

Because x and y are independent variables the relation (II.3.6) can only be fulfilled if

$$f_m(x) = \text{const.} \int_{-A}^{A} f_m(x')\, e^{ik\frac{xx'}{R}}\, dx' \qquad \text{(II.3.7)}$$

and a corresponding relation for g_n holds.

According to SLEPIAN and POLLAK[12] the solutions of (II.3.7) are given by the angular wavefunctions in prolate spherical coordinates as defined by FLAMMER[13].

Provided the field has not too many modes in the $x-y$-plane, it is well concentrated around the axis, so that $y/A \ll 1$. In this case the solution of (II.3.6) and thus the field simplifies to

$$E(x, y) = E_0\, H_m(X)\, H_n(Y)\, e^{-\frac{1}{2}(X^2+Y^2)} \quad \text{where} \quad \left.\begin{aligned} X &= x\sqrt{\frac{k}{R}} \\ Y &= y\sqrt{\frac{k}{R}} \end{aligned}\right\} \qquad \text{(II.3.8)}$$

where H_m are Hermite polynomials, $m, n = 0, 1, 2, \ldots$

This solution can also be immediately obtained by letting the Fresnel number $N = \frac{1}{2\pi} \cdot \frac{A^2 k}{R} \to \infty$ in (II.3.6) after a suitable transformation.

It follows from (II.3.8) that the field in the lowest-mode falls to $1/e$, of its value at a

$$\text{spot radius } w_S = \sqrt{\frac{R\lambda}{\pi}} \qquad \text{(II.3.9)}$$

which is independent of the reflector aperture.

The *field* over an arbitrary plane $z = z_0$ *inside or outside the resonator* can be obtained again by the use of Huygen's principle.

a) Field outside the resonator. For large Fresnel numbers the travelling wave field resulting from the field at one of the reflectors is represented by

$$\left.\begin{aligned} E(x, y, z_0) = c_t E_0 \sqrt{\frac{2}{1+\xi^2}}\, H_m\!\left(X\sqrt{\frac{2}{1+\xi^2}}\right) H_n\!\left(Y\sqrt{\frac{2}{1+\xi^2}}\right) \times \\ \times\, e^{-(X^2+Y^2)\frac{1}{1+\xi^2}}\, e^{-i\varphi(X,Y)} \end{aligned}\right\} \qquad \text{(II.3.10)}$$

where

$$\left.\begin{aligned} \varphi(X, Y) &= k\left[\frac{R}{2}(1+\xi) + \frac{\xi}{(1+\xi^2)k}(X^2+Y^2)\right] - (1+m+n)\left(\frac{\pi}{2} - \varphi_0\right), \\ \xi &= \frac{2z_0}{R}, \\ \tan\varphi_0 &= \frac{1-\xi}{1+\xi} \end{aligned}\right\} \qquad \text{(II.3.11)}$$

c_t is the transmission coefficient of the reflector.

---

[12] D. SLEPIAN and H. O. POLLAK: Bell System Tech. J. **40**, 43 (1961).
[13] C. FLAMMER: Spheroidal Wave Functions. Palo Alto: Stanf. Univ. Press 1957.

b) Field inside the resonator. The factor c_t is omitted, and $e^{-i\varphi(X,Y)}$ is replaced by $\sin\varphi(X, Y)$. The surfaces of constant phase are approximately spherical with radii of curvature

$$R' = \left|\frac{1+\xi^2}{2\xi}\right| R. \tag{II.3.12}$$

c) Far field pattern of the confocal resonator. As can be seen from (II.3.10), the spot size of the TEM_{00q} mode is at a distance z:

$$w_S = \left[\frac{R\lambda}{2\pi}(1+\xi^2)\right]^{\frac{1}{2}}. \tag{II.3.13}$$

The angular beam width Θ can be defined as the ratio w/z for $z\to\infty$, which yields

$$\Theta = \sqrt{\frac{R\lambda}{2\pi}}\,\frac{\xi}{z} = \sqrt{\frac{R\lambda}{2\pi}}\,\frac{2}{R} = \sqrt{\frac{2\lambda}{\pi R}}. \tag{II.3.14}$$

In order to obtain the beam width between the half-power points, (II.3.14) is to be multiplied with $(2 \ln 2)^{\frac{1}{2}}$.

The beam width of the confocal resonator is larger than that of the Fabry-Perot resonator by a factor $(\lambda/R)^{-\frac{1}{2}}$.

d) Phase shifts and losses. We return now to the integral Eq. (II.3.6) or (II.3.7) and their eigenvalues. They are

$$\sigma_m\,\sigma_n = \chi_m\,\chi_n\,i\,e^{-ikR} \tag{II.3.15}$$

where (SLEPIAN and POLLAK[12])

$$\left.\begin{aligned} \chi_m &= \sqrt{\frac{2c}{\pi}}\, i^m R_{0m}^{(1)}(c, 1), \qquad m = 0, 1, 2, \ldots \\ &\left(c = \frac{A^2 k}{R} = 2\pi\cdot\text{Fresnel number}\right) \end{aligned}\right\} \tag{II.3.16}$$

$R_{0m}^{(1)}(c, 1)$ is the radial wavefunction in prolate spherical coordinates (FLAMMER[13]).

Because in the resonance case the phase factor of the field must be reproduced at one surface after a full round trip, it follows that

$$(\sigma_m\,\sigma_n)^2 \quad \text{must be equal} \quad e^{2\pi i q}\, r_0 \tag{II.3.17}$$

where q is an integer and r_0 is real.

Because the functions $R_{0m}^{(1)}$ are real, it follows from (II.3.15)—(II.3.17)

$$2\pi q = 2\left(\frac{\pi}{2} - kR + (m+n)\frac{\pi}{2}\right) \tag{II.3.18}$$

or, with $k = 2\pi/\lambda$, we obtain as *resonant condition*

$$\frac{4\pi}{\lambda} R = 2q + (1+m+n). \tag{II.3.19}$$

Note that the mode spectrum is highly degenerate.

The fractional energy loss per reflection due to diffraction is given by

$$1 - |\sigma_m \sigma_n|^2. \tag{II.3.20}$$

A plot is given in Fig. 16, together with results of FOX and LI[3] for the plane parallel resonator with circular reflectors. The diffraction losses for the confocal resonators are orders of magnitude smaller than for the plane parallel resonator. BOYD and KOGELNIK have considered also more general configurations.

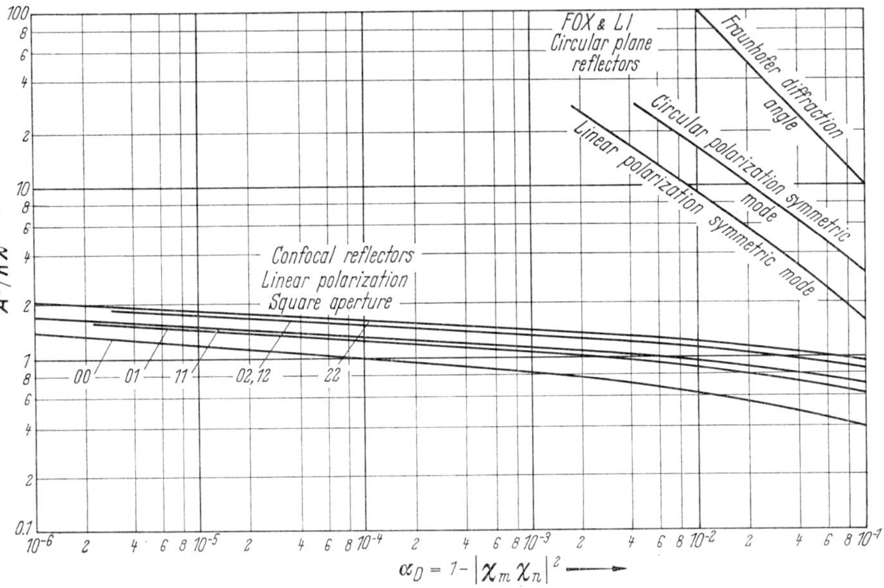

Fig. 16. Diffraction losses for confocal and plane-parallel resonator. (From Boyd and GORDON, l.c. [11] on p. 19. The American Telephone and Telegraph Co., 1961, reprinted by permission.)

II.4. More general configurations[14]. When the system of the mirrors becomes asymmetric, e.g. by unequal apertures or curvatures, one must require that the field pattern is reproduced (besides a constant factor) after a *full* round trip from S_1 to S_2 back to S_1. One then has to apply Huygen's principle twice. The integral equation which replaces (II.3.6) then reads

$$\sigma_m^2\,\sigma_n^2 E(x,y) = -\frac{k^2\,e^{-2ikD}}{4\pi^2 D^2}\int\limits_{S_1} d\,x'\,d\,y'\,E(x',y')\,K(x,x';y,y') \qquad \text{(II.4.1)}$$

where, with the same approximation as (II.3.2)

$$K(x,x';y,y') = \int d\,x''\,d\,y''\,\exp\left(i\,\frac{k}{D}\,[x''(x+x')+y''(y+y')]\right). \qquad \text{(II.4.2)}$$

This kernel depends on the shape and size of the reflector S, and has been evaluated by BOYD and KOGELNIK for various situations. We do not give the details of their calculations, but list merely their main results:

a) Confocal resonators with unequal square and rectangular apertures. The kernel and thus the integral equation factorize. The mirrors $j(=1,2)$ are of dimensions $2a_j$ in x-direction, and $2A_j$ in y-direction. The spot sizes are then:

$$\left.\begin{aligned} &\text{on reflector } S_1\text{:}\quad x_S = \sqrt{\frac{a_1}{a_2}}\sqrt{\frac{R\lambda}{\pi}}, \qquad y_S = \sqrt{\frac{A_1}{A_2}}\sqrt{\frac{R\lambda}{\pi}}\\ &\text{on reflector } S_2\text{:}\quad x'_S = \sqrt{\frac{a_2}{a_1}}\sqrt{\frac{R\lambda}{\pi}}, \qquad y'_S = \sqrt{\frac{A_2}{A_1}}\sqrt{\frac{R\lambda}{\pi}}. \end{aligned}\right\} \qquad \text{(II.4.3)}$$

The resonance condition (II.3.19) is unaltered.

[14] G. D. BOYD and H. KOGELNIK: Bell System Tech. J. **41**, 1347 (1962).

The diffraction losses are equal to that of a confocal system with equal aperture dimensions a_0; A_0 if $a_0^2 = a_1 a_2$ and $A_0^2 = A_1 A_2$.

b) Resonators with reflectors of unequal curvature. The resonance condition reads

$$\frac{2D}{\lambda} = q + \frac{1}{\pi}(1+m+n)\cos^{-1}\sqrt{\left(1-\frac{D}{R_1}\right)\left(1-\frac{D}{R_2}\right)}. \qquad \text{(II.4.4)}$$

For square apertures of dimensions $2A_1$ and $2A_2$, respectively, the diffraction losses at each reflector ($i=1, 2$) can be obtained from that of the confocal resonator by replacing, e.g. in diagram 16,

$$\frac{A^2}{R\lambda} \quad \text{by} \quad \frac{A_i^2}{D_i\lambda}\left[2\frac{D_i}{R_i} - \left(\frac{D_i}{R_i}\right)^2\right]^{\frac{1}{2}} \qquad \text{(II.4.5)}$$

where

$$D_i = 2D\,\frac{R_1 + R_2 - R_i - D}{R_1 + R_2 - 2D}.$$

A special case is the *resonator with equal spherical mirrors but with nonconfocal spacing*:

α) Large circular apertures. One obtains for the spot size of the fundamental TEM_{00q} mode:

$$w_S' = \sqrt{\frac{\lambda R}{\pi}}\left(\frac{D}{2R-D}\right)^{\frac{1}{4}}. \qquad \text{(II.4.6)}$$

The relative field-distribution at the reflectors of a TEM_{plq} mode is given in polar coordinates r, φ by

$$E(r,\varphi) = E_0\left(\frac{r}{w_S'}\sqrt{2}\right)^l L_p^l\left(2\frac{r^2}{w_S'^2}\right)e^{-r^2/w_S'^2}\cos l\varphi \qquad \text{(II.4.7)}$$

where L_p^l are the associated Laguerre polynomials. The resonant condition reads:

$$\frac{2D}{\lambda} = q + \frac{1}{\pi}(2p+l+1)\cos^{-1}\left(1-\frac{D}{R'}\right). \qquad \text{(II.4.8)}$$

β) Large square aperture.

$$E(x,y) = E_0 H_m\left(\frac{x\sqrt{2}}{w_S'}\right)H_n\left(\frac{y\sqrt{2}}{w_S'}\right)e^{-(x^2+y^2)/w_S'^2} \qquad \text{(II.4.9)}$$

where w_S is given under α, and the H_m's are Hermite polynomials. The resonant condition can be obtained from (II.4.10) (see below) by putting $R_1 = R_2 = R$.

A second special case is the resonator consisting of a spherical mirror $R_1 = 2D$ and a plane mirror ($R_2 = \infty$) located at $D = \frac{1}{2}R_1$ where the surface of constant phase is a plane. This system is in a stable region (see below). The resonant condition is

$$\frac{2D}{\lambda} = q + \frac{1}{2\pi}(1+m+n)\cos^{-1}\left(1-\frac{2D}{R_1}\right). \qquad \text{(II.4.10)}$$

The spot sizes are

$$\left.\begin{aligned} &\text{on the curved reflector} \quad w_1 = \left(\frac{\lambda D}{\pi}\right)^{\frac{1}{2}}\left[\frac{R_1^2}{D(R_1-D)}\right]^{\frac{1}{4}},\\ &\text{on the plane reflector} \quad w_2 = \left(\frac{\lambda D}{\pi}\right)^{\frac{1}{2}}\left[\frac{R_1-D}{D}\right]^{\frac{1}{4}}.\end{aligned}\right\} \qquad \text{(II.4.11)}$$

II.5. Stability. Depending on the relative size of R_1/D and R_2/D, there exist stable (or „low-loss") and unstable (or „high-loss") regions*, as has been shown

* It should be noted, however, that laser action can also occur in the "high-loss" region provided the gain in the active material is high enough.

by BOYD and KOGELNIK[14] by a „translation" of PIERCE'S[15] treatment of an infinite sequence of lenses. The boundaries of the stable regions are given by

$$\left(\frac{D}{R_1}-1\right)\left(\frac{D}{R_2}-1\right) \leqq 1 \tag{II.4.12}$$

and

$$\left(\frac{D}{R_1}-1\right)\left(\frac{D}{R_2}-1\right) \geqq 0 .$$

As is seen from (II.4.12), the confocal resonator $R_1=R_2=D$ is just on the border of an unstable region, although it has minimum diffraction losses. By deviations from the ideal confocal resonator, one can possibly obtain a system of high loss. By making *both* radii slightly smaller or larger, one can move away from the unstable region.

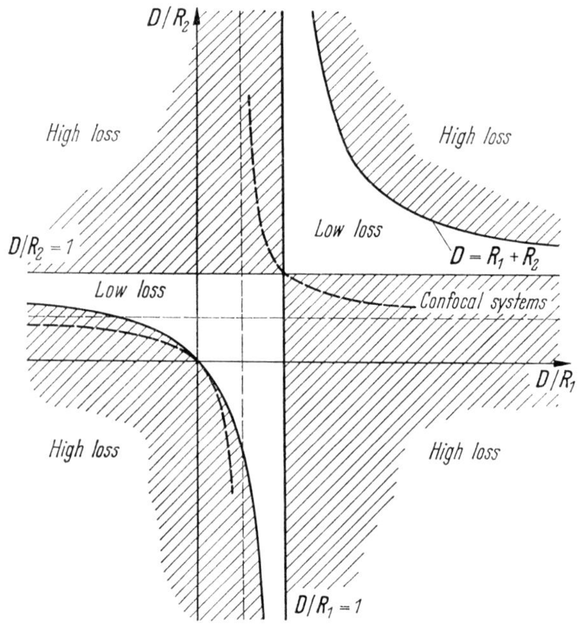

Fig. 17. Two-dimensional diagram of stable and unstable regions. (From BOYD and KOGELNIK, l.c. [14] on p. 22. The American Telephone and Telegraph Co, 1962, reprinted by permission.)

III. Quantum mechanical equations of the light field and the atoms without losses.

III.1. Quantization of the light field[1]. The radiation field is represented by Maxwell's equations

$$\left.\begin{aligned} \operatorname{curl} \boldsymbol{H} &= \frac{4\pi}{c}\boldsymbol{j} + \frac{1}{c}\frac{d\boldsymbol{E}}{dt} + \frac{4\pi}{c}\frac{d\boldsymbol{P}}{dt} \\ \operatorname{curl} \boldsymbol{E} &= -\frac{1}{c}\frac{d\boldsymbol{H}}{dt} \\ \operatorname{div} \boldsymbol{B} &= 0, \quad \operatorname{div} \boldsymbol{D} = 0 \end{aligned}\right\} \tag{III.1.1}$$

[15] J. R. PIERCE: Theory and Design of Electron Beams. New York: D. van Nostrand Co. 1954.

[1] For more details about the quantization process see e.g. W. HEITLER: The Quantum Theory of Radiation. Oxford: Clarendon Press 1954.

where we use in the following the cgs-system (and assume $\mu=1$). $\boldsymbol{E}$ and $\boldsymbol{H}$ can be obtained from the scalar potential φ and the vector potential $\boldsymbol{A}$ in the well known way

$$\boldsymbol{H} = \operatorname{curl} \boldsymbol{A}, \tag{III.1.2}$$

$$\boldsymbol{E} = -\frac{1}{c}\frac{d}{dt}\boldsymbol{A} - \operatorname{grad}\varphi. \tag{III.1.3}$$

In the following we choose for $\boldsymbol{A}$ the Coulomb gauge

$$\operatorname{div} \boldsymbol{A} = 0. \tag{III.1.4}$$

The current density $\boldsymbol{j}$ and the polarization $\boldsymbol{P}$ stem from the motion (or displacement) of the electrons and molecules, which we neglect for the time being, however. We consider first the quantization of the radiation field. Because we want to consider the laser, first of all we suppose the radiation field to be transversal, i.e. that (III.1.4) holds, and $\varphi=0$. (In cavities and wave guides the field has a longitudinal component also, so that the quantization must be done in these cases in a somewhat different way). We decompose the vector potential into a series of orthogonal normalized functions, $\boldsymbol{u}_\lambda(\boldsymbol{x})$, which are suitable for the problem under consideration. In the Fabry-Perot resonator these expansion functions can be chosen for instance as functions (II.2.6) multiplied by a normalization constant and a polarization vector, where it is merely necessary to define the $\boldsymbol{u}(\boldsymbol{x})$'s in the region of the Fabry-Perot interferometer. One should bear in mind, however, that due to the slight nonorthogonality of decaying modes this expansion is not completely exact.

We write the expansion in the form

$$\boldsymbol{A}(\boldsymbol{x}) = \sum_\lambda \sqrt{\frac{2\pi c^2 \hbar}{\omega_\lambda}}\,(b_\lambda^+ + b_\lambda)\,\boldsymbol{u}_\lambda(\boldsymbol{x}) \tag{III.1.5}$$

where in the normalization volume

$$\int \boldsymbol{u}_\lambda(\boldsymbol{x})\,\boldsymbol{u}_\lambda(\boldsymbol{x})\,dV = 1 \tag{III.1.6}$$

(III.1.5) can be split into (III.1.7) and (III.1.8) where:

$$\boldsymbol{A}^{(-)} = \sum_\lambda \sqrt{\frac{2\pi c^2 \hbar}{\omega_\lambda}}\, b_\lambda^+\,\boldsymbol{u}_\lambda(\boldsymbol{x}) \tag{III.1.7}$$

and

$$\boldsymbol{A}^{(+)} = \sum_\lambda \sqrt{\frac{2\pi c^2 \hbar}{\omega_\lambda}}\, b_\lambda\,\boldsymbol{u}_\lambda(\boldsymbol{x}) \tag{III.1.8}$$

$\boldsymbol{A}^{(-)}$ is called the negative frequency part, while $\boldsymbol{A}^{(+)}$ is the positive frequency part.

This notation becomes clear, when we replace b_λ in the interaction representation by

$$b_\lambda\, e^{-i\omega_\lambda t} \tag{III.1.9}$$

and take into account that, in quantum theory, exponential functions with positive frequencies are written in the form (III.1.9). While (III.1.5) still represents a classical expansion the quantization is performed by the requirement that the amplitudes b, b^+ become operators with the following commutation relations

$$[b_\lambda, b_\varrho^+] \equiv b_\lambda b_\varrho^+ - b_\varrho^+ b_\lambda = \delta_{\lambda\varrho}, \tag{III.1.10}$$

$$[b_\lambda, b_\varrho] = 0, \tag{III.1.11}$$

$$[b_\lambda^+, b_\varrho^+] = 0. \tag{III.1.12}$$

The energy of the field now becomes an operator, namely the Hamiltonian H_L which reads

$$H_L = \sum \hbar\omega_\lambda b_\lambda^+ b_\lambda. \tag{III.1.13}$$

From $\boldsymbol{A}$ we can obtain again the electric and magnetic field strength by the rule that the time derivative of an operator O is given by its commutator with the Hamiltonian

$$\frac{dO}{dt} = \frac{i}{\hbar}[H, O], \tag{III.1.14}$$

as is usual in quantum theory.

We thus obtain for the operator of the electric field strength

$$\left.\begin{aligned} \boldsymbol{E} &= -\frac{1}{c}\frac{i}{\hbar}[H_L, \boldsymbol{A}] = \sum_\lambda \boldsymbol{E}_\lambda(\boldsymbol{x}) \\ &\text{with} \\ \boldsymbol{E}_\lambda(\boldsymbol{x}) &= -i(b_\lambda^+ - b_\lambda)\sqrt{2\pi\hbar\omega_\lambda}\,\boldsymbol{u}_\lambda(\boldsymbol{x}) \end{aligned}\right\} \tag{III.1.15}$$

and for the operators of the magnetic field strength

$$\boldsymbol{H} = \text{curl}\,\boldsymbol{A}. \tag{III.1.16}$$

The state of the light field is described by eigenfunctions which are also often called Hilbert space vectors. When the energy of the radiation field is measured, the eigenstates must obey the Schrödinger equation

$$H_L\Phi = E\Phi. \tag{III.1.17}$$

They can be obtained from the vacuum state Φ_0, which is defined by

$$b_\lambda \Phi_0 = 0 \tag{III.1.18}$$

for all λ's by a multiple application of the creation operators b_λ^+:

$$\Phi_{n_1, n_2, \ldots, n_N} = \frac{1}{\sqrt{n_1!\, n_2! \ldots n_N!}}(b_1^+)^{n_1} \ldots (b_N^+)^{n_N}\Phi_0. \tag{III.1.19}$$

The n's are positive integers.

This function represents a state in which the specific waves $\lambda_1, \lambda_2, \ldots$ are occupied with $n_1, n_2, \ldots$ quanta, respectively. The corresponding energy reads

$$E_{n_1, n_2, \ldots, n_N} = n_1\hbar\omega_1 + n_2\hbar\omega_2 + \cdots + n_N\hbar\omega_N. \tag{III.1.20}$$

An arbitrary state of the light field can be represented as a linear combination of the states (III.1.19).

One finds generally

$$\langle\Phi^* b_\lambda^+ b_\lambda \Phi\rangle\hbar\omega_\lambda = \frac{1}{4\pi}\int\langle E_\lambda^2\rangle\, dV. \tag{III.1.21}$$

For the following we need also to know the selection rules

$$\langle\Phi^*_{n_1,\ldots,n_\lambda,\ldots,n_N} b_\lambda^+ \Phi_{m_1,\ldots,m_\lambda,\ldots,m_N}\rangle = \delta_{n_\lambda, m_\lambda+1}\sqrt{m_\lambda+1}\prod_{\lambda'\neq\lambda}\delta_{n_{\lambda'}, m_{\lambda'}}, \tag{III.1.22}$$

$$\langle\Phi^*_{n_1,\ldots,n_\lambda,\ldots,n_N} b_\lambda \Phi_{m_1,\ldots,m_\lambda,\ldots,m_N}\rangle = \delta_{n_\lambda, m_\lambda-1}\sqrt{m_\lambda}\prod_{\lambda'\neq\lambda}\delta_{n_{\lambda'}, m_{\lambda'}}. \tag{III.1.23}$$

III.2. Second quantization of the electron wave field[1]. Within the second quantization the electronic wave functions $\psi(\boldsymbol{x})$ become operators in the same way as the vector potential above. We decompose ψ into a complete set of normalized and orthogonal functions φ_j

$$\psi(\boldsymbol{x}) = \sum_j a_j \varphi_j(\boldsymbol{x}) \tag{III.2.1}$$

and correspondingly

$$\psi^+(\boldsymbol{x}) = \sum_j a_j^+ \varphi_j^*(\boldsymbol{x}). \tag{III.2.2}$$

Because ψ is supposed to describe electrons which obey the Pauli principle, ψ, ψ^+ must obey the +commutation relations

$$\psi(\boldsymbol{x})\,\psi^+(\boldsymbol{x}') + \psi^+(\boldsymbol{x}')\,\psi(\boldsymbol{x}) = \delta(\boldsymbol{x}-\boldsymbol{x}'), \tag{III.2.3}$$

$$\psi(\boldsymbol{x})\,\psi(\boldsymbol{x}') + \psi(\boldsymbol{x}')\,\psi(\boldsymbol{x}) = 0, \tag{III.2.4}$$

$$\psi^+(\boldsymbol{x})\,\psi^+(\boldsymbol{x}') + \psi^+(\boldsymbol{x}')\,\psi^+(\boldsymbol{x}) = 0 \tag{III.2.5}$$

which are equivalent to the following commutation relations for a^+, a

$$a_j a_l^+ + a_l^+ a_j = \delta_{jl}, \qquad a_j a_l + a_l a_j = 0, \qquad a_j^+ a_l^+ + a_l^+ a_j^+ = 0. \tag{III.2.6}$$

If the interaction between the electrons can be neglected, the Hamiltonian H_{El} of the second quantization can be obtained from the Hamiltonian of the usual Schrödinger equation

$$H_0\psi \equiv \left(-\frac{\hbar^2}{2m}\Delta + V(\boldsymbol{x})\right)\psi = E\psi \tag{III.2.7}$$

as follows

$$H_{El} = \int \psi^+(\boldsymbol{x})\, H_0\, \psi(\boldsymbol{x})\, dV. \tag{III.2.8}$$

Choosing for φ especially the eigenfunctions of (III.2.7), (III.2.8) simplifies immediately to

$$H_{El} = \sum E_j a_j^+ a_j \tag{III.2.9}$$

where use of the orthogonality of the φ_j's has been made. H_{El} acts now on functions χ, which describe electron quantum states in a many-body description. The eigenstates of H_{El} read

$$\chi = (a_1^+)^{n_1} (a_2^+)^{n_2} \dots (a_N^+)^{n_N} \chi_0 \tag{III.2.10}$$

where the vacuum state is defined by

$$a_j \chi_0 = 0 \tag{III.2.11}$$

for all j's. On account of the Pauli principle [compare also (III.2.6)] the n's may have only the values 0 or 1. By means of ψ and χ the one-particle expectation values can be represented as follows

$$\left.\begin{aligned} &\langle \chi^+ \left(\int \psi^+(\boldsymbol{x})\, O(\boldsymbol{x})\, \psi(\boldsymbol{x})\, dV\right) \chi \rangle \\ &\quad = \left\langle \chi^+ \left(\sum_{jl} a_j^+ a_l \int \varphi_j^* O \varphi_l \, dV\right) \chi \right\rangle. \end{aligned}\right\} \tag{III.2.12}$$

For the whole article we adopt the convention that all *circular* atomic transition frequencies are denoted by ν, e.g. the frequency connected with the transition from level i to k by

$$\nu_{ik} = \frac{1}{\hbar}(E_i - E_k).$$

Note that ν_{ik} is negative if $E_i < E_k$.

[1] See footnote [1] on p. 24.

III.3. Interaction between radiation field and electron wave field[1]. Replacing H_0 in (III.2.8) by

$$H = \frac{1}{2m}\left(\boldsymbol{p} - \frac{e}{c}\boldsymbol{A}\right)^2 + V(\boldsymbol{x}) \quad \text{where} \quad \boldsymbol{p} = \frac{\hbar}{i}\boldsymbol{\nabla} \tag{III.3.1}$$

we obtain with respect to (III.2.9) the following additional term, which describes the interaction between electron motion and light field:

$$H_I = H_{I,1} + H_{I,2} \tag{III.3.2}$$

with

$$H_{I,1} = \int \psi^+(\boldsymbol{x})\left(-\frac{e}{mc}\right)\boldsymbol{A}\,\boldsymbol{p}\,\psi(\boldsymbol{x})\,dV \tag{III.3.3}$$

and

$$H_{I,2} = \frac{e^2}{2mc^2}\int \psi^+(\boldsymbol{x})\,\boldsymbol{A}^2(\boldsymbol{x})\,\psi(\boldsymbol{x})\,dV. \tag{III.3.4}$$

Inserting for $\psi(\boldsymbol{x})$ and $\boldsymbol{A}(\boldsymbol{x})$ the expansion (III.2.1) and (III.1.5), respectively, we obtain

$$H_{I,1} = -\frac{e}{m}\sum_{\lambda,j,l}\sqrt{\frac{2\pi\hbar}{\omega_\lambda}}\,(b_\lambda^+ + b_\lambda)\left(\left(\int \varphi_j^*\,\boldsymbol{u}_\lambda(\boldsymbol{x})\,\boldsymbol{p}\,\varphi_l\,dV\right) a_j^+\,a_l\right), \tag{III.3.5}$$

$$\left.\begin{aligned} H_{I,2} = \frac{e^2}{m}\sum_{\lambda,\lambda',j,l}(\pi\hbar)\,\frac{1}{\sqrt{\omega_\lambda\,\omega_\lambda}}\int \varphi_j^*\,\boldsymbol{u}_\lambda(\boldsymbol{x})\,\boldsymbol{u}_{\lambda'}(\boldsymbol{x})\,\varphi_l\,dV \times \\ \times (b_\lambda^+ + b_\lambda)\,(b_{\lambda'}^+ + b_{\lambda'})\,a_j^+\,a_l. \end{aligned}\right\} \tag{III.3.6}$$

In the following we will consider mainly $H_{I,1}$, because $H_{I,2}$ is usually much smaller.

If the φ's represent functions of an electron which is localized around a nucleus at the space point $\boldsymbol{x}_0$, we can put in the lowest approximation[2] (dipole approximation) $\boldsymbol{u}_\lambda$ in front of the integral

$$H_{I,1} = -\frac{e}{m}\sum_{\lambda,j,l}\sqrt{\frac{2\pi\hbar}{\omega_\lambda}}\,(b_\lambda^+ + b_\lambda)\,\boldsymbol{u}_\lambda(\boldsymbol{x}_0)\int \varphi_j^*\,\boldsymbol{p}\,\varphi_l\,dV\cdot a_j^+\,a_l. \tag{III.3.7}$$

The matrix element $e\int \varphi_j^*\,\boldsymbol{p}\,\varphi_l\,dV$ can also be written* as dipole matrix element

$$m\,i\,\nu_{jl}\int \varphi_j^*\,(e\boldsymbol{x})\,\varphi_l\,dV \tag{III.3.8}$$

where

$$\nu_{jl} = \frac{1}{\hbar}\,(E_j - E_l). \tag{III.3.9}$$

Using (III.1.13), (III, 2.9) and (III.3.2) we find for the total Hamiltonian of the light field and the electrons

$$H_{Tot} = H_{El} + H_L + H_I. \tag{III.3.10}$$

This Hamiltonian may refer not only to electrons at one atom but also to electrons at all the atoms. In the following we will assume that we have N atoms each

* This relation follows in the well-known manner:

$$\boldsymbol{p}_{21} = m\,\dot{\boldsymbol{x}}_{21} = \frac{mi}{\hbar}\,([H,\boldsymbol{x}])_{21} = \frac{mi}{\hbar}\,(E_2 - E_1)\,\boldsymbol{x}_{21} = i\,m\,\nu_{21}\,\boldsymbol{x}_{21}.$$

[1] See footnote [1] on p. 24.

[2] For a detailed investigation of the dipole approximation and the possibility of replacing $\boldsymbol{A}$ by a scalar potential in addition to $V(\boldsymbol{x})$, see M. Goeppert-Mayer: Ann. Physik **9**, 273 (1931). The question of replacing $\boldsymbol{A}$ is treated quite generally by J. Fiutak: Canad. J. Phys. **41**, 12 (1963).

with a single electron. In this case $(a_j^+ a_l)$ has to be supplemented by the further index μ

$$(a_{j,\mu}^+ a_{l,\mu}) \quad \text{or} \quad (a_j^+ a_l)_\mu. \tag{III.3.11}$$

If the electronic wave functions of the different atoms do not overlap, we can consider the electrons as distinguishable. Because the electronic wave functions are in this case orthogonal, it follows immediately that the pairs (III.3.11) commute for different μ's. If there are many atoms within the wavelength of the electromagnetic oscillation under consideration one can average in many cases over this ensemble although one loses information about linewidth etc., as will be shown below.

As with the usual quantum mechanics, the wave functions which refer to the total system consisting of electrons and light waves must obey the Schrödinger equation

$$H_{Tot}\,\Phi = i\hbar \frac{d\Phi}{dt}. \tag{III.3.12}$$

III.4. The interaction representation[1] and the rotating wave approximation[2]. According to (III.1.19), (III.1.20), (III.2.9) and (III.2.10), the problem

$$H_0 \Phi = i\hbar \frac{d\Phi}{dt}, \qquad H_0 = H_{El} + H_L \tag{III.4.1}$$

can be considered as solved. For this reason we can split off the unperturbed time dependence of Φ by means of the transformation

$$\Phi = U\,\Phi_I \quad \text{with} \quad U = e^{-iH_0 t/\hbar}. \tag{III.4.2}$$

Φ_I then obeys the equation

$$\tilde{H}_I \Phi_I = i\hbar \frac{d\Phi_I}{dt} \tag{III.4.3}$$

where

$$\tilde{H}_I = U^+ H_I U. \tag{III.4.4}$$

This representation is called the interaction representation.

On account of the form of H_0 and due to the commutation relations (III.1.10) to (III.1.12), (III.2.6), we find

$$\left.\begin{aligned}\tilde{H}_{I,1} = -\frac{e}{m} \sum_{\lambda,j,l} \sqrt{\frac{2\pi\hbar}{\omega_\lambda}}\,(b_\lambda^+ e^{i\omega_\lambda t} + b_\lambda e^{-i\omega_\lambda t}) \left(\int \varphi_j^* \boldsymbol{u}_\lambda(\boldsymbol{x})\, \boldsymbol{p} \varphi_l \, dV\right) \times \\ \times a_j^+ e^{iE_j t/\hbar} a_l e^{-iE_l t/\hbar}\end{aligned}\right\} \tag{III.4.5}$$

and a corresponding expression for $\tilde{H}_{I,2}$.

In (III.4.5) exponential functions of the form

$$\exp\{i(\omega_\lambda + \nu_{jl})\,t\} \tag{III.4.6}$$

occur.

When $\tilde{H}_I$ in Eq. (III.4.3) is treated by time dependent perturbation theory, the factor of t in (III.4.6) occurs in the denominator of the transition amplitude.

[1] See footnote [1] on p. 24.

[2] For the application of the rotating wave approximation in spin-resonance, see e.g.: F. Bloch and A. J. Siegert: Phys. Rev. **57**, 522 (1940). — J. J. Rabi, N. F. Ramey, and J. Schwinger: Rev. Mod. Phys. **26**, 107 (1954). — A. Abragam: The Principles of Nuclear Magnetism. Oxford: Clarendon Press 1962.

In the case of approximate resonance

$$\omega_\lambda + \nu_{jl} \approx 0 \tag{III.4.7}$$

the corresponding contributions are very large, whereas in the opposite case ($\omega_\lambda \to -\omega_\lambda$):

$$-\omega_\lambda + \nu_{jl} \approx -2\omega_\lambda . \tag{III.4.8}$$

Due to the high frequency, the corresponding contribution is extremely small. Thus one can often completely neglect those terms in (III.4.5) which show the behavior (III.4.8). This approximation is called the *rotating wave approximation*[2]. As we will see below, it possesses a very simple meaning for spins in a magnetic field.

III.5. The equations of motion in the Heisenberg picture. In this picture the time dependence is completely transferred to the operators from the eigenfunctions (wave functions) by putting

$$\Phi = U_H \, \Phi_H(0) \quad \text{with} \quad U_H = e^{-\frac{i}{\hbar} H_{Tot} t} \tag{III.5.1}$$

(the explicit form of U_H holds as long as H_{Tot} is time independent). U_H obeys the equation

$$H_{Tot} \, U_H = i\hbar \dot{U}_H \tag{III.5.2}$$

and is unitary:

$$U_H^+ \, U_H = 1 \tag{III.5.3}$$

because H_{Tot} is hermitian:

$$H_{Tot}^+ = H_{Tot} . \tag{III.5.4}$$

The operators O are then transformed according to

$$O_H = U_H^+ \, O \, U_H \tag{III.5.5}$$

and obey the equation

$$\dot{O}_H = \frac{i}{\hbar} [H_{Tot}, O_H] + \frac{\partial}{\partial t} O_H \tag{III.5.6}$$

as can be checked immediately by differentiation with respect to the time. The differentiation $\partial O_H / \partial t$ refers to the explicit time dependence of O_H.

Because in quantum theory the operators correspond to observables of the classical theory, and because these operators are fully time dependent in the Heisenberg picture as the classical observables, the transition from quantum theory to the classical theory can be most easily done in the Heisenberg picture. This is one of the reasons why this picture is preferably used in the laser theory.

Choosing especially b_λ^+ for O_H, we obtain by means of (III.3.10), (III.1.10)

$$\dot{b}_\lambda^+ = i\omega_\lambda \, b_\lambda^+ + i \sum_{j,l,\mu} g_{j,l,\lambda,\mu} \, a_{j,\mu}^+ \, a_{l,\mu} \tag{III.5.7}$$

where

$$g_{j,l,\lambda,\mu} = -\frac{1}{\hbar} \frac{e}{m} \sqrt{\frac{\hbar \, 2\pi}{\omega_\lambda}} \int \varphi_{j,\mu}^* \, \boldsymbol{u}_\lambda(\boldsymbol{x}) \, \boldsymbol{p} \, \varphi_{l,\mu} \, dV . \tag{III.5.8}$$

With respect to the equations of motion for the electronic operators one could try to derive those for $\dot{a}_{j,\mu}^+$. In these cases, however, one obtains on the right hand side of the corresponding equations expressions bilinear in $a_{j,\mu}^+$, $a_{l,\mu}$, so that one would now have to determine equations for their time derivatives again. On the other hand the $a_{j,\mu}^+$'s and $a_{l,\mu}$'s always jointly occur in all expectation values.

It is therefore advisable to investigate the temporal change of $a_{j,\mu}^{+} a_{l\mu}$ from the very beginning. In this way we find

$$\left.\begin{aligned}(a_{j'}^{+} a_{l'})_{\mu}^{\cdot} = i\, \nu_{j' l'} (a_{j'}^{+} a_{l'})_{\mu} + i \sum_{j,l,\lambda} g_{j,l,\lambda,\mu} (b_{\lambda}^{+} + b_{\lambda}) \times \\ \times \{(a_{j}^{+} a_{l'})_{\mu}\, \delta_{l j'} - (a_{j'}^{+} a_{l})_{\mu}\, \delta_{j l'}\}.\end{aligned}\right\} \tag{III.5.9}$$

The field Eq. (III.5.7), together with the material Eq. (III.5.9), completely describe the coupled system of light waves and electrons as long as there are no external perturbations like the pumping and loss mechanisms. Because the behavior of the laser depends decisively on these latter mechanisms, we have to investigate their influence (this will be done in Chap. IV).

Let us consider as a special example a two level system in the rotating wave approximation:

Two level system.

The indices j and l can assume only the values 1 and 2:

$$(a_1^{+} a_1)_{\mu}^{\cdot} = -i \sum_{\lambda} \left(g_{1,2,\lambda,\mu}\, b_{\lambda}^{+} (a_1^{+} a_2)_{\mu} - g_{2,1,\lambda,\mu}\, b_{\lambda} (a_2^{+} a_1)_{\mu}\right), \tag{III.5.10}$$

$$(a_2^{+} a_2)_{\mu}^{\cdot} = i \sum_{\lambda} \left(g_{1,2,\lambda,\mu}\, b_{\lambda}^{+} (a_1^{+} a_2)_{\mu} - g_{2,1,\lambda,\mu}\, b_{\lambda} (a_2^{+} a_1)_{\mu}\right), \tag{III.5.11}$$

$$(a_1^{+} a_2)_{\mu}^{\cdot} = i\, \nu_{12} (a_1^{+} a_2)_{\mu} + i \sum_{\lambda} g_{2,1,\lambda,\mu}\, b_{\lambda} \left((a_2^{+} a_2) - (a_1^{+} a_1)_{\mu}\right), \tag{III.5.12}$$

where we assume $g_{l,l,\lambda,\mu} = 0$ for parity reasons

$$(a_2^{+} a_1)_{\mu}^{\cdot} = -i\, \nu_{12} (a_2^{+} a_1)_{\mu} - i \sum_{\lambda} g_{1,2,\lambda,\mu}\, b_{\lambda}^{+} \left((a_2^{+} a_2)_{\mu} - (a_1^{+} a_1)_{\mu}\right). \tag{III.5.13}$$

On account of $g_{2,1,\lambda,\mu} = g_{1,2,\lambda,\mu}^{*}$ this equation is the hermitian conjugate of Eq. (III.5.12). Because only the occupation number difference $N_2 - N_1$ occurs in (III.5.12) and (III.5.13), one can introduce this difference, the so called inversion, as a new variable

$$\Delta N = a_2^{+} a_2 - a_1^{+} a_1. \tag{III.5.14}$$

From (III.5.10)—(III.5.12) we then obtain the following equations

$$(\Delta N)_{\mu}^{\cdot} = 2i \sum_{\lambda} \left(g_{1,2,\lambda,\mu}\, b_{\lambda}^{+} (a_1^{+} a_2)_{\mu} - g_{2,1,\lambda,\mu}\, b_{\lambda} (a_2^{+} a_1)_{\mu}\right), \tag{III.5.15}$$

$$(a_1^{+} a_2)_{\mu}^{\cdot} = i\, \nu_{12} (a_1^{+} a_2)_{\mu} + i \sum_{\lambda} g_{2,1,\lambda,\mu}\, b_{\lambda}\, \Delta N_{\mu}. \tag{III.5.16}$$

Because the third equation is merely the hermitian conjugate of Eq. (III.5.16), we drop it.

III.6. The formal equivalence of the system of atoms each having 2 levels with a system of $\frac{1}{2}$ spins[1]. Because the spins as well as the electrons belonging to different atoms can be considered as distinguishable, it is sufficient to prove the equivalence for a single spin and a single electron. For this purpose we need only show that the following correspondence exists:

$$\left.\begin{array}{lcl} \text{spin-operators} & & \text{electron-operators} \\ \sigma^{+} = \sigma_x + i\,\sigma_y & \leftrightarrow & a_2^{+} a_1 \\ \sigma^{-} = \sigma_x - i\,\sigma_y & \leftrightarrow & a_1^{+} a_2 \\ \sigma_z & \leftrightarrow & \frac{1}{2}(a_2^{+} a_2 - a_1^{+} a_1) \end{array}\right\} \tag{III.6.1}$$

[1] For more details with regard to the spin system see: A. Abragam: The principles of nuclear magnetism. Oxford: Clarendon Press 1962. — Charles P. Slichter: Principles of magnetic resonance. New York: Harper & Row 1964.

$\sigma_x, \sigma_y, \sigma_z$ are the components of the spin-operator (without the factor $\hbar$) and can be represented by Pauli matrices. σ^+ and σ^- are spin-flip operators, so that σ^+ flips the spin from the negative z direction to the positive one, while σ^- causes the inverse process.

The operators σ^+, σ^- and σ_z obey the commutation relations:

$$\left.\begin{array}{ll} \sigma^{+2} = \sigma^{-2} = 0, & \sigma^+\sigma^- + \sigma^-\sigma^+ = 1, \\ [\sigma^\pm, \sigma_z] = \mp\sigma^\pm & \sigma^+\sigma^- - \sigma^-\sigma^+ = 2\sigma_z. \end{array}\right\} \tag{III.6.2}$$

Using the commutation relations (III.2.6) for the electron-operators and the fact that $a_1^+ a_1 + a_2^+ a_2 = 1$ in the two-level system, it may be easily shown that the combinations of electron operators on the right-hand side of (III.6.1) obey the same commutation relations (III.6.2). In order to make the correspondence complete we identify the quantum states:

$$\left.\begin{array}{ll} \Phi(\downarrow)\text{: spin down} & \Phi_1\text{: electron in the lower state} \\ \Phi(\uparrow)\text{: spin up} & \Phi_2\text{: electron in the upper state.} \end{array}\right\} \tag{III.6.3}$$

The equivalence (III.6.1)—(III.6.3) is of a purely mathematical and not of a physical nature. It allows us to transcribe calculations made with the spin $\frac{1}{2}$ system to an electronic 2-level system and vice versa. Thus, phenomena known from one field must have corresponding ones in the other field. An important application will be described in VII.12 in connection with the spin or photo-echo experiment. In particular, one finds the following equivalence between the Hamiltonians:

$$\left.\begin{array}{lll} \text{Spin Hamiltonian} & & \text{Electron Hamiltonian*} \\ H_S = \hbar(\nu_0\sigma_z + g_S^* F_m^+ \sigma^+ + g_S F_m \sigma^-) & \leftrightarrow & \frac{\Delta E}{2}(a_2^+ a_2 - a_1^+ a_1) + \\ & & + \hbar(g_{el}^* F_{el}^+ a_2^+ a_1 + \\ & & + g_{el} F_{el} a_1^+ a_2) \\ & & \text{with} \\ & & \Delta E = E_2 - E_1. \end{array}\right\} \tag{III.6.4}$$

Note that the physical meanings of $\hbar\nu_0$ and ΔE as well as of $g_S F_m$ and $g_{el} F_{el}$ are completely different:

$\hbar\nu_0$ and g_S are determined by the product of *the magnetic field* $\boldsymbol{H}$ *and the magnetic moment* connected with the spin**:

$$\left.\begin{array}{l} H_S = -\frac{e\hbar}{mc}\boldsymbol{\sigma H} = \underbrace{-\frac{e\hbar}{mc} H_z}_{\hbar\nu_0}\sigma_z \underbrace{-\frac{e\hbar}{2mc}(H_x - iH_y)}_{\hbar g_S^* F_m^+}\sigma^+ - \\ \underbrace{-\frac{e\hbar}{2mc}(H_x + iH_y)}_{\hbar g_S F_m}\sigma^-. \end{array}\right\} \tag{III.6.5}$$

* The notation g for the coupling constant is chosen in accordance with our general notation and has nothing to do with atomic g-factors.

** The Hamiltonian H_S shows clearly what is meant by the rotating wave approximation: For a rotating transverse magnetic field

$$H_x = H_t \cos\omega t, \quad H_y = H_t \sin\omega t$$

so that

$$\left.\begin{array}{ll} & \\ g_S F_m \sim e^{+i\omega t}, & g_S^* F_m^+ \sim e^{-i\omega t}. \end{array}\right\} \tag{III.6.8}$$

Because σ^+ is connected with a factor $e^{i\nu_0 t}$ (and σ^- with $e^{-i\nu_0 t}$) in the interaction representas tion, in the resonant case $\omega \approx \nu_0$ the time modulation vanishes, so that the interaction has it-

$g_{el} F_{el}$ in the electron Hamiltonian stems (at least in the dipole-approximation) from the product of the electric dipole moment of the bound electron and the electric field (because the electric field component $\boldsymbol{E}_\lambda$ is proportional to the corresponding component of the vector potential).

On account of the correspondence (III.6.1) and (III.6.4) we can readily transcribe the equations of motion (III.5.15), (III.5.16) into

$$\dot{\sigma}_z = i \sum_\lambda (g_\lambda b_\lambda^+ \sigma^- - g_\lambda^* b_\lambda \sigma^+), \tag{III.6.6}$$

$$\dot{\sigma}^+ = i \nu_0 \sigma^+ - i\, 2 \sum_\lambda b_\lambda^+ g_\lambda \sigma_z. \tag{III.6.7}$$

IV. Dissipation and fluctuation of quantum systems. The realistic laser equations.

IV.1. Some remarks on homogeneous and inhomogeneous broadening.

a) Natural linewidth. The optical transition energy of an atom is not exactly sharp for several reasons. Let us consider first an excited atom in the vacuum. Then the electron makes a spontaneous transition to its lower state emitting a light quantum. The finite lifetime of the upper state (if the lower state is the ground state) gives rise to the spontaneous natural linewidth, derived first by the theory of WIGNER and WEISSKOPF[1]. This width is directly connected with the transition probability into the lower state or, in other words, with the radiative lifetime. An outline of the theory of this process is given in V.3 and V.5.

When the atoms are embedded in a solid (or a liquid) an *additional* broadening[2] occurs. The same happens, through different mechanisms, for gas atoms.

b) Inhomogeneous broadening.

α) *Impurity atoms in solids.* If the impurity atoms are inserted at non-equivalent lattice sites, so that the atoms experience different crystal fields, the optical transition energy can be split (in the case of degeneracies) or shifted in a *different way for the different sites.* If there is a continuum of energetically different time-independent positions, the total line will be *inhomogeneously broadened* or, more exactly, an inhomogeneous broadening is superimposed upon the homogeneous spontaneous broadening. Such a broadening can be expected for instance in strained crystals. (This will generally be internal strain, inherent in the crystals

---

full strength. If, on the other hand, there is a linearly polarized transverse a.c. field,

$$H_x = H_t \cos \omega t, \qquad H_y = 0 \tag{III.6.9}$$

the $g_S F_m$ contains *both* $e^{-i\omega t}$ and $e^{i\omega t}$. The term $e^{i\omega t}$ gives together with $e^{i\nu_0 t}$ a very rapid modulation, so that the net interaction which stems from this term is very small. The physical reason is as follows: under the action of H_z the spin precesses. In the case (III.6.8) the transverse field precesses in the same direction as the spin (and for $\nu_0 = \omega$ with the same angular velocity). Thus the magnetic field acts all the time with its full strength. In the case (III.6.9) the field can be thought of as being decomposed into two circularly polarized fields. The one field then precesses in the same direction, whereas the other precesses in the opposite direction. When we transform to the rotating coordinate system (rotating with frequency ν_0) the spin is subject to a constant field (if $\omega \approx \nu_0$) and a very rapidly changing field.

It is interesting to note, that the introduction of the rotating coordinate system is equivalent to the interaction representation.

[1] V. WEISSKOPF and E. WIGNER: Z. Physik **63**, 54 (1930); **65**, 18 (1930).

[2] A review of several broadening mechanisms is given e.g. by D. L. DEXTER: Solid State Phys. **6**, 353 (1958). — J. E. GEUSIC, H. M. MARCOS, and L. G. VAN UITERT: [Appl. Phys. Letters **4**, 182 (1964)] assume that the line broadening is due to local strain at the impurity site and in addition due to lattice vibrations. See also R. H. SILSBEE: Quantum Electron. **3**, 773 (1964).

from the very beginning due to the growing process). In many cases one may assume that the additional broadening has a Gaussian distribution, so that the total distribution of transition energy is a folded Lorentz-Gaussian distribution[3]. A similar behavior can be expected in glass.

β) Gases. If we neglect collisions in gases, the atoms have different, but fixed individual velocities v. A running light wave with wave-vector $\boldsymbol{k}$ thus sees from each atom a DOPPLER shifted frequency, $\omega - \boldsymbol{k}\boldsymbol{v}$. The broadening with wave vector $\boldsymbol{k}$ is governed by the Maxwell velocity distribution.

γ) Semiconductors. A further example is provided by optical transitions between the energy bands of crystals. Due to the $\boldsymbol{k}$-selection rule for periodic structures, the transitions occur in a vertical direction between different bands, where they cover again a total energy range ΔE. The distribution of transition energies is governed here by the density of states of the two energy bands. Also if the transitions occur between impurity states and energy bands an inhomogeneous broadening must be taken into account.

c) Homogeneous broadening[4].

α) Impurity atoms in solids. We now consider atoms embedded in a solid at energetically equivalent lattice sites. The crystal field fluctuates due to the thermal motion, which steadily causes changes of the transition energy of the atom. In the course of time, all atoms suffer the same changes. Because these changes take place in a time characteristic of the lattice vibrations, 10^{-14}—10^{-15} sec., a time short compared to the laser process, the light field is subjected to a "homogeneous broadening". In this case no real electronic transitions occur, and no real phonons are emitted or absorbed, although a double phonon process (absorption and reemission) may occur (see e.g. MCCUMBER and STURGE[5]).

A similar situation appears, if spin-spin interaction plays a role. The individual spin then is influenced by the fluctuating field of all other spins (compare e.g. KUBO[6]). On the other hand, the interaction with lattice vibrations can also lead to real transitions by means of emission of phonons, which shortens the lifetime of the atomic state under consideration.

If the electron-phonon interaction is weak this case can be treated in a manner similar to the natural broadening, so that a Lorentzian line-shape results. For a strong coupling the line-shape is, on the other hand, Gaussian[7].

β) Gases. Homogeneous broadening is caused in gases by collisions between atoms or with the walls, or with electrons.

γ) Semiconductors. If the electrons belong to energy bands in solids, their lifetime within a band state is shortened due to collisions with phonons, other carriers, and impurities, so that a broadening again results.

A clearcut separation between homogeneous and inhomogeneous broadening is not always possible. For instance in gases the collisions between atoms reshuffle the atoms steadily within their Maxwellian distribution.

---

[3] A. C. G. MITCHELL and M. W. ZEMANSKY: Resonance Radiation and Excited Atoms. London and New York: Cambridge University Press 1961.

[4] For details see: R. G. BREENE: The Shift and Shape of Spectral Lines. New York: Pergamon Press 1961.

[5] D. E. MCCUMBER and M. D. STURGE: J. Appl. Phys. **34**, 1682 (1962).

[6] For a review see: R. KUBO: In Fluctuation, Relaxation and Resonance in Magnetic Systems. Edinburgh: Oliver & Boyd 1962.

[7] For treatments of the strong coupling case see e.g.: S. I. PEKAR: Elektronentheorie der Kristalle. Berlin: Akademie-Verlag 1954. — K. HUANG and A. RHYS: Proc. Roy. Soc. (London) A **204**, 406 (1951). — M. WAGNER: Z. Naturforsch. **15**a, 889 (1960); **16**a, 302, 410 (1961), who treat predominantly F-centers in polar crystals.

As we will see below, the laser action in an inhomogeneously broadened line must be treated differently from that of a homogeneously broadened line, which leads to basically different results.

The inhomogeneous broadening can be taken into account simply by summing up over all individual atoms which can be replaced by a suitable weighted average at the end of the calculation (see e.g. VII.6).

The homogeneous broadening can be treated in an adequate (quantum mechanically consistent) way only if the corresponding "heatbaths" are either included explicitly or are represented both by decay constants and fluctuating forces as will be demonstrated in the Chap. IV.2 to IV.11.

IV.2. A survey of IV.2.—IV.11.

a) Definition of heatbaths (reservoirs). In the laser the field modes and electrons (or atoms or molecules) not only interact with one another in the way described in III.3, but also with their surroundings. Thus the electrons of the individual atoms interact with the external pump, with lattice vibrations, with nonlasing light modes, etc. The field interacts with the mirrors, with scattering centers etc. All these external sources acting on electrons and fields are called heatbaths (or reservoirs). In general they are large systems compared to the subsystems of the laser, so that one may assume that these heatbaths are kept at their individual temperatures, which may differ from each other appreciably. Thus, the temperature of the pump source is much higher than that of the lattice vibrations etc.

Survey III

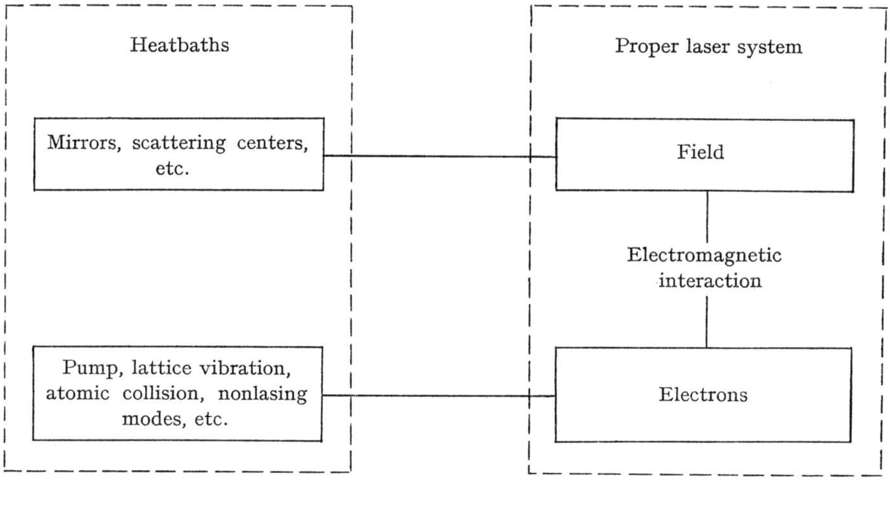

b) The role of heatbaths. In general the heatbaths act in a random fashion on electrons and fields and cause both damping and fluctuations of the latter systems. An analogy to Brownian motion in classical physics may be established:

The field (or atom) plays the role of the Brownian particle, whereas the heatbaths correspond to the gas or liquid in which the Brownian particle moves in thermal equilibrium. There are, however, two main differences:

1. The fields and atoms must be treated in a fully quantum mechanical manner;

2. These systems are far away from thermal equilibrium.

In spite of these difficulties it is possible to develop quantum mechanical methods to deal in an adequate way with these problems. Since these methods bear some formal resemblance with the Langevin and Fokker-Planck procedures of classical physics, we briefly remind the reader of these methods.

c) Classical Langevin and Fokker-Planck equations[1].

α) *Langevin equations.* A set of physical quantities $u = (u_1(\tau), \ldots, u_n(\tau))$ may be given, which obey the equations of motion

$$\frac{du_i}{dt} = \sum_k M_{ik}(u)\, u_k + \Gamma_i(t). \tag{IV.2.1}$$

The quantities M_{ik} may describe damping or, more generally, relaxation processes, as well as external "forces".

If the M's are independent of u, Eqs. (IV.2.1) are linear, otherwise they are nonlinear equations. In laser theory we have mainly to deal with nonlinear equations. $\Gamma_i(t)$ are stochastic forces.

A well-known example is that of a Brownian particle with velocity v:

$$\frac{d}{dt}\boldsymbol{v} = -\gamma\, \boldsymbol{v} + \boldsymbol{F}(t) + \boldsymbol{\Gamma}(t). \tag{IV.2.2}$$

$-\gamma\boldsymbol{v}$ is a friction force, representing the damping of the particle motion, $\boldsymbol{F}(t)$ is an externally prescribed force, e.g. an electric field acting on a charged particle, $\boldsymbol{\Gamma}(t)$ is the stochastic force just mentioned. The damping or relaxation constants together with the stochastic forces describe the action of the above-mentioned heatbaths on the particle. Because of the Γ_i's, the u's also become stochastic functions. Their statistical properties can be determined, if those of the Γ_i's are given. We call equations of the type (IV.2.1) *Langevin equations.*

In the following we denote the statistical average by E. Important classes of problems are characterized by one (or both) of the following properties of the Γ's: The Γ's are called Markoffian if

$$\mathsf{E}(\Gamma_i(t)\,\Gamma_k(t')) = G_{ik}\,\delta(t-t') \tag{IV.2.3}$$

(IV.2.3) applies to a physical situation, in which the "memory" of the heatbaths is much shorter than all other time constants of the system.

They are called Gaussian if

$$\left.\begin{aligned} \mathsf{E}(\Gamma_{i_1}(t_1)\,\Gamma_{i_2}(t_2)\ldots\Gamma_{i_n}(t_n)) &= 0 \qquad \text{for } n \text{ odd}\\ &= \sum_p \mathsf{E}(\Gamma_{i\lambda_1}(t_{\lambda_1})\,\Gamma_{i\lambda_2}(t_{\lambda_2})) \qquad \text{for } n \text{ even}\\ &\ldots\, \mathsf{E}(\Gamma_{i\lambda_{n-1}}(t_{\lambda_{n-1}})\,\Gamma_{i\lambda_n}(t_{\lambda_n})) \end{aligned}\right\} \tag{IV.2.4}$$

where $\sum_p$ runs over all permutations $(\lambda_1, \ldots, \lambda_n)$ of $(1, \ldots, n)$. (We assume here and in the following that $\mathsf{E}(\Gamma_j(t)) = 0$).

β) *The Fokker-Planck equation.* Because the u_j's are stochastic functions, their values are distributed over a certain range, which may change in the course

[1] For a detailed discussion of these equations see e.g.: S. Chandrasekkar: Rev. Mod. Phys. **15**, 1 (1943). — G. E. Uhlenbeck and L. S. Ornstein: Phys. Rev. **36**, 823 (1930). — R. L. Stratonovich: Topics in the Theory of Random Noise. New York: Gordon & Breach 1963. — A. T. Bharucha Reid: Theory of Markov Processes. Toronto-London: McGraw Hill Book Co. 1960. — J. L. Doob: Stochastic Processes. New York, London: John Wiley & Sons 1963. — L. Takacs: Stochastic Processes. New York: John Wiley & Sons 1962. London, Methuen & Co. Ltd. — N. Wax: Noise and Stochastic Processes. New York: Dover Publ. 1954.

of time. The time-dependent distribution of (u) is described by a distribution function $f(u, t)$. *If the Langevin forces are Markoffian,* $f(u, t)$ obeys the Fokker-Planck equation

$$\frac{\partial f(u,t)}{\partial t} = -\sum_i \frac{\partial}{\partial u_i}(B_i(u) f) + \frac{1}{2}\sum_{i,j}\frac{\partial^2 (Q_{ij}(u) f)}{\partial u_i \, \partial u_j} \tag{IV.2.5}$$

where

$$B_i(u) = \lim_{t\to 0}\frac{1}{t}\,\mathsf{E}\left[u_i(t) - u_i(0)\right], \tag{IV.2.6}$$

$$Q_{ij}(u) = \lim_{t\to 0}\frac{1}{t}\,\mathsf{E}\left[\left(u_i(t) - u_i(0)\right)\left(u_j(t) - u_j(0)\right)\right]. \tag{IV.2.7}$$

The coefficients B_i, Q_{ij} are directly connected with the M's and G's [see (IV.2.3)], as may be seen from the following: We represent the solution of (IV.2.1) in the form*

$$u_i = \int_0^t \sum_k K_{ik}(t,\tau)\, \Gamma_k(\tau)\, d\tau + u_i^h(t), \tag{IV.2.8}$$

where u_i^h and K_{ik} obey the homogeneous equation and $K_{ik}(t, t) = \delta_{ik}$.

Inserting (IV.2.8) into (IV.2.6) we obtain

$$\left.\begin{aligned} B_i(u) &= \frac{1}{t}\,\mathsf{E}\left[\int_0^t \sum_k K_{ik}(t,\tau)\,\Gamma_k(\tau) + u_i^h(t) - u_i^h(0)\right] \\ &= \lim \frac{1}{t}\left(u_i^h(t) - u_i^h(0)\right) = \dot{u}_i^h\big|_0 = \sum_k M_{ik}\, u_k^h(0). \end{aligned}\right\} \tag{IV.2.9}$$

When we insert (IV.2.8) into (IV.2.7), we find:

$$\left.\begin{aligned} &\lim_{t\to 0}\frac{1}{t}\,\mathsf{E}\Bigg[\left(\int_0^t d\tau \sum_k K_{ik}(t,\tau)\,\Gamma_k(\tau) + u_i^h(t) - u_i^h(0)\right)\times \\ &\qquad\times\left(\int_0^t d\sigma \sum_l K_{jl}(t,\sigma)\,\Gamma_l(\sigma) + u_j^h(t) - u_j^h(0)\right)\Bigg] \\ &= \frac{1}{t}\left(\int_0^t d\tau \sum_k K_{ik}(t,\tau) \sum_l K_{jl}(t,\tau)\, G_{kl} + O(t^2)\right) \\ &= \sum_{k,l} K_{ik}(t,t)\, K_{jl}(t,t)\, G_{kl} = G_{ij} \end{aligned}\right\} \tag{IV.2.10}$$

i.e.

$$\boxed{Q_{ij} \equiv G_{ij}}\,. \tag{IV.2.11}$$

d) Quantum mechanical formulation: the total Hamiltonian. If one tries to treat laser action starting from first principles, the interaction of field and electrons with their respective heatbaths must be taken into account explicitly.

---

* This form of the solution is exact if (IV.2.1) are linear equations. If the Eqs. (IV.2.1) are nonlinear, the form (IV.2.8) can still be used for the derivation of the formulas (IV.2.9) and (IV.2.11) if the time t is small compared to the time constants inherent in averaged motion of the system but large compared to the time constants of the fluctuating forces. (IV.2.8) has in this case the character of an iterative solution.

The Hamiltonian of the total system then reads:

$$H = H_{laser} + H_{laser\text{-}bath} + H_{bath} \tag{IV.2.12}$$

where H_{laser} is the Hamiltonian (III.3.10) of the proper laser system. H_{bath} contains only the bath variables.

$H_{laser\text{-}bath}$ describes the interaction of electrons and fields with their respective heatbaths. It consists of two parts

$$H_{laser\text{-}bath} = H_{El\text{-}B} + H_{L\text{-}B}. \tag{IV.2.13}$$

The interaction Hamiltonian for the electrons may be written in the form

$$H_{El\text{-}B} = \sum_{\mu} H_{El\mu\text{-}B\mu} \equiv \sum_{i<k,\mu} (B_{ik,\mu}\, a_{i\mu}^{+}\, a_{k\mu} + B_{ik,\mu}^{+}\, a_{k\mu}^{+}\, a_{i\mu}) + \sum_{k,\mu} B_{k,k,\mu}\, a_{k\mu}^{+}\, a_{k\mu}. \tag{IV.2.14}$$

The corresponding Hamiltonian for the light fields is taken in the form

$$H_{L\text{-}B} = \sum_{\lambda} (B_{\lambda}\, b_{\lambda}^{+} + B_{\lambda}^{+} b_{\lambda}) + \sum_{\lambda} \Theta_{\lambda}\, b_{\lambda}^{+} b_{\lambda}. \tag{IV.2.15}$$

The B's and Θ's in (IV.2.14) and (IV.2.15) are operators only acting on the variables of the corresponding heatbaths but not on electron or field variables.

Those sums in (IV.2.14) and (IV.2.15) which contain $B_{ik} (i \neq k)$ and B_{λ}, describe interactions causing real transitions, whereas the remaining sums destroy only the phases of the individual states. This difference of action is decisive for the difference between "longitudinal" and "transversal" relaxation times as will turn out later. The total Hamiltonian forms the basis for the derivation of the fundamental laser equations.

e) Quantum mechanical Langevin equations, Fokker-Planck equation and density matrix equation[2]. In view of the enormous number of degrees of freedom of the total system described by the Hamiltonian it seems hopeless to attempt to find exact solutions. The situation is rather analogous to that of statistical mechanics. The heatbaths cause certain stochastic effects, which manifest themselves by damping and fluctuations of the motion of field and electrons. We will therefore try to eliminate the heatbath coordinates so that we end up with equations which contain only the variables of the proper system explicitly. There are three different known ways:

α) *Langevin equations.* We start from the Heisenberg equations of motion for operators O. Elimination of the bath variables leads to equations of the form

$$\frac{dO}{dt} = \left(\frac{\partial O}{\partial t}\right)_{coh} + \left(\frac{\partial O}{\partial t}\right)_{incoh}, \tag{IV.2.16}$$

where

$$\left(\frac{\partial O}{\partial t}\right)_{coh} = \frac{i}{\hbar} [(H_{El} + H_L + H_I), O] \tag{IV.2.17}$$

$(\partial O/\partial t)_{inc}$ has a form completely analogous to the classical Langevin equations. For an explicit form of (IV.2.16) see (IV.6.1) and (IV.6.2).

β) *Density matrix equation.* We start from the density matrix equations of the total system and eliminate the bath variables by second order perturbation theory. We then find an effective density matrix equation which refers only to the variables of the proper laser system, but which describes besides the

[2] For a comparison of these methods with respect to the laser see also: HAKEN, H., and W. WEIDLICH: Proc. of the Varenna Summer School "Enrico Fermi", ed. R. J. GLAUBER, 1967.

coherent laser process also incoherent relaxation processes:

$$\frac{d\varrho}{dt} = \left(\frac{\partial \varrho}{\partial t}\right)_{coh} + \left(\frac{\partial \varrho}{\partial t}\right)_{incoh} \tag{IV.2.18}$$

where in the interaction representation:

$$\left(\frac{\partial \varrho}{\partial t}\right)_{coh} = -\frac{i}{\hbar}[\widetilde{H}_I, \varrho]. \tag{IV.2.19}$$

The matrix elements of $(\partial\varrho/\partial t)_{incoh}$ obey equations of the form

$$\left(\frac{\partial \varrho}{\partial t}\right)_{nm} = \sum_{lk} A_{nm\,lk}\,\varrho_{lk} \tag{IV.2.20}$$

where the coefficients A are given constants describing relaxation effects. The explicit form of (IV.2.20) will be given in (IV.7.70) and (IV.7.71).

γ) *Generalized Fokker-Planck equation.* A classical distribution function f of macroscopic laser variables may be defined in such a way that all quantum mechanical information is retained. f obeys an equation of the form

$$\frac{df}{dt} = \left(\frac{\partial f}{\partial t}\right)_{coh} + \left(\frac{\partial f}{\partial t}\right)_{incoh} \tag{IV.2.21}$$

with

$$\left(\frac{\partial f}{\partial t}\right)_{coh} = L_{coh}\,f \tag{IV.2.22}$$

and

$$\left(\frac{\partial f}{\partial t}\right)_{incoh} = L_{incoh}\,f. \tag{IV.2.23}$$

L_{coh}, L_{incoh} are (generalized) Liouville differential operators.

L_{coh} stems from the Hamiltonian (III.3.10) of the proper laser system, whereas L_{incoh} describes dissipation and fluctuations of the proper laser variables caused by the heatbaths. The explicit form of (IV.2.21) is presented in (IV.10.32) to (IV.10.34).

IV.3. Quantum mechanical Langevin equations: origin of quantum mechanical Langevin forces (the effect of heatbaths). We first show how quantum mechanical Langevin forces arise by means of two examples which can be treated exactly. For this purpose we couple the light field or the electron field to a heatbath which consists of an infinite number of harmonic oscillators. Thus H_{bath} which occurs in (IV.2.12) has the form

$$\sum_{\omega} \hbar\,\omega\,B_{\omega}^{+}\,B_{\omega}. \tag{IV.3.1)*$$

The bath variable occurring in the interaction Hamiltonian (IV.2.15) is chosen as a superposition

$$B = \hbar \sum_{\omega} g_{\omega}\,B_{\omega} \tag{IV.3.2)*$$

or in the interaction representation

$$B(t) = \hbar \sum_{\omega} g_{\omega}\,B_{\omega}\,e^{-i\omega t}. \tag{IV.3.3}$$

We now distinguish between field and electrons (atoms).

* More accurately one should write $B_{\lambda} = \hbar\sum g_{\lambda,\omega}\,B_{\lambda,\omega}$ etc. Because we treat only a single field mode, we drop the corresponding index λ everywhere, however.

a) The field (one mode). The total interaction-Hamiltonian (IV.2.15) has the form *[1]

$$\hat{H}_{L\text{-}B} = b^+ \sum_\omega g_\omega B_\omega e^{-i\omega t} + b \sum_\omega g_\omega^* B_\omega^+ e^{-i\omega t}. \tag{IV.3.4}$$

The Hamiltonian of the free field is given by

$$\hbar\,\omega_0\, b^+ b\,. \tag{IV.3.5}$$

The Heisenberg equations for b^+ and B_ω^+ read

$$\frac{d b^+}{dt} = i\,\omega_0\, b^+ + i \sum_\omega g_\omega^* B_\omega^+ e^{+i\omega t}, \tag{IV.3.6}$$

$$\frac{d B_\omega^+}{dt} = i\, b^+ g_\omega\, e^{-i\omega t} \tag{IV.3.7}$$

(IV.3.7) immediately allows for the solution

$$B_\omega^+ = i \int_{t_0}^{t} b^+(\tau)\, g_\omega\, e^{-i\omega t}\, d\tau + B_\omega^+(t_0) \tag{IV.3.8}$$

where $B_\omega^+(t_0)$ is the operator at time t_0. Inserting (IV.3.8) into (IV.3.6) we arrive at

$$\frac{d b^+}{dt} = i\,\omega_0\, b^+ - \int_{t_0}^{t} b^+(\tau) \sum_\omega |g_\omega|^2 e^{i\omega(t-\tau)}\, d\tau + i \sum_\omega g_\omega^* B_\omega^+(t_0)\, e^{i\omega t}. \tag{IV.3.9}$$

We now assume that the g_ω's are of about equal amplitudes so that we have

$$\left.\begin{aligned} \int_0^\infty |g|^2 e^{i\omega(t-\tau)}\, d\omega &= \varkappa\left(\delta(t-\tau) + \frac{i}{\pi}\,\frac{P}{t-\tau}\right), \\ \varkappa &= \pi |g|^2. \end{aligned}\right\} \tag{IV.3.10}$$

If we would allow also for negative frequencies that would mean for harmonic oscillators with an inverted energy scale we would find instead of (IV.3.10)

$$\int_{-\infty}^{+\infty} |g|^2 e^{i\omega(t-\tau)}\, d\omega = 2\varkappa\, \delta(t-\tau)\,. \tag{IV.3.11}$$

For the evaluation of the integrals it must be noted that the δ-function has the property

$$\int_0^t \delta(t-\tau)\, d\tau = \tfrac{1}{2} \tag{IV.3.12}$$

(IV.3.10) corresponds to the choice made by SENITZKY[2], whereas the choice (IV.3.11) was made by LAX[3]. We see that (IV.3.10) is more physical but (IV.3.11) has certain computational advantages. The actual difference between (IV.3.10) and (IV.3.11) is in practical calculations negligible. Under these assumptions

* We drop the term containing b^+b of (IV.2.15) because in general the terms kept in (IV.3.4) are the important ones. For a discussion of the terms dropped see R. BONIFACIO and F. HAAKE: Z. Physik **200**, 526 (1967).

[1] For a coupling of the harmonic oscillator to a general heatbath (treated approximately) see J. R. SENITZKY: Phys. Rev. **119**, 670 (1960); **124**, 642 (1961). The coupling of an oscillator to many others has been treated by a number of authors in different ways. Our present treatment seems best suited to explain how quantum mechanical Langevin forces arise.

[2] l.c. [1].

[3] M. LAX: Phys. Rev. **129**, 2342 (1963).

(IV.3.9) becomes

$$\frac{db^+}{dt} = i\omega_0 b^+ - \varkappa b^+ + \underbrace{i\sum_\omega g_\omega^* B_\omega^+(t_0)\, e^{i\omega t}}_{F^+(t)} \tag{IV.3.13}$$

where the last term is evidently the fluctuating force.

We determine the properties of this fluctuating force and evaluate

$$\left.\begin{aligned}\langle F^+(t)\, F(t')\rangle_B &= \sum_\omega |g_\omega|^2\, e^{i\omega(t-t')}\, \text{trace}\left(B_\omega^+ B_\omega\, e^{-H_B^{(\omega)}/kT}\right) Z_\omega^{-1} \\ &= \sum_\omega |g_\omega|^2\, e^{-i\omega(t-t')}\, \bar{n}_\omega(T)\end{aligned}\right\} \tag{IV.3.14}$$

where

$$Z_\omega = \text{trace}\,(e^{-H_B^{(\omega)}/kT})$$

and correspondingly

$$\langle F(t)\, F^+(t')\rangle_B = \sum_\omega |g_\omega|^2\, e^{-i\omega(t-t')}\, (\bar{n}_\omega(T) + 1) \tag{IV.3.15}$$

$\bar{n}_\omega$ is the occupation number of the mode corresponding to frequency ω of the *heatbath*. The bath average over the commutator has the form

$$\langle [F(t), F^+(t')]\rangle_B = \left\{\begin{matrix}\varkappa\left(\delta(t-t') + \dfrac{i}{\pi}\dfrac{P}{t'-t}\right), \\ 2\varkappa\, \delta(t-t')\end{matrix}\right\} \tag{IV.3.16}$$

(IV.3.14) and (IV.3.15) can be directly expressed by the damping constant $\varkappa$ only for the temperature $T=0$. For higher temperatures it must be noted that in practical calculations

$$\langle F^+(t)\, F(t')\rangle = \sum_\omega |g_\omega|^2\, e^{i\omega(t-t')}\, \langle B_\omega^+ B_\omega\rangle \tag{IV.3.17}$$

and

$$\langle F(t)\, F^+(t')\rangle = \sum_\omega |g_\omega|^2\, e^{i\omega(t-t')}\, \langle B_\omega\, B_\omega^+\rangle \tag{IV.3.18}$$

always appear under an integral which contains a factor $e^{i\omega_0 t}$, where ω_0 is the frequency of the harmonic oscillator (light field). If this is taken into account, (IV.3.14) may be expressed in the form

$$\sum_\omega |g_\omega|^2\, e^{i\omega(t-t')}\, \bar{n}_\omega(T) = 2\varkappa\, \bar{n}_{\omega_0}(T)\, \delta(t-t') \tag{IV.3.19}$$

and similarly:

$$\langle F(t)\, F^+(t')\rangle = 2\varkappa\, (\bar{n}_{\omega_0}(T) + 1)\, \delta(t-t'). \tag{IV.3.20}$$

The δ-function in (IV.3.16) and (IV.3.19) expresses the fact that the heatbath has a very short memory. As may be seen from (IV.3.9), (IV.3.10), it is essential for the derivation that the heatbath frequencies have a spread which covers a whole range around the oscillator frequency ω_0. The present example is also very illustrative in order to show how the fluctuating forces restore the quantum mechanical consistency (see also below, p. 44). For this end we solve Eq. (IV.3.13) explicitly

$$b^+ = b^+(0)\, e^{(i\omega_0 - \varkappa)t} + \int_0^t e^{(i\omega_0-\varkappa)(t-\tau)}\, F^+(\tau)\, d\tau \tag{IV.3.21}$$

and form

$$\left.\begin{aligned}\frac{d}{dt}\langle b^+ b\rangle_B = \langle \dot{b}^+ b + b^+ \dot{b}\rangle_B = -2\varkappa\langle b^+ b\rangle_B + \\ + \int_0^t \langle F^+(t)\, F(\tau)\rangle_B\, e^{-(i\omega_0+\varkappa)(t-\tau)}\, d\tau + \\ + \int_0^t \langle F^+(\tau)\, F(t)\rangle_B\, e^{(i\omega_0-\varkappa)(t-\tau)}\, d\tau .\end{aligned}\right\} \tag{IV.3.22}$$

The integrals may be evaluated using (IV.3.19) and (IV.3.20) so that we find

$$\frac{d}{dt}\langle b^+ b\rangle_B = -2\varkappa\langle b^+ b\rangle_B + 2\varkappa\bar{n}. \tag{IV.3.23}$$

Expression (IV.3.23) agrees with the corresponding expression (IV.7.50) of the density matrix treatment of WEIDLICH and HAAKE[4] or with the Langevin treatment of SENITZKY. In a similar way we show how the fluctuating forces restore the commutation relation. In analogy to (IV.3.22) we form

$$\frac{d}{dt}\langle [b, b^+]\rangle_B = -2\varkappa\langle [b, b^+]\rangle_B + 2\varkappa. \tag{IV.3.24}$$

The general solution reads

$$\langle [b, b^+]\rangle_{B,t} = e^{-2\varkappa t}\, C + 1. \tag{IV.3.25}$$

If we insert the initial condition

$$\langle [b, b^+]\rangle_{B,0} = 1 \tag{IV.3.26}$$

we find

$$C = 0 \tag{IV.3.27}$$

which means that the commutation relation (III.1.10) is preserved for all times. We now indicate how Langevin forces arise for the atomic system.

b) Electrons ("atoms"). Because this problem becomes rather complex, we treat the example of a two-level atom in which the transition $1 \rightarrow 2$ is coupled to a heatbath[5]. The Hamiltonians H_{El} and H_{bath} have the form (III.2.9) and (IV.3.1) respectively, whereas the interaction Hamiltonian takes the form

$$H_{El-B} = \hbar\left\{a_1^+ a_2 \sum_\omega g_\omega^* B_\omega^+ + a_2^+ a_1 \sum_\omega g_\omega B_\omega\right\}. \tag{IV.3.28}$$

We only include resonant terms. We proceed to the interaction representation doing away the motion of the free atom and the free bath so that the Heisenberg equations read as follows:

$$\frac{d(a_1^+ a_2)}{dt} = -i\sum_\omega g_\omega (a_1^+ a_1 - a_2^+ a_2)\, B_\omega\, e^{-i\omega t + i\nu t}, \tag{IV.3.29}$$

$$\frac{d(a_1^+ a_1)}{dt} = -i\, a_1^+ a_2 \sum_\omega g_\omega^* B_\omega^+ e^{i\omega t - i\nu t} + i\sum_\omega g_\omega B_\omega e^{-i\omega t + i\nu t} a_2^+ a_1, \tag{IV.3.30}$$

$$\frac{d(a_2^+ a_2)}{dt} = i\, a_1^+ a_2 \sum_\omega g_\omega^* B_\omega^+ e^{i\omega t - i\nu t} - i\sum_\omega g_\omega B_\omega a_2^+ a_1 e^{-i\omega t + i\nu t}, \tag{IV.3.31}$$

$$\frac{dB_\omega}{dt} = -i\, g_\omega^*\, a_1^+ a_2\, e^{i\omega t - i\nu t}, \tag{IV.3.32}$$

$$\nu = \nu_{21} \tag{IV.3.33}$$

[4] W. WEIDLICH, and F. HAAKE. Z. Physik **185**, 30 (1965).

[5] For a different treatment of this example, see H. SAUERMANN: Z. Physik **188**, 480 (1965). For a different introduction of the atomic noise sources, see H. HAKEN: Z. Physik **181**, 96 (1964) and **182**, 346 (1965), (compare footnote * on p. 45).

(IV.3.32) immediately allows for the solution

$$B_\omega(t) = -i \int_{t_0}^{t} g_\omega^* (a_1^+ a_2)_\tau e^{i(\omega-\nu)\tau} d\tau + B_\omega(t_0). \tag{IV.3.34}$$

Inserting it into (IV.3.29) gives

$$\left.\begin{aligned}\frac{d(a_1^+ a_2)}{dt} = -\int_{t_0}^{t} \sum_\omega |g_\omega|^2 e^{-i(\omega-\nu)(t-\tau)} (a_1^+ a_1 - a_2^+ a_2)_t (a_1^+ a_2)_\tau d\tau - \\ -i \sum_\omega g_\omega (a_1^+ a_1 - a_2^+ a_2)_t B_\omega(t_0) e^{-i(\omega-\nu)t}.\end{aligned}\right\} \tag{IV.3.35}$$

Again using the Markoff property (IV.3.11) we end up with

$$\left.\begin{aligned}\frac{d(a_1^+ a_2)}{dt} = -\gamma (a_1^+ a_1 - a_2^+ a_2)_t (a_1^+ a_2)_t - \\ -i \sum_\omega g_\omega (a_1^+ a_1 - a_2^+ a_2) B_\omega(t_0) e^{-i(\omega-\nu)t}\end{aligned}\right\} \tag{IV.3.36}$$

where γ is defined in analogy to $\varkappa$ in (IV.3.11).

When we return from the interaction representation to the full Heisenberg picture (IV.3.36) takes the form

$$\frac{d(a_1^+ a_2)}{dt} = -i\nu a_1^+ a_2 - \gamma a_1^+ a_2 \underbrace{- i \sum_\omega g_\omega (a_1^+ a_1 - a_2^+ a_2) B_\omega(t_0) e^{-i\omega t}}_{\Gamma_{12}(t)}. \tag{IV.3.37}$$

A corresponding equation can be obtained for $a_2^+ a_1$ by simply taking the hermitian conjugate of Eq. (IV.3.37). A closer inspection of the above procedure shows that the sequence of the operators $a_1^+ a_2$, $a_1^+ a_1$, $a_2^+ a_2$ in Eq. (IV.3.35) is essential for the correct form of the damping term $-\gamma a_1^+ a_2$.

The last sum in (IV.3.37) is again to be identified with the fluctuating force. Our treatment shows that the new Langevin forces are explicitly dependent on the atomic operators. The evaluation of the correlation functions between the Γ's turns out to be rather complicated. We therefore give only one example. We immediately find

$$\langle \Gamma_{12}(t)\, \Gamma_{12}(t')\rangle = 0 \tag{IV.3.38}$$

$$\big(\text{because } \langle B(t_0)\, B(t_0)\rangle = 0\big)$$

and after some algebra

$$\left.\begin{aligned}\langle \Gamma_{12}(t)\, \Gamma_{21}(t')\rangle &= \sum_\omega |g_\omega|^2 \langle (a_1^+ a_1 - a_2^+ a_2) B_\omega(t_0) \times \\ &\quad \times B_\omega^+(t_0) (a_1^+ a_1 - a_2^+ a_2)\rangle e^{i\omega(t'-t)} \\ &= \langle (a_1^+ a_1 - a_2^+ a_2)^2\rangle\, 2\gamma\, \delta(t-t') \\ &= \langle (N_1 + N_2)\rangle\, 2\gamma\, \delta(t-t').\end{aligned}\right\} \tag{IV.3.39}$$

Note, that in the one-electron subspace

$$N_1^2 \equiv a_1^+ a_1 a_1^+ a_1 = a_1^+ a_1 \equiv N_1$$

and

$$N_2 N_1 \equiv a_2^+ a_2 a_1^+ a_1 = 0$$

where use was made of the relation

$$a_i^+ a_k a_l^+ a_m = a_i^+ a_m \delta_{kl}.$$

For the proof of this relation and a more general one see (IV.7.57).

IV.4. The requirement of quantum mechanical consistency.

a) The field[1]. Let us consider a general heatbath (not necessarily consisting of harmonic oscillators) which causes damping of the field amplitude. In a purely classical treatment one may write:

$$E = E_0 \, e^{-i\omega t - \varkappa t}. \tag{IV.4.1}$$

It is, however, not sufficient within the framework of a fully quantum mechanical treatment to take the damping of the electromagnetic field into account only by supplementing ω with an imaginary part, $\varkappa$. In this case the field operators would have the form

$$b^+(t) = b^+(0) \, e^{i\omega t - \varkappa t}, \qquad b(t) = b(0) \, e^{-i\omega t - \varkappa t} \tag{IV.4.2}$$

so that the commutation relation would read:

$$[b(t), b^+(t)] = e^{-2\varkappa t}. \tag{IV.4.3}$$

This is in contradiction to one of the fundamental postulates of quantum theory namely that the commutation relation

$$[b, b^+] = 1 \tag{IV.4.4}$$

holds for all times. Our earlier example (see p. 42) has shown that (IV.4.4) can be restored if not only the dissipation of the field caused by the heatbath, but also the fluctuations of b are taken into account. Therefore b^+ and b must obey the following differential equations:

$$\dot{b}^+ = (i\omega - \varkappa) \, b^+ + F^+(t), \tag{IV.4.5}$$

$$\dot{b} = (-i\omega - \varkappa) \, b + F(t) \tag{IV.4.6}$$

where $F(t)$ is an operator which still depends on the heatbath coordinates and represents fluctuations. The explicit dependence of $F(t)$ on the heatbath variables need not be known, however. All that is needed for explicit calculations are the correlation functions of F.

If the F's are Gaussian [see (IV.2.4)], only the correlation functions of first and second order need be given. The examples of Chap. (IV.3) suggest that it is reasonable to assume that the F's are also Markoffian, i.e.

$$\langle F^+(t) \, F(t') \rangle = C_1 \, \delta(t - t'), \qquad \langle F(t) \, F^+(t') \rangle = C_2 \, \delta(t - t'), \; \ldots \tag{IV.4.7}$$

A simple calculation similar to that of formulas (IV.3.22), (IV.3.24) of Chap. IV.3 shows that the coefficients C_i are uniquely determined by the requirement of quantum mechanical consistency and by $\langle b^+ b \rangle = n_{th}$ in thermal equilibrium (n_{th} is the number of thermal quanta of the mode described by b^+, b). The properties of the F's which we use in the following are thus given by (IV.3.19), (IV.3.20):

$$\left.\begin{aligned} \langle F^+(t) \, F(t') \rangle &= 2\varkappa \, \bar{n}_\omega(T) \, \delta(t - t'), \\ \langle F(t) \, F^+(t') \rangle &= 2\varkappa \left(\bar{n}_\omega(T) + 1\right) \delta(t - t'), \\ \langle F(t) \, F(t') \rangle &= \langle F^+(t) \, F^+(t') \rangle = 0. \end{aligned}\right\} \tag{IV.4.8}$$

[1] For a discussion see also SENITZKY, l.c.[1] on p. 40.

b) Dissipation and fluctuations of the atoms[2]*. The phase factor of the atomic dipole moment which is described by the operator $a_2^+ a_1$ decreases exponentially ($\sim e^{-\gamma t}$) due to the interaction of the atom with its surroundings: lattice interaction, atomic collision in gases, the external pump light, but also by the spontaneous atomic decay with emission of light. In a phenomenological treatment, which considers the operators as classical quantities, one would merely supplement the dipole oscillation frequency ν_{21} with an imaginary part $+i\gamma$:

$$(a_2^+ a_1)^{\cdot} = (i\nu_{21} - \gamma)\, a_2^+ a_1 . \tag{IV.4.9}$$

The above mentioned "reservoirs" simultaneously cause transitions between the atomic levels, so that the occupation number changes according to transition rates w_{ik}. Thus, e.g. in a two-level system, these changes are described by equations of the form:

$$(a_1^+ a_1)^{\cdot} = -w_{12}(a_1^+ a_1) + w_{21}(a_2^+ a_2), \tag{IV.4.10}$$

$$(a_2^+ a_2)^{\cdot} = -w_{21}\, a_2^+ a_2 + w_{12}(a_1^+ a_1). \tag{IV.4.11}$$

Again it follows that the solutions of Eqs. (IV.4.9)—(IV.4.11) are not consistent with the commutation relations (III.2.6).

Although the commutation relations of the b's and those of the $(a_i^+ a_k)$'s are completely different, the introduction of fluctuation forces is also possible in this case (see p. 43), and allows the quantum mechanical properties of the $(a_i^+ a_k)$'s to be restored. The Eqs. (IV.4.9)—(IV.4.11) become for the example of two levels:

$$(a_2^+ a_1)^{\cdot} = (i\nu_{21} - \gamma)\, a_2^+ a_1 + \Gamma_{21}, \tag{IV.4.12}$$

$$(a_1^+ a_1)^{\cdot} = -w_{12}(a_1^+ a_1) + w_{21}(a_2^+ a_2) + \Gamma_{11}, \tag{IV.4.13}$$

$$(a_2^+ a_2)^{\cdot} = w_{12}(a_1^+ a_1) - w_{21}(a_2^+ a_2) + \Gamma_{22}. \tag{IV.4.14}$$

The fluctuating forces F and Γ which occur in an analogous way in the Langevin equation of Brownian motion are distinguished from their classical counterparts by their property that they are in general non-commuting operators. Below, we determine the properties of these atomic Langevin operators by the method of HAKEN and WEIDLICH[2].

---

[2] H. HAKEN and W. WEIDLICH: Z. Physik **189**, 1 (1966).

* The fluctuating forces F are of fundamental importance for the determination of the phase and amplitude fluctuations of the maser (microwave region), because they represent *thermal* noise. In the optical region the *spontaneous emission noise* plays the decisive role. This effect was first taken care of in the following way [HAKEN, H.: Z. Physik **181**, 96 (1964)]: The Eqs. (III.5.7) and (III.5.9) alone are not sufficient to treat spontaneous emission by the Weißkopf-Wigner-method (see Chap. V.3), but must be supplemented by the initial condition that the atom is in its upper state. In the case of the ammonia maser this "preparation" of the atoms is done when they are shot into the cavity at random times t_ν. In the case of fixed atoms the preparation of the initial state is achieved by the pumping processes. In a completely symmetric way the initial state is also prepared e.g. by the spontaneous emission into the nonlasing modes (which form a continuum). Because Eq. (III.5.9) is a first order differential equation, such initial conditions can be built in by adding on the right hand side a term

$$\Gamma_{\mu,21}(t) = \sum_\nu \delta(t - t_{\mu\nu})\{(a_2^+ a_1)_\mu (t_{\mu\nu} + 0) - (a_2^+ a_1)_\mu (t_{\mu\nu} - 0)\} \tag{IV.4.15}$$

where e.g. $(a_2^+ a_1)_\mu (t_{\mu\nu} + 0)$ is the "value" of the operator immediately after the pumping processes [we have specialized (III.5.9) for a two-level system (see Chap. III, p. 31)]. One may show that the Γ's are non-commuting operators with the property

$$\langle \Gamma_\mu \rangle = 0, \ \langle \Gamma_{\mu, i_1 i_2}(t_1)\, \Gamma_{\mu, j_1 j_2}(t_2) \rangle = G_{i_1 i_2 j_1 j_2}\, \delta(t_1 - t_2) \tag{IV.4.16}$$

where the δ-function expresses the fact that the external perturbations ("heatbaths" or "reservoirs") which "prepare" the initial state have a short memory (compared to the other time-constants in the system), or in other words, that they are "Markoffian".

IV.5. The explicit form of the correlation functions of Langevin forces.

a) The field[1]. The results of the foregoing chapter allow us to write down the properties of the fluctuating forces for an arbitrary set of modes which are in contact with the statistically independent reservoirs at temperature T and which have decay constants $\varkappa_\lambda$. The fluctuating forces have the following properties

$$\langle F_\lambda^+(t)\, F_{\lambda'}^+(t')\rangle = \langle F_\lambda(t)\, F_{\lambda'}(t')\rangle = 0, \tag{IV.5.1}$$

$$\langle F_\lambda^+(t)\, F_{\lambda'}(t')\rangle = \bar{n}_\lambda(T)\, 2\varkappa_\lambda\, \delta(t-t')\, \delta_{\lambda\lambda'}, \tag{IV.5.2}$$

$$\langle F_\lambda(t)\, F_{\lambda'}^+(t')\rangle = (\bar{n}_\lambda(T)+1)\, 2\varkappa_\lambda\, \delta(t-t')\, \delta_{\lambda\lambda'}. \tag{IV.5.3}$$

The occurrence of the Kronecker symbol $\delta_{\lambda\lambda'}$ stems from the fact that the reservoirs are assumed to be statistically independent. The F's may be assumed Gaussian.

b) The N-level atom[2]*. As we have mentioned at the beginning of Chap. IV.3 the effect of the heatbaths on an atomic N-level system can be fully taken into account by transition probabilities w_{ik}, decay constants γ_{ik}, and by fluctuating forces. w_{ik} and γ_{ik} will be determined in IV.7. In the following, we assume that these quantities are given, either by a calculation from first principles or, which is of great practical importance, phenomenologically. We require further, that the fluctuating forces have two properties:

1. They have a short memory, so that their correlation function is proportional to a δ-function of the time difference (or, in other words, they have a Markoff character).

2. They are determined in such a way that the electron operators form an algebraic ring with the property

$$\langle a_i^+ a_k\, a_l^+ a_m\rangle = \delta_{kl}\, \langle a_i^+ a_m\rangle \tag{IV.5.4}$$

where the a's are the time-dependent operators under the influence of the full Hamilton operator, which comprises also the effect of the heatbaths [for a proof of (IV.5.4) see (IV.7.57)]. We show that the fluctuating forces can then be determined uniquely and explicitly by a rather simple formula [see (IV.5.22) and (IV.5.26)]. They secure also the quantum mechanical consistency of the equations of motion for the operators $a_i^+ a_k$ in complete analogy to the harmonic oscillator.

We start with the averaged equations of the type (IV.7.59-60) which we write in the general form

$$\frac{d}{dt}\langle a_{i_1}^+ a_{i_2}\rangle = \sum_{j_1, j_2} \langle M_{i_1 i_2,\, j_1 j_2}\, a_{j_1}^+ a_{j_2}\rangle. \tag{IV.5.5}$$

The number of levels 1, 2, ... may be arbitrary.

Denoting the vector formed of $a_{i_1}^+ a_{i_2}$ by A and the matrix formed of $M_{i_1 i_2;\, j_1 j_2}$ by M (note that i_1, i_2 counts as a single index) we can represent Eq. (IV.5.5) as

$$\frac{d}{dt}\langle A\rangle = \langle M A\rangle. \tag{IV.5.6}$$

---

* Actually the results of this chapter are not confined to atoms but apply to arbitrary quantum systems. This rests on the fact that the projection operators P_{ik} of a quantum system obey just the relation (IV.5.4). We are grateful to W. Weidlich for drawing our attention to this fact.

[1] See Senitzky, l.c.[1] on p. 40.

[2] H. Haken and W. Weidlich: Z. Physik **189**, 1 (1966). We closely follow that paper. For a different treatment with equivalent results see Ch. Schmid and H. Risken: Z. Physik **189**, 365 (1966).

In it, M contains the transition probabilities (especially the losses), the non-diagonal phase-destroying elements produced by the heatbath, and the "coherent" driving field. Because these driving fields, if treated quantum mechanically, may still depend in an implicit way on the heatbath, the average over the heatbath has to comprise both M and A. We now go one step back and consider the motion of the unaveraged operators A by adding fluctuating driving forces, Γ.

$$\frac{d}{dt} A = MA + \Gamma. \tag{IV.5.7}$$

The driving forces can have any form and in particular can still depend on $a_i^+ a_k$, for instance, in the following form

$$\Gamma(t) = L(t)\, A + N(t) \tag{IV.5.8}$$

where L and N do not depend on A.

In order to come back from (IV.5.7) to (IV.5.6) we must assume

$$\langle \Gamma \rangle = 0. \tag{IV.5.9}$$

It should be noted that we have to impose as an additional condition the conservation of the total occupation number, expressed by

$$\sum_{i=1}^{N} a_i^+ a_i = 1 \tag{IV.5.10}$$

The formal solution of (IV.5.7) can be written as an integral representation

$$A = \int^{t} K(t, \tau)\, \Gamma(\tau)\, d\tau + A_h \tag{IV.5.11}$$

where A_h is a solution of the homogeneous part of (IV.5.7). We postulate in the well known way that the kernel has the property

$$K(t, t) = E. \tag{IV.5.12}$$

The initial condition requires

$$A(0) = \left(a_{k_1}^+(0)\, a_{k_2}(0)\right) \tag{IV.5.13}$$

where the a's and a^+'s are the Schrödinger operators. We now construct

$$\langle \tilde{A} B A \rangle = \left\langle \left(\int^{t} \tilde{\Gamma}(\tau) \tilde{K}(t, \tau)\, d\tau + \tilde{A}_h \right) B \left(\int^{t} K(t, \tau')\, \Gamma(\tau')\, d\tau' + A_h \right) \right\rangle \tag{IV.5.14}$$

where $\sim$ denotes the adjoint matrix.

B is assumed as a constant matrix

$$B = (B_{k\,l}) \tag{IV.5.15}$$

having only one nonvanishing element $(i, j) = (i_1, i_2; j_1, j_2)$:

$$B_{k_1, k_2;\, l_1, l_2} = \delta_{k_1, i_1}\, \delta_{k_2 i_2}\, \delta_{l_1 j_1}\, \delta_{l_2 j_2}. \tag{IV.5.16}$$

Using the property (IV.5.4) we find for the left hand side of Eq. (IV.5.14) by differentiation

$$\delta_{i_2 j_1} \frac{d}{dt} \langle a_{i_1}^+ a_{j_2} \rangle = \delta_{i_2 j_1} \sum_{m,n} \langle M_{i_1 j_2; m,n}\, a_m^+ a_n \rangle. \tag{IV.5.17}$$

By differentiation of the right hand side of Eq. (IV.5.14) and using the properties of (IV.5.11) we obtain the following terms

$$\langle \tilde{\Gamma}(t)\, B A\rangle_{i,j}, \tag{IV.5.18}$$

$$\langle \tilde{A}\, B \Gamma(\tau)\rangle_{i,j}, \tag{IV.5.19}$$

$$\langle \tilde{A}\, \tilde{M} B A\rangle_{i,j}, \tag{IV.5.20}$$

$$\langle \tilde{A}\, B M A\rangle_{i,j}. \tag{IV.5.21}$$

For the evaluation of expressions (IV.5.18), (IV.5.19) we use the Markoffian property i.e.

$$\langle \Gamma_{i_1,i_2}(t)\, \Gamma_{j_1,j_2}(t')\rangle = G_{i_1,i_2;j_1,j_2}\, \delta(t-t') \tag{IV.5.22}$$

and

$$\langle \Gamma_{i_1,i_2}(t)\rangle = 0. \tag{IV.5.23}$$

We now have to perform the average over the heatbath where we know that the fluctuations have only a very short memory. We assume that the response of the particle (or spin) is slow enough for K and A_h to be taken out of the average. [Note that M contains at most the heatbath coordinates indirectly over the quantum mechanical ("coherent") fields, but in all loss terms and phase memory destroying parts they *do not* appear.] (IV.5.18) then gives the contribution

$$\tfrac{1}{2} G_{i,j} \tag{IV.5.24}$$

and the same quantity similarly follows from (IV.5.19).

In order to evaluate (IV.5.20) we contract the product of the A's by means of the rule (IV.5.2), where we make use of the commutativity of the operators a_k^+, a_k with the operators of the driving fields within M, which holds in the whole Heisenberg picture. We thus obtain

$$\Big\langle \sum_{m,n} a_m^+ a_n\, \tilde{M}_{m,n;i_1,i_2}\, a_{j_1}^+ a_{j_2} \Big\rangle = \sum_m \langle a_m^+ a_{j_2} M_{i_1,i_2;m,j_1}\rangle \tag{IV.5.25}$$

from Eq. (IV.5.20) and a similar one from Eq. (IV.5.21).

We now compare the results of both sides of (IV.5.14), thus obtaining the following equation for the G's which occur in (IV.5.18), (IV.5.19)

$$G_{i_1,i_2;j_1,j_2} = \sum_{m,n} \langle\{\delta_{i_2,j_1} M_{i_1,j_2;m,n} - \delta_{n,j_2} M_{i_1 i_2;m j_1} - \delta_{m,i_1} M_{j_1 j_2;i_2 n}\}\rangle\, \langle a_m^+ a_n\rangle. \tag{IV.5.26}$$

This equation represents the required result. It allows us to calculate all correlation functions of the type (IV.5.22) if the coefficients M are given, and the solutions $\langle a_i^+ a_k\rangle$ of the *averaged* equations are known. We now show that in (IV.5.26) all terms containing external fields cancel each other* so that only transition rates and damping constants need to be used for the M's:

The coherent part of M stems from the commutation of $a_{i_1}^+ a_{i_2}$ with a Hamiltonian which is of the form

$$H = \sum_{mn} c_{mn}\, a_m^+ a_n .$$

The evaluation of the commutator

$$\frac{i}{\hbar}[H, a_{i_1}^+ a_{i_2}] = \frac{i}{\hbar}\Big(\sum_{mn} c_{mn}\, a_m^+ \delta_{n i_1} a_{i_2} - \sum_{mn} c_{mn}\, a_{i_1}^+ \delta_{i_2 m}\, a_n\Big)$$

yields immediately

$$\sum_{j_1 j_2} \underbrace{\frac{i}{\hbar}\,(c_{j_1 i_1}\delta_{j_2 i_2} - c_{i_2 j_2}\delta_{j_1 i_1})}_{M^{coh}_{i_1,i_2;j_1,j_2}}\, a_{j_1}^+ a_{j_2}.$$

* WEIDLICH, W.: Private communication.

A comparison of this expression with the right hand side of equation (IV.5.5) shows that

$$M^{coh}_{i_1,i_2;j_1,j_2} = \frac{i}{\hbar}\,(c_{j_1 i_1}\,\delta_{j_2 i_2} - c_{i_2 j_2}\,\delta_{j_1 i_1}).$$

When this explicit form is inserted into the right hand side of (IV.5.26) these terms cancel each other.

In the case of the laser, the incoherent part of the atomic equations often reads:

$$\frac{d}{dt}\,(a_j^+ a_j) = \sum_k w_{kj}\, a_k^+ a_k - \sum_k w_{jk}\, a_j^+ a_j + \Gamma_{jj}(t), \tag{IV.5.27}$$

$$\frac{d}{dt}\,(a_j^+ a_k) = -\gamma_{jk}(a_j^+ a_k) + \Gamma_{jk}(t), \quad j \neq k. \tag{IV.5.28}$$

By specializing formula (IV.5.26) we obtain

$$\left.\begin{aligned} G_{ii,jj} = \delta_{ij}\Big\{\sum_k w_{ki}\langle a_k^+ a_k\rangle + \sum_k w_{ik}\langle a_i^+ a_i\rangle\Big\} - \\ - w_{ij}\langle a_i^+ a_i\rangle - w_{ji}\langle a_j^+ a_j\rangle. \end{aligned}\right\} \tag{IV.5.29}$$

$$G_{ij,ij} = 0 \quad i \neq j, \tag{IV.5.30}$$

$$G_{ij,ji} = \sum_k w_{ki}\langle a_k^+ a_k\rangle - \sum_k w_{ik}\langle a_i^+ a_i\rangle + (\gamma_{ij} + \gamma_{ji})\langle a_i^+ a_i\rangle, \quad i \neq j. \tag{IV.5.31}$$

For many applications, e.g. laser theory, a knowledge of the second moments of the fluctuating forces is completely sufficient, because the fields interact with many independent atoms. Thus the results, e.g. on the linewidth, depend only on a sum over very many independently fluctuating forces, which possesses a Gaussian distribution (the fluctuating forces of a single atom are not Gaussian, however*).

So far we have shown that (IV.5.26) follows from condition (IV.5.4). It is rather simple to show that the opposite also holds. Thus all commutation relations between the $a_i^+ a_k$'s are also satisfied.

IV.6. The complete laser equations[1] (*quantum mechanical Langevin equations*). The results of (IV.2)—(IV.5) can be summarized as follows:

The equations of motion for the light field and the atoms, as they are known from quantum electrodynamics and represented in (III.5.7)—(III.5.9) are not sufficient, because they neglect the losses of the light field as well as the various perturbations which act on the atoms. The losses of the light field as caused by the mirrors and by other absorptive processes were therefore taken into account by an appropriate heatbath. Similarly the pumping and the decay, e.g. by spontaneous emission into the nonlasing modes and other phase-destroying processes acting on the atoms (such as lattice vibrations, collisions in gases, etc.) have been represented by a set of heatbaths.

These heatbaths can then be eliminated and replaced in the case of:

a) the light field:

by decay constants $\varkappa_\lambda$ for the modes λ, *and* by fluctuating forces, F_λ.

b) the atoms:

by transition rates w_{ik} (from state i to state k), by decay constants γ_{ik} for the relative phases between states i and k, *and* by fluctuating forces Γ.

* J. R. SENITZKY: Phys. Rev. **161**, 165 (1967).

[1] These complete equations are presented here for the first time.

Putting these results together* we obtain the following

a) Quantum-mechanically consistent equations for the operators b_λ^+ and $(a_i^+ a_k)_\mu$.

α) The field equations:

$$\frac{d}{dt} b_\lambda^+ = (i\omega_\lambda - \varkappa_\lambda)\, b_\lambda^+ + i \sum_{j,l,\mu} g_{j,l;\lambda,\mu} (a_j^+ a_l)_\mu + F_\lambda(t) \tag{IV.6.1}$$

where b_λ^+ is the creation operator of the cavity mode λ, ω_λ its frequency and $t_\lambda = \frac{1}{2\varkappa_\lambda}$ its lifetime (both in the unloaded cavity).

$(a_j^+ a_l)_\mu$ is an abbreviation for $a_{j,\mu}^+ a_{l,\mu}$ where $a_{j,\mu}^+$ is the creation operator of an electron in the level j at an atom μ (μ denotes for instance the lattice site) and $a_{l,\mu}$ is correspondingly the annihilation operator.

$g_{j,l;\lambda,\mu}$ describes the coupling of the mode λ with the electron at atom μ, which causes a real or virtual transition between levels l and j. g is given in (III.5.8).

The fluctuating forces $F_\lambda(t)$ are characterized by the properties (IV.5.1) to (IV.5.3).

β) The matter equations:

$$\left.\begin{aligned}
\frac{d}{dt}(a_{j'}^+ a_{l'})_\mu &= \nu_{j'l',\mu} (a_{j'}^+ a_{l'})_\mu + \\
&+ i \sum_{j,l,\lambda} g_{j,l,\lambda,\mu} (b_\lambda^+ + b_\lambda) \{(a_j^+ a_{l'})_\mu \delta_{lj'} - (a_{j'}^+ a_l)_\mu \delta_{jl'}\} - \\
&- \gamma_{j'l'} (a_{j'}^+ a_{l'})_\mu (1 - \delta_{j'l'}) + \\
&+ \delta_{j'l'} \left\{\sum_k w_{kj'} (a_k^+ a_k)_\mu - (a_{j'}^+ a_{j'})_\mu \sum_k w_{jk}\right\} + \Gamma_{j'l',\mu}(t), \\
\nu_{jl,\mu} &= \frac{1}{\hbar} (E_j - E_l)_\mu .
\end{aligned}\right\} \tag{IV.6.2}$$

$E_{j,\mu}$ is the energy of an electron in the level j at atom μ.

The inhomogeneous broadening is taken into account by letting the E's explicitly depend on μ.

δ_{lj} is Kronecker's symbol ($=1$ for $l=j$; $=0$ for $l \neq j$). The Γ's are the fluctuating forces. It is assumed that they are uncorrelated for different μ's and have the properties (IV.5.22) and (IV.5.23), where the coefficients G in (IV.5.22) are given by (IV.5.26). The M's occurring in (IV.5.26) are defined in (IV.5.5). Eqs. (IV.6.1) and (IV.6.2) are the basis of Chap. VI.

If we take averages with respect to the heatbath variables which still occur in the F's and Γ's, the fluctuating forces vanish. We take further averages $\langle\ldots\rangle$ with respect to the electronic states and the light-field and make the *approximation*, that [2]

$$\langle b_\lambda a_l^+ a_j \rangle_{total} = \langle b_\lambda \rangle_{field} \langle a_l^+ a_j \rangle_{el} . \tag{IV.6.3}$$

---

* Because the heatbaths of the light-field and those of the atoms are independent of each other, this putting together is possible in a straightforward manner, as could be shown also by a more detailed analysis.

[2] H. Paul: Ann. Phys. **16**, 403 (1965); Fortschr. Physik **14**, 141 (1966), has pointed out that the assumption (IV.6.3) has the far-reaching consequence that the light field must be in a coherent state (compare V.1.35). Because one of the essential objectives of a laser theory is, in fact, to determine the coherence properties of the field, this assumption is not satisfactory, although it can be justified by the rigorous theory (compare VI.7), if the (small) phase and amplitude fluctuations are neglected.

Using further

$$\langle (a_l^+ a_j)_\mu \rangle_{el} = \varrho_{j,l,\mu}, \qquad \langle b_\lambda \rangle_{field} = \langle b_\lambda \rangle \tag{IV.6.4}$$

we obtain finally from (IV.6.1) and (IV.6.2) the so called

b) "Semiclassical" equations[3].

α) The field equations

$$\frac{d}{dt} \langle b_\lambda^+ \rangle = (i\omega_\lambda - \varkappa_\lambda) \langle b_\lambda^+ \rangle + i \sum_{j,l,\mu} g_{j,l,\lambda,\mu}\, \varrho_{l,j;\mu}. \tag{IV.6.5}$$

β) The matter equations

$$\left.\begin{aligned} \frac{d}{dt} \varrho_{l',j';\mu} &= \nu_{j'l',\mu} \cdot \varrho_{l',j';\mu} + \\ &+ i \sum_{j,l,\lambda} g_{j,l,\lambda,\mu} (\langle b_\lambda^+ \rangle + \langle b_\lambda \rangle) \{ \varrho_{l',j,\mu}\, \delta_{l,j'} - \varrho_{l,j';\mu}\, \delta_{j\,l'} \} - \\ &- \gamma_{j'l'}\, \varrho_{l',j';\mu} (1 - \delta_{j',l'}) + \\ &+ \delta_{j'l'} \left\{ \sum_k w_{kj'}\, \varrho_{k,k;\mu} - \varrho_{j',j';\mu} \sum_k w_{j'k} \right\}. \end{aligned}\right\} \tag{IV.6.6}$$

Note, that $\langle b_\lambda \rangle$, $\langle b_\lambda^+ \rangle$ and $\varrho_{l,j}$ are (time-dependent) c-numbers. Eqs. (IV.6.5) and (IV.6.6) are the basis of Chap. VII.

IV.7. The density matrix equation[1].

a) General derivation. We consider the interaction of a free field (e.g. the light mode or the electron of an atom) with a heatbath or a set of heatbaths. The density matrix W of the total system obeys the equation

$$\frac{dW}{dt} = -\frac{i}{\hbar} [H, W]. \tag{IV.7.1}$$

where

$$H = H_0 + H_{F\text{-}B} \tag{IV.7.2)*}$$

and

$$H_0 = H_F + H_B. \tag{IV.7.3}$$

H_F is the Hamiltonian of the free field and reads explicitly

$$H_F = \begin{cases} \hbar\,\omega\, b^+ b & \text{light-mode,} \qquad \text{(IV.7.4)} \\ \sum \hbar\, \nu_i\, a_i^+ a_i & \text{atom.} \qquad \text{(IV.7.5)} \end{cases}$$

* In this chapter we abbreviate $H_{field\text{-}bath}$ and $H_{electron\text{-}bath}$ by $H_{F\text{-}B}$, H_{field} or $H_{electron}$ by H_F, and H_{bath} by H_B.

[3] Derivations of these equations and some discussions of the range of validity have been given by W. Weidlich and F. Haake: Z. Physik **185**, 30 (1965); **186**, 203 (1965). — C. R. Willis and P. G. Bergmann: Phys. Rev. **128**, 391 (1962). — C. R. Willis: J. Math. Phys. **5**, 1241 (1964).

[1] Density matrix equations for dissipative quantum systems were derived by many authors. The first derivations referred to spin-systems: R. K. Wangsness and F. Bloch: Phys. Rev. **89**, 728 (1953). — P. N. Argyres: In: Magnetic and Electric Resonance and Relaxation, p. 555, ed. by J. Smidt. Amsterdam: North Holland Publ. Co. 1963. — A. Abragam: The Principles of Nuclear Magnetism. Oxford: Clarendon Press 1961. — P. N. Argyres and P. L. Kelley: Phys. Rev. **134**, A 98 (1964); — For nonlinear optics see e.g. N. Bloembergen: Nonlinear Optics. New York: W. A. Benjamin Inc. 1965. — For the laser see: W. Weidlich and F. Haake: Z. Physik **185**, 30 (1965); **186**, 203 (1965). — C. R. Willis and P. G. Bergmann: Phys. Rev. **128**, 371 (1962). — C. R. Willis: J. Math. Phys. **5**, 1241 (1964); — We follow the treatment of Weidlich and Haake, who included in particular reservoirs at *different* temperatures and treat the field quantum mechanically. We extend their treatment in a straightforward manner to atoms with an arbitrary number of levels.

The Hamiltonian of the heatbath H_B may consist of a sum of several Hamiltonians corresponding to different heatbaths

$$H_B = \sum_j H_B^{(j)} \tag{IV.7.6}$$

$H_{F\text{-}B}$ represents the interaction between the two systems. For our analysis it is convenient to proceed to the interaction representation:

$$\widetilde{W} = e^{\frac{i}{\hbar} H_0 t} W e^{-\frac{i}{\hbar} H_0 t}, \tag{IV.7.7}$$

$$\widetilde{H}_{F\text{-}B} = e^{\frac{i}{\hbar} H_0 t} H_{F\text{-}B}\, e^{-\frac{i}{\hbar} H_0 t}. \tag{IV.7.8}$$

The interaction Hamiltonian has the form

$$\widetilde{H}_{F\text{-}B}(t) = \hbar \sum_k V_k(t)\, B_k(t) = \begin{cases} \underbrace{b\, e^{-i\omega t}}_{V_1}\ \underbrace{B^+(t)}_{\hbar B_1} + \underbrace{b^+\, e^{i\omega t}}_{V_2}\ \underbrace{B(t)}_{\hbar B_2} & \text{light-mode,} \quad \text{(IV.7.9)} \\ \sum_{ij} \underbrace{a_i^+\, a_j\, e^{i(\nu_{ij})t}}_{V_k}\ \underbrace{B_{ij}(t)}_{\hbar B_k} & \text{atom.} \quad \text{(IV.7.10)} \end{cases}$$

The explicit form of the V's suggests that we write $V_k(t) = V_k\, e^{i\Delta \nu_k t}$ (IV.7.11)

We treat the interaction by perturbation theory up to second order which yields

$$\widetilde{W} = W_0 + \left(-\frac{i}{\hbar}\right) \int_0^t d\tau\, [\widetilde{H}_{F\text{-}B}(\tau); W_0] + \left(-\frac{i}{\hbar}\right)^2 \int_0^t d\tau_2 \int_0^{\tau_2} d\tau_1 \cdot \{\cdots\}, \tag{IV.7.12}$$

where

$$\left.\begin{aligned} \{\cdots\} = \{&\widetilde{H}_{F\text{-}B}(\tau_2)\, \widetilde{H}_{F\text{-}B}(\tau_1)\, W_0 - \widetilde{H}_{F\text{-}B}(\tau_2)\, W_0\, \widetilde{H}_{F\text{-}B}(\tau_1) - \\ &- \widetilde{H}_{F\text{-}B}(\tau_1)\, W_0\, \widetilde{H}_{F\text{-}B}(\tau_2) + W_0\, \widetilde{H}_{F\text{-}B}(\tau_1)\, \widetilde{H}_{F\text{-}B}(\tau_2)\}. \end{aligned}\right\} \tag{IV.7.13}$$

Since we are ultimately interested only in the variables of the field we eliminate the bath variables from (IV.7.12) by taking the trace over the heatbath

$$\tilde{\varrho} = \text{trace}_B\, \widetilde{W}. \tag{IV.7.14}$$

It is essential that at time $t = 0$ the total density matrix factorizes in that of the free field, $\varrho(0)$ and that of the heatbath, ϱ_B:

$$W_0 = \widetilde{W}(t=0) = \varrho(0)\, \varrho_B. \tag{IV.7.15}$$

ϱ_B may itself factorize into the density matrix of the individual baths

$$\varrho_B = \varrho_{B,1} \cdot \varrho_{B,2} \cdot \cdots. \tag{IV.7.16}$$

We allow that these baths are kept at different temperatures T_j, so that

$$\varrho_{Bj} = \frac{1}{Z}\, e^{-\frac{H_B^{(j)}}{kT_j}} \tag{IV.7.17}$$

where

$$Z = \text{trace}_{Bj}\, e^{-\frac{H_B^{(j)}}{kT_j}}. \tag{IV.7.18}$$

For simplification of the further analysis, we may assume

$$\text{trace}_B (B_k\, \varrho) = 0 \tag{IV.7.19}$$

because the corresponding terms in (IV.7.12) would lead to mere energy shifts, but no damping effects. After taking the trace over Eq. (IV.7.12) we find

$$\left.\begin{aligned}
&\tilde{\varrho}(t)=\varrho(0)-\int_0^t d\tau_2\int_0^{\tau_2} d\tau_1\cdot\{\cdots\},\\
&\{\cdots\}=\sum_{kk'}\{V_k(\tau_2)\,V_{k'}(\tau_1)\,\varrho(0)\,\mathrm{trace}_B\,(B_k(\tau_2)\,B_{k'}(\tau_1)\,\varrho_B(0))-\\
&\quad-V_k(\tau_2)\,\varrho(0)\,V_{k'}(\tau_1)\,\mathrm{trace}_B\,(B_k(\tau_2)\,\varrho_B(0)\,B_{k'}(\tau_1))-\\
&\quad-V_k(\tau_1)\,\varrho(0)\,V_{k'}(\tau_2)\,\mathrm{trace}_B\,(B_k(\tau_1)\,\varrho_B(0)\,B_{k'}(\tau_2))+\\
&\quad+\varrho(0)\,V_k(\tau_1)\,V_{k'}(\tau_2)\,\mathrm{trace}_B\,(\varrho_B(0)\,B_k(\tau_1)\,B_{k'}(\tau_2))\}.
\end{aligned}\right\}\qquad\text{(IV.7.20)}$$

The second term on the right hand side of Eq. (IV.7.12) was now dropped because of Eq. (IV.7.19).

In order to further simplify Eq. (IV.7.20), we discuss the expressions

$$\int_0^t d\tau_2\, e^{i\Delta\nu_k\tau_2}\int_0^{\tau_2} d\tau_1\, e^{i\Delta\nu_{k'}\tau_1}\,K_{kk'}(\tau_2-\tau_1)\qquad\text{(IV.7.21)}$$

where K is an abbreviation for

$$K_{kk'}(\tau_2,\tau_1)=K_{kk'}(\tau_2-\tau_1)=\mathrm{trace}_B\big(B_k(\tau_2)\,B_{k'}(\tau_1)\,\varrho_B(0)\big).\qquad\text{(IV.7.22)}$$

The first part of Eq. (IV.7.22) expresses the fact that the correlation function K depends only on the time difference, provided, that the interaction is stationary[2].

A simple coordinate transformation

$$\tau_2-\tau_1=\tau'\qquad\text{(IV.7.23)}$$

yields for (IV.7.21)

$$\int_0^t d\tau_2\, e^{i(\Delta\nu_k+\Delta\nu_{k'})\tau_2}\int_0^{\tau_2} e^{-i\Delta\nu_{k'}\tau_1}\,K_{kk'}(\tau)\,d\tau.\qquad\text{(IV.7.24)}$$

[2] The proof runs as follows: Stationary means that in the Schrödinger picture B_k, B_k and H_B are time independent. We distinguish between two cases:

a) B_k and $B_{k'}$ belong to different heatbaths. The trace (IV.7.22) then factorizes into $K_k\cdot K_{k'}$, where e.g.

$$K_k=\mathrm{trace}_{B,k}\left(e^{\frac{i}{\hbar}H_B^{(k)}\tau_2}B(0)\,e^{-\frac{i}{\hbar}H_B^{(k)}\tau_2}e^{-\frac{H_B^{(k)}}{kT_k}}\right)\cdot\frac{1}{Z_k}\qquad\text{(IV.7.22a)}$$

where (IV.7.17) was used.

Because the trace is cyclic: trace $(AB)=$ trace (BA) we immediately find

$$K_k=\mathrm{trace}_{B,k}\left(B_k(0)\,e^{-\frac{H_B^{(k)}}{kT_k}}\right)=0$$

due to (IV.7.19).

b) B_k and $B_{k'}$ belong to the same heatbath. Using (IV.7.17) we write (IV.7.22) in the form

$$\frac{1}{Z_k}\,\mathrm{trace}_B\left(e^{\frac{i}{\hbar}H_B^{(k)}\tau_2}B_k(0)\,e^{-\frac{i}{\hbar}H_B^{(k)}\tau_2}e^{\frac{i}{\hbar}H_B^{(k)}\tau_1}B_{k'}(0)\,e^{-\frac{i}{\hbar}H_B^{(k)}\tau_1}e^{-\frac{H_B^{(k)}}{kT_k}}\right).$$

Because the trace is cyclic (IV.7.22) transforms into

$$\frac{1}{Z_k}\,\mathrm{trace}_B\left(e^{\frac{i}{\hbar}H_B^{(k)}(\tau_2-\tau_1)}B_k(0)\,e^{-\frac{i}{\hbar}H_B^{(k)}(\tau_2-\tau_1)}B_{k'}(0)\,e^{-\frac{H_B^{(k)}}{kT_k}}\right)$$

which depends only on $(\tau_2-\tau_1)$.

We assume that the heatbath has a short memory so that (IV.7.22) is only nonvanishing for

$$|\tau_2 - \tau_1| < \tau_0 . \tag{IV.7.25}$$

In the following we consider times t which are large compared to the correlation time τ_0. Then we may simply replace the upper limit τ_2 by infinity, so that we finally find for (IV.7.21)

$$\int_0^t d\tau_2 \, e^{i(\Delta\nu_k + \Delta\nu_{k'})\tau_2} \int_0^\infty e^{-i\Delta\nu_{k'}\tau} K_{kk'}(\tau)\, d\tau . \tag{IV.7.26}$$

If

$$\Delta\nu_k + \Delta\nu_{k'} = 0 \tag{IV.7.27}$$

the expression (IV.7.26) becomes

$$t \int_0^\infty e^{-i\Delta\nu_{k'}\tau} K_{kk'}(\tau)\, d\tau . \tag{IV.7.28}$$

In the following we may simply assume that (IV.7.27) holds because otherwise there appear rapidly oscillating terms which cancel the corresponding expressions. We now make the second essential assumption, namely, that the interaction with the heatbath is so small that during the time $\tau > \tau_0$ the density matrix has changed very little. This allows us to make the replacement

$$\frac{\tilde{\varrho}(t) - \tilde{\varrho}(0)}{t} = \frac{d\tilde{\varrho}(t)}{dt} .$$

If we further use the cyclic property of traces

$$\text{trace}_B\left(B_k(\tau_1)\, \varrho_B(0)\, B_{k'}(\tau_2)\right) = \text{trace}_B\left(B_{k'}(\tau_2)\, B_k(\tau_1)\, \varrho_B(0)\right) \tag{IV.7.29}$$

we may rewrite (IV.7.20) in the form

$$\left.\begin{aligned} \frac{d\tilde{\varrho}(t)}{dt} = -\sum_{kk'} \Big\{ & V_k V_{k'}\, \varrho(0)\, e^{i(\Delta\nu_k + \Delta\nu_{k'})t} \int_0^\infty e^{-i\Delta\nu_{k'}\tau}\, \text{trace}_B\left(B_k(\tau)\, B_{k'}(0)\, \varrho_B(0)\right) d\tau - \\ & - V_k\, \varrho(0)\, V_{k'}\, e^{i(\Delta\nu_{k'} + \Delta\nu_k)t} \int_0^\infty e^{-i\Delta\nu_k\tau}\, \text{trace}_B\left(B_{k'}(\tau)\, B_k(0)\, \varrho_B(0)\right) d\tau + \\ & + \varrho(0)\, V_k V_{k'}\, e^{i(\Delta\nu_{k'} + \Delta\nu_k)t} \int_0^\infty e^{-i\Delta\nu_k\tau}\, \text{trace}_B\left(B_k(0)\, B_{k'}(\tau)\, \varrho_B(0)\right) d\tau - \\ & - V_k\, \varrho(0)\, V_{k'}\, e^{i(\Delta\nu_k + \Delta\nu_{k'})t} \int_0^\infty e^{-i\Delta\nu_{k'}\tau}\, \text{trace}_B\left(B_{k'}(0)\, B_k(\tau)\, \varrho_B(0)\right) d\tau \Big\} . \end{aligned}\right\} \tag{IV.7.30}$$

We now assume that the iteration step from time $t=0$ to the time t may be repeated at consecutive times so that we may replace the initial time $t=0$ by the arbitrary time t. Because the integral expressions of the first and second sum agree provided k is exchanged with k', and the same is true for the third and fourth term, Eq. (IV.7.30) takes the very simple form

$$\frac{d\tilde{\varrho}(t)}{dt} = \sum_{kk'} \{[V_{k'}\tilde{\varrho}(t), V_k]\, A_{kk'} + [V_{k'}, \tilde{\varrho}(t)\, V_k]\, A'_{kk'}\} \tag{IV.7.31}$$

where the coefficients A and A' are defined by

$$A_{kk'} = e^{i(\Delta\nu_k + \Delta\nu_{k'})t} \int_0^\infty e^{-i\Delta\nu_{k'}\tau} \operatorname{trace}_B\big(B_k(\tau)\, B_{k'}(0)\, \varrho_B(0)\big)\, d\tau \qquad \text{(IV.7.32)}$$

$$A'_{kk'} = e^{i(\Delta\nu_k + \Delta\nu_{k'})t} \int_0^\infty e^{-i\Delta\nu_k\tau} \operatorname{trace}_B\big(B_k(0)\, B_{k'}(\tau)\, \varrho_B(0)\big)\, d\tau\,. \qquad \text{(IV.7.33)}$$

A and A' are in a close internal connection: We consider the particular case that

$$\Delta\nu_k + \Delta\nu_{k'} = 0; \qquad B_{k'}^+ = B_k \qquad \text{(IV.7.34)}$$

holds. This is suggested by the detailed consideration of the interaction Hamiltonian (IV.7.9), (IV.7.10), since it must be Hermitian. We find, e.g. in the atomic case, always the following combination

$$a_i^+ a_j\, e^{i(\nu_i - \nu_j)t}\, B_{ij} + a_j^+ a_i\, e^{-i(\nu_i - \nu_j)t}\, B_{ij}^+\,. \qquad \text{(IV.7.35)}$$

It follows that k and k' have the meaning

$$(i, j) = k; \qquad (j, i) = k'. \qquad \text{(IV.7.36)}$$

Under this assumption we obtain

$$\left.\begin{aligned} A'_{kk'} &= \int_0^\infty e^{-i\Delta\nu_k\tau} \operatorname{trace}_B\big(\varrho_B(0)\, B_{k'}^+(0)\, B_{k'}(\tau)\big)\, d\tau \\ &= \left(\int_0^\infty e^{-i\Delta\nu_{k'}\tau} \operatorname{trace}_B\big(B_{k'}^+(\tau)\, B_{k'}(0)\, \varrho_B(0)\big)\, d\tau\right)^* = A_{kk'}^*. \end{aligned}\right\} \qquad \text{(IV.7.37)}$$

It is therefore sufficient to discuss the properties of the A's only. We evaluate the trace over the heatbath in the energy representation of the heatbath alone. This yields

$$\left.\begin{aligned} A_{k,k'} &= \int_0^\infty e^{-i\Delta\nu_{k'}\tau} \sum_{nm} \langle n|B_k|m\rangle \langle m|B_{k'}|n\rangle\, e^{i(\omega_n - \omega_m)\tau}\, e^{-\hbar\omega_n/kT_k}\, d\tau \cdot \frac{1}{Z} \\ &= \sum_{nm} |\langle n|B_k|m\rangle|^2 \cdot e^{-\hbar\omega_n/kT_k} \Big\{\pi\,\delta(\omega_n - \omega_m - \Delta\nu_{k'}) + \\ &\quad + \frac{iP}{\omega_m - \omega_n - \Delta\nu_{k'}}\Big\} \cdot \frac{1}{Z} \end{aligned}\right\} \qquad \text{(IV.7.38)}$$

where Z is the normalization of the heatbath trace. In the same way we obtain

$$\left.\begin{aligned} A_{k',k} = {} & \sum_{nm} |\langle n|B_k|m\rangle|^2 \cdot \frac{1}{Z}\, e^{-\hbar\omega_m/kT_k} \Big\{\pi\,\delta(\omega_m - \omega_n + \Delta\nu_{k'}) + \\ & + \frac{iP}{\omega_m - \omega_n - \Delta\nu_k}\Big\} \end{aligned}\right\} \qquad \text{(IV.7.39)}$$

where use was made of (IV.7.17). A comparison of (IV.7.38) and (IV.7.39) leads to the important relation

$$\boxed{\operatorname{Re} A_{k',k} = \operatorname{Re} A_{k,k'} \cdot e^{\hbar\Delta\nu_{k'}/kT_{k'}}} \qquad \text{(IV.7.40)}$$

which we discuss both for the light field and the atom:

1. light field: with (IV.7.9) we have e.g. $k = 1$, $k' = 2$, $\Delta\nu_{k'} = \omega$

$$\operatorname{Re} A_{2,1} = \operatorname{Re} A_{1,2} \cdot e^{\hbar\omega/kT} \qquad \text{(IV.7.41)}$$

2. atom: with (IV.7.10), (IV.7.35), and (IV.7.36), we have

$$\mathrm{Re}\, A_{\underbrace{ij}_{k'},\, \underbrace{ji}_{k}} = \mathrm{Re}\, A_{\underbrace{ji}_{k},\, \underbrace{ij}_{k'}}\, e^{\hbar\nu_{ij}/kT(ij)}\,. \tag{IV.7.42}$$

b) Specialization of Eq. (IV.7.31).

α) *Light mode*[3]. We make the following identifications

$$\left.\begin{aligned} V_1 &= b, & V_2 &= b^+ \\ \Delta\nu_1 &= -\omega, & \Delta\nu_2 &= \omega, \\ B_1 &= B^+, & B_2 &= B. \end{aligned}\right\} \tag{IV.7.43}$$

Eq. (IV.7.31) then takes the form

$$\frac{d\tilde{\varrho}}{dt} = [b\tilde{\varrho}, b^+]A_{21} + [b^+\tilde{\varrho}, b]A_{12} + [b, \tilde{\varrho} b^+]A_{21}^* + [b^+, \tilde{\varrho} b]A_{21}^*. \tag{IV.7.44}$$

Because the imaginary parts of A give rise to mere frequency shifts which can be absorbed into the frequency of the actual oscillator we keep only the real parts and put

$$A_{21} = \xi, \tag{IV.7.45}$$

$$A_{12} = \delta = e^{-\hbar\omega/kT}\,\xi \tag{IV.7.46}$$

where WEIDLICH's notation is used.

The final equation for the density matrix of the light field alone thus takes the form

$$\frac{d\varrho}{dt} = \delta([b^+\varrho, b] + [b^+, \varrho b]) + \xi([b\varrho, b^+] + [b, \varrho b^+]). \tag{IV.7.47}$$

We now show how this equation allows us to treat the mean motion of the light field amplitude and intensity, where we define the average by

$$\langle O\rangle = \mathrm{trace}_F(O\varrho)\,. \tag{IV.7.48}$$

The average light field amplitude obeys the equation

$$\frac{d\langle b^+\rangle}{dt} = \{i\omega - (\xi - \delta)\}\langle b^+\rangle. \tag{IV.7.49}$$

The photon number obeys the equation

$$\frac{d\langle b^+ b\rangle}{dt} = 2\delta - 2(\xi - \delta)\langle b^+ b\rangle \tag{IV.7.50}$$

(IV.7.49) allows for the solution

$$\langle b^+\rangle = \langle b^+\rangle_0\, e^{i\omega t - (\xi-\delta)t} \tag{IV.7.51}$$

whereas (IV.7.50) is solved by

$$\langle b^+ b\rangle = \langle b^+ b\rangle_0\, e^{-2(\xi-\delta)t} + \frac{\delta}{\xi-\delta}\left(1 - e^{-2(\xi-\delta)t}\right). \tag{IV.7.52}$$

Evidently the difference $\xi - \delta$ must be identified with the decay constant which is introduced in the classical theory

$$\xi - \delta = \varkappa \tag{IV.7.53}$$

[3] W. WEIDLICH and F. HAAKE: l.c.[1] on p. 51.

whereas the expression $\frac{\delta}{\xi-\delta}$ can be evaluated using (IV.7.45) and (IV.7.46)

$$\frac{\delta}{\xi-\delta}=\frac{1}{e^{\hbar\omega/kT}-1}=\bar{n}. \tag{IV.7.54}$$

From Eq. (IV.7.52) we see that the photon number tends towards the thermal photon number at the temperature of the heatbath.

β) *Atom.* Here we make the following identification

$$V_k \rightarrow a_i^+ a_j,$$
$$\Delta\nu_k \rightarrow \nu_i - \nu_j,$$
$$B_k = B_{ij}; \qquad B_{ji} = B_{ij}^+.$$

The density matrix equation takes the simple form

$$\frac{d\tilde{\varrho}}{dt}=\sum_{ij}[a_i^+ a_j\tilde{\varrho},\, a_j^+ a_i]\,A_{ji,ij}+\sum_{ij}[a_i^+ a_j,\,\tilde{\varrho}\, a_j^+ a_i]\,A_{ji,ij}^*. \tag{IV.7.55}$$

We derive the equation for the average of the operator $a_m^+ a_n$ by multiplying both sides of Eq. (IV.7.55) by $a_m^+ a_n$ and taking the trace:

$$\left.\begin{aligned}\frac{d\langle a_m^+ a_n\rangle}{dt} &= \mathrm{trace}_F\Big\{\sum_{ij}(a_m^+ a_n a_i^+ a_j\tilde{\varrho}\, a_j^+ a_i - a_m^+ a_n a_j^+ a_i a_i^+ a_j\tilde{\varrho})\,A_{ji,ij}+\\ &+(a_m^+ a_n a_i^+ a_j\tilde{\varrho}\, a_j^+ a_i - a_m^+ a_n\tilde{\varrho}\, a_j^+ a_i a_i^+ a_j)\,A_{ji,ij}^*\Big\}.\end{aligned}\right\} \tag{IV.7.56}$$

For the further evaluation of the trace we use its cyclic property and arrange the operators in such a way that the density matrix ϱ stands on the right hand side. We then use the theorem

$$a_{i_1}^+ a_{j_1} a_{i_2}^+ a_{j_2}\dots a_{i_n}^+ a_{j_n} = a_{i_1}^+ a_{j_n}\,\delta_{j_1 i_2}\,\delta_{j_2 i_3}\dots\delta_{j_{n-1} i_n} \tag{IV.7.57}$$

which holds if only one electron is present. The relation (IV.7.57) can be proved as follows: consider within the product the terms which bear the indices λ and $\lambda+1$. Exchanging $a_{j\lambda}$ with $a_{j\lambda+1}$, we find using the commutation relations

$$\left.\begin{aligned}& a_{i_1}^+ a_{j_1}\dots a_{j\lambda}\, a_{i\lambda+1}^+ a_{i\lambda+1}\dots a_{i_n}^+ a_{j_n}\\ &\qquad = a_{i_1}^+ a_{j_1}\dots(\delta_{j\lambda,\, i\lambda+1} - a_{i\lambda+1}^+ a_{j\lambda})\, a_{i\lambda+1}\dots a_{i_n}^+ a_{j_n}.\end{aligned}\right\} \tag{IV.7.58}$$

The term which still contains $a_{i\lambda+1}^+ a_{j\lambda}$ contains two subsequent annihilation operators. These operators are applied to a state which contains only one electron. The corresponding state vector vanishes.

If (IV.7.58) is performed for all indices λ, (IV.7.57) results. Using (IV.7.57) we transform Eq. (IV.7.56) into

$$\frac{d}{dt}\langle a_m^+ a_n\rangle = \langle a_m^+ a_n\rangle(i\Delta\Omega_{mn}-\gamma_{mn}) \quad \text{for} \quad m\neq n, \tag{IV.7.59}$$

$$\frac{d}{dt}\langle a_m^+ a_m\rangle = \sum_j \langle a_j^+ a_j\rangle\, w_{jm} - \langle a_m^+ a_m\rangle \sum_j w_{mj} \tag{IV.7.60}$$

where we have used the abbreviations

$$w_{jm} = A_{jmmj} + A_{jmmj}^*, \tag{IV.7.61}$$

$$\gamma_{mn} = \sum_i \mathrm{Re}(A_{miim}) + \mathrm{Re}(A_{niin}^*) = \tfrac{1}{2}\sum_i (w_{ni}+w_{mi}), \tag{IV.7.62}$$

$$\Delta\Omega_{mn} = -\sum_i \mathrm{Im}(A_{niin} + A_{miim}^*) \tag{IV.7.63}$$

w_{jm} is evidently the transition rate from the state j to the state m, as may be seen by considering Eq. (IV.7.60). γ_{mn} is the phase halfwidth from the second part of Eq. (IV.7.59). We see that the halfwidth is determined by half the sum of the transition rates out of the two states n and m which are considered. (IV.7.62) contains also an additional term which stems from phase fluctuations alone which are not accompanied by real transitions, i.e. w_{nn}. Eq. (IV.7.59) contains frequency shifts which we neglect, however, in the following. We now show that the Eq. (IV.7.59), (IV.7.60) can also be interpreted as an equation for the density matrix itself if the latter one is written in the occupation number representation. We put

$$\varrho(t) = \sum_{m'n'} \varrho_{m'n'}(t)\, a_{m'}^+ a_{n'} \qquad \text{(IV.7.64)}$$

and consider

$$\text{trace}_F\,(a_m^{+\prime} a_n\, \varrho(t)) = \sum_l \langle \Phi_0\, a_l\, a_m^+ a_n \sum_{m'n'} \varrho_{m'n'}(t)\, a_{m'}^+ a_{n'}\, a_l^+\, \Phi_0 \rangle \qquad \text{(IV.7.65)}$$

where Φ_0 is the vacuum state and $a_l^+ \Phi_0$ a one-electron state.

Using (IV.7.57) we find immediately

$$\text{trace}_F\,(a_m^+ a_n\, \varrho(t)) = \varrho_{nm}(t) \qquad \text{(IV.7.66)}$$

so that we can identify

$$\langle a_m^+ a_n \rangle = \varrho_{nm}(t)\,. \qquad \text{(IV.7.67)}$$

γ) *The density matrix equation of the complete system of M laser modes and N atoms.* We return from Eq. (IV.7.47) and (IV.7.59), (IV.7.60) which were written in the interaction representation to the Schrödinger equation and assume that the dissipation and fluctuation mechanism is not altered by the interaction between the lightfields and the atoms. For field strengths that are not too high this is an excellent approximation*.

Under these assumptions the density matrix equation has the form

$$\frac{d\varrho}{dt} = -\frac{i}{\hbar}[H_S, \varrho] + \left(\frac{\partial \varrho}{\partial t}\right)_{\textit{incoh, light}} + \left(\frac{\partial \varrho}{\partial t}\right)_{\textit{incoh, atoms}}. \qquad \text{(IV.7.68)}$$

H_S is the Hamiltonian of the loss-less system given by (III.3.10) with (III.1.13), (III.2.9), and (III.3.7), (III.5.8) and reads

$$H_S = \sum_\lambda \hbar\,\omega_\lambda\, b_\lambda^+ b_\lambda + \sum_{\mu,j,l} \hbar \nu_{j,\mu} (a_j^+ a_j)_\mu + \hbar \sum_{\lambda,j,l,\mu} g_{j,l;\lambda,\mu} (b_\lambda^+ + b_\lambda)\,(a_j^+ a_l)_\mu\,. \qquad \text{(IV.7.69)}$$

The incoherent decay of the light field is given according to (IV.7.47) by

$$\left(\frac{\partial \varrho}{\partial t}\right)_{\textit{incoh, light}} = \xi_\lambda([b_\lambda\, \varrho, b_\lambda^+] + [b_\lambda, \varrho\, b_\lambda^+]) + \delta_\lambda([b_\lambda^+ \varrho, b_\lambda] + [b_\lambda^+, \varrho b_\lambda]) \qquad \text{(IV.7.70)}$$

and that of the atoms according to (IV.7.55) by

$$\left(\frac{\partial \varrho}{\partial t}\right)_{\textit{incoh, atoms}} = \sum_{ij,\mu} ([(a_i^+ a_j)_\mu\, \varrho, (a_j^+ a_i)_\mu] + [(a_i^+ a_j)_\mu, \varrho\,(a_j^+ a_i)_\mu])\, A_{jiij;\mu}\,. \qquad \text{(IV.7.71)}$$

The coefficients ξ, δ and A were assumed to be real. The physical meaning of the A's follows from the identification (IV.7.61), (IV.7.63). Eq. (IV.7.68) is the basic master equation which governs the whole laser statistics. We shall discuss its direct solution in Chap. VI.11.

---

* It should be noted, however, that very strong fields cause a partial quenching of the spontaneous emission linewidth. In this region the incoherent terms of the density matrix equation should be derived by using the coupled system: field + atoms from the very beginning.

IV.8. The evaluation of multi-time correlation functions by the single-time density matrix[1]. We now show that the density matrix which contains only one time variable may be used to calculate multi-time correlation functions. As we will show in Chap. V.4, the statistical properties (or in other words, the coherence properties) of laser light are determined by correlation functions of the form

$$\langle b^+(t_1') \ldots b^+(t_m')\, b(t_n) \ldots b(t_2)\, b(t_1)\rangle \tag{IV.8.1}$$

where we consider for simplicity a single mode. All the following considerations apply equally well to the multimode case, however. We assume that the b's occur ordered with respect to their time

$$t_{j+1} \geqq t_j; \quad t_{j+1}' \geqq t_j'. \tag{IV.8.2}$$

In order to show how (IV.8.1) can be calculated by the density matrix we represent (IV.8.1) in the form

$$\mathrm{tr}\{b(t_n)\, b(t_{n-1}) \ldots b(t_1)\, \varrho(t_0)\, b^+(t_1') \ldots b^+(t_m')\} \tag{IV.8.3}$$

where $\varrho(t_0)$ is the density matrix at time $t_0 = \min(t_j, t_j')$. We write expression (IV.8.3) in a new fashion

$$\left.\begin{aligned}(\text{IV.8.1}) = {} & \frac{1}{(-i)^n}\,\frac{1}{i^m} \times \\ & \times \left[\frac{\delta^n}{\delta\alpha(t_n)\,\delta\alpha(t_{n-1}) \ldots \delta\alpha(t_1)}\,\frac{\delta^m}{\delta\gamma^*(t_1') \ldots \delta\gamma^*(t_m')}\,\varrho(\alpha,\gamma)\right]_{\alpha=0,\gamma=0}\end{aligned}\right\} \tag{IV.8.4}$$

where

$$\left.\begin{aligned}\varrho(\alpha,\gamma) = {} & \Big\langle T e^{-i\int_{t_0}^{t} H_\tau d\tau - i\int_{t_0}^{t} (\alpha(\tau) b_\tau + \alpha^*(\tau) b_\tau^+)\, d\tau} \times \\ & \times \varrho(t_0)\, \tilde{T} e^{i\int_{t_0}^{t} H_{\tau'} d\tau' + i\int_{t_0}^{t} (\gamma(\tau') b_{\tau'} + \gamma^*(\tau') b_{\tau'}^+)\, d\tau'}\Big\rangle_{\text{reservoir}}\end{aligned}\right\} \tag{IV.8.5}$$

where H is the total Hamiltonian of the system which comprises the light field and its interaction with the atoms as well as the interaction between the light field and heatbaths and the interaction of the atoms with heatbaths:

$$\left.\begin{aligned}H = H_S + H_{S\text{-}B}, \quad \text{where} \quad H_S &= H_{\text{field}} + H_{\text{atoms}} + H_{\text{field-atom}}, \\ H_{S\text{-}B} &= H_{\text{field-bath}} + H_{\text{atoms-bath}}.\end{aligned}\right\} \tag{IV.8.6}$$

It is assumed that the free motion of the heatbaths is split off, so that H_τ contains the heatbath coordinates in the interaction representation. $T, \tilde{T}$ are time ordering operators, where T and $\tilde{T}$ order the terms with latest time to the left and right, respectively.

The problem is now reduced to the determination of the *extended* density matrix (IV.8.5). By derivation of both sides of (IV.8.5) with respect to time we obtain an equation for $\varrho(\alpha, \gamma)$:

$$\left.\begin{aligned}\frac{d\varrho(\alpha,\gamma)}{dt} = {} & -i[H_S, \varrho] - i(\alpha(t)\, b + \alpha^*(t)\, b^+)\,\varrho + \varrho\, i(\gamma(t)\, b + \gamma^*(t)\, b^+) - \\ & - i\langle [H_{S\text{-}B}, \tilde{\varrho}]\rangle_R\end{aligned}\right\} \tag{IV.8.7}$$

$\tilde{\varrho}$ is defined by the right-hand side of (IV.8.5) but without reservoir average.

[1] H. Haken and W. Weidlich: Z. Physik **205**, 96 (1967). $\hbar$ is put equal 1.

The last term was calculated in the foregoing and coincides with

$$\left(\frac{\partial \varrho}{\partial t}\right)_{incoh., light} + \left(\frac{\partial \varrho}{\partial t}\right)_{incoh., atoms}$$

on the right hand side of Eq. (IV.7.68).

Therefore the only difference between the usual density matrix and our new Eq. (IV.8.7) consists in the additional terms containing the α's and γ's which are arbitrary functions (which may, however, be small). In order to determine the correlation function (IV.8.1), the system must be "probed" by classical external fields*. The b's in (IV.8.7) are, of course, time independent.

If one wants to calculate only a limited number of correlation functions it is sufficient to formulate an initial value problem. To this end we put

$$\alpha(t) = \Sigma \alpha_i \,\delta(t-t_i), \qquad \gamma^* = \Sigma \gamma_i^* \,\delta(t-t_i') \tag{IV.8.8}$$

and replace the functional derivative by an ordinary derivative with respect to α. Obviously the solution of (IV.8.7) is now replaced by solving the usual density matrix Eq. (IV.7.68) with the additional jump conditions

$$\left.\begin{aligned} \left.\frac{d\varrho}{dt}\right|_{t=t_i} &= \varrho(t_i+0) - \varrho(t_i-0) = -i\alpha_i b \int\limits_{t_i-\varepsilon}^{t_i+\varepsilon} \delta(t'-t_i)\,\varrho(t')\,dt' \\ &= -i\alpha_i b\,\varrho(t_i-0) + \text{corrections of order } \alpha^2. \end{aligned}\right\} \tag{IV.8.9}$$

The corrections of order α^2 and higher order may be neglected as long as the differentiation in Eq. (IV.8.9) is performed only once with respect to a single time. The case where several operators occur in (IV.8.1) at the same time may then be obtained as a special case by letting the time difference go to zero.

So far we have been considering correlation functions with respect to photon operators. In a similar way, we may do this, of course, with atomic operators or with mixed terms, for instance, with

$$\langle a_{i_1'}^+ a_{k_1'}(t_1') \ldots a_{i_m'}^+ a_{k_m'}(t_m')\,(a_{i_n}^+ a_{k_n})(t_n) \ldots (a_{i_1}^+ a_{k_1})(t_1)\rangle \tag{IV.8.10}$$

where the a_i^+ and a_j are the creation and annihilation operators of electrons. Obviously one has now a density matrix equation of the form

$$\left.\begin{aligned} \frac{d\varrho}{dt} &= -i[H_S, \varrho] + \left(\frac{d\varrho}{dt}\right)_{incoh.} - i\sum_{i,k} D_{ik}(t)\,a_i^+ a_k\,\varrho + \\ &\quad + \varrho\sum_{i,k} D_{i,k}'(t)\,a_i^+ a_k - i\left(\alpha(t)\,b + \alpha^*(t)\,b^+\right)\varrho + i\varrho\left(\gamma(t)\,b + \gamma^*(t)\,b^+\right) \end{aligned}\right\} \tag{IV.8.11}$$

and the correlation function (IV.8.10) may be found by functional derivatives with respect to D.

IV.9. Generalized Fokker-Planck equation: definition of distribution functions.

The distribution functions of this chapter are classical functions of classical variables. They allow us to calculate all *quantum mechanical* averages (see IV.9a, γ below) and time ordered correlation functions (see IV.11 below) by pure c-number procedures, i.e. evaluation of *classical* averages, without any loss of the quantum mechanical information.

* This establishes a connection with Kubo's point of view[2].

[2] R. Kubo: J. Phys. Soc. Japan **12**, 570 (1957), and in: Fluctuation, Relaxation and Resonance in Magnetic Systems. Edinburgh: Oliver & Boyd 1962.

a) Field.

α) *Wigner distribution function and related representations.* The Wigner distribution function[1] establishes a connection between the quantum mechanical wave function, or more generally, the density matrix, and a c-number distribution function of (classical) variables.

The Wigner function reads:

$$\hat{W}(p, q) = \frac{1}{(2\pi)^2} \iint e^{-ip\mu - iq\nu} \operatorname{tr}(e^{i\mu\hat{p} + i\nu\hat{q}} \varrho) \, d\mu \, d\nu, \qquad \text{(IV.9.1)}$$

where $\hat{q}$ and $\hat{p}$ are coordinate and moment operators, respectively, of a harmonic oscillator and ϱ the density matrix, tr is the abbreviation for trace. [$\hat{q}$ and $\hat{p}$ are defined in such a way that the Hamilton operator $\hat{H}$ is given by $\hat{H} = \frac{1}{2}(\hat{p}^2 + \omega^2 \hat{q}^2)$.]

Eq. (IV.9.1) establishes a connection between the quantum mechanical operators $\hat{q}$ and $\hat{p}$ and classical variables q and p

$$\left.\begin{array}{lcc} \text{classical} & \text{transformation (IV.9.1)} & \text{quantum mechanical} \\ q & \leftrightarrow & \hat{q} \\ p & \leftrightarrow & p = \dfrac{\hbar}{i} \dfrac{d}{d\hat{q}} \end{array}\right\} \qquad \text{(IV.9.2)}$$

Eq. (IV.9.1) corresponds to the definition of the distribution function in probability theory of classical variables where q and p would be random functions and $\operatorname{tr}(e^{i\mu\hat{p}+i\nu\hat{q}}\varrho)$ would just be the characteristic function.

The definition is not confined to an harmonic oscillator, but can also be used for other quantum systems (for a discussion see HUSIMI, K., Proc. Phys. Math. Soc. Japan **22**, 264 (1940)]. In the following we leave it open whether ϱ or $(\hat{q}, \hat{p})$ are time-dependent operators, because (IV.9.1) holds for both the Schrödinger and Heisenberg representations.

In our present context it is advantageous to proceed from $(\hat{q}, \hat{p})$ of the harmonic oscillator to the corresponding Bose creation and annihilation operators b^+, b:

$$\hat{q} = \sqrt{\frac{\hbar}{2\omega}} (b^+ + b); \quad \hat{p} = i \sqrt{\frac{\hbar\omega}{2}} (b^+ - b). \qquad \text{(IV.9.3)}$$

Writing (IV.9.1) in terms of the creation and annihilation operator b^+, b we obtain

$$\left.\begin{aligned} \hat{W}(p, q) &= \frac{1}{2\hbar} W(u, u^*), \\ W(u, u^*) &= \frac{1}{\pi^2} \int e^{-i\beta u - i\beta^* u^*} \operatorname{tr}(e^{i\beta b + i\beta^* b^+} \varrho) \, d^2\beta, \end{aligned}\right\} \qquad \text{(IV.9.4)}$$

where we have used the variables

$$\left.\begin{aligned} \beta = \sqrt{\frac{\hbar}{2}} \left[\frac{\nu}{\sqrt{\omega}} - i\sqrt{\omega}\,\mu\right]; \quad u = \frac{1}{\sqrt{2\hbar\,\omega}} (\omega q + ip); \\ d^2\beta = d(\operatorname{Re}\beta)\, d(\operatorname{Im}\beta). \end{aligned}\right\} \qquad \text{(IV.9.5)}$$

[1] E. P. WIGNER: Phys. Rev. **40**, 749 (1932). — Wigner defined his distribution function for a N-particle system; — For a single particle it can be shown that (IV.9.1) is equivalent to the original definition; — For a detailed investigation of $\hat{W}$ and its associate equation see J. E. MOYAL: Proc. Cambridge Phil. Soc. **45**, 99 (1949).

The difference between the Wigner distributions $\hat{W}(p, q)$ and $W(u, u^*)$ ist that $\hat{W}(p, q)$ is normalized with respect to the $p-q$-plane, whereas $\hat{W}(u, u^*)$ is normalized with respect to the u-plane.

Using Feynman's disentangling techniques[2] we may write

$$e^{i\beta b+i\beta^* b^+} = e^{|\beta|^2/2}\, e^{i\beta b}\, e^{i\beta^* b^+}. \tag{IV.9.6}$$

The sequence of the operators $e^{i\beta b}$ and $e^{i\beta^* b^+}$ is rather unimportant because

$$e^{i\beta^* b^+} e^{i\beta b} = e^{i\beta b}\, e^{i\beta^* b^+}\, e^{+|\beta|^2}. \tag{IV.9.7}$$

Thus there are two equivalent forms of (IV.9.4), namely:

$$W(u, u^*) = \pi^{-2} \int e^{-i\beta u - i\beta^* u^*} \operatorname{tr}(e^{i\beta b} e^{i\beta^* b^+} \varrho)\, e^{|\beta|^2/2}\, d^2\beta \tag{IV.9.8}$$

$$= \pi^{-2} \int e^{-i\beta u - i\beta^* u^*} \operatorname{tr}(e^{i\beta^* b^+} e^{i\beta b} \varrho)\, e^{-|\beta|^2/2}\, d^2\beta \tag{IV.9.9}$$

Somewhat different definitions for a distribution function are obtained if the factor $e^{-|\beta|^2/2}$ in (IV.9.9) or if the factor $e^{|\beta|^2/2}$ in (IV.9.8) are omitted:*

$$P(u, u^*) = \pi^{-2} \int e^{-i\beta u - i\beta^* u^*} \operatorname{tr}(e^{i\beta^* b^+} e^{i\beta b} \varrho)\, d^2\beta, \tag{IV.9.10a}$$

$$Q(u, u^*) = \pi^{-2} \int e^{-i\beta u - i\beta^* u^*} \operatorname{tr}(e^{i\beta b} e^{i\beta^* b^+} \varrho)\, d^2\beta. \tag{IV.9.10b}$$

If we introduce coherent states defined by the equation

$$b|\alpha\rangle = \alpha|\alpha\rangle, \tag{IV.9.11}$$

it turns out that $P(\alpha, \alpha^*)$ is the Glauber and Sudarshan[3] diagonal representation of the density operator as given by

$$\varrho = \int P(\alpha, \alpha^*)\, |\alpha\rangle\langle\alpha|\, d^2\alpha, \tag{IV.9.12}$$

whereas $Q(u, u^*)$ is the matrix element of the density operator with respect to the coherent states:

$$Q(u, u^*) = \pi^{-1} \langle u|\varrho|u\rangle. \tag{IV.9.13}$$

To prove that P defined in (IV.9.10a) is identical with the definition (IV.9.12) we insert the representation (IV.9.12) for ϱ. We obtain for the right hand side of (IV.9.10a) using the cyclic properties of traces $(\operatorname{tr}(AB) = \operatorname{tr}(BA))$

$$\int \left\{ \frac{1}{\pi^2} \int e^{-i\beta(u-\alpha) - i\beta^*(u^*-\alpha^*)}\, d^2\beta \right\} P(\alpha, \alpha^*)\, d^2\alpha,$$

which is equal to $P(u, u^*)$ because the curly bracket represents the δ-function.

---

* The definitions of W, P and Q in this chapter ensure that they are real, provided ϱ is a hermitian operator. Note, however, that W and P are not necessarily positive (Q is always positive). W and P therefore do not have the usual meaning of a probability distribution, but they allow all quantum mechanical expectation values to be calculated purely by c-number procedures (see below).

[2] R. P. Feynman: Phys. Rev. **84**, 108 (1951). (IV.9.6) also follows by means of the Baker-Hausdorff theorem.

[3] R. J. Glauber: Phys. Rev. Letters **10**, 84 (1963); — Phys. Rev. **130**, 2529 (1963); **131**, 2766 (1963). — Photon statistics, in: Fundamental Problems in Statistical Mechanics II, ed. by E. G. D. Cohen, Amsterdam: North-Holland Publ. Comp. 1968. — E. C. G. Sudarshan: Phys. Rev. Letters **10**, 277 (1963); — For a different introduction of this representation see J. R. Klauder: Phys. Rev. Letters **16**, 534 (1966).

To prove the equivalence of (IV.9.10b) and (IV.9.13) we use the coherent state representation of the unit operator

$$\mathbf{1} = \pi^{-1} \int |\alpha\rangle \langle\alpha| \, d^2\alpha$$

under the trace in the following way

$$\begin{aligned} \operatorname{tr}\{e^{i\beta b} e^{i\beta^* b^+} \varrho\} &= \frac{1}{\pi} \int \operatorname{tr}\{e^{i\beta b} |\alpha\rangle\langle\alpha| e^{i\beta^* b^+} \varrho\} \, d^2\alpha \\ &= \frac{1}{\pi} \int e^{i\beta\alpha + i\beta^*\alpha^*} \langle\alpha|\varrho|\alpha\rangle \, d^2\alpha . \end{aligned}$$

Thus we obtain for the right hand side of (IV.9.10b)

$$\frac{1}{\pi} \int \left\{ \frac{1}{\pi^2} \int e^{-i\beta(u-\alpha) - i\beta^*(u^*-\alpha^*)} \, d^2\beta \right\} \langle\alpha|\varrho|\alpha\rangle \, d^2\alpha ,$$

which is equal to (IV.9.13) again because of the δ-function. According to GLAUBER[3] the relations between P, W and Q are the following ones

$$\begin{aligned} Q(u, u^*) &= \frac{2}{\pi} \int e^{-2|u-\alpha|^2} W(\alpha, \alpha^*) \, d^2\alpha \\ W(u, u^*) &= \frac{2}{\pi} \int e^{-2|u-\alpha|^2} P(\alpha, \alpha^*) \, d^2\alpha \\ Q(u, u^*) &= \frac{1}{\pi} \int e^{-|u-\alpha|^2} P(\alpha, \alpha^*) \, d^2\alpha . \end{aligned}$$

The following connection was worked out by LAX and LOUISELL[4]. Assuming that the density operator can be expanded in a Taylor series of the creation and annihilation operator b^+, b we may use the commutation relation $b b^+ - b^+ b = 1$ and bring this series either in an antinormal or normal ordered form

$$\begin{aligned} \varrho &= \sum_{n,m} c_{nm} \, b^n (b^+)^m \\ &= \sum_{n,m} d_{nm} (b^+)^n b^m . \end{aligned}$$

The Taylor series of the $P(u, u^*)$ and $Q(u, u^*)$ functions defined by (IV.9.10) are then given by

$$P(u, u^*) = \frac{1}{\pi} \sum_{n,m} c_{nm} \, u^n (u^*)^m , \tag{IV.9.10c}$$

$$Q(u, u^*) = \frac{1}{\pi} \sum_{n,m} d_{nm} (u^*)^n u^m . \tag{IV.9.10d}$$

The proof of (IV.9.10d) follows immediately from relation (IV.9.13). To prove (IV.9.10c) we make use of the unit operator in the coherent state representation

$$\begin{aligned} \varrho &= \sum_{n,m} c_{nm} \, b^n \frac{1}{\pi} \int |u\rangle \langle u| \, d^2u \, (b^+)^m \\ &= \int \left\{ \frac{1}{\pi} \sum_{n,m} c_{nm} u^n (u^*)^m \right\} |u\rangle \langle u| \, d^2u . \end{aligned}$$

Because of (IV.9.12), the curly bracket is equal to $P(u, u^*)$.

β) *Transforms of the distribution functions: characteristic functions.* Depending on the problem, one may use the Fourier-transform of the distribution functions $\hat{W}(p, q)$, $W(u, u^*)$, $P(u, u^*)$, $Q(u, u^*)$.

[4] M. LAX and W. LOUISELL: IEEE J. Quant. Electronics **QE 3**, 47 (1967).

These new functions may be called the characteristic functions (in analogy to classical statistics). They read

$$\left.\begin{aligned} \chi_{\hat{W}}(\mu,\nu) &= \mathrm{tr}\,(e^{i\mu\hat{p}+i\nu\hat{q}}\,\varrho)\,,\\ \chi_W(\beta,\beta^*) &= \mathrm{tr}\,(e^{i\beta b+i\beta^* b^+}\,\varrho)\,,\\ \chi_P(\beta,\beta^*) &= \mathrm{tr}\,(e^{i\beta^* b^+}\,e^{i\beta b}\,\varrho)\,,\\ \chi_Q(\beta,\beta^*) &= \mathrm{tr}\,(e^{i\beta b}\,e^{i\beta^* b^+}\,\varrho)\,. \end{aligned}\right\} \tag{IV.9.14}$$

γ) *Calculation of expectation values by means of the distribution functions.* In quantum mechanics one is interested in expectation values like

$$\langle g(b,b^+)\rangle = \mathrm{tr}\,\big(g(b,b^+)\,\varrho\big)\,.$$

The advantage of using the distribution functions is to obtain these expectation values by simple integrations:

$$\left.\begin{aligned} \mathrm{tr}\,\big(g_N(b,b^+)\,\varrho\big) &= \iint g_N(u,u^*)\,P(u,u^*)\,d^2u\,,\\ \mathrm{tr}\,\big(g_A(b,b^+)\,\varrho\big) &= \iint g_A(u,u^*)\,Q(u,u^*)\,d^2u\,. \end{aligned}\right\} \tag{IV.9.15}$$

In (IV.9.15) g_N and g_A means that $g(b,b^+)$ has to be arranged by use of the commutation relation $b\,b^+ - b^+ b = 1$ in such a manner that $g_N(g_A)$ is in normal (antinormal) order. To prove (IV.9.15) we assume that $g_N(g_A)$ can be expanded in a Taylor series. Therefore we have to prove the relations only for

$$\left.\begin{aligned} \mathrm{tr}\,\big((b^+)^n\,b^m\,\varrho\big) &= \iint (u^*)^n\,u^m\,P(u,u^*)\,d^2u\,,\\ \mathrm{tr}\,\big(b^n\,(b^+)^m\,\varrho\big) &= \iint u^n\,u^{*m}\,Q(u,u^*)\,d^2u\,. \end{aligned}\right\} \tag{IV.9.15a}$$

The proof of (IV.9.15a) follows from the fact that the integration of P and Q in (IV.9.15a) corresponds to a differentiation of the Fourier transforms, i.e. χ_P and χ_Q

$$\iint (u^*)^n\,u^m\,P(u,u^*)\,d^2u = \left(\frac{\partial}{\partial i\beta^*}\right)^n \left(\frac{\partial}{\partial i\beta}\right)^m \chi_P(\beta,\beta^*)\Big|_{\beta=\beta^*=0}$$

$$\iint u^n\,(u^*)^m\,Q(u,u^*)\,d^2u = \left(\frac{\partial}{\partial i\beta}\right)^n \left(\frac{\partial}{\partial i\beta^*}\right)^m \chi_Q(\beta,\beta^*)\Big|_{\beta=\beta^*=0}.$$

For the corresponding use of the Wigner distribution function, the reader is referred to ref. [1] on p. 61.

In Chap. IV.11 we will show how the distribution functions allow us to calculate *multi-time correlation functions.*

b) Electrons.

α) *Distribution functions for a single electron*: SCHMID and RISKEN[5] have defined the distribution function

$$f(v_{jk}) = N'\int\ldots\int e^{-i\sum_{jk} v_{jk}\xi_{jk}}\;\mathrm{tr}\,\Big(e^{i\sum_{jk}\xi_{jk}a_j^+a_k}\,\varrho\Big)\,d\{\xi\} \tag{IV.9.16}$$

with $v_{ik} = v_{ki}^*$, N' is a normalization constant, so that $\int\ldots\int f(v)\,d\{v\} = 1$. According to HAKEN, RISKEN, and WEIDLICH[6] another possibility is given by

$$f(v_{jk}) = N'\int\ldots\int e^{-i\sum_{jk} v_{jk}\xi_{jk}}\cdot\mathrm{tr}\,\Big(\prod_{j<k} e^{i\xi_{jk}a_j^+a_k}\prod_k e^{i\xi_{kk}a_k^+a_k}\prod_{j>k} e^{i\xi_{jk}a_j^+a_k}\,\varrho\Big)\,d\{\xi\} \tag{IV.9.17}$$

which has great advantages in practical calculations (see Chap. IV.10).

[5] CH. SCHMID and H. RISKEN: Z. Physik **189**, 365 (1966). See also J. P. GORDON: Phys. Rev. **161**, 367 (1967).

[6] H. HAKEN, H. RISKEN, and W. WEIDLICH: Z. Physik **206**, 355 (1967).

When an equation of the Fokker-Planck type for the f's is derived in the case of a single atom, it turns out that the f's are singular and that diffusion coefficients of high order are needed. In the actual laser one has to deal, however, with a great number of equivalent atoms. It is then possible to introduce macroscopic variables which show only small fluctuations, so that only the first terms of the generalized Fokker-Planck equation are important (see below). For a system of two-level atoms these macroscopic variables are defined as follows:

$$\left.\begin{array}{lll} \text{Macroscopic classical variables} & & \text{quantum mechanical operators} \\ \text{total dipole moment} & v^* \leftrightarrow & S^+ = \sum_\mu (a_2^+ a_1)_\mu, \\ & v \leftrightarrow & S^- = \sum_\mu (a_1^+ a_2)_\mu, \\ \text{total inversion} & D \leftrightarrow & 2S_z = \sum_\mu (a_2^+ a_2 - a_1^+ a_1)_\mu. \end{array}\right\} \quad \text{(IV.9.18)}$$

The above correspondence is established by the distribution function

$$f(v, v^*, D) = N' \int \ldots \int e^{-i\left\{v\xi + v^*\xi^* + \zeta\frac{D}{2}\right\}} \chi(\xi, \xi^*, \zeta)\, d^2\xi\, d\zeta \quad \text{(IV.9.19)}$$

where

$$\chi = \mathrm{tr}\{e^{i\xi^* S^+} e^{i\zeta S_z} e^{i\xi S^-} \varrho\}. \quad \text{(IV.9.20)}$$

The correspondence (IV.9.18) and (IV.9.19) may be generalized in an obvious way (for the introduction of collective atomic coordinates see p. 125).

A trivial generalization consists in replacing ζS_z by $\zeta_1 N_1 + \zeta_2 N_2$ where N_1, N_2 are the total occupation numbers of levels 1 and 2.

β) *Characteristic functions.* They are defined by

$$\left.\begin{array}{l} \chi(\xi_{jk}) = \mathrm{tr}\left(e^{i\sum_{jk}\xi_{jk} a_j^+ a_k} \varrho\right) \\ \text{or} \\ \chi(\xi_{jk}) = \mathrm{tr}\left\{\prod_{j<k} e^{i\xi_{jk} a_j^+ a_k} \prod_k e^{i\xi_{kk} a_k^+ a_k} \prod_{j>k} e^{i\xi_{jk} a_j^+ a_k} \varrho\right\}. \end{array}\right\} \quad \text{(IV.9.21)}$$

γ) *Electrons and fields.* In this case it is possible to take any combinations of the definitions of α) and β). In the following we will treat the example:

$$\left.\begin{array}{l} f(u, u^*, v, v^*, D) = N' \int \ldots \int e^{-i\left(v\xi + v^*\xi^* + \zeta\frac{D}{2} + u\beta + u^*\beta^*\right)} \times \\ \qquad \times \chi(\xi, \xi^*, \zeta, \beta, \beta^*)\, d^2\xi\, d\zeta\, d^2\beta \end{array}\right\} \quad \text{(IV.9.22)}$$

where

$$\chi = \mathrm{tr}\{O(\xi, \xi^*, \zeta, \beta, \beta^*)\, \varrho(t)\}$$

and

$$O = e^{i\zeta^* S^+} e^{i\zeta S_z} e^{i\xi S^-} e^{i\beta^* b^+} e^{i\beta b}. \quad \text{(IV.9.23)}$$

χ is obviously the characteristic function.

IV.10. Equation for the laser distribution function (IV.9.22). We confine our derivation to the example of a two-level laser with one running mode, because the calculations would otherwise be rather lengthy.

a) Comparison of the advantages of the Heisenberg and the Schrödinger representations.

α) *The Heisenberg representation.* The operators $V_j = b^+, b, a_i^+ a_k$ are timedependent, the density matrix, ϱ, is time-independent. We consider

$$f(u, t) = N' \int \ldots \int e^{-i\sum_j u_j \xi_j}\, \mathrm{tr}\left(\prod_j e^{i\xi_j V_j(t)} \varrho\right) d\{\xi\}. \quad \text{(IV.10.1)}$$

In order to derive a differential equation for f, we form as usual

$$\Delta f(u,t) = f(u, t+\Delta t) - f(u,t). \tag{IV.10.2}$$

Because only the operators $V_j(t)$ are time-dependent, we first investigate the product

$$O(t) \equiv \prod_j e^{i\xi_j V_j(t)}: \tag{IV.10.3}$$

$$O(t+\Delta t) = \prod_j e^{i\xi_j V_j(t+\Delta t)} \equiv \prod_j e^{i\xi_j (V_j(t)+\Delta V_j(t))} \tag{IV.10.4}$$

which contains the new definition:

$$\Delta V_j(t) = V_j(t+\Delta t) - V_j(t). \tag{IV.10.5}$$

If the V's were classical quantities, we could easily expand $O(t+\Delta t)$ into a power series of $\Delta V_j(t)$:

$$\left.\begin{aligned}\prod_j e^{i\xi_j(V_j(t)+\Delta V_j(t))} \approx \prod_j e^{i\xi_j V_j(t)} + \sum_k \Delta V_k(t)\, i\xi_k \prod_j e^{i\xi_j V_j(t)} - \\ - \tfrac{1}{2}\sum_{kl} \Delta V_k\, \Delta V_l\, \xi_k\, \xi_l \prod_j e^{i\xi_j V_j(t)}\end{aligned}\right\} \tag{IV.10.6}$$

$$\approx \Big(\underbrace{1}_{\mathrm{I}} + \underbrace{\sum_k \Delta V_k(t)\, i\xi_k}_{\mathrm{II}} - \underbrace{\tfrac{1}{2}\sum_{kl} \Delta V_k\, \Delta V_l\, \xi_l\, \xi_k}_{\mathrm{III}} + \cdots\Big) \prod_j e^{i\xi_j V_j(t)}. \tag{IV.10.7}$$

The terms II and III would then lead to the drift and diffusion terms of the usual Fokker-Planck Eq. (IV.2.5).

In a completely quantum mechanical treatment the operators V_j as well as ΔV_j in general do not commute. Therefore, strictly speaking, it is *not permissible* to write simply

$$e^{i\xi(V+\Delta V)} = \left(1 + i\xi\,\Delta V + \tfrac{1}{2}(i\xi\,\Delta V)^2 + \cdots\right) e^{i\xi V} \tag{IV.10.8}$$

if $[V, \Delta V] \neq 0$. Similarly, a rearrangement of factors [as in (IV.10.7)] is not admitted. There are restricted classes of problems in which these commutators may be neglected giving an excellent approximation:

1. An important example is the Risken-Schmid-Weidlich[1] treatment of the laser. It refers to macroscopic observables, which may be treated classically, whereas the fluctuations are treated on a quantum mechanical level by using quantum mechanically defined dissipation and fluctuation coefficients

$$B_i(u) = \lim_{t\to 0} \frac{1}{t} \operatorname{tr}\{(V_i(t) - V_i(0))\,\varrho\} \equiv \lim_{t\to 0} \frac{1}{t} \operatorname{tr}\{\Delta V_i\, \varrho\}, \tag{IV.10.9*}$$

$$\left.\begin{aligned}Q_{ij}(u) &= \lim_{t\to 0} \frac{1}{t} \operatorname{tr}\{(V_i(t) - V_i(0))\,(V_j(t) - V_j(0))\,\varrho\} \\ &\equiv \lim_{t\to 0} \frac{1}{t} \operatorname{tr}\{\Delta V_i\, \Delta V_j\, \varrho\}.\end{aligned}\right\} \tag{IV.10.10*}$$

Note that u_i is the classical quantity which corresponds to the operator V_i. As may be seen by direct calculation (see below), the right-hand sides of (IV.10.9) are expressible by the classical quantities u. In their treatment, RISKEN, SCHMID and WEIDLICH take into account all relevant laser variables (IV.9.18).

* These definitions correspond to the classical ones (IV.2.6) and (IV.2.7).

[1] H. RISKEN, CH. SCHMID, and W. WEIDLICH: Phys. Letters **29**, 489 (1966); — Z. Physik **193**, 37 (1966); **194**, 337 (1966).

2. In rather restricted problems it may be shown that $[\varDelta V_j, V_j]=0$. This case was treated by LAX[2] for the examples of a light mode (which is nonlinearly coupled to itself) and of a light mode coupled to atoms, where only occupation numbers are considered explicitly.

In general an exact treatment which takes into account all commutators and all higher diffusion coefficients seems to be at least very complicated, if the Heisenberg picture is used, and has not been performed so far.

β) *The Schrödinger representation.* The operators $V_j=b^+, b, a_i^+ a_k$ are time independent, but the density matrix, ϱ, is time dependent. In this case it is possible to derive a completely exact equation for f. HAKEN, RISKEN and WEIDLICH[3] have chosen for this purpose f in the form (IV.9.17)*. If the operators $a_i^+ a_k$ in (IV.9.17) are replaced by projection operators P_{ik}, it is even possible to derive a completely exact equation of the Fokker-Planck type for arbitrary dissipative quantum systems [4]. Because the resulting equation and its derivation are lengthy, we postpone its representation to Chap. IX. In the present chapter we exhibit the characteristic features of such a theory by treating a laser of two-level atoms interacting with one running light wave (mode). We follow closely the paper of HAKEN, RISKEN and WEIDLICH (l.c. [3]). The procedure is as follows: We differentiate both sides of (IV.9.22) with respect to the time:

$$\frac{df}{dt}=N'\int\cdots\int e^{-i\left(\xi v+\xi^* v^*+\zeta\frac{D}{2}+\beta^* u^*+\beta u\right)}\frac{\partial\chi}{\partial t}\,d^2\xi\,d\zeta\,d^2\beta \qquad \text{(IV.10.11)}$$

where

$$\frac{d\chi}{dt}=\frac{d}{dt}\operatorname{tr}(O\varrho)=\operatorname{tr}\left(O\frac{d\varrho}{dt}\right). \qquad \text{(IV.10.12)}$$

Because ϱ obeys the density matrix Eq. (IV.7.68) with (IV.7.69), $\frac{d\varrho}{dt}$ consists of three parts referring to the electrons (atoms), the fields, and the interaction between electrons and field. We decompose $\frac{d\chi}{dt}$ [see (IV.10.12)] correspondingly:

$$\frac{d\chi}{dt}=\left(\frac{\partial\chi}{\partial t}\right)_A+\left(\frac{\partial\chi}{\partial t}\right)_L+\left(\frac{\partial\chi}{\partial t}\right)_{A\text{-}L}$$

where

$$\left(\frac{\partial\chi}{\partial t}\right)_A=\operatorname{tr}\left\{O\sum_{\mu=1}^{N}\left(\frac{\partial\varrho}{\partial t}\right)_{A\mu}\right\}. \qquad \text{(IV.10.13)}$$

The sum runs over all atoms which are distinguished by an index μ.

$$\left(\frac{\partial\chi}{\partial t}\right)_L=\operatorname{tr}\left\{O\left(\frac{\partial\varrho}{\partial t}\right)_L\right\}, \qquad \text{(IV.10.14)}$$

$$\left(\frac{\partial\chi}{\partial t}\right)_{A\text{-}L}=\operatorname{tr}\left\{O\frac{-i}{\hbar}[H_{A\text{-}L},\varrho]\right\}. \qquad \text{(IV.10.15)}$$

We first evaluate (IV.10.13)—(IV.10.15). We introduce the abbreviations

$$a_2^+ a_1=s^+,\quad a_1^+ a_2=s^-,\quad \tfrac{1}{2}(a_2^+ a_2-a_1^+ a_1)=s_z;\quad a_1^+ a_1+a_2^+ a_2=1. \qquad \text{(IV.10.16)}$$

* GORDON l.c. [5] on p. 64 chose the form (IV.9.16) for N three-level atoms, as had been done by SCHMID and RISKEN l.c. [5] on p. 64 for a single multi-level atom (the latter authors used the Heisenberg representation, however). In both papers the derivation is confined to the drift and diffusion coefficients.

[2] M. LAX, in: Dynamical Processes in Solid State Optics (ed. R. KUBO and H. KAMIMURA). Tokyo: Syokabo 1967.

[3] H. HAKEN, H. RISKEN, and W. WEIDLICH: l.c.[6] on p. 64. In the present representation we have exchanged the sequence of the operators S^+ and S^- in (IV.10.11) which leads to a simplification in the interaction operator (IV.10.33). We wish to thank H. D. VOLLMER for doing the corresponding calculations.

[4] H. HAKEN: Z. Physik **219**, 411 (1969).

Note that s^+, s^- and s_z have the formal meaning of spin operators [compare (III.6)].

We further put

$$\left.\begin{aligned} &O_{A\mu} = e^{i\xi^* s_\mu^+} e^{i\zeta s_{z\mu}} e^{i\xi s_\mu^-}; && O'_{A\mu} = \prod_{\nu\neq\mu} O_{A\nu} O_F, \\ &O = O_{A\mu} O'_{A\mu}; && O_F = e^{i\beta^* b^+} e^{i\beta b}, \\ &O_A = \prod_{\mu=1}^{N} O_{A\mu}; && O = O_A O_F. \end{aligned}\right\} \tag{IV.10.17}$$

Because we are treating a two-level system, we specialize in Eq. (IV.7.61) $i, j, = 1, 2$, and use the relations

$$\left.\begin{aligned} w_{12} &= 2A_{1221} \\ w_{21} &= 2A_{2112} \end{aligned}\right\} \tag{IV.10.18}$$

where the A's are assumed to be real.

We drop terms which are not connected with real transitions caused by the heatbaths. For the evaluation of (IV.10.13) we use

$$\left.\begin{aligned} \left(\frac{\partial\varrho}{\partial t}\right)_A = \sum_{\mu=1}^{N}\left(\frac{\partial\varrho}{\partial t}\right)_{A,\mu} = \sum_{\mu=1}^{N}\Big[\frac{w_{21}}{2}\{[s_\mu^-, \varrho s_\mu^+] + [s_\mu^- \varrho, s_\mu^+]\} + \\ + \frac{w_{12}}{2}\{[s_\mu^+, \varrho s_\mu^-] + [s_\mu^+ \varrho, s_\mu^-]\} \end{aligned}\right\} \tag{IV.10.19}$$

and find after a slight rearrangement of operators under the traces:

$$\left.\begin{aligned} \left(\frac{\partial\chi}{\partial t}\right)_A &= \sum_{\mu=1}^{N} \operatorname{tr}\left\{O_{A\mu} O'_{A\mu}\left(\frac{\partial\varrho}{\partial t}\right)_{A\mu}\right\} \\ &= \sum_{\mu=1}^{N}\left\{\frac{w_{21}}{2}\left[2\operatorname{tr}\{s_\mu^+ O_{A\mu} s_\mu^- O'_{A\mu}\varrho\} - \operatorname{tr}\{s_\mu^+ s_\mu^- O_{A\mu} O'_{A\mu}\varrho\} -\right.\right. \\ &\quad - \operatorname{tr}\{O_{A\mu} s_\mu^+ s_\mu^- O'_{A\mu}\varrho\}\Big] + \frac{w_{12}}{2}\left[2\operatorname{tr}\{s_\mu^- O_{A\mu} s_\mu^+ O'_{A\mu}\varrho\} -\right. \\ &\quad - \operatorname{tr}\{s_\mu^- s_\mu^+ O_{A\mu} O'_{A\mu}\varrho\} - \operatorname{tr}\{O_{A\mu} s_\mu^- s_\mu^+ O'_{A\mu}\varrho\}\Big]. \end{aligned}\right\} \tag{IV.10.20}$$

Because s^-, s^+ and s_z are composed of Fermi operators according to (IV.10.16) which act in the one-particle subspace we may express their products by the single operators:

$$\left.\begin{aligned} &s_\mu^- s_\mu^+ = \tfrac{1}{2} - s_{z\mu}; && s_\mu^+ s_\mu^- = \tfrac{1}{2} + s_{z\mu}, \\ &s_{z\mu} s_\mu^- = -\tfrac{1}{2} s_\mu^-; && s_\mu^- s_{z\mu} = \tfrac{1}{2} s_\mu^-, \\ &s_{z\mu} s_\mu^+ = \tfrac{1}{2} s_\mu^+; && s_\mu^+ s_{z\mu} = -\tfrac{1}{2} s_\mu^+, \\ &(s_\mu^-)^2 = (s_\mu^+)^2 = 0; && (s_{z\mu})^2 = \tfrac{1}{4}. \end{aligned}\right\} \tag{IV.10.21}$$

For our further analysis it is most important that the operators which occur in front of the density matrix under the trace can be expressed as linear combinations of derivatives of the operator $O_{A\mu}$ with respect to the variables ξ, ξ^*, ζ and of the operator $O_{A\mu}$ defined in (IV.10.17)

$$\left.\begin{aligned} \frac{\partial O_{A\mu}}{\partial(i\xi^*)} &= s_\mu^+ e^{i\xi^* s_\mu^+} e^{i\zeta s_{z\mu}} e^{i\xi s_\mu^-}, \\ \frac{\partial O_{A\mu}}{\partial(i\zeta)} &= e^{i\xi^* s_\mu^+} s_{z\mu} e^{i\zeta s_{z\mu}} e^{i\xi s_\mu^-}, \\ \frac{\partial O_{A\mu}}{\partial(i\xi)} &= e^{i\xi^* s_\mu^+} e^{i\zeta s_{z\mu}} s_\mu^- e^{i\xi s_\mu^-}. \end{aligned}\right\} \tag{IV.10.22}$$

We exhibit these linear combinations explicitly by the following formulas

$$\left.\begin{aligned}
s_\mu^- O_{A\mu} s_\mu^+ &= \left[\frac{1}{2} e^{+i\zeta} + \frac{1}{2} e^{-i\zeta} (i\xi)^2 (i\xi^*)^2 + (i\xi)(i\xi^*)\right] O_{A\mu} + \\
&\quad + i\xi\left[-(i\xi)(i\xi^*) - e^{+i\zeta}\right] \frac{\partial O_{A\mu}}{\partial(i\xi)} + i\xi^*\left[-(i\xi)(i\xi^*) - e^{+i\zeta}\right] \frac{\partial O_{A\mu}}{\partial(i\xi^*)} - \\
&\quad - \left[e^{+i\zeta} - e^{-i\zeta}(i\xi)^2 (i\xi^*)^2\right] \frac{\partial O_{A\mu}}{\partial(i\zeta)}, \\
s_\mu^+ O_{A\mu} s_\mu^- &= \frac{e^{-i\zeta}}{2} O_{A\mu} + e^{-i\zeta} \frac{\partial O_{A\mu}}{\partial(i\zeta)}, \\
s_{z\mu} O_{A\mu} &= i\xi^* \frac{\partial O_{A\mu}}{\partial(i\xi^*)} + \frac{\partial O_{A\mu}}{\partial(i\zeta)}, \\
O_{A\mu} s_{z\mu} &= i\xi \frac{\partial O_{A\mu}}{\partial(i\xi)} + \frac{\partial O_{A\mu}}{\partial(i\zeta)}, \\
s_{z\mu} O_{A\mu} s_{z\mu} &= \left(\frac{1}{4} + \frac{1}{2}(i\xi)(i\xi^*) e^{-i\zeta}\right) O_{A\mu} - \frac{1}{2} i\xi^* \frac{\partial O_{A\mu}}{\partial(i\xi^*)} - \\
&\quad - \frac{1}{2} i\xi \frac{\partial O_{A\mu}}{\partial(i\xi)} + (i\xi)(i\xi^*) e^{-i\zeta} \frac{\partial O_{A\mu}}{\partial(i\zeta)}.
\end{aligned}\right\} \quad \text{(IV.10.23)}$$

Inserting (IV.10.23) into (IV.10.20) yields

$$\left.\begin{aligned}
\left(\frac{\partial\chi}{\partial t}\right)_A &= \frac{w_{12}}{2}\Big\{\left[e^{-i\zeta} - 1 + e^{i\zeta}(i\xi)^2(i\xi^*)^2 + 2i\xi\, i\xi^*\right] N\chi + \\
&\quad + 2i\xi\left[-(i\xi)(i\xi^*) - e^{-i\zeta} + \frac{1}{2}\right] \frac{\partial\chi}{\partial(i\xi)} + \\
&\quad + 2i\xi^*\left[-(i\xi)(i\xi^*) - e^{-i\zeta} + \frac{1}{2}\right] \frac{\partial\chi}{\partial(i\xi^*)} - \\
&\quad - \left[2(e^{+i\zeta} - 1) - 2e^{-i\zeta}(i\xi)^2(i\xi^*)^2\right] \frac{\partial\chi}{\partial(i\zeta)}\Big\} + \\
&\quad + \frac{w_{21}}{2}\left\{(e^{-i\zeta} - 1) N\chi - i\xi \frac{\partial\chi}{\partial(i\xi)} - i\xi^* \frac{\partial\chi}{\partial(i\xi^*)} + 2(e^{-i\zeta} - 1) \frac{\partial\chi}{\partial(i\zeta)}\right\}.
\end{aligned}\right\} \quad \text{(IV.10.24)}$$

We now calculate the field term (IV.10.14) which reads explicitly:

$$\left.\begin{aligned}
\left(\frac{\partial\chi}{\partial t}\right)_L &= \varkappa \operatorname{tr}\left\{\left[2b^+ O_F b - O_F b^+ b - b^+ b O_F\right] O_A \varrho\right\} + \\
&\quad + 2\varkappa n_{th} \operatorname{tr}\left\{\left[b^+ O_F b - b b^+ O_F - O_F b^+ b + b O_F b^+\right] O_A \varrho\right\}.
\end{aligned}\right\} \quad \text{(IV.10.25)}$$

Using the operator relations*

$$\left.\begin{aligned}
b^+ O_F b &= \frac{\partial^2 O_F}{\partial(i\beta^*)\,\partial(i\beta)}, \\
b O_F b^+ &= \left[\frac{\partial^2}{\partial(i\beta^*)\,\partial(i\beta)} + 1 + i\beta^* \frac{\partial}{\partial(i\beta^*)} + i\beta \frac{\partial}{\partial(i\beta)} + (i\beta^*)(i\beta)\right] O_F, \\
O_F b^+ b &= \left[\frac{\partial^2}{\partial(i\beta^*)\,\partial(i\beta)} + i\beta \frac{\partial}{\partial(i\beta)}\right] O_F, \\
b^+ b O_F &= \left[\frac{\partial^2}{\partial(i\beta^*)\,\partial(i\beta)} + i\beta^* \frac{\partial}{\partial(i\beta^*)}\right] O_F
\end{aligned}\right\} \quad \text{(IV.10.26)}$$

---

* The operator techniques used in Eqs. (IV.10.26) are described in the appendix, X. 2, or e.g. in the book of LOUISELL: "Radiation and Noise in Quantum Electronics". New York: McGraw Hill 1964.

we immediately obtain

$$\left(\frac{\partial \chi}{\partial t}\right)_L = -\varkappa\left[i\beta\frac{\partial}{\partial(i\beta)} + i\beta^*\frac{\partial}{\partial(i\beta^*)}\right]\chi + 2\varkappa\, n_{th}(i\beta^*)(i\beta)\chi. \quad \text{(IV.10.27)}$$

Finally the interaction term (IV.10.15) is to be evaluated. We assume that the running wave has infinite wavelength. In Chap. VI.4 we will show that in laser theory a single running wave with finite wavelength is mathematically equivalent to one with infinite wavelength.

Under this assumption the interaction Hamiltonian depends only on the total dipole moment

$$\left.\begin{aligned} S^+ &= \sum_\mu s_\mu^+ , \\ S^- &= \sum_\mu s_\mu^- \end{aligned}\right\} \quad \text{(IV.10.28)}$$

so that we obtain

$$\left(\frac{\partial \chi}{\partial t}\right)_{AL} = -ig\left\{\mathrm{tr}\left\{[O S^+ b + O S^- b^+]\,\varrho\right\} - \mathrm{tr}\{[S^+ b O + S^- b^+ O]\,\varrho\}\right\}. \quad \text{(IV.10.29)}$$

g is the coupling constant

$$g = g_{21,\mu}$$

which may be assumed real.

Using the operator relations

$$\left.\begin{aligned}
O_A S^+ &= e^{i\zeta}\frac{\partial O_A}{\partial(i\xi^*)} - (i\xi)^2\frac{\partial O_A}{\partial(i\xi)} - 2i\xi\frac{\partial O_A}{\partial(i\zeta)}, \\
O_A S^- &= \frac{\partial O_A}{\partial(i\xi)}, \\
S^+ O_A &= \frac{\partial O_A}{\partial(i\xi^*)}, \\
S^- O_A &= -(i\xi^*)^2\frac{\partial O_A}{\partial(i\xi^*)} + e^{i\zeta}\frac{\partial O_A}{\partial(i\xi)} - 2i\xi^*\frac{\partial O_A}{\partial(i\zeta)}, \\
O_F b &= \frac{\partial O_F}{\partial(i\beta)}, \\
O_F b^+ &= \left(\frac{\partial}{\partial(i\beta^*)} + i\beta\right) O_F, \\
b^+ O_F &= \frac{\partial O_F}{\partial(i\beta^*)}, \\
b O_F &= \left(\frac{\partial}{\partial(i\beta)} + i\beta^*\right) O_F
\end{aligned}\right\} \quad \text{(IV.10.30)}$$

we find

$$\left.\begin{aligned}
\left(\frac{\partial \chi}{\partial t}\right)_{AL} = -ig\Bigg\{&\left[e^{i\zeta}\frac{\partial}{\partial(i\xi^*)} - (i\xi)^2\frac{\partial}{\partial(i\xi)} - 2i\xi\frac{\partial}{\partial(i\zeta)}\right]\frac{\partial}{\partial(i\beta)} - \\
&-\left[-(i\xi^*)^2\frac{\partial}{\partial(i\xi^*)} + e^{i\zeta}\frac{\partial}{\partial(i\xi)} - 2i\xi^*\frac{\partial}{\partial(i\zeta)}\right]\frac{\partial}{\partial(i\beta^*)} + \\
&+\frac{\partial}{\partial(i\xi)}\left(\frac{\partial}{\partial(i\beta^*)} + i\beta\right) - \frac{\partial}{\partial(i\xi^*)}\left(\frac{\partial}{\partial(i\beta)} + i\beta^*\right)\Bigg\}\chi.
\end{aligned}\right\} \quad \text{(IV.10.31)}$$

b) Final form of the generalized Fokker-Planck equation. According to the definition of f this distribution function is a Fourier transform of χ. Thus the expressions (IV.10.24), (IV.10.27) and (IV.10.31) may be reexpressed by f and its derivatives. We write this resulting equation for f in the form

$$\frac{\partial f}{\partial t} = L f$$

where the Liouville operator L consists of the atomic part, the interaction and the field part

$$L = L_A + L_{AL} + L_L.$$

The different contributions are defined as follows

$$\left.\begin{aligned} L_A = \frac{w_{12}}{2} \Big\{ & N\left(e^{-2\frac{\partial}{\partial D}} - 1\right) + N e^{2\frac{\partial}{\partial D}} \frac{\partial^4}{\partial v^2\, \partial v^{*2}} + 2N \frac{\partial^2}{\partial v\, \partial v^*} + \\ & + \frac{\partial}{\partial v}\left[2\frac{\partial^2}{\partial v\, \partial v^*} + 2 e^{-2\frac{\partial}{\partial D}} - 1\right] v + \\ & + \frac{\partial}{\partial v^*}\left[2\frac{\partial^2}{\partial v\, \partial v^*} + 2 e^{-2\frac{\partial}{\partial D}} - 1\right] v^* - \\ & - 2\left[\left(e^{-2\frac{\partial}{\partial D}} - 1\right) - e^{2\frac{\partial}{\partial D}} \frac{\partial^4}{\partial^2 v\, \partial^2 v^*}\right] \frac{D}{2}\Big\} + \\ & + \frac{w_{21}}{2}\left\{N\left(e^{2\frac{\partial}{\partial D}} - 1\right) + \frac{\partial}{\partial v} v + \frac{\partial}{\partial v^*} v^* + 2\left(e^{2\frac{\partial}{\partial D}} - 1\right)\frac{D}{2}\right\}, \end{aligned}\right\} \quad \text{(IV.10.32)†}$$

$$\left.\begin{aligned} L_{AL} = -ig \Big\{ & \left[e^{-2\frac{\partial}{\partial D}} v^* - \frac{\partial^2}{\partial v^2} v + \frac{\partial}{\partial v} D\right] u - \\ & - \left[-\frac{\partial^2}{\partial v^{*2}} v^* + e^{-2\frac{\partial}{\partial D}} v + \frac{\partial}{\partial v^*} D\right] u^* + \\ & + \left[-\frac{\partial}{\partial u} + u^*\right] v - \left[-\frac{\partial}{\partial u^*} + u\right] v^* \Big\}, \end{aligned}\right\} \quad \text{(IV.10.33)}$$

$$L_L = \varkappa\left[\frac{\partial}{\partial u} u + \frac{\partial}{\partial u^*} u^*\right] + 2\varkappa\, n_{th} \frac{\partial^2}{\partial u\, \partial u^*}\,. \quad \text{(IV.10.34)}$$

IV.11. The calculation of multi-time correlation functions by means of the distribution function[1]. So far we have shown that we can derive a well defined equation for the distribution function (IV.9.22). We have now to show how the properties of the laser can be calculated by means of this function f. The laser properties, in particular its coherence properties, are determined by correlation functions of the form

$$\langle b^+(t_1') \ldots b^+(t_m')\, b(t_n) \ldots b(t_1)\rangle \quad \text{(IV.11.1)}$$

as will be shown in Chap. V.

† If the phases of the atomic dipole moments are destroyed not only by real transitions but also by virtual transitions, the following term must be added to the r.h.s. of Eq. (IV.10.32):

$$\frac{\eta}{2}\left\{\frac{\partial}{\partial v} v + \frac{\partial}{\partial v^*} v^* + 2\frac{\partial^2}{\partial v\, \partial v^*} e^{2\frac{\partial}{\partial D}} \cdot \frac{D}{2} + N \cdot \frac{\partial^2}{\partial v\, \partial v^*} e^{2\frac{\partial}{\partial D}}\right\} \quad \text{(VI.10.32a)}$$

η is connected with the total phase width by

$$2\gamma = w_{12} + w_{21} + \eta$$

(compare the paper of Haken, Risken, and Weidlich, l.c.[6] on p. 64).

[1] H. Haken, H. Risken, and W. Weidlich: l.c.[6] on p. 64. For two-time correlation functions see also R. Bonifacio and F. Haake: Z. Physik **200**, 526 (1967).

In (IV.8) we have shown that (IV.11.1) can be evaluated by the single time density matrix. Because f is intimately connected with ϱ, we start with the form (IV.8.4).

$\varrho(\alpha, \gamma)$ is defined as solution of Eq. (IV.8.7), where the last term is explicitly given by (IV.7.70) and (IV.7.71) with $\alpha^* = \gamma = 0$.

We define a generalized distribution function by[2]

$$\left.\begin{aligned} f(u, u^*, v, v^*, D; \alpha(t), \gamma(t); t) = N' \int \cdots \int e^{-iv\xi - iv^*\xi^* - i\zeta\frac{D}{2} - iu\beta - iu^*\beta^*} \times \\ \times \chi(\xi, \xi^*, \zeta, \beta, \beta^*; \alpha(t), \gamma(t))\, d^2\xi\, d\zeta\, d^2\beta \end{aligned}\right\} \quad \text{(IV.11.2)}$$

where

$$\chi = \operatorname{tr}\{e^{i\xi^* S^+} e^{i\zeta S_z} e^{i\xi S^-} e^{i\beta^* b^+} e^{i\beta b} \varrho(\alpha, \gamma; t)\}. \quad \text{(IV.11.3)}$$

Evidently the trace over $\varrho(\alpha, \gamma)$ may be expressed by

$$\operatorname{tr}(\varrho(\alpha, \gamma, t)) = \int f(u, u^*, v, v^*; D, \alpha, \gamma; t)\, d^2u\, d^2v\, dD. \quad \text{(IV.11.4)}$$

All steps which were done for the derivation of an equation for f may be simply repeated. We then find immediately

$$\frac{df}{dt} - Lf = -i\alpha(t)\, uf - i\gamma^*(t)\, u^* f. \quad \text{(IV.11.5)}$$

We again introduce the decomposition (IV.8.8), so that (IV.11.5) becomes

$$\frac{df}{dt} - Lf = 0 \quad \text{(IV.11.6)}$$

for $t \neq t_i$, and additional jump conditions

$$f(t_i + 0) - f(t_i - 0) = -i\alpha_i\, uf(t_i - 0) + \text{corrections of order } \alpha^2. \quad \text{(IV.11.7)}$$

We define a Green's function by

$$\frac{dP}{dt} - LP = \delta(t - t_1)\, \delta(v - v_1)\, \delta(u - u_1)\, \delta(D - D_1) \quad \text{(IV.11.8)}$$

and use the notation

$$\left.\begin{aligned} P(u_2, u_2^*, v_2, v_2^*, D_2; u_1, u_1^*, v_1, v_1^*, D_1; t_2, t_1) = P(2, 1); \\ P(1, 0) \equiv f(\alpha = 0; \gamma = 0). \end{aligned}\right\} \quad \text{(IV.11.9)}$$

According to the jumps at time $t_1, t_2, t_3, \ldots$ we decompose the solution of Eq. (IV.11.6) into these time intervals, where we find for

$$t < t_1 : f = P(1, 0) \quad \text{(IV.11.10)}$$

where $P(1, 0)$ is the stationary state solution of Eq. (IV.11.6). For the next time interval we find

$$f = P(1, 0) + \int (-i\alpha_1 u_1 + i\gamma_1^* u_1^*)\, P(2, 1)\, P(1, 0)\, dV_1 \quad \text{(IV.11.11)}$$

where

$$dV_l = d^2u_l\, d^2v_l\, dD_l \quad \text{(IV.11.12)}$$

---

[2] The present method is also applicable to other distribution functions, e.g. the Q-representation (IV.9.10b). See: R. GRAHAM, F. HAAKE, H. HAKEN and W. WEIDLICH, Z. Physik **213**, 21 (1967).

and for a time t after the last jump at t_n we find the expression

$$\left.\begin{aligned} f = P(1,0) + \sum_{\nu=1}^{n} \underbrace{\int \cdots \int}_{\nu} \left\{ \prod_{\mu=1}^{\nu} P(\mu+1,\mu)\,(-i\alpha_\mu u_\mu + i\gamma_\mu^* u_\mu^*) \right\} \times \\ \times P(1,0)\, dV_1 \ldots dV_\nu\,. \end{aligned}\right\} \quad \text{(IV.11.13)}$$

Because we must differentiate f finally with respect to all α's or γ^*'s, we need only retain the term with $\nu = n$, so that we need only consider

$$[\int f\, dV]_n = \underset{n}{\int \cdots \int} \prod_{\mu=1}^{n} (-i\alpha_\mu u_\mu + i\gamma_\mu^* u_\mu^*)\, P(\mu, \mu-1)\, dV_1 \ldots dV_n \qquad \text{(IV.11.14)}$$

where we have integrated both sides over $V_{n+1} = V$ and have used the relation

$$\int P(n+1, n)\, dV_{n+1} = 1\,. \qquad \text{(IV.11.15)}$$

When we differentiate with respect to α and γ^* and make, if necessary, the transition to equal times (e.g. $t_i = t_{i+1}$) we find our final result in the following form

$$\left.\begin{aligned} &\langle (b^+(t_1))^{\nu_1} (b^+(t_2))^{\nu_2} \ldots (b^+(t_n))^{\nu_n} (b(t_n))^{\mu_n} \ldots (b(t_1))^{\mu_1} \rangle \\ &= \underset{n}{\int \cdots \int} u_n^{*\,\nu_n} u_n^{\mu_n} \ldots u_1^{*\,\nu_1} u_1^{\mu_1} \prod_{\mu=1}^{n} P(\mu, \mu-1)\, dV_1 \ldots dV_n\,. \end{aligned}\right\} \quad \text{(IV.11.16)}$$

Note that (IV.11.16) is the most general formula, because by putting some of the μ's or ν's equal to zero, we may obtain all desired expectation values provided the operators are time ordered and in normal order.

V. Properties of quantized electromagnetic fields.

V.1. Coherence properties of the classical and the quantized electromagnetic field.

a) Classical description[1]*: definitions.*

α) *The complex analytical signal.* We consider a real classical wave function $V^{(r)}(\boldsymbol{x}, t)$ which describes the field at the point $\boldsymbol{x}$ at time t. This function may represent, for example, a component of the electric field, the magnetic field or the vector potential. We assume that $V^{(r)}(\boldsymbol{x}, t)$ can be expressed by a Fourier integral with respect to the time variable:

$$V^{(r)}(\boldsymbol{x}, t) = \frac{1}{2\pi} \int_{-\infty}^{+\infty} v(\boldsymbol{x}, \omega)\, e^{-i\omega t}\, d\omega\,. \qquad \text{(V.1.1)}$$

If the Fourier representation of $V^{(r)}$ does not exist, one can consider in place of $V^{(r)}$ the truncated function:

$$\begin{aligned} V_T^{(r)}(\boldsymbol{x}, t) &= V^{(r)}(\boldsymbol{x}, t) \quad \text{when} \quad |t| < T \\ &= 0 \qquad\qquad\quad \text{when} \quad |t| > T. \end{aligned} \qquad \text{(V.1.2)}$$

Because $V^{(r)}$ is real,

$$v(\boldsymbol{x}, -\omega) = v^*(\boldsymbol{x}, \omega)\,. \qquad \text{(V.1.3)}$$

[1] For more details see H. Born and E. Wolf: Principles of Optics (Oxford: Pergamon Press, Ltd.; New York: The Macmillan Comp., 2nd ed. 1964) and L. Mandel and E. Wolf: Rev. Mod. Phys. **37**, 231 (1965), whom we will follow closely. The latter paper contains also a very detailed list of references both of classical and quantum theoretical treatments of coherence up to 1964. — M. Beran and G. B. Parrent: Theory of Partial Coherence. Englewood Cliffs, New Jersey: Prentice Hall 1964.

Consequently the negative frequency components ($\omega < 0$) do not provide any information which is not already contained in the positive ones ($\omega > 0$) and may be omitted. This leads us to introduce the "complex analytical signal"[2]*

$$V(\boldsymbol{x}, t) = \frac{1}{2\pi} \int_0^\infty v(\boldsymbol{x}, \omega)\, e^{-i\omega t}\, d\omega. \tag{V.1.4}$$

As we will see below, the complex field is most appropriate for the transition to the quantum theoretical treatment.

β) The average. We define the time average of a function $F(t)$ (or, more exactly, a stationary process in the statistical sense) by**

$$\langle F(t)\rangle_t = \lim_{T\to\infty} \frac{1}{2T} \int_{-T}^{T} F_T(t)\, dt. \tag{V.1.5}$$

For the later connection with the quantum theory we treat also the average over an ensemble

$$\langle F\rangle_e \tag{V.1.6}$$

where the ensemble depends on statistical properties of the field source. Provided the process is stationary and ergodic:

$$\langle \cdots \rangle_t = \langle \cdots \rangle_e. \tag{V.1.7}$$

We define

$$\langle V^{(r)2}(t)\rangle_t = \frac{1}{2\pi} \int_{-\infty}^{+\infty} G(\omega)\, d\omega \tag{V.1.8}$$

and find readily

$$G(\omega) = \lim_{T\to\infty} \frac{|v_T(\omega)|^2}{2T} \tag{V.1.9}$$

where $v_T(\omega)$ is the Fourier coefficient of the truncated function $V_T^{(r)}$.

From (V.1.3) and (V.1.4) we obtain further, using $V = \frac{1}{2}\left(V^{(r)}(t) + iV^{(i)}(t)\right)$

and

$$\left.\begin{aligned} \langle V^{(r)2}(t)\rangle_t &= \langle V^{(i)2}(t)\rangle_t \\ \langle V^{(r)2}(t)\rangle_t &= 2\langle V(t)\, V^*(t)\rangle_t. \end{aligned}\right\} \tag{V.1.10}$$

This equation holds approximately also for the ensemble average, if quasi-monochromatic light is used and an average over a few oscillations of the light is taken.

γ) The mutual coherence function. We consider an example which leads us in a natural way to the definition of the mutual coherence function, which describes the correlation of the amplitudes $V_1(\boldsymbol{x}_1, t_1')$ and $V_2(\boldsymbol{x}_2, t_2')$ at space points $\boldsymbol{x}_1, \boldsymbol{x}_2$ and at times t_1', t_2'. The amplitudes may stem either from two different light sources or from the same one. The amplitudes V_1 and V_2 are first singled out at two different space-points P_1 and P_2 for example by two pinholes in an opaque screen (compare Fig. 18).

Then a time difference is generated by two delay lines of different lengths, in our example by the optical paths s_1 and s_2. Finally, the two amplitudes are

* We follow the definition of Mandel and Wolf, l.c.[1] and omit a factor 2 in front of the integral (V.1.4), which occurs in the customary definition.

** We omit the space-coordinate $\boldsymbol{x}$ if it is not explicitly needed.

[2] D. Gabor: J. Inst. Elec. Eng. (London) **93**, 429 (1946).

superimposed and their total intensity is measured. The total amplitude at point P with the coordinate $\boldsymbol{x}$ is given by

$$V(\boldsymbol{x}, t) = K_1 V_1(\boldsymbol{x}_1, t-t_1) + K_2 V_2(\boldsymbol{x}_2, t-t_2) \tag{V.1.11}$$

where $t_1 = s_1/c$ and $t_2 = s_2/c$ are the times needed for the light to travel from P_1 to P and P_2 to P respectively. K_1 and K_2 are constant factors, which depend on the geometry. Besides a factor 2, the intensity $I(\boldsymbol{x}, t)$ is given by

$$V^*(\boldsymbol{x}, t)\, V(\boldsymbol{x}, t) \tag{V.1.12}$$

so that we define directly

$$I(\boldsymbol{x}, t) = V^*(\boldsymbol{x}, t)\, V(\boldsymbol{x}, t). \tag{V.1.13}$$

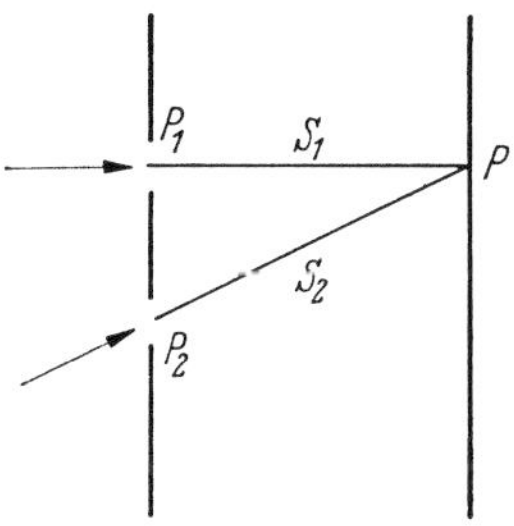

Fig. 18.

Inserting the expression (V.1.11) into (V.1.13) and taking the ensemble average we obtain

$$\left.\begin{aligned}\langle I(\boldsymbol{x}, t)\rangle_e = |K_1|^2 \langle I_1(\boldsymbol{x}_1, t-t_1)\rangle_e + |K_2|^2 \langle I_2(\boldsymbol{x}_2, t-t_2)\rangle_e + \\ +2\,\mathrm{Re}\{K_1^* K_2 \Gamma_{12}(\boldsymbol{x}_1, \boldsymbol{x}_2, t-t_1, t-t_2)\}\end{aligned}\right\} \tag{V.1.14}$$

where

$$\Gamma_{12}(\boldsymbol{x}_1, \boldsymbol{x}_2, t_1, t_2) = \langle V_1^*(\boldsymbol{x}_1, t_1)\, V_2(\boldsymbol{x}_2, t_2)\rangle_e; \tag{V.1.15}$$

$$I_j(\boldsymbol{x}_j, t) = V_j^*(\boldsymbol{x}_j, t)\, V_j(\boldsymbol{x}_j, t). \tag{V.1.16}$$

Re denotes the real part.

If the process is stationary, Γ_{12} depends on the time difference $\tau = t_1 - t_2$, but not on the single times, so that

$$\left.\begin{aligned}\langle I(\boldsymbol{x}, t)\rangle_e = |K_1|^2 \langle I_1(\boldsymbol{x}_1)\rangle_e + |K_2|^2 \langle I_2(\boldsymbol{x}_2)\rangle_e + \\ +2\,\mathrm{Re}\{K_1^* K_2 \Gamma_{12}(\boldsymbol{x}_1, \boldsymbol{x}_2, t_1 - t_2)\}.\end{aligned}\right\} \tag{V.1.17}$$

$\Gamma_{12}(\boldsymbol{x}_1, \boldsymbol{x}_2, \tau)$ is the socalled *mutual coherence function**.

If V_1 and V_2 stem from the same light source, $\Gamma_{12}(\boldsymbol{x}_1, \boldsymbol{x}_2, \tau)$** is sometimes called the auto-coherence function.

The *normalized* mutual coherence function

$$\gamma_{12}(\boldsymbol{x}_1, \boldsymbol{x}_2, \tau) = \frac{\Gamma_{12}(\boldsymbol{x}_1, \boldsymbol{x}_2, \tau)}{\sqrt{\langle I(\boldsymbol{x}_1)\rangle_e \langle I(\boldsymbol{x}_2)\rangle_e}} \tag{V.1.18}$$

is called the *complex degree of mutual coherence.*

* On account of the missing factor 2 (compare footnote * on p. 74) Γ_{12} differs from the usual definition by a factor $\frac{1}{4}$.

** A more consequent notation for the auto-coherence function is Γ_{11}.

From Schwarz' inequality it follows that

$$|\gamma_{12}| \leqq 1 . \tag{V.1.19}$$

For $|\gamma_{12}| = 1$ the two signals are completely coherent, for
$|\gamma_{12}| = 0$ completely incoherent.

$|\gamma_{12}|$ is connected with Michelson's "fringe visibility", $v(\boldsymbol{x})$, which is the usual measure of the sharpness of interference fringes.

$v(\boldsymbol{x})$ is defined as

$$v(\boldsymbol{x}) = \frac{\langle I_{max}\rangle - \langle I_{min}\rangle}{\langle I_{max}\rangle + \langle I_{min}\rangle} \tag{V.1.20}$$

where $\langle I_{max}\rangle$ and $\langle I_{min}\rangle$ are the intensity maxima and minima in the immediate neighbourhood of $\boldsymbol{x}$. One then finds, in a good approximation

$$v(\boldsymbol{x}) = 2(\mu + \mu^{-1})^{-1} |\gamma_{12}(\boldsymbol{x}_1, \boldsymbol{x}_2, (s_1 - s_2)/c)| \tag{V.1.21}$$

where

$$\mu = [\langle I^{(1)}(\boldsymbol{x}, t)\rangle_e / \langle I^{(2)}(\boldsymbol{x}, t)\rangle_e]^{\frac{1}{2}}$$

and where

$$I^{(j)} = |K_j|^2 I_j .$$

When the intensities of the two beams are equal, then $\mu = 1$ and

$$v(\boldsymbol{x}) = |\gamma_{12}(\boldsymbol{x}_1, \boldsymbol{x}_2, (s_1 - s_2)/c)| . \tag{V.1.22}$$

From the mathematical point of view, it is quite simple to define higher order correlation functions by

$$\langle V_{i_1}^*(\boldsymbol{x}_1, t_1) \ldots V_{in}^*(\boldsymbol{x}_n, t_n) V_{j_1}(\boldsymbol{x}_1', t_1') \ldots V_{j_n}(\boldsymbol{x}_n', t_n')\rangle_e \tag{V.1.23}$$

although in general it is difficult to give an experimental setup which measures such a quantity. (V.1.23) is written in a way, which remains valid also in a quantum theoretical treatment. An important case is, however, the *intensity correlation function* of the Hanbury Brown-Twiss[3] experiment

$$\left.\begin{aligned} &\langle V_1^*(\boldsymbol{x}_1, t_1) V_2^*(\boldsymbol{x}_2, t_2) V_2(\boldsymbol{x}_2, t_2) V_1(\boldsymbol{x}_1, t_1)\rangle_e - \\ &\quad - \langle V_1^*(\boldsymbol{x}_1, t_1) V_1(\boldsymbol{x}_1, t_1)\rangle_e \langle V_2^*(\boldsymbol{x}_2, t_2) V_2(\boldsymbol{x}_2, t_2)\rangle_e . \end{aligned}\right\} \tag{V.1.24}$$

It can be determined experimentally by measuring two intensities at points $\boldsymbol{x}_1$ and $\boldsymbol{x}_2$, then inserting a delay line, and then giving the intensities to an electronic device which multiplies them and averages over a time interval. Usually, the product of the averaged intensities is then subtracted, as is done in (V.1.24).

b) Quantum theoretical coherence functions[4].

α) *Elementary introductions.* In quantum theory the classical quantities ("observables") become operators. Examples were given in Chap. III.1, where the electric field or the vector potential were transformed into such operators. We are now concerned with the adequate description of measurements in quantum theory. As is well known from quantum mechanics, in general one must perform

[3] R. Hanbury Brown and R. O. Twiss: Nature **177**, 27 (1956). This experiment has recently been repeated by W. Martienssen and E. Spiller: Am. J. Phys. **32**, 919 (1964) using their "pseudothermal source".

[4] For more details see: R. J. Glauber, in: Quantum Optics and Electronics (C. de Witt, A. Blandin, and C. Cohen-Tannoudji, eds.). New York: Gordon & Breach 1965. — L. Mandel and E. Wolf: Rev. Mod. Phys. **37**, 231 (1965). — J. R. Klauder and E. C. G. Sudarshan: Fundamentals of Quantum Optics. New York and Amsterdam: W. A. Benjamin, Inc. 1968.

measurements with an ensemble. The ensemble average over the measured values of such a quantity, e.g. $E(\boldsymbol{x},t)$, can be calculated by forming the expectation value $\langle \Phi^+ E(\boldsymbol{x},t)\,\Phi\rangle$ of the operator E, if the system is in the pure state Φ, or by forming the expression $\sum_i w_i \langle \Phi_i^+ E(\boldsymbol{x},t)\,\Phi_i\rangle$, if the system belongs to an ensemble described by a mixture of states Φ_i. (The w_i's are statistical weights).

Let us illustrate this by an example from quantum mechanics: One wishes to measure the distribution of the squared momentum of the electron of a hydrogen atom in a certain energy eigenstate (n, l, m). Then one has to calculate

$$\overline{p^2_{n,l,m}} = \langle \varphi^*_{n,l,m}\, p^2\, \varphi_{n,l,m}\rangle\,. \tag{V.1.25}$$

If the atom is subjected to a heatbath, $\varphi_{n,l,m}$ is only a member of a statistical ensemble of states, and the expectation value is given by

$$\overline{p^2} = \frac{\sum\limits_{n,l,m} \overline{p^2_{n,l,m}}\, e^{-E_n/kT}}{\sum\limits_{n,l,m} e^{-E_n/kT}}\,. \tag{V.1.26}$$

We now consider

some examples in quantum electrodynamics.

We use throughout the Heisenberg picture (III.5), so that the states are time independent, whereas the operators depend fully on time (free fields):

1. $$\langle \Phi(0)\, \boldsymbol{A}(\boldsymbol{x},t)\, \Phi(0)\rangle \tag{V.1.27}$$

where $\boldsymbol{A}$ denotes the vector potential.

a) $\Phi(0)$ contains a definite number of photons

$$\Phi(0) = \frac{1}{\sqrt{n_1!\ldots n_M!}}\,(b_1^+)^{n_1}\ldots(b_M^+)^{n_M}\,\Phi_0 \tag{V.1.28}$$

[compare (III.1.19)]. Using the decomposition of $\boldsymbol{A}(\boldsymbol{x},t)$ (III.1.5) in interaction representation*, we find immediately

$$(\text{V.1.27}) = 0\,. \tag{V.1.29}$$

Because the electric and magnetic fields are obtained from $\boldsymbol{A}(\boldsymbol{x},t)$ by differentiation processes, (V.1.29) holds also for their expectation values.

b) If $\Phi(0)$ is a superposition of states with different numbers of photons, we can obtain nonvanishing expectation values, however, (see below).

2. We consider next the energy density of the electric field:

$$\frac{1}{4\pi}\,\langle \Phi(0)\, \boldsymbol{E}(\boldsymbol{x},t)\, \boldsymbol{E}(\boldsymbol{x},t)\, \Phi(0)\rangle\,. \tag{V.1.30}$$

This quantity *diverges* for any state, even for the vacuum state, Φ_0. In the latter case it represents just one half the zero point energy. In quantum electrodynamics this (trivial) divergence is avoided by the prescription that annihilation operators must always act first ("Normal product"). In the present context, the use of the normal product is a natural consequence of the light detection process, as we will discuss in the following.

β) Coherence functions. As just outlined, any measurable quantity now becomes an operator, especially the mutual coherence functions (V.1.15), (where

* Note, that we are considering *free fields*, so that in the "interaction representation" b_λ, b_λ^+ are merely to be supplemented with factors $e^{-i\omega_\lambda t}$ and $e^{i\omega_\lambda t}$ respectively.

we omit for the moment the average)

$$V_1^*(\boldsymbol{x}_1, t_1)\, V_2(\boldsymbol{x}_2, t_2) \tag{V.1.31}$$

where V is the positive frequency part, e.g. of a component of the vector potential. When we use the same sequence of operators, we avoid divergencies (see above) and we have then immediately the form encountered in the photon counting experiment. Thus we consider*

$$\Gamma_{12}(\boldsymbol{x}_1, \boldsymbol{x}_2, t_1, t_2) = \langle \Phi(0)\, V_1^{(-)}(\boldsymbol{x}_1, t_1)\, V_2^{(+)}(\boldsymbol{x}_2, t_2)\, \Phi(0)\rangle \tag{V.1.32}$$

as the *quantum theoretical coherence function.*

In the following we identify V with a component of $\boldsymbol{A}$. All higher order correlation functions can be defined in a similar way.

The interpretation of (V.1.32) can be done exactly as in the classical case by the consideration of an experimental setup as shown in Fig. 18. This is a consequence of the fact that the spatial propagation of quantized electromagnetic fields is completely determined by classical physics (wave equation with boundary conditions).

The *complex degree of mutual coherence* γ_{12} is defined exactly as in the classical case (V.1.18).

γ) Coherent wave functions. We define the *complex degree of coherence of a single mode* by

$$\gamma_{11} = \frac{\langle \Phi^+ b\, \Phi\rangle\, \langle \Phi^+ b^+\, \Phi\rangle}{\langle \Phi^+ b^+ b\, \Phi\rangle} \tag{V.1.33}$$

where we have dropped the index λ and the time dependence $\sim e^{i\omega t}$. One may show quite generally, that

$$|\gamma_{11}| \leqq 1. \tag{V.1.34}$$

There is an infinity of states Φ, for which $|\gamma_{11}| = 1$. They can be found by a variational principle, which chooses the maximum of γ_{11}.

These "*coherent states*" read explicitly[5]

$$\Phi_\alpha = e^{-\frac{1}{2}|\alpha|^2}\, e^{\alpha b^+}\, \Phi_0 \equiv e^{-\frac{1}{2}|\alpha|^2} \sum_{n=0}^{\infty} \alpha^n \frac{1}{n!}\, (b^+)^n\, \Phi_0 \tag{V.1.35}$$

where α can be any complex number.

If the field is produced by a source (laser), α can even contain the operators of the source-atoms (see below, p. 81). Because the normalization factor of $(b^+)^n \Phi_0$ is $\frac{1}{\sqrt{n!}}$, the probability of finding in a state Φ_α a given number of photons, n_0 is

$$p_\alpha(n_0) = e^{-|\alpha|^2}\, \frac{|\alpha|^{2 n_0}}{n_0!} \tag{V.1.36}$$

which is called the Poisson distribution.

Φ_α is an eigenstate to b:

$$b\, \Phi_\alpha = \alpha\, \Phi_\alpha. \tag{V.1.37}$$

---

* Note that $\Phi(0)$ can be an arbitrary initial state and is *not* the vacuum state, which is denoted by Φ_0.

[5] These coherent states were first treated as minimum uncertainty states by E. Schrödinger: Ann. Phys. **87**, 570 (1928), and used by J. Schwinger: Phys. Rev. **91**, 728 (1953) in his formulation of quantum electrodynamics. They were introduced into maser theory with respect to coherence by J. R. Senitzky: Phys. Rev. **111**, 3 (1958); **115**, 227 (1959) and have been investigated in great detail by R. J. Glauber: Phys. Rev. Letters **10**, 84 (1963); Phys. Rev. **130**, 2529 (1963); **131**, 2766 (1963). An introduction to the states (V.1.35) is also given by W. H. Louisell: Radiation and Noise in Quantum Electronics. New York: McGraw-Hill Book Co. 1964.

Thus $\langle \Phi_\alpha^+ b \Phi_\alpha \rangle = \alpha$ so that there exists a nonvanishing expectation value for b and also for the electric field strength $\boldsymbol{E}$. Returning to many, or even infinitely many modes, we can now construct eigenfunctions of

$$\boldsymbol{A}^{(+)}(\boldsymbol{x}, t) = \sum_\lambda \sqrt{\frac{2\pi c^2 \hbar}{\omega_\lambda}}\, \boldsymbol{u}_\lambda(\boldsymbol{x})\, e^{-i\omega_\lambda t}\, b_\lambda. \tag{V.1.38}$$

Putting

$$\Phi = e^{-\sum_\lambda \frac{1}{2}|\alpha_\lambda|^2}\, e^{\sum_\lambda \alpha_\lambda b_\lambda^+}\, \Phi_0 \tag{V.1.39}$$

one readily verifies, that

$$\boldsymbol{A}^{(+)}(\boldsymbol{x}, t)\, \Phi = \Big(\sum_\lambda \sqrt{\frac{2\pi c^2 \hbar}{\omega_\lambda}}\, \boldsymbol{u}_\lambda(\boldsymbol{x})\, e^{-i\omega_\lambda t}\, \alpha_\lambda\Big)\, \Phi \tag{V.1.40}$$

where the quantity in brackets is the *eigenvalue* to the operator $\boldsymbol{A}^+(\boldsymbol{x}, t)$ (III.1.8).

If such coherent states are used in the quantum theoretical mutual coherence function (V.1.32) it reduces to a product of the two eigenvalues, i.e. it is identical with the classical mutual coherence function (V.1.15). The same is true if, besides the quantum mechanical expectation value, an average is to be taken over a statistical ensemble.

These properties of the coherent functions suggest expanding any wave function into a series of functions Φ_α[6]. This is feasible because the Φ_α's form a complete set, if α takes all values of the complex plane. The Φ's are, however, not orthogonal. One has instead

$$|\langle \Phi_\alpha^+ \Phi_\beta \rangle| = e^{-\frac{1}{2}|\alpha-\beta|^2}. \tag{V.1.41}$$

Provided a wave function is represented by

$$\Phi = f(b^+)\, \Phi_0 \tag{V.1.42}$$

where

$$f(b^+) = \sum c_n \frac{(b^+)^n}{\sqrt{n!}}$$

this expansion reads

$$\Phi = \frac{1}{\pi} \iint f(\alpha^*)\, e^{-\frac{1}{2}|\alpha|^2}\, \Phi_\alpha\, d^2\alpha \tag{V.1.43}$$

where $d^2\alpha = d(\mathrm{Re}\,\alpha)\, d(\mathrm{Im}\,\alpha)$.

In the same way, any operator, especially the density matrix, can be expanded into Φ_α's and thus expressed, in general, by a nondiagonal matrix.

Besides this, the socalled "*P representation*" of the density matrix, ϱ, has been introduced by GLAUBER[6] and SUDARSHAN[7]*:

$$\varrho = \int P(\alpha, \alpha^*)\, |\alpha\rangle\langle\alpha|\, d^2\alpha \tag{V.1.44}$$

where $|\alpha\rangle\langle\alpha|$ is a projection operator. Obviously, this representation is diagonal.

---

* This representation is applicable to a broad class of radiation fields although there are counter examples where (V.1.44) seems not to exist. [See e.g. the parametric amplifier treated by B. R. MOLLOW and R. J. GLAUBER: Phys. Rev. **160**, 1076 (1967); **160**, 1097 (1967)].

[6] R. J. GLAUBER: l.c. [5] on p. 78.

[7] E. C. G. SUDARSHAN: Phys. Rev. Letters **10**, 277 (1963); — For a different introduction see: J. R. KLAUDER: Phys. Rev. Letters **16**, 534 (1966); — The mathematical properties of this and related functions were investigated by a number of authors: R. J. GLAUBER: l.c. [5]. — E. C. G. SUDARSHAN: l.c. [7]. — K. E. CAHILL: Phys. Rev. **138**, B 1566 (1965). — D. HOLLIDAY and M. L. SAGE: Phys. Rev. **138**, B 485 (1965). — R. BONIFACIO, L. M. NARDUCCI, and E. MONTALDI: Nuovo Cimento **47**, 890 (1967). — M. M. MILLER: J. Math. Phys. **9**, 1270 (1968) (This latter paper contains a criticism of the forementioned two papers). — K. E. CAHILL, and R. J. GLAUBER: Phys. Rev. **177**, 1857, 1882 (1969). — For other representations and extensions see: C. L. METHA: Phys. Rev. Letters **18**, 752 (1967). — G. S. AGARWAL and E. WOLF: Phys. Rev. Letters **21**, 180, 656 E (1968).

$P(\alpha, \alpha^*)$ cannot be consistently interpreted as a probability, because it may take on negative values, even for positive density operators. $P(\alpha, \alpha^*)$ is closely related to the Wigner distribution which has also been successfully applied to the maser-laser problem in the subthreshold region[8], and which we describe therefore briefly (see also Chap. IV.9).

If $\psi(q)$ is a quantum mechanical wave function which depends on the space coordinate, q, then the Wigner distribution is defined as

$$\hat{W}(p, q) = \frac{1}{2\pi\hbar} \int \psi^* \left(q + \frac{1}{2} y\right) e^{\frac{ipy}{\hbar}} \psi\left(q - \frac{1}{2} y\right) dy \tag{V.1.45}$$

where p and q are numbers (and not operators).

(V.1.45) is the Fourier transform

$$\hat{W}(p, q) = \frac{1}{(2\pi)^2} \int e^{-i\mu p - i\nu q} \chi_{\hat{W}}(\mu, \nu)\, d\mu\, d\nu \tag{V.1.46}$$

of the characteristic function

$$\chi_{\hat{W}}(\mu, \nu) = \int \psi^*(\hat{q})\, e^{i(\mu\hat{p} + \nu\hat{q})}\, \psi(\hat{q})\, d\hat{q} \tag{V.1.47}$$

where $\hat{p}$ and $\hat{q}$ are the momentum and position operators respectively. We can also take both in (V.1.45) and (V.1.47) instead of $\psi^*(q) \ldots \psi(q)$ a mixture

$$\sum c_n \psi_n^*(q) \ldots \psi_n(q) \tag{V.1.48}$$

so that we write (V.1.47) more generally as

$$\chi_{\hat{W}}(\mu, \nu) = \langle e^{i(\mu\hat{p} + \nu\hat{q})} \rangle . \tag{V.1.49}$$

Because for the harmonic oscillator $\hat{p}$ and $\hat{q}$ can be expressed by b^+, b, (V.1.49) can also be written as

$$\chi_{\hat{W}}(\beta, \beta^*) = \langle e^{i(\beta^* b^+ + \beta b)} \rangle = \mathrm{tr}\,(e^{i\beta^* b^+ + i\beta b}\, \varrho) \tag{V.1.50}$$

where

$$\beta = \sqrt{\frac{\hbar}{2}} \left(\frac{\nu}{\sqrt{\omega}} - i \sqrt{\omega}\, \mu \right), \tag{V.1.51}$$

$$b = \frac{1}{\sqrt{2\hbar\omega}} (\omega\, \hat{q} + i\hat{p}); \quad b^+ = \frac{1}{\sqrt{2\hbar\omega}} (\omega\, \hat{q} - i\hat{p}) . \tag{V.1.52}$$

The form (V.1.50) has been successfully used by GORDON, WALKER and LOUISELL[9] for the maser amplifier. For further discussions and the connection with other distribution functions see Chap. IV.9.a.

δ) Generation of coherent fields by classical sources[10] *(the forced harmonic oscillator).* In the radio frequency range the electromagnetic fields are generated by prescribed, macroscopic currents. These fields, treated quantum theoretically, provide an example of coherent states.

---

[8] A. E. GLASSGOLD and D. HOLLIDAY: Phys. Rev. **139**, A 1717 (1965).

[9] J. P. GORDON, L. R. WALKER, and W. H. LOUISELL: Phys. Rev. **129**, 481 (1963); **130**, 807 (1963). — W. H. LOUISELL: Radiation and Noise in Quantum Electronics, Chap. 7. New York: McGraw Hill Book Co. 1964.

[10] The forced harmonic oscillator has been treated by G. LUDWIG: Z. Physik **130**, 468 (1951). — R. J. GLAUBER: l.c.[4] on p. 76 and other authors; — In our representation we follow a method used in the exercises in the lectures on quantum mechanics (1960) by the present author.

We start with the Schrödinger equation

$$H\Phi = i\hbar\dot{\Phi} \tag{V.1.53}$$

where H is given by (III.3.10). We neglect as usual $H_{I,2}$ (III.3.6) and keep for simplicity only one mode λ. We use $H_{I,1}$ in the form (III.3.7) where

$$-\frac{e}{mc}\sqrt{\frac{2\pi\hbar c^2}{\omega_\lambda}}\,\boldsymbol{u}_\lambda(\boldsymbol{x}_0)\sum_{\mu\nu}\int\varphi_\mu^*\,\boldsymbol{p}\,\varphi_\nu\,dV\,a_\mu^+\,a_\nu = \hbar g(t) \tag{V.1.54}$$

is treated as a given classical and real function of time. H_{El} can be dropped, because it does not belong to the system now. Thus we are left with the following Schrödinger equation:

$$H'\Phi \equiv \{\hbar\omega\, b^+ b + (b^+ + b)\,\hbar g(t)\}\,\Phi = i\hbar\dot{\Phi}. \tag{V.1.55}$$

(Because we are dealing with a single mode, we drop the mode index λ completely.) Putting

$$\Phi = e^{-iF(t)}\,e^{f(t)b^+}\,\Phi_0 \tag{V.1.56}$$

and inserting it into (V.1.55) yields

$$\{\hbar\omega\, b^+ f(t) + \hbar g(t)\, f(t) + \hbar g(t)\, b^+\}\,\Phi = (i\hbar\, b^+\dot{f} + \hbar\dot{F})\,\Phi \tag{V.1.57}$$

where use was made of the formal relation*

$$[b, G(b^+)] = \frac{dG(b^+)}{db^+}. \tag{V.1.58}$$

The comparison of the coefficients of the linearly independent functions Φ and $b^+\Phi$ yields:

$$\omega f(t) + g(t) = i\dot{f} \tag{V.1.59}$$

and

$$\dot{F} = g(t)\, f(t) \tag{V.1.60}$$

with the solution

$$f(t) = -i\int_0^t e^{-i\omega(t-\tau)}\,g(\tau)\,d\tau + f(0)\,e^{-i\omega t} \tag{V.1.61}$$

and

$$F(t) = \int_0^t g(\sigma)\,f(\sigma)\,d\sigma + F(0). \tag{V.1.62}$$

(V.1.56) has just the form required for a coherent state with $\alpha = f(t)$. The factor $e^{-iF(t)}$ serves for the normalization. Although (V.1.56) is a coherent function, it mirrors on account of (V.1.61) the noise properties of $g(t)$. Because all essential mathematical features of the solution can be seen, if the initial state $\Phi(0)$ is the oscillator ground state, Φ_0, we choose in the following part of this chapter this initial condition and put correspondingly $f(0) = F(0) = 0$.

A particularly simple solution of (V.1.55) can be obtained in the Heisenberg picture (compare III.5).

The equation of motion for b^+ reads:

$$\dot{b}^+(t) = i\omega\, b^+(t) + i g(t) \tag{V.1.63}$$

and the one for $b(t)$ is just the conjugate complex of (V.1.63). (V.1.63) is a linear first order differential equation with the solution

$$b^+(t) = i\int_0^t e^{i\omega(t-\tau)}\,g(\tau)\,d\tau + b^+(0)\,e^{i\omega t} \equiv f^*(t) + b^+(0)\,e^{i\omega t}, \tag{V.1.64}$$

* For a proof see the appendix.

where $b^+(0)$ is the Schrödinger operator b^+. Because g commutes with b^+, b; $b^+(t), b(t)$ fulfill the commutation relations at all times. In contrast to Eq. (V.1.61), $b^+(0)$ and $b(0)$ cannot be put equal to zero because otherwise the commutation relations would no longer hold.

It is instructive to study the unitary transformation U (compare III.5.1), which connects the Heisenberg and Schrödinger pictures, in detail.

According to (III.5.5), U must have the properties:

$$b^+(t) = U^+ b^+(0)\, U \quad \text{and} \quad b(t) = U^+ b(0)\, U, \tag{V.1.65}$$

where, according to (III.5.2), U must satisfy the equation

$$H'\, U = i\hbar \dot{U}, \tag{V.1.66}$$

where H' is given in (V.1.55).

A comparison of the right hand side of (V.1.65) with that of (V.1.64) shows that U must have the properties

$$U^+ b^+(0)\, U = b^+(0)\, e^{i\omega t} + f^*(t); \qquad U^+ b(0)\, U = b(0)\, e^{-i\omega t} + f(t). \tag{V.1.67}$$

Thus U must cause a *displacement* of b^+, b by an amount $f^*(t), f(t)$ respectively (and a multiplication with factors $e^{i\omega t}$, $e^{-i\omega t}$ respectively).

We first solve Eq. (V.1.66) and then show how (V.1.67) and the wave function in the Schrödinger picture (V.1.56) with (V.1.61) and (V.1.62) follow. The Eq. (V.1.66), which is a first order differential equation in time, has the formal solution

$$\left.\begin{aligned} U &= \exp\left\{\frac{1}{i\hbar}\int_0^t H'(\tau)\, d\tau\right\} \\ &\equiv \exp\left\{-i\int_0^t \omega\, b_\tau^+ b_\tau\, d\tau - i\int_0^t g(\tau)\,(b_\tau^+ + b_\tau)\, d\tau\right\}, \end{aligned}\right\} \tag{V.1.68}$$

where $H'(\tau)$, b_τ^+ and b_τ are time ordered operators in the sense of FEYNMAN[11]: If the exponential function is expanded into a power series, operators with a lower index τ must be applied prior to those with a higher index.

We now use a "disentangling" theorem of FEYNMAN, which allows us to transform the exponential function of a sum of operators A_i into a product of exponential functions of the single operators A_i (with additional factors). Using this theorem and some algebraic transformations, we find for (V.1.68) the form*

$$U = U_1\, U_2, \tag{V.1.69}$$

where

$$U_1 = e^{-iF(t)}\, e^{f(t)\, b^+}\, e^{-f^*(t)\, b} \tag{V.1.70}$$

and

$$U_2 = e^{-i\omega t b^+ b}. \tag{V.1.71}$$

The detailed treatment shows, that $f(t)$ and $F(t)$ are given by (V.1.61) and (V.1.62) respectively [with $f(0) = F(0) = 0$]. Furthermore, U_1 and U_2 are unitary operators. We now show how the form of U [(V.1.69)—(V.1.71)] establishes the connection (V.1.67): [We write everywhere $b^+(0) = b^+$, $b(0) = b$].

$$U^+ b\, U \equiv U_2^+\, U_1^+ b\, U_1\, U_2 \tag{V.1.72}$$

* A derivation of this theorem and the further algebraic transformations are given in the appendix, X. 5.

[11] R. P. FEYNMAN: Phys. Rev. **84**, 108 (1951).

We first look what is achieved by the transformation U_1: Using again the theorem (V.1.58) and $U_1^+ U_1 = 1$, we find immediately

$$U_1^+ b\, U_1 = b + f(t). \tag{V.1.73}$$

Thus the unitary operator U_1 (V.1.70) provokes a "displacement" of the operator b by an amount $f(t)$.

This remark will be of importance in the quantum theoretical treatment of the laser (see Chap. VI.4 and VI.7). We now investigate

$$U_2^+ (U_1^+ b\, U_1)\, U_2 \equiv U_2^+ (b + f(t))\, U_2. \tag{V.1.74}$$

Because U_2^+ commutes with the c-function $f(t)$, and $U_2^+ U_2 = 1$, we obtain

$$U_2^+ f(t)\, U_2 = f(t). \tag{V.1.75}$$

As is shown in the appendix

$$U_2^+ b\, U_2 = b\, e^{-i\omega t}. \tag{V.1.76}$$

Putting (V.1.73), (V.1.76), (V.1.75) together, we find the second Eq. (V.1.67). [The other one in (V.1.67) is just the hermitian conjugate].

So far we have investigated how U operates on b in the Heisenberg picture. In the Schrödinger picture the time-dependent wave function $\Phi(t)$ arises from the initial wave function $\Phi(0)$ by the unitary transformation $U(t)$, which in our case is given by (V.1.69) with (V.1.70), (V.1.71). Again we choose the initial state $\Phi(0)$ as the oscillator ground state Φ_0. We thus have

$$\Phi(t) = U(t)\, \Phi_0 \equiv e^{-iF(t)}\, e^{f(t)\, b^+}\, e^{-if^* b}\, e^{-i\omega t b^+ b}\, \Phi_0. \tag{V.1.77}$$

Since $b\Phi_0 = 0$ we find immediately for $\Phi(t)$ the expression (V.1.56), which is our original wave function in the Schrödinger picture.

V.2. Uncertainty relations and limits of measurability. In quantum mechanics operators Ω_1, Ω_2, which do not commute, cannot be measured simultaneously with absolute exactness. There is always an uncertainty left, the degree of which is expressed by uncertainty relations. A measure for the uncertainty in the measurement of the observable with the operator Ω is the root mean square deviation:

$$\Delta\Omega = [\langle(\Omega - \langle\Omega\rangle)^2\rangle]^{\frac{1}{2}} \tag{V.2.1}$$

where the average refers to a single wave function:

$$\langle\Omega\rangle = \langle\Phi^+ \Omega\, \Phi\rangle. \tag{V.2.2}$$

For hermitian operators one can show quite generally[1]

$$\Delta\Omega_1 \cdot \Delta\Omega_2 \geqq \tfrac{1}{2} |\langle[\Omega_1, \Omega_2]\rangle| \tag{V.2.3}$$

where $[\Omega_1, \Omega_2] = \Omega_1\Omega_2 - \Omega_2\Omega_1$.

A famous example is in quantum mechanics $\Omega_1 = p$ (momentum), $\Omega_2 = q$ (position). Because $[p, q] = \hbar/i$ one finds: $\Delta p\, \Delta q \geqq \hbar/2$. We are here primarily concerned with the electromagnetic fields, the photon number and the phase of light.

a) Field and photon number. The question arises how accurately can one measure the field strength and the number of photons in a certain mode λ_0

[1] For the proof see e.g. A. S. Davydov: Quantum Mechanics, p. 35. Oxford-London-Edinburgh-New York-Paris-Frankfurt: Pergamon Press.

simultaneously? Since the b_λ^+'s, b_λ's commute for different λ's, we need only consider the same mode λ_0 in the expansion of E or H which can be obtained from (III.1.5) using (III.1.2) and (III.1.3). The time-dependent mode amplitude is represented by the operators b_{λ_0} and $b_{\lambda_0}^+$, so that we have merely to determine the commutation relation between $b^+ + b$ and $\frac{b^+ - b}{i}$ on the one hand and $n = b^+ b$ on the other.

On account of (V.2.1), and

$$[b^+, b^+ b] = -b^+, \tag{V.2.4}$$

$$[b, b^+ b] = b \tag{V.2.5}$$

we find

$$\left.\begin{aligned} \Delta(b^+ + b)\,\Delta n &\geqq \frac{1}{2}\,|\langle -b^+ + b\rangle|, \\ \Delta\left(\frac{b^+ - b}{i}\right)\Delta n &\geqq \frac{1}{2}\,|\langle b^+ + b\rangle|. \end{aligned}\right\} \tag{V.2.6}$$

If the averages on the right hand side are evaluated for states, for which $\langle \pm b^+ + b\rangle$ does not vanish, it is evident, that a small uncertainty of n neccessitates a big one of $\Delta(b^+ \pm b)$ and vice versa. Because the electric (or magnetic) field strength is proportional to b^+ and b, a precise measurement of the photon number is connected with an uncertainty of the field amplitudes. (This is the case even if no light quanta at all are present).

If we use states for which the right hand sides of Eq. (V.2.6) vanish, at a first glance it would seem possible to determine $(b^+ \pm b)$ and n exactly in a simultaneous measurement. We want to show that this conclusion is misleading, because in these cases one factor on the left hand side of Eq. (V.2.6) vanishes:

1. If we use eigenstates of the number operator $b^+ b$, both sides vanish,

2. The same is true, if we use certain coherent states. Thus in these two cases the relations (V.2.6) fail to give us information about the uncertainty of n, if a coherent state is measured and vice versa. Therefore one is forced for these two cases, at least, to calculate Δn or $\Delta(b^+ + b)$ directly:

a) A coherent state Φ_α (V.1.35) is given.

Inserting Φ_α into $(\Delta n)^2$ we obtain according to the definition (V.2.1)

$$\left.\begin{aligned} (\Delta n)^2 &= \langle \Phi_\alpha^+ b^+ b b^+ b\, \Phi_\alpha\rangle - \langle \Phi_\alpha^+ b^+ b\, \Phi_\alpha\rangle^2 \\ &= \langle \Phi_\alpha^+ b^+ b^+ b b\, \Phi_\alpha\rangle + \langle \Phi_\alpha^+ b^+ b\, \Phi_\alpha\rangle - \langle \Phi_\alpha^+ b^+ b\, \Phi_\alpha\rangle^2 \\ &= \langle \Phi_\alpha^+ b^+ b\, \Phi_\alpha\rangle = |\alpha|^2 = \bar{n}. \end{aligned}\right\} \tag{V.2.7}$$

The mean square deviation of n, determined for a coherent state equals the average number of photons in that state.

The relative fluctuations decrease, however, with $\bar{n}$: $\frac{\Delta n}{\bar{n}} = \frac{1}{\sqrt{\bar{n}}}$.

b) An eigenstate of $b^+ b$ is given: Φ_n (III.1.19).

We obtain

$$(\Delta(b^+ + b))^2 = \langle \Phi_n^+ (b^{+2} + (b b^+ + b^+ b) + b^2)\, \Phi_n\rangle - \langle \Phi_n^+ (b^+ + b)\, \Phi_n\rangle^2. \tag{V.2.8}$$

Due to (III.1.22), (III.1.23) the second expectation value vanishes while the first reduces to

$$(\text{V.2.8}) = 2n + 1. \tag{V.2.9}$$

The same is found for

$$\left(\Delta\left(\frac{b^+ - b}{i}\right)\right)^2.$$

The mean square deviation of $(b^+ + b)$ and $\left(\frac{b^+ - b}{i}\right)$ determined for a state with a definite photon number, goes with the photon number.

b) Phase and photon number.

α) *Heuristic considerations.* Since the classical analogue of b, b^+ is a complex amplitude, one may try to decompose these operators into a real amplitude and a phase factor[2]:

$$b^+ = \sqrt{n}\, e^{-i\varphi} \tag{V.2.10}$$

and

$$b = e^{i\varphi} \sqrt{n} \tag{V.2.11}$$

where φ and $\sqrt{n}$ are operators.

We assume for the time being that such a representation does exist. Then it follows from the commutation relation (III.1.10) that

$$e^{i\varphi} n - n e^{i\varphi} = e^{i\varphi}. \tag{V.2.12}$$

This relation is satisfied if φ and n satisfy the commutation relation

$$\varphi n - n \varphi = -i \tag{V.2.13}$$

and thus on account of (V.2.1) the uncertainty relation $\Delta n \, \Delta \varphi \geqq \frac{1}{2}$ follows.

There are, however, two serious difficulties:

1. the decomposition (V.2.10), (V.2.11) does not exist,
2. it is necessary to derive (V.2.13) from (V.2.12) and not vice versa.

That the decomposition (V.2.10), (V.2.11) does not exist, can be seen in a qualitative way as follows (a rigorous proof has been given by GLOWGOWER and SUSSKIND[3]).

It is obviously required that

$$e^{-i\varphi} e^{i\varphi} = 1. \tag{V.2.14}$$

Expressing now $e^{-i\varphi}$ by $\frac{1}{\sqrt{n}} b^+$ and $e^{i\varphi}$ by $b \frac{1}{\sqrt{n}}$, we immediately run into difficulties: If we calculate these exponentials in the n-representation, $\sqrt{n}$ vanishes in those matrix-elements which contain the vacuum.

This difficulty can easily be overcome, however, by replacing $\sqrt{n}$ by $\sqrt{n+1}$. One is thus led to a new decomposition of b, b^+ which also allows the uncertainty relation (V.2.13) to be replaced by one which can be rigorously derived:

β) *Exact treatment.* We write according to GLOWGOWER and SUSSKIND[3]

$$b^+ = E_+ (b^+ b + 1)^{\frac{1}{2}}, \qquad b = (b^+ b + 1)^{\frac{1}{2}} E_- . \tag{V.2.15*}$$

Because $(b^+ b + 1)^{\frac{1}{2}}$ possesses an inverse, E_+ and E_- are uniquely defined by

$$E_+ = b^+ (b^+ b + 1)^{-\frac{1}{2}}, \qquad E_- = (b^+ b + 1)^{-\frac{1}{2}} b \tag{V.2.16}$$

where $E_\pm$ are obviously the substitutes for the former $e^{\mp i\varphi}$.

The E's have the property to create normalized eigenfunctions in the n-representation

$$\Phi_{n\pm 1} = E_\pm \Phi_n . \tag{V.2.17}$$

---

* The operator introduced in (V.2.15) is not to be mixed up with the electric field.

[2] W. HEITLER: The Quantum Theory of Radiation, 3rd ed., p. 65. Oxford: At the Clarendon Press 1954. For a recent detailed review of the quantum-mechanical description of phase and angle variables, with emphasis on the proper mathematical treatment of these coordinates, see: P. CARRUTHERS and M. M. NIETO: Rev. Mod. Phys. **40**, 411 (1968).

[3] L. SUSSKIND and J. GLOWGOWER: Physics **1**, 49 (1964).

Neither b^+, b^+ nor E_-, E_+ are hermitian operators, however, for which an uncertainty relation of the form (V.2.1) can be deduced[4].

We therefore define

$$S = \frac{1}{2i}(E_- - E_+), \quad (\sim \sin\varphi!) \tag{V.2.18}$$

and

$$C = \frac{1}{2}(E_+ + E_-) \quad (\sim \cos\varphi!). \tag{V.2.19}$$

The operators S and C have a continuous eigenvalue spectrum in the interval from -1 to $+1$.

Using (V.2.3) one derives the following uncertainty relations:

$$\left.\begin{aligned} \Delta n\, \Delta S &\geqq \frac{1}{2}|\langle C\rangle|, \\ \Delta n\, \Delta C &\geqq \frac{1}{2}|\langle S\rangle|. \end{aligned}\right\} \tag{V.2.20}$$

From (V.2.20) the relation

$$U \equiv (\Delta n)^2 \frac{[(\Delta S)^2 + (\Delta C)^2]}{[\langle S\rangle^2 + \langle C\rangle^2]} \geqq \frac{1}{4} \tag{V.2.21}$$

follows.

In particular one can show for coherent states that $\frac{1}{2} \geqq U \geqq \frac{1}{4}$. The Eqs. (V.2.20) are no use, if eigenstates of n are involved, and again one has to evaluate ΔC by the given wave-function:

$$(\Delta C)^2 = \frac{1}{4}\langle \Phi_n^+ (E_-^2 + E_- E_+ + E_+ E_- + E_+^2)\, \Phi_n\rangle = \frac{1}{2} \tag{V.2.22}$$

and $(\Delta S)^2 = \frac{1}{2}$.

Note, that in a classical interpretation

$$\frac{1}{2\pi}\int_0^{2\pi} \sin^2\varphi\, d\varphi = \frac{1}{2}.$$

For a definite value of n, C and S can take any value between -1 and $+1$, so that the "phase" is completely undetermined.

We mention some useful results[4] for expectation values taken with respect to the coherent states (V.1.35), indicated by an index α

$$\left.\begin{aligned} \langle S^2\rangle_\alpha &= \frac{1}{2} - \frac{1}{4}e^{-n} - \frac{1}{2}e^{-n}\, n(1-2\xi)\sum_{\nu=0}^{\infty}\frac{n^\nu}{\nu!\,[(\nu+1)(\nu+2)]^{\frac{1}{2}}} \\ \text{where} \qquad & \\ \xi &= \frac{(\operatorname{Im}\alpha)^2}{|\alpha|^2}; \qquad n = |\alpha|^2. \end{aligned}\right\} \tag{V.2.23}$$

$$\langle S\rangle_\alpha = e^{-n}(\operatorname{Im}\alpha)\sum_{\nu=0}^{\infty}\frac{n^\nu}{\nu!\,(\nu+1)^{\frac{1}{2}}}, \tag{V.2.24}$$

$$\langle C^2 + S^2\rangle_\alpha = 1 - \frac{1}{2}e^{-n}, \tag{V.2.25}$$

$$(\langle C\rangle_\alpha)^2 + (\langle S\rangle_\alpha)^2 = n\, e^{-2n}\left(\sum_{\nu=0}^{\infty}\frac{n^\nu}{\nu!\,(\nu+1)^{\frac{1}{2}}}\right)^2 \approx 1 - \frac{1}{4n}. \tag{V.2.26}$$

[4] For this and the following see: P. Carruthers and M. M. Nieto: Phys. Rev. Letters **14**, 387 (1965).

Here n always means the mean photon number [see (V.2.23)].

$$(\Delta S)_\alpha (\Delta C)_\alpha \geqq \frac{1}{4} e^{-n}. \tag{V.2.27}$$

From these considerations it is evident that for photon numbers $n \gg 1$ (say 10) the old "heuristic" considerations hold, whereas for photon numbers of order 1 at least quantitative differences occur. In the maser or laser process a great number of photons is involved, so that one may safely use the more elementary treatment. On the other hand, when the measurement of phases of single light quanta is discussed, for instance in counting experiments, the exact treatment must be used.

c) Field strength[5]. Because the electric and magnetic fields $\boldsymbol{E}$ and $\boldsymbol{H}$ are linear combinations of the photon operators b_λ^+, b_λ, which do not all commute, $\boldsymbol{E}$ and $\boldsymbol{H}$ do not commute either. Particularly one finds for free fields:

$$[E_i(\boldsymbol{x}_1, t_1), H_i(\boldsymbol{x}_2, t_2)] = 0, \tag{V.2.28}$$

$$\left.\begin{aligned} [E_i(\boldsymbol{x}_1, t_1), H_k(\boldsymbol{x}_2, t_2)] &= -i\hbar \frac{\partial^2}{\partial x_{l,2}\, \partial t_1} \left\{\frac{\delta(x-ct)-\delta(x+ct)}{x}\right\}; \\ x = |\boldsymbol{x}_1 - \boldsymbol{x}_2|&, \; t = |t_1 - t_2| \end{aligned}\right\} \tag{V.2.29}$$

where $k \neq i$ and $l \neq i$, $k. (i, k, l)$ form an even permutation of x, y, z.

$$\left.\begin{aligned} [H_i(\boldsymbol{x}_1, t_1), H_k(\boldsymbol{x}_2, t_2)] &= [E_i(\boldsymbol{x}_1, t_1), E_k(\boldsymbol{x}_2, t_2)], \\ &= i\hbar c \left(\frac{1}{c^2} \frac{\partial^2}{\partial t_1\, \partial t_2} \delta_{ik} - \frac{\partial^2}{\partial x_{i1}\, \partial x_{k2}}\right) \left\{\frac{\delta(x-ct)-\delta(x+ct)}{x}\right\}. \end{aligned}\right\} \tag{V.2.30}$$

Because two times t_1 and t_2 appear and because the action propagates with the finite light velocity, the usual terminology of "simultaneous measurements of two quantities" cannot be used in the sense "measurement at the same time", but means that the reciprocal influence of the two measurements is taken into account.

We mention some consequences of (V.2.28) to (V.2.30). The field strengths at two space-time-points which cannot be connected by a light-signal, commute, and can be "measured simultaneously".

Due to the singular functions on the right hand side one has to average the field strengths over finite space-time regions $L_1^3 T_1$ and $L_2^3 T_2$ at the points $\boldsymbol{x}_1, t_1$ and $\boldsymbol{x}_2$, t_2 respectively. We denote such an averaged field strength by $\boldsymbol{E}_{L_1, T_1}$ etc. and quote the essential results:

a) $t_1 = t_2$ and $T_1 = T_2 = T$ i.e. the time regions coincide

$$\Delta E_{i, L_1 T}\, \Delta E_{k, L_2 T} = \Delta H_{i L_1 T}\, \Delta H_{k, L_2 T} = 0. \tag{V.2.31}$$

Thus the average values of the components of the electric field or those of the magnetic field *over the same time region* and the same *or* different space regions can be measured simultaneously.

b) $\boldsymbol{x}_1 = \boldsymbol{x}_2$, $L_1 = L_2 = L$ i.e. the space regions coincide:

$$\Delta E_{i, L T_1}\, \Delta H_{k L, T_2} = 0. \tag{V.2.32}$$

The average values of the electric field and the magnetic field *over the same space region* and the same or different time regions can be measured simultaneously.

[5] N. Bohr and L. Rosenfeld: Det. Kgl. dansk. Vid. Selskab. **12**, 8 (1933); see also Heitler, l.c. [2] on p. 85.

c) The two regions (I) and (II) are situated so that light signals emitted from points of (I) can reach (II) but not vice versa during the time T_1. We assume $L_1 \approx L_2 = L$, $T_1 \approx T_2 = T$ and find for the order of magnitude:

$$\left.\begin{aligned} \Delta E_{xL}(\mathrm{I})\,\Delta H_{yL}(\mathrm{II}) &\sim \hbar/x^2 L T^2 \quad \text{for} \quad (L \gg cT) \\ &\sim \hbar/x^2 c T^2 \quad \text{for} \quad (L \ll cT) \end{aligned}\right\} \tag{V.2.33}$$

x is the distance between (I) and (II).

The fields can be considered as classical ones, when they are so large that their product is large as compared to the right hand side of (V.2.33).

Putting $H \approx E$ and the distance $x \approx L$ we obtain

$$E^2 L^3 T \gg \hbar \qquad (\text{for } L > cT). \tag{V.2.34}$$

Since $E^2 L^3 \sim n\hbar\omega$ where n is the number of light quanta in the volume L^3 and since the time T of the measurement must be smaller than $1/\omega$ (otherwise the average value of E vanishes), we obtain from (V.2.34)

$$n \gg 1.$$

In connection with the generation of ultra-short optical pulses (compare Chap. VII.10. and VII.11.), it is hoped that duration times of the order of $1/\omega$ or perhaps even less can be reached. Clearly in that region the uncertainty relations just discussed may play an important role.

V.3. Spontaneous and stimulated emission and absorption[1]. For the treatment of these problems it is convenient to start with the Schrödinger equation in the interaction representation (compare III.4.). We consider a single atom with two levels, 1 and 2, whose upper level, 2, is occupied in the beginning ($t=0$). Because the spontaneous and stimulated emission as well as the absorption are processes in which only one light quantum is created or annihilated, it suffices to take into account only that part (III.4.5) of the Hamiltonian, (III.3.2), which is linear in the light-field operators b_λ^+, b_λ. Thus we have to solve the Schrödinger equation:

$$i\hbar\dot{\Phi}_I = \widetilde{H}_{I,1}\Phi_I. \tag{V.3.1}$$

Keeping only *resonant terms* (compare III.4.) we find from (III.4.5)

$$\widetilde{H}_{I,1} = \hbar\sum_\lambda \{a_1^+ a_2 b_\lambda^+ g_\lambda e^{i(\omega_\lambda - \nu_{21})t} + a_2^+ a_1 b_\lambda g_\lambda^* e^{-i(\omega_\lambda - \nu_{21})t}\} \tag{V.3.2}$$

where, according to (III.5.8)

$$g_\lambda \equiv g_{12,\lambda} = -\frac{1}{\hbar}\frac{e}{m}\sqrt{\frac{\hbar\cdot 2\pi}{\omega_\lambda}}\int \varphi_1^*(\boldsymbol{x})\,\boldsymbol{u}_\lambda(\boldsymbol{x})\,\boldsymbol{p}\,\varphi_2(\boldsymbol{x})\,dV. \tag{V.3.3}$$

In order to treat the

a) Spontaneous emission we assume an initial state for the total system: light field + one electron, in which there is no photon present, and the electron in its excited state:

$$\Phi_I(0) = a_2^+\Phi_0 \tag{V.3.4}$$

Φ_0: vacuum state.

As usual in first order perturbation theory, we insert (V.3.4) as unperturbed wave function into the right hand side of (V.3.1) and determine the improved wave function by integration over the time interval, t.

[1] These processes are treated to a greater or smaller extent in most modern textbooks on quantum theory.

Using (V.3.2), we obtain

$$\left.\begin{aligned} \Phi_I(t) &= -\sum_\lambda g_\lambda \frac{e^{i\Delta\omega_\lambda t}-1}{\Delta\omega_\lambda}\, a_1^+ b_\lambda^+ \Phi_0 + a_2^+ \Phi_0 \\ \text{where} \quad & \\ \Delta\omega_\lambda &= \omega_\lambda - \nu_{21}. \end{aligned}\right\} \tag{V.3.5}$$

The absolute square of the coefficient connected with the normalized wave-function $a_1^+ b_\lambda^+ \Phi_0$ gives the probability of finding a photon of the mode λ and the electron in the ground state. In general, it makes sense to single out a specific mode connected with spontaneous emission only if the dimensions of the cavity (with closed walls) are of the same order as the wavelengths of the emitted electro-magnetic wave.

On the other hand, in infinite space or in a cavity with open sides, there is a continuum of modes, so that we have to sum up (integrate) over a total range of final states:

$$\sum_\lambda |g_\lambda|^2 \left|\frac{e^{i\Delta\omega_\lambda t}-1}{\Delta\omega_\lambda}\right|^2. \tag{V.3.6}$$

By differentiation of this expression with respect to the time we obtain for the transition probability:

$$W = 2\pi \sum_\lambda |g_\lambda|^2\, \delta(\omega_\lambda - \nu_{21}) \tag{V.3.7}$$

where we used

$$\lim_{t\to\infty} \frac{\sin \Delta\omega t}{\Delta\omega} = \pi\,\delta(\Delta\omega). \tag{V.3.8}$$

δ: Dirac's function.

We calculate first the sum over λ considering only those photons, which are emitted into a certain group of modes. In order to be specific, we treat plane waves, so that

$$\boldsymbol{u}_\lambda(\boldsymbol{x}) = \frac{1}{\sqrt{V}}\, e^{i\boldsymbol{k}_\lambda \boldsymbol{x}} \cdot \boldsymbol{e} \tag{V.3.9}$$

$\boldsymbol{e}$: vector of polarization. V = volume of the box, with $V\to\infty$. We consider photons emitted into a space angle $d\Omega$.

Using

$$\sum_\lambda = \frac{V}{(2\pi)^3} \int \dots k^2\, dk\, d\Omega \tag{V.3.10}$$

we find from (V.3.7)

$$W(d\Omega) = a^1_{2,e}\, d\Omega = \frac{1}{2\pi\hbar} \frac{e^2}{m^2} \frac{\nu_{21}}{c^3} \left| \int \varphi_1^*(\boldsymbol{x})\, e^{i\boldsymbol{k}_{\lambda_0}\boldsymbol{x}}\, \boldsymbol{e}\,\boldsymbol{p}\, \varphi_2(\boldsymbol{x})\, dV \right|^2 d\Omega \tag{V.3.11}$$ *

where

$$|k_{\lambda_0}| = \frac{\nu_{21}}{c}$$

$a^1_{2,e}$ in (V.3.11) is the Einstein coefficient for spontaneous emission of a light quantum of polarization $\boldsymbol{e}$ into $d\Omega$.

In the *dipole approximation* [compare (III.3.7), (III.3.8)], (V.3.11) simplifies to

$$W(d\Omega) = \frac{1}{2\pi\hbar} \frac{\nu_{21}^3}{c^3} |\boldsymbol{e}\,\boldsymbol{\vartheta}_{21}|^2\, d\Omega \tag{V.3.12}$$

* Note that ν_{21} is a circular frequency in accordance with the notation used throughout this article.

where

$$\boldsymbol{\vartheta}_{21} = e \int \varphi_2^* \, \boldsymbol{x} \, \varphi_1 \, dV \tag{V.3.13}$$

and we have used

$$\boldsymbol{p}_{21} = i \, m \, \nu_{21} \, \boldsymbol{x}_{21} \, .$$

In order to get the total transition probability for the emission of any photon, we integrate (V.3.12) over $d\Omega$ and sum up over the two directions of polarization:

$$W = \frac{4}{3} \frac{\nu_{21}^3}{\hbar \, c^3} |\boldsymbol{\vartheta}_{21}|^2 . \tag{V.3.14}$$

As we will show in V.5, $W = 1/\tau$, where τ is the lifetime of the upper state.

b) Stimulated emission. We assume first that there is a definite number of light quanta in a definite mode λ_0 initially present.

The normalized initial state is then

$$\Phi_I(0) = a_2^+ \frac{1}{\sqrt{n!}} (b_{\lambda_0}^+)^n \, \Phi_0 \, . \tag{V.3.15}$$

We insert again $\Phi_I(0)$ into the right hand side of Eq. (V.3.1) and find immediately that there are *two kinds of final states*, depending whether in $\sum_\lambda$ the index λ equals λ_0 or not:

a) $$\lambda \neq \lambda_0 : \quad a_1^+ \frac{1}{\sqrt{n!}} (b_{\lambda_0}^+)^n \, b_\lambda^+ \, \Phi_0 , \tag{V.3.16}$$

b) $$\lambda = \lambda_0 : \quad a_1^+ \sqrt{n+1} \frac{1}{\sqrt{(n+1)!}} (b_{\lambda_0}^+)^{n+1} \, \Phi_0 \, . \tag{V.3.17}$$

In the case a) a quantum of another mode λ has been emitted spontaneously.

In case b) a light quantum has been added to the mode under consideration. Under the action of the light-field the atom is forced to emit a quantum by "stimulated emission".

In order to obtain the transition probability, we repeat first naively the above steps, which yields:

$$W = 2\pi |g_{\lambda_0}|^2 \, \delta(\nu_{21} - \omega_\lambda) \, (n+1) + 2\pi \sum_{\lambda \neq \lambda_0} |g_\lambda|^2 \, \delta(\nu_{21} - \omega_\lambda) \, . \tag{V.3.18}$$

The first term stems from b), while the second one comes from a). We split the first term $(n+1)$ into n and 1 and combine the expressions connected with "1" with the second expression in (V.3.18), so that we obtain with this combination just the *spontaneous emission into all modes.*

The remaining expression:

$$W_{st} = 2\pi \, n \, |g_{\lambda_0}|^2 \, \delta(\nu_{21} - \omega_{\lambda_0}) \tag{V.3.19}$$

is then the stimulated emission rate. In it there appears no summation over λ. On the other hand it is necessary to integrate over a continuum in order that the δ-function can be evaluated. Thus the formalism forces us to start with a more realistic initial state, which is formed as a wavepacket. We assume, that it is built up out of plane waves within a region $\Delta k_x \, \Delta k_y \, \Delta k_z$. If there are M modes in it the normalized wave function reads:

$$\Phi_I(0) = \frac{1}{\sqrt{M}} \frac{1}{\sqrt{n!}} \sum_{\Delta k} a_2^+ (b_{\boldsymbol{k}}^+)^n \, \Phi_0 \tag{V.3.20}$$

(in order to be quite clear we use $\boldsymbol{k}$ instead of λ).

With this initial state we obtain readily

$$W_{st} = 2\pi \frac{n}{M} \sum_{\Delta k} |g_k|^2 \, \delta(\nu_{21} - \omega_k). \tag{V.3.21}$$

With

$$M = m_x \cdot m_y \cdot m_z \quad \text{and} \quad \Delta k_i = \frac{2\pi m_i}{L}$$

(L: length of normalization box)

we find

$$M = \frac{L^3}{(2\pi)^3} \Delta k_x \Delta k_y \Delta k_z = \frac{V}{(2\pi)^3} k^2 \Delta k \, d\Omega. \tag{V.3.22}$$

Using further (V.3.3) and (V.3.10), we obtain for the transition probability for stimulated emission of photons into a space angle $d\Omega$:

$$W_{st}(d\Omega) = \frac{4\pi^2 e^2}{\hbar^2 m^2 \nu_{21}^2} \cdot \frac{n \hbar \nu_{21}}{\Delta\omega \, d\Omega \, V} \left| \int \varphi_1^*(\boldsymbol{x}_1) \, e^{i \boldsymbol{k}_{\lambda_0} \boldsymbol{x}} \, \boldsymbol{e}\boldsymbol{p} \, \varphi_2(\boldsymbol{x}) \, dV \right|^2 d\Omega. \tag{V.3.23}$$

We write $W_{st}(d\Omega)$ in the form

$$W_{st}(d\Omega) = \varrho_{\boldsymbol{e}}(\nu_{21}, d\Omega) \cdot b_{2,\boldsymbol{e}}^1 \, d\Omega \tag{V.3.24}$$

where

$$b_{2,\boldsymbol{e}}^1 = \frac{4\pi^2 e^2}{\hbar^2 m^2 \nu_{21}^2} \left| \int \varphi_1^*(\boldsymbol{x}_1) \, e^{i \boldsymbol{k}_{\lambda_0} \boldsymbol{x}} \, \boldsymbol{e}\boldsymbol{p} \, \varphi_2(\boldsymbol{x}) \, dV \right|^2 \tag{V.3.25}$$

is the Einstein coefficient for stimulated emission of photons with polarization $\boldsymbol{e}$ into $d\Omega$.

$$\varrho_{\boldsymbol{e}}(\nu_{21}, d\Omega) = \frac{n \hbar \nu_{21}}{\Delta\omega \, d\Omega \, V} \tag{V.3.26}$$

is the total energy of the n photons, divided by the frequency spread, the space angle, and the volume, or in other words: ϱ is the energy density per unit frequency interval, unit space angle and unit volume.

A comparison between (V.3.11) and (V.3.25) yields one of Einstein's relations:

$$\frac{a_{2,\boldsymbol{e}}^1}{b_{2,\boldsymbol{e}}^1} = \frac{\hbar \nu^3}{(2\pi)^3 c^3}. \tag{V.3.27}$$

In the *dipole approximation* the transition probability $W_{st}(d\Omega)$ becomes:

$$W_{st}(d\Omega) = \varrho_{\boldsymbol{e}}(\nu_{21}, d\Omega) \cdot b_{2,\boldsymbol{e}}^1 \, d\Omega \tag{V.3.28}$$

where ϱ is given by (V.3.26) and

$$b_{2,\boldsymbol{e}}^1 = \frac{4\pi^2}{\hbar^2} |\boldsymbol{e} \boldsymbol{\vartheta}_{21}|^2. \tag{V.3.29}$$

c) Comparison between spontaneous and stimulated emission rates: We already know of one connection, namely the one between Einstein's coefficients (V.3.27). This connection can be given another appearance. We determine the spontaneous emission rate, $\widetilde{W}$, per number of *modes* (not photons!) in the volume V, the angle $d\Omega$ and the frequency range $\Delta\omega \, (\nu_{21} \ldots \nu_{21} + \Delta\omega)$ which we had considered just now. Dividing (V.3.11) by this number

$$N_m = \frac{k^2 \Delta k V d\Omega}{8\pi^3} = \frac{\nu^2 \Delta\omega V d\Omega}{8\pi^3 c^3}$$

we find

$$\left.\begin{aligned}\tilde{W} = \frac{W}{N_m} &= \frac{1}{2\pi\hbar}\frac{e^2}{m^2}\frac{\nu_{21}}{c^3}\left|\int \varphi_1^*(\boldsymbol{x}_1)\, e^{i\boldsymbol{k}_{\lambda_0}\boldsymbol{x}}\,\boldsymbol{e}\boldsymbol{p}\,\varphi_2(\boldsymbol{x})\, dV\right|^2 \frac{8\pi^3 c^3\, d\Omega}{\nu_{21}^2\, \Delta\omega\, d\Omega\, V}\\ &= \frac{4\pi^2 e^2}{\hbar m^2 \nu_{21}\, V\Delta\omega}\left|\int \varphi_1^*(\boldsymbol{x}_1)\, e^{i\boldsymbol{k}_{\lambda_0}\boldsymbol{x}}\,\boldsymbol{e}\boldsymbol{p}\,\varphi_2(\boldsymbol{x})\, dV\right|^2 = \frac{1}{n} W_{st}\end{aligned}\right\} \quad \text{(V.3.30)}$$

so that the ratio of stimulated emission rate to the spontaneous emission rate is $= n$ ($=$ total number of photons in this range).

d) Absorption. The calculation of the transition probability W_{abs} for the atom going from its ground state to the excited state by absorption of a quantum out of a wave-packet which propagates within a space angle $d\Omega$ can be done in complete analogy to the stimulated emission. The only difference is that we start now with an electron in its ground state instead of one in its excited state:

$$W_{abs} = \varrho_e(\nu_{21}, d\Omega)\, b_{1,\boldsymbol{e}}^{(2)}\, d\Omega \quad \text{(V.3.31)}$$

where $b_{1,\boldsymbol{e}}^{(2)} = b_{2,\boldsymbol{e}}^{1}$, so that the *Einstein coefficients for absorption is equal to that for stimulated emission.* The absorption rate is proportional to the energy density, ϱ, of the incident light.

In order to derive the absorption coefficient α, we introduce the energy flux density $I(\omega) = \varrho_{\boldsymbol{e}}(\nu_{21}, d\Omega)\, c\, d\Omega$ (energy flux per sec., per unit area) into (V.3.31), so that

$$W_{abs} = I(\omega)\, \frac{b_{1,\boldsymbol{e}}^{(2)}}{c}. \quad \text{(V.3.32)}$$

The decrease of the photon number, n, per second is equal W_{abs} for a single atom. If there are N atoms, we find

$$\frac{dn}{dt} = -W_{abs} N. \quad \text{(V.3.33)}$$

Introducing

$$I(\omega) = \frac{\bar{n}\hbar\omega c}{\Delta\omega} \quad \text{(V.3.34)}$$

($\bar{n} = n/V$: photon density) into (V.3.33) yields

$$\frac{dI(\omega)}{dt} = -I(\omega)\, \frac{b_{1,\boldsymbol{e}}^{(2)}}{c}\, \frac{N}{V}\, \frac{\hbar\omega}{\Delta\omega}. \quad \text{(V.3.35)}$$

Writing $dx = c\,dt$ we obtain the spatial absorption equation

$$\frac{dI(\omega)}{dx} = -I(\omega)\,\alpha \quad \text{(V.3.36)}$$

where the absorption coefficient

$$\alpha = \frac{b_{1,\boldsymbol{e}}^{(2)}}{c}\, \frac{N}{V}\, \frac{\hbar\omega}{\Delta\omega} \quad \text{(V.3.37)}$$

(as usual, we put $\Delta\omega = 1$).

Because W_{abs} and W_{st} play a completely symmetrical role, we find quite generally for a system of noninteracting, partially inverted atoms:

$$\frac{dI}{dt} = I\, \frac{(N_2 - N_1)}{N}\, \alpha. \quad \text{(V.3.38)}$$

We conclude with an important *remark about coherence properties of light created by spontaneous or stimulated emission.*

The above transition rates were calculated for an experimental setup, in which the *photon-number* is measured. Such a measurement, however, destroys any phase information (compare V.2.). An appropriate treatment which retains phase information is given in V.5.

V.4. Photon counting.

a) Quantum mechanical treatment, correlation functions. We use the same equations and notations, as in the preceding Chap. V.3, but we treat the light quantum absorption in still another way for the purposes of VI.8. and VI.9. Because a photon is annihilated, only the "positive frequency" part of $\tilde{H}_{I,1}$,

$$\tilde{H}_{I,1}^{(+)} = \hbar \sum_{\lambda} a_2^+ a_1 b_\lambda g_\lambda^* e^{-i(\omega_\lambda - \nu_{21})t} \tag{V.4.1}$$

is of importance in calculating the disturbed wave-function, so that we write

$$\Phi_I(t) - \Phi_I(0) = \frac{i}{\hbar} \int_0^t H_{I,1}^{(+)} d\tau\, \Phi_I(0). \tag{V.4.2}$$

We assume as usual that the interaction between the detector atom and the field is switched on at $t=0$. Thus the initial wave function factorizes:

$$\Phi_I(0) = \Phi_{atom}(0)\, \Phi(0), \tag{V.4.2a}$$

where $\Phi(0)$ contains the field variables and, if the field is coupled to sources, *also the source variables.*

We further assume that the detector atom is initially in its ground state:

$$\Phi_{atom}(0) = a_1^+ \Phi_{0,\,atom} \tag{V.4.2b}$$

The probability of finding a quantum absorbed (and thus the electron in its upper state) is given by the expectation value

$$\langle \Phi_I^+(t)\, a_2^+ a_2\, \Phi_I(t) \rangle = \frac{1}{\hbar^2} \left\langle \Phi_I^+(0) \int_0^t \tilde{H}_{I,1}^{(-)} d\tau \int_0^t \tilde{H}_{I,1}^{(+)} d\tau'\, \Phi_I(0) \right\rangle \tag{V.4.3}$$

where the quantum mechanical average $\langle \cdots \rangle$ refers to all variables. In going from the left hand side of Eq. (V.4.3) to its right hand side, use has been made of (V.4.1), (V.4.2), (V.4.2a), (V.4.2b).

We now go one step back in the representation of $\tilde{H}_{I,1}^{(+)}$ in that we introduce the explicit from of g_λ (V.3.3), so that we write, using (III.1.8)

$$\tilde{H}_{I,1}^{(+)} = \int \varphi_2^*(\boldsymbol{x}) \left(-\frac{e}{mc}\right) \boldsymbol{A}^{(+)}(\boldsymbol{x}, \tau)\, \boldsymbol{p}\, \varphi_1(\boldsymbol{x})\, dV e^{i\nu_{21}\tau} a_2^+ a_1. \tag{V.4.4}$$

Inserting (V.4.4) into (V.4.3) and taking all the integrations out of the quantum mechanical expectation value we find:

$$\left.\begin{aligned} &\frac{e^2}{m^2c^2} \frac{1}{\hbar^2} \sum_{i,j=x,y,z} \int_0^t d\tau\, e^{i\nu_{12}\tau} \int_0^t d\tau'\, e^{i\nu_{21}\tau'} \int dV \int dV' \varphi_2(\boldsymbol{x})\, p_i^*\, \varphi_1^*(\boldsymbol{x}) \times \\ &\quad \times \varphi_2^*(\boldsymbol{x}')\, p_j'\, \varphi_1(\boldsymbol{x}')\, \langle \Phi^+(0) A_i^{(-)}(\boldsymbol{x}, \tau) A_j^{(+)}(\boldsymbol{x}', \tau')\, \Phi(0) \rangle. \end{aligned}\right\} \tag{V.4.5}$$

The photon counting rate thus depends on an electronic part containing the atomic wave functions and on a part, which refers to the light field only:

$$\langle \Phi^+(0) A_i^{(-)}(\boldsymbol{x}, t) A_j^{(+)}(\boldsymbol{x}', t')\, \Phi(0) \rangle. \tag{V.4.6}$$

Note, that the A's are the *free fields* in the interaction representation. "Free" refers to the interaction with the detector. *When there is an interaction of A with a source* on the other hand, *the A's are the full Heisenberg operators with respect to the source,* but they are still *free* with respect to the detector. The quantum mechanical averages $\langle\cdots\rangle$ in (V.4.5) and (V.4.6) refer to the field (+ source), but no longer to the detector. The electronic part is responsible for the sensitivity of the detector.

The classical analogue of (V.4.6) is the so-called mutual coherence function (compare V.2). In the present case $\boldsymbol{x}$ and $\boldsymbol{x}'$ are very close together, within the atomic dimensions. Using a system of atoms with macroscopic dimensions, one encounters also macroscopic differences of $\boldsymbol{x}, \boldsymbol{x}'$ in (V.4.6). In the dipole approximation we can replace $\boldsymbol{x}$ and $\boldsymbol{x}'$ in the A's by $\boldsymbol{x}_0$ (position of the atom), so that (V.4.5) reduces to

$$\left.\begin{aligned}&\frac{e^2}{m^2c^2}\frac{1}{\hbar^2}\sum_{ij=x,y,z}\int_0^t d\tau\int_0^t d\tau'\, e^{i\nu_{12}\tau}e^{i\nu_{21}\tau'}(p_i)^*_{21}(p_j)_{21}\times\\ &\qquad\qquad\times\langle\Phi^+(0)A_i^{(-)}(\boldsymbol{x}_0,\tau)A_j^{(+)}(\boldsymbol{x}_0,\tau')\Phi(0)\rangle.\end{aligned}\right\}\quad\text{(V.4.7)}$$

When there is a successive annihilation of photons to be described one has to apply $\widetilde{H}^{(+)}_{I,1}$ several times on $\Phi_I(0)$, with a corresponding sequence of time integrations. The probability that n photons are absorbed is thus determined by[1] correlation functions of the form

$$\langle\Phi^+(0)A_{i_1}^{(-)}(\boldsymbol{x}_1,t_1)\ldots A_{in}^{(-)}(\boldsymbol{x}_n,t_n)A_{jn}^{(+)}(\boldsymbol{x}'_n,t'_n)\ldots A_{j_1}^{(+)}(\boldsymbol{x}'_1,t'_1)\Phi(0)\rangle.\quad\text{(V.4.8)}$$

In general the counter consists of many atoms, each of which has more than two states. One has then, as usual, to average over the initial and to sum up over the final states of all these atoms. Whether the double time integration $(t_i\, t'_i)$ can be replaced by a single one (t_i) depends on the spread of atomic energies (bandwidth $\Delta\omega$ of the counter) involved in the counting process. If $t > \frac{1}{\Delta\omega}$, this can be achieved (see e.g. GLAUBER[1]).

We refrain from a further discussion, as to how expressions of the type (V.4.7) can be further simplified. We would rather present a classical treatment of photon counting, since the fully quantum mechanical theory gives the same final results as the classical one.

b) Classical treatment of photon counting[2]. The problem is to connect the light intensity distribution $I(t')$ of the light source at time t' with the number of photoelectrons counted within a time interval $t<t'<t+T$. In the following we assume for simplicity that each photon instantaneously generates an electron, so that we may identify the photoelectron distribution with the photon distribution. Because the number of laser photons is high (even at threshold there are several thousand photons present), $I(t)$ is virtually a continuous function. On the other hand the number of photoelectrons counted is very small in the relevant experiments, say of the order of 5, so that the probability $p(n, T, t)$ of finding n photons in the interval $t\ldots t+T$ is a discrete function. The connection between

[1] This general process has been treated by P. L. KELLEY and W. H. KLEINER: Phys. Rev. **136**, A 316 (1964). — R. J. GLAUBER, in: Quantum Optics and Electronics (DE WITT, BLANDIN, and COHEN-TANNOUDJI, eds.). New York: Gordon & Breach 1965. — L. MANDEL, E. C. G. SUDARSHAN, and E. WOLF: Proc. Phys. Soc. (London) **84**, 435 (1964). — R. J. GLAUBER and U. M. TITULAER: Phys. Rev. **140**, 676 (1965). G. BÉDARD: J. Opt. Soc. Am. **57**, 1201 (1967); B. R. MOLLOW: Phys. Rev. **168**, 1896 (1968).

[2] L. MANDEL: Proc. Phys. Soc. (London) **71**, 1037 (1958); — Fluctuations of Light Beams, in: Progr. in Optics, p. 227 (ed. E. WOLF). Amsterdam 1963.

$I(t')$ and p can be established by the usual methods of probability theory: The probability of measuring a single photon in the infinitesimal time interval dt is proportional to the intensity:

$$p(1, dt, t) = \alpha I(t)\, dt. \tag{V.4.9}$$

The probability $p(n, T, t)$ is then given by the POISSON distribution (for a derivation see MANDEL[2])

$$p(n, T, t) = \frac{1}{n!}\left[\alpha \int_t^{t+T} I(t')\, dt'\right]^n \exp\left(-\alpha \int_t^{t+T} I(t')\, dt'\right). \tag{V.4.10}$$

The light intensity is not a fixed quantity, but shows fluctuations, i.e. it is a random function. (Its explicit statistical properties will be treated in VI.12, but are at present of no concern to us.) The experimentally observed photon distribution is obtained by averaging over the distribution of I. This average will be denoted by:

$$\langle p(n, T, t)\rangle_I = p(n, T). \tag{V.4.11}$$

Let us first consider the case where the time T of the measurement is small compared to the time constants of $I(t)$ (as we will see in VI.12, these time constants are given by the relaxation times of the laser). Then we may put

$$\int_t^{t+T} I(t')\, dt' = T I(t)$$

and the distribution of $I(t)$ is just given by the stationary distribution $\widetilde{W}(I)$. This leaves us with

$$p(n, T, t) = \frac{1}{n!}(\alpha T I)^n e^{-\alpha T I} \tag{V.4.12}$$

and, after performing the average over I, with

$$p(n, T) = \int_0^\infty \frac{(\alpha T I)^n}{n!} e^{-\alpha T I}\, \widetilde{W}(I)\, dI. \tag{V.4.13}$$

Explicit expressions for $\widetilde{W}(I)$ will be given below (VI.13.3).

In the *general case* the explicit evaluation of (V.4.13) may be difficult but there exists a relation between the cumulants k_l of the photon count distribution und the correlation functions of the intensity distribution which can be seen as follows:

The cumulants k_l of a random variable ξ are defined by

$$\sum_{l=1}^\infty \frac{s^l}{l!} k_l = \ln \langle e^{s\xi}\rangle_\xi = k(s) \tag{V.4.14}$$

where $\langle \ldots \rangle_\xi$ is the average over the distribution of ξ. The expression of (V.4.14) is the generating function of the cumulants k_l. In our case the random variable is n, the distribution function of n is given by $\langle p(n, T, t)\rangle_I$. Then

$$\langle e^{sn}\rangle_n \equiv \sum_{n=0}^\infty e^{sn} \langle p(n, T, t)\rangle_I, \tag{V.4.15}$$

and the cumulants follow by

$$\sum_{l=1}^{\infty} \frac{s^l}{l!} k_l = \ln \left\{ \sum_{n=0}^{\infty} e^{s n} \langle p(n, T, t) \rangle_I \right\} \equiv k(s). \tag{V.4.16}$$

We further consider the random variable

$$E(T, t) = \int_t^{t+T} I(t')\, dt'. \tag{V.4.17}$$

The cumulants of the distribution of $E(T, t)$ are denoted by K_l, thus the generating function of these cumulants is

$$\sum_{l=1}^{\infty} \frac{s^l}{l!} K_l = \ln \langle e^{sE} \rangle_E \equiv K(s). \tag{V.4.18}$$

Inserting (V.4.10) into (V.4.16) yields

$$k(s) = \ln \left\{ \left\langle \sum_{n=0}^{\infty} \frac{(\alpha\, e^s E)^n}{n!} e^{-\alpha E} \right\rangle_I \right\}. \tag{V.4.19}$$

The evaluation of the sum leads to

$$k(s) = \ln \{ \langle \exp(\alpha (e^s - 1) E) \rangle_I \}. \tag{V.4.20}$$

This, is, however, of the form (V.4.18), when we replace s in (V.4.18) by $\alpha(e^s - 1)$; so that

$$k(s) = K(\alpha(e^s - 1)). \tag{V.4.21}$$

This equation establishes the relation between the cumulants, k, of the discrete distribution $\langle p(n, T, t) \rangle_I$ and those of the continuous intensity distribution. A comparison of the coefficients of the powers of s allows us to determine the cumulants k explicitly.

$$\left.\begin{aligned} k_1 &= \alpha K_1, \\ k_2 &= \alpha K_1 + \alpha^2 K_2, \\ k_3 &= \alpha K_1 + 3\alpha^2 K_2 + \alpha^3 K_3, \\ k_4 &= \alpha K_1 + 7\alpha^2 K_2 + 6\alpha^3 K_3 + \alpha^4 K_4 \quad \text{etc.} \end{aligned}\right\} \tag{V.4.22}$$

It remains to show how the K's can be determined. For this purpose we raise the last two expressions of Eq. (V.4.18) into the exponent of the exponential function, expand e^{sE} into a power series of E and insert (V.4.17) for E.

This leaves us with

$$\sum_{m=0}^{\infty} \frac{1}{m!} s^m \int_t^{t+T} dt_m \int_t^{t+T} dt_{m-1} \int_t^{t+T} dt_1 \langle I(t_m) \ldots I(t_1) \rangle = \exp \left\{ \sum_{l=1}^{\infty} \frac{s^l}{l!} K_l \right\}. \tag{V.4.23}$$

After expanding the right hand side into a power series and comparing the coefficients of powers of s, we find (see e.g. STRATONOVICH, R. L.: "Topics in the Theory of Random Noise", Vol. 1, Gordon and Breach, New York (1963), where more general relations are also discussed)

$$K_1 = \int_t^{t+T} \langle I(t_1) \rangle \, dt_1, \tag{V.4.24}$$

$$K_2 = \int_t^{t+T} \int_t^{t+T} \langle (I(t_2) - \langle I(t_2) \rangle)(I(t_1) - \langle I(t_1) \rangle) \rangle \, dt_1 \, dt_2, \tag{V.4.25}$$

$$\left. \begin{aligned} K_3 = \int_t^{t+T} \int_t^{t+T} \int_t^{t+T} \langle (I(t_3) - \langle I(t_3) \rangle)(I(t_2) - \langle I(t_2) \rangle) \times \\ \times (I(t_1) - \langle I(t_1) \rangle) \rangle \, dt_1 \, dt_2 \, dt_3. \end{aligned} \right\} \tag{V.4.26}$$

The higher K's are somewhat more complicated (see STRATONOVICH) e.g.:

$$K_4 = \int_t^{t+T} \cdots \int_t^{t+T} \left\langle \prod_{j=1}^{4} (I(t_j) - \langle I(t_j) \rangle) \right\rangle dt_1 \ldots dt_4 - 3K_2^2. \tag{V.4.27}$$

The expectation values under the integrals are correlation functions of 1., 2., 3.,... order of the intensity distribution so that by means of (V.4.22) and (V.4.24) to (V.4.27) we can express the cumulants k of the discrete photo count distribution by the correlation functions of the intensity distribution.

V.5. Coherence properties of spontaneous and stimulated emission. The spontaneous linewidth. In V.3 we solved the Schrödinger equation of a single atom interacting with the radiation field by first order perturbation theory. In order to obtain the finite linewidth, we proceed now more exactly[1] by putting, for the wave function in the *full* Schrödinger representation

$$\Phi(t) = A(t)\, a_2^+ \Phi_0 \, e^{-\frac{i}{\hbar} E_2 t} + \sum_\lambda c_\lambda(t)\, b_\lambda^+ a_1^+ \Phi_0 \, e^{-i\omega_\lambda t - i\frac{1}{\hbar} E_1 t}. \tag{V.5.1}$$

Inserting it into (V.3.1) (using the interaction representation) and comparing the coefficients of the linearly independent components on both sides yields:

$$\dot{A} = \frac{1}{i} \sum_\lambda c_\lambda g_\lambda^* \, e^{i(\nu_{21} - \omega_\lambda)t}, \tag{V.5.2}$$

$$\dot{c}_\lambda = \frac{1}{i} g_\lambda A \, e^{i(\omega_\lambda - \nu_{21})t}. \tag{V.5.3}$$

Making the hypothesis $A = e^{-\gamma t}$ we find immediately

$$c_\lambda = \frac{1}{i} g_\lambda \frac{1 - e^{-\gamma t + i(\omega_\lambda - \nu_{21})t}}{\gamma - i(\omega_\lambda - \nu_{21})} \tag{V.5.4}$$

and

$$\gamma = \sum_\lambda |g_\lambda|^2 \frac{e^{\gamma t - i(\omega_\lambda - \nu_{21})t} - 1}{\gamma - i(\omega_\lambda - \nu_{21})}. \tag{V.5.5}$$

For small γ we can use

$$\lim_{t\to\infty} \frac{1 - e^{i(\nu-\omega)t}}{\nu - \omega} = \frac{P}{\nu - \omega} - i\pi\,\delta(\nu - \omega). \tag{V.5.6}$$

Inserting the δ-function part into (V.5.5) yields exactly the same expression for the total transition probability W (V.3.14) besides a factor of 2 if we sum over

[1] V. WEISSKOPF and E. WIGNER: Z. Physik **63**, 54 (1930); **65**, 18 (1930); see also W. HEITLER: The Quantum Theory of Radiation, 3rd ed. Oxford: At the Clarendon Press 1954.

all λ, as in the elementary theory, so that

$$2\gamma = W. \tag{V.5.7}$$

The principal part $P/(\nu - \omega)$ gives rise to a frequency shift, which is very small, however.

We wish to show now explicitly, how the formalism of the coherence functions (V.1.32) allows the coherence properties of the spontaneously emitted light field to be determined directly[2]. To this end we calculate

$$\Gamma(t+\tau, t) = \langle \Phi^+(0)\, \boldsymbol{E}^{(-)}(\boldsymbol{x}, t+\tau)\, \boldsymbol{E}^{(+)}(\boldsymbol{x}, t)\, \Phi(0)\rangle. \tag{V.5.8}$$

It is most important that the atom should serve as source and not as a counter. Consequently, $\boldsymbol{E}^{(-)}$ and $\boldsymbol{E}^{(+)}$ (or $\boldsymbol{A}^{(-)}$ and $\boldsymbol{A}^{(+)}$) must be the full Heisenberg operators:

$$E_x^{(-)}(\boldsymbol{x}, t) = e^{iHt}\Big\{-\frac{i}{\sqrt{V}}\sum_\lambda \sqrt{2\pi\hbar\omega_\lambda}\,(\boldsymbol{e}_\lambda)_x\, e^{-i\boldsymbol{k}_\lambda \boldsymbol{x}}\, b_\lambda^+\Big\}\, e^{-iHt} \tag{V.5.9}$$

where H is the complete Hamiltonian (III.3.10) divided by $\hbar$.

Because $\boldsymbol{E}^{(-)}$ and $\boldsymbol{E}^{(+)}$ are decomposed into a linear combination of operators b^+, b we investigate first:

$$\langle \Phi(0)\, b_\lambda^+(t+\tau)\, b_{\lambda'}(t)\, \Phi(0)\rangle \tag{V.5.10}$$

where

$$b_\lambda^+(t) = e^{iHt}\, b_\lambda^+(0)\, e^{-iHt}, \qquad b_{\lambda'}(t) = e^{iHt}\, b_{\lambda'}(0)\, e^{-iHt}, \tag{V.5.11}$$

$\Phi(0)$ is the initial state (one atom excited, no photon present).

We insert (V.5.11) into (V.5.10) and observe that

$$e^{-iHt}\Phi(0) = \Phi(t) \tag{V.5.12}$$

so that

$$(\text{V.5.10}) = \langle \Phi^+(t+\tau)\, b_\lambda^+\, e^{-iH\tau}\, b_{\lambda'}\, \Phi(t)\rangle. \tag{V.5.13}$$

We use now the explicit form (V.5.1) of $\Phi(t)$ and find after some elementary algebra:

$$(\text{V.5.10}) = c_\lambda^*(t+\tau)\, e^{i\omega_\lambda(t+\tau)}\, c_{\lambda'}(t)\, e^{-i\omega_{\lambda'}t + \frac{i}{\hbar}E_1\tau} \langle (a_1^+\Phi_0)^+ e^{-iH\tau} a_1^+\Phi_0\rangle. \tag{V.5.14}$$

Because $a_1^+\Phi_0$ represents a state in which the electron is in its ground state and no photon is present, H_I can cause only virtual transitions. Because they give rise only to renormalization effects, we need not consider it, so that of $e^{-iH\tau}$ only $e^{-iH_{at}\tau}$ is left.

Therefore we find finally

$$(\text{V.5.10}) = c_\lambda^*(t+\tau)\, e^{i\omega_\lambda(t+\tau)}\, c_{\lambda'}(t)\, e^{-i\omega_{\lambda'}t}. \tag{V.5.15}$$

When we insert (V.5.15) into (V.5.8) we obtain a double sum (λ, λ') of expressions (V.5.15), still multiplied with coefficients occurring in (V.5.9). This evaluation is rather tedious. Because it sheds no light on the quantum-theoretical treatment, which has been completely done above, we quote merely the result:

$$\Gamma(t+\tau, t) = \frac{2\nu_{21}^4}{c^4}\, e^{-2\gamma\left(t-\frac{r}{c}\right)}\, \frac{\sin^2\Theta}{r^2}\, |\vartheta_{21}|^2\, e^{+i\nu_{21}\tau - \gamma\tau} \tag{V.5.16}$$

[2] H. Haken, R. Hübner, and K. Zeile: Unpublished manuscript.

where ϑ_{21} is the (electric) dipole matrix element, and it was assumed, that

$$\gamma \ll \nu_{21}, \tag{V.5.17}$$

$$r = |\boldsymbol{x}| \gg \frac{c}{\nu_{21}} = \frac{\lambda}{2\pi}, \tag{V.5.18}$$

$$r < ct. \tag{V.5.19}$$

The condition (V.5.17) is always well fulfilled. (V.5.18) means that we confine ourselves to the wave region of a Hertzian dipole, while Γ would vanish (on account of causality) if (V.5.19) is not fulfilled. For the explicit derivation of (V.5.16) the dipole approximation has been used.

It is most remarkable that the classical treatment of a damped Hertzian oscillator leads to exactly the same expression. When we normalize Γ in order to find the complex degree of coherence [compare (V.1.18)] we obtain immediately:

$$\gamma(t+\tau, t) = e^{+i\nu_{21}\tau - \gamma\tau}. \tag{V.5.20}$$

This clearly shows, that the spontaneously emitted wave-track has a coherence time $1/\gamma$.

Although $\Gamma(t+\tau, t)$ does not vanish, $\langle \boldsymbol{E}^{(+)}(\boldsymbol{x}, t)\rangle = \langle \boldsymbol{E}^{(-)}(\boldsymbol{x}, t)\rangle$ does. This can be interpreted as meaning, that the initial phase of $\boldsymbol{E}(\boldsymbol{x}, t)$ is unknown. This phase cancels out, however, in (V.5.8).

The build up of coherent laser light due to stimulated emission is treated e.g. in VI.12. The question, whether the coherence properties of a *coherent state* are changed by (partial) absorption, stimulated emission and scattering has been treated by BRUNNER, PAUL and RICHTER[3]. They have found by first order perturbation theory, that in these processes a *coherent state* remains a coherent state (where only the parameter α changes) if the contribution due to spontaneous emission is neglected.

VI. Fully quantum mechanical solutions of the laser equations.

VI.1. Disposition. We have three different, but essentially equivalent sets of equations at our disposal:

1. quantum mechanical Langevin equations,
2. density matrix equation,
3. generalized Fokker-Planck equation.

They were derived in Chap. IV. In the present chapter, we adopt and specialize these equations. In order to enable the reader to start immediately with these equations, we again explain all quantities which appear.

The Langevin equations are particularly well suited for giving a quick survey of laser action (see VI.4. below): There exists a certain threshold with respect to the pump: Below threshold, laser light consists of many statistically independent wavetracks and its amplitude thus possesses a Gaussian distribution. In this region the linewidth becomes narrower and narrower with increasing pump power. Somewhat above threshold, laser light possesses (in analogy to radio waves) a stable real amplitude with small residual fluctuations. Its phase still diffuses in an undamped manner. The Langevin equations allow for a simple solution from which we can easily obtain the expressions for the total linewidth below threshold and the amplitude and phase fluctuations above threshold.

[3] W. BRUNNER, H. PAUL, and G. RICHTER: Ann. Physik **14**, 384 (1964); **15**, 17 (1965).

The *density matrix equations* can also be solved in the threshold region and then give in particular the photon statistics which changes continuously from a Bose-Einstein to a Poisson distribution when threshold is passed.

The *Fokker-Planck methods* are suitable for calculating linewidth formulas, even at threshold, for the photon distribution, and for the determination of correlation functions.

Survey IV

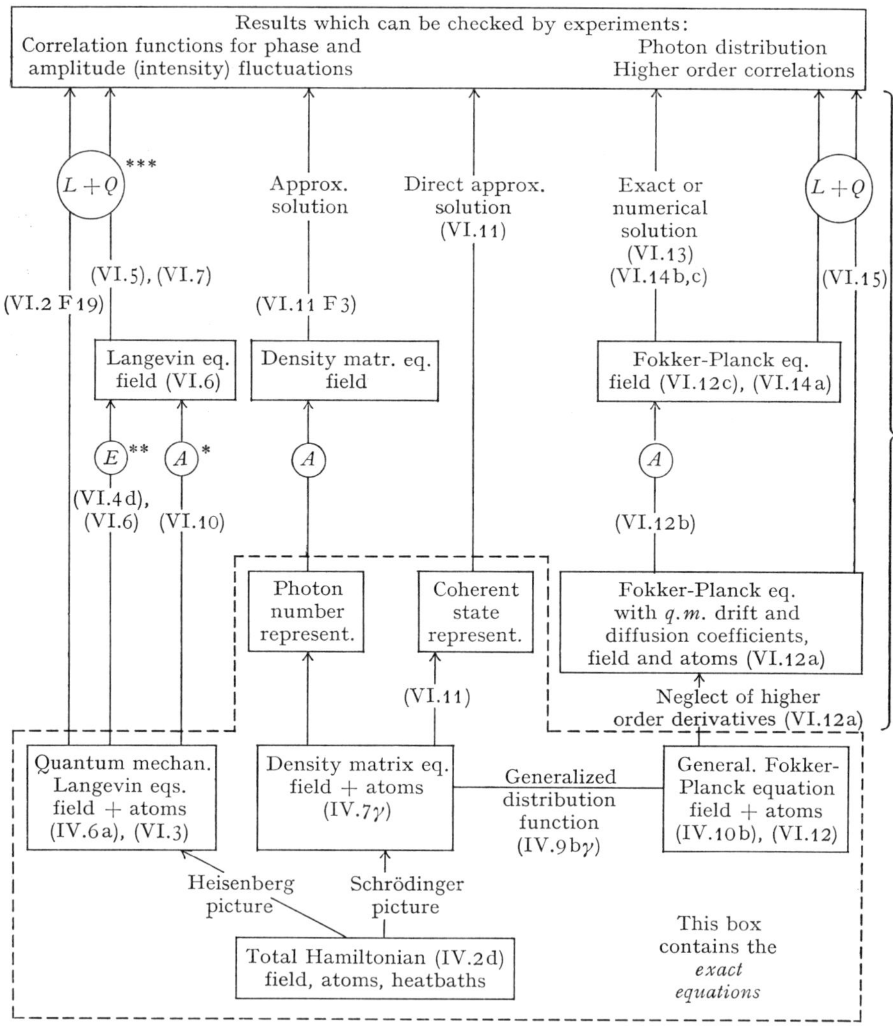

* A: adiabatic elimination of atomic variables.
** E: exact elimination of atomic variables.
*** $L+Q$: linearization and quantum mechanical quasi-linearization.

Survey IV. Family tree of the quantum theory of the laser. (A similar representation was given by H. HAKEN and W. WEIDLICH in lectures at the Varenna Summer School 1967.) The numbers in the above scheme refer to the chapters of the present article where the corresponding calculations or equations are given. Those numbers which contain an F refer to footnotes where reference is given to the corresponding original paper which is not represented in detail in the present article.

In talking about linewidths and photon numbers we make use of the dualism between the wave and particle picture which stems from the quantum aspects of light. Clearly, at high photon numbers these quantum numbers may be replaced by the light intensity.

VI.2. Summary of theoretical results and comparison with the experiments. The measurable quantities of laser light which will be calculated in this Chap. VI are:

1. the spectroscopic linewidth,
2. the intensity correlation function,
3. the photon statistics.

These quantities will be determined for solid-state, gas, and semiconductor lasers.

With respect to the mathematical treatment, several overlapping regions must be distinguished:

<table>
<tr><th></th><th>Photons in the lasing mode</th><th>Notation of region with respect to threshold</th><th>Validity of adiabatic elimination of atomic variables</th><th>Possible mathematical methods</th></tr>
<tr><td rowspan="15">Increasing laser output ↓</td><td>10</td><td>well below</td><td>no</td><td rowspan="2">↑
linearization
↓</td></tr>
<tr><td>10^2</td><td rowspan="2">somewhat below</td><td rowspan="7">yes
↑
van der Pol-equation valid
↓</td></tr>
<tr><td>10^3</td><td rowspan="2">fully nonlinear treatment</td></tr>
<tr><td>10^4</td><td>at threshold</td></tr>
<tr><td>10^5</td><td rowspan="9">↑
well above
↓</td><td rowspan="11">↑
quasilinear
↓</td></tr>
<tr><td>10^6</td></tr>
<tr><td>10^7</td></tr>
<tr><td>10^8</td></tr>
<tr><td>10^9</td><td rowspan="7">no</td></tr>
<tr><td>10^{10}</td></tr>
<tr><td>10^{11}</td></tr>
<tr><td>10^{12}</td></tr>
<tr><td>10^{13}</td></tr>
<tr><td>10^{14}</td><td rowspan="2">high above
↓</td></tr>
<tr><td>10^{15}
⋮</td></tr>
</table>

The photon numbers only serve as a first indication. The exact limitations of the different approximations may be found in the following chapters.

The photon numbers refer to the example of a gas laser. The scheme clearly shows that a fully nonlinear treatment (e.g. by the Fokker-Planck equation) is only necessary for the rather narrow threshold region; we therefore quote the results of the other two regions (linear and quasilinear) first and then show how they are continuously connected with each other.

a) Qualitative discussion of the characteristic features of the laser output: homogeneously broadened line. Well below threshold the spontaneous emission predominates. Far below threshold we expect the usual fluorescent spectrum: a Lorentzian line of half width γ.

With increasing pump-power, stimulated emission supports the cavity modes which increase in intensity but decrease in linewidth, $\Delta\omega \sim \frac{1}{P}$, where P is the output power. *Below* threshold all these modes show up (the situation may be completely different above threshold, see e. g. p. 143). The mode envelope is still a Lorentzian, as is the line shape of each mode. (Compare Fig. 7 on p. 6.) The light output is a random superposition of the output of the single mode. The output of each mode again consists of many *statistically independent* wavetracks originating from single spontaneous emission processes but enhanced by stimulated emission. Thus the ensemble average of the intensity correlation function (compare Chap. V.1).

$$\langle E^{(-)}(t)\, E^{(-)}(t')\, E^{(+)}(t')\, E^{(+)}(t)\rangle \tag{VI.2.1}$$

may be reduced to the product of amplitude correlation functions

$$|\langle E^{(-)}(t)\, E^{(+)}(t')\rangle|^2. \tag{VI.2.2}$$

Both functions may be directly measured (see VI.1). $E^{(+)}(t)$ and $E^{(-)}(t)$ respectively are the positive and negative frequency parts of the complex amplitude of the mode under consideration. With increasing pump power (increasing atomic inversion) the light of the mode with the highest gain (and thus the highest output) undergoes a phase transition (in the present context "phase" is used in the thermodynamic sense). This phase transition at threshold will be discussed in detail below.

It is possible that several modes exist above threshold. The conditions are discussed in VI.9, VII.3—VII.7 and VIII.2. With respect to the statistical properties we confine our following discussion to *a single mode*.

b) Quantitative results: single mode action.

α) *The spectroscopic linewidth well above threshold.* This linewidth $\Delta\omega$ (half width at half power) is defined by

$$\langle E^{(-)}(\tau)\, E^{(+)}(0)\rangle = \langle E^{(-)}(0)\, E^{(+)}(0)\rangle\, e^{-\Delta\omega\tau} \tag{VI.2.3}$$

(provided the line has a Lorentzian shape). Experimentally, it is measured by an interferometer.

Well above threshold (i.e. for photon numbers which are bigger than about 10 times the photon number at threshold) it has for all kinds of lasers the following form:

$$\Delta\omega = \frac{\hbar\omega}{P}\, \varkappa^2 f(\varkappa, \gamma)\, (n_{sp} + n_{th})\, (1 + \Delta^2) \tag{VI.2.4}$$

ω is the mode frequency, P the output power $(= 2\varkappa\langle n\rangle\hbar\omega)$. $\langle n\rangle$ average number of photons in the mode, $\varkappa$ the cavity half width. f is a correction factor which is of the order of unity, n_{sp} is the number of spontaneously emitted quanta, n_{th} the number of thermal quanta, Δ depends on the detuning and vanishes for exact resonance. γ is the homogeneous half-width of the atomic optical transition, which includes contributions from spontaneous emission into all modes, the pumping, and other linebroadening mechanisms.

The explicit expressions for f, n_{sp} and Δ depend on the different laser systems, as is shown in the following scheme:

Kind of laser	$f(\varkappa, \gamma)$	n_{sp}	Δ
Homogeneous solid state *[1,2]	$\frac{\gamma^2}{(\varkappa+\gamma)^2}$	$\frac{N_{2,s}}{(N_2-N_1)_{thr}}$	$\left(\frac{\nu_0-\omega}{\varkappa+\gamma}\right)$
Inhomogeneous solid state **[3]	1	$\frac{1}{2}\,\frac{\overline{(N_2+N_1)_s}}{(N_2-N_1)_{thr}}+\frac{1}{2}\approx\frac{\overline{N}_{2,s}}{(N_2-N_1)_{thr}}$	$\frac{2\gamma}{\alpha}\;\frac{(\nu_0-\omega)}{\alpha}$
Gas **[4]	1	(as above)	$\frac{\gamma(\nu_0-\omega)}{(2\gamma^2+(\omega-\nu_0)^2)}$
Semiconductor **[5] (inhomogeneous)	1	$\frac{1}{\varkappa}\sum_{k k'}\frac{\lvert g_{k k'}\rvert^2 f_{k c}(1-f_{k' v})\gamma_{k k'}}{(\Omega-\nu_{k k'})^2+\gamma_{k k'}^2}$	$\left.\frac{\frac{d}{dn}\operatorname{Im} S(n)}{\frac{d}{dn}\operatorname{Re} S(n)}\right\rvert_{n=n_0}$

* Valid up to arbitrary photon numbers.
** The approximations made imply $\varkappa \ll \gamma$, and the validity of the adiabatic approximation.

The levels 2 and 1 may be arbitrarily situated in a multi-level atom. N_i is the occupation number of the corresponding level i, s stands for saturated, thr for threshold. $\overline{N_2+N_1}$ is an average over the full line, for its exact definition see (VI.10.16). $g_{kk'}$ are coupling constants (of the semiconductor laser) proportional to optical matrix elements, which connect the states k, k' in the conduction band C and the valence band, V. f_{kc} and f_{kv} are the saturated quasi-Fermi distribution functions. ν_0 is the atomic frequency at line center, α the half width of the inhomogeneous line. S is the difference between the saturated and the unsaturated complex gain function.

β) *The spectroscopic linewidth somewhat below threshold* differs from that well above threshold only by the factor 2 (for vanishing detuning)[6,7].

$$\Delta\omega_{below}\,P_{below} = 2\Delta\omega_{above}\,P_{above}. \tag{VI.2.5}$$

[1] H. Haken: Z. Physik **182**, 346 (1965) (two-level atoms, running wave mode, n_{th} neglected). — H. Sauermann: Z. Physik **189**, 312 (1966) (two-level atoms, $\Delta=0$, running wave mode). — H. Haken: Z. Physik **190**, 327 (1966) (three-level atoms, lower or upper transition showing laser action, standing or running wave mode, $\Delta=0$). — H. Risken, Ch. Schmid, and W. Weidlich: Z. Physik **194**, 337 (1966); — Phys. Letters **20**, 489 (1966) (two-level atoms, $\Delta\neq 0$ admitted, running wave mode).

[2] Results for the maser (i.e. infinite wavelength) are derived by M. Lax, in: Physics of Quantum Electronics (eds. P. L. Kelley, B. Lax, and P. E. Tannenwald). New York: McGraw-Hill Book Co. 1966 ($\Delta\neq 0$ is included). — M. Scully and W. E. Lamb: Phys. Rev. Letters **16**, 853 (1966); — Phys. Rev. **166**, 246 (1968); — In the references [1] to [5] we include only those papers, which give a fully quantum theoretical treatment.

[3] V. Arzt, H. Haken, H. Risken, H. Sauermann, Ch. Schmid, and W. Weidlich: Z. Physik **197**, 207 (1966).

[4] l.c. [3] above.

[5] H. Haug and H. Haken: Z. Physik **204**, 262 (1967).

[6] This factor 2 was noted by P. Grivet and A. Blaquiere: Proc. on the Symposium on Opt. Masers (ed. Jerome Fox), New York: Polytechnic Press 1963, in the framework of the classical theory of the van der Pol-oscillator driven by noise.

[7] Quantum mechanical treatments of the (linear) region below threshold were given by D. E. McCumber: Phys. Rev. **130**, 675 (1962). — W. H. Wells: Ann. Phys. (N.Y.) **12**, 1 (1961). — G. Kemeny: Phys. Rev. **133**, A 96, 1—5 (1964), and by the authors of references [1–5].

The linewidth changes smoothly from below threshold $(\Delta\omega)_{below}$ to above threshold $(\Delta\omega)_{above}$

$$\Delta\omega(a)\,P(a) = \Delta\omega_{above}\,P_{above}\cdot\alpha(a). \tag{VI.2.6}$$

α varies from 2 somewhat below threshold to 1 well above threshold[8]. Its dependence on the pump parameter a is given in Fig. 19 [see also Chap. VI.13, Eq. (VI.13.28)].

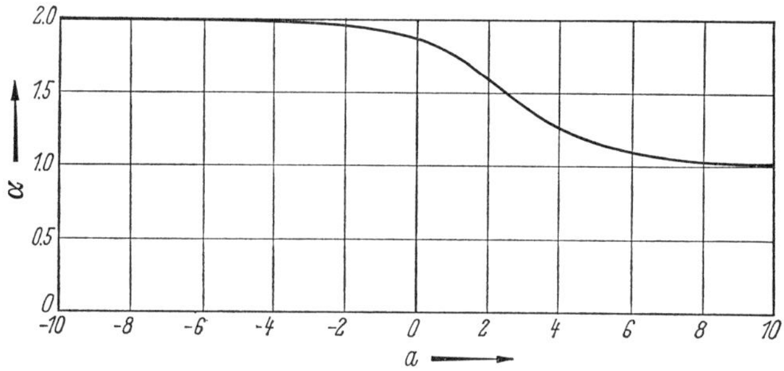

Fig. 19. The linewidth factor $\alpha(a) = \lambda_{10}$ as a function of the pump parameter a. (After H. RISKEN, l.c. [8] on p. 104).

The pump parameter a is defined by [see also (VI.13.1)]

$$a = \sqrt{\beta/Q}\,\tilde{d} \tag{VI.2.7}$$

where [compare also (VI.12.24) to (VI.12.26), (VI.12.33), assuming $\varkappa \ll \gamma_\perp$]

$$\beta = \frac{4\varkappa^2 T}{D_{thr}}; \quad Q = \frac{1}{2}(n_{sp}+n_{th})\,\varkappa; \quad \tilde{d} = \frac{1}{4\varkappa T}(D_0 - D_{thr}). \tag{VI.2.8}$$

$D_{thr} = N(N_2 - N_1)_{thr}$ is the total inversion at threshold, whereas $D_0 = N(N_2 - N_1)_0$ is the total unsaturated inversion, i.e. the inversion which would be present without the coupling of the atoms to the lasing mode. N is the total number of laser atoms, T is the relaxation time after which the population inversion comes to its equilibrium value under the action of pump and incoherent decay processes.

The observed linewidth is at least 10 times bigger than the theoretically predicted one, which is presumably due to mirror fluctuations etc[9].

γ) *The intensity (or amplitude) fluctuations* are described by the correlation function

$$K(\tau) = \langle b^+(t)\,b^+(t+\tau)\,b(t+\tau)\,b(t)\rangle - \langle b^+ b\rangle\langle b^+ b\rangle \tag{VI.2.9*}$$

where b^+, b are photon creation and annihilation operators of the mode under consideration. For a classical interpretation we note that $b^+ \sim E^{(-)}(t)$, $b \sim E^{(+)}(t)$; the proportionality factors are time independent c-numbers.

---

* This correlation function is the quantum mechanical analogue of the classical correlation function for intensities as defined in Eq. (V.1.24). For the quantum mechanical formulation the field amplitude V is substituted by the quantized vector potential which then occurs in the correlation function in the general form (V.4.8). The final form (VI.2.9) is obtained by expanding the quantized vector potential into cavity modes [see (III.1.7) and (III.1.8)] and retaining only a single mode. The spatial dependence is eliminated by integration over a certain volume.

[8] This smooth transition was calculated by H. RISKEN: Z. Physik **191**, 302 (1966). — See also R. D. HEMPSTEAD and M. LAX: Phys. Rev. **161**, 350 (1967).

[9] See e.g. A. E. SIEGMAN and R. ARRATHOON: Phys. Rev. Letters **20**, 901 (1968).

γ1) Below threshold $K(\tau)$ may be reduced to[10]

$$\left.\begin{aligned} K(\tau) &= |\langle b^+(t+\tau)\, b(t)\rangle|^2 \sim |\langle E^{(-)}(t+\tau)\, E^{(+)}(t)\rangle|^2 \\ &= \langle E^{(-)}(t)\, E^{(+)}(t)\rangle^2\, e^{-2\Delta\omega_a \tau}\,. \end{aligned}\right\} \quad \text{(VI.2.10)}$$

$K(0)$ thus increases with the square of the intensity $\langle E^{(-)}E^{(+)}\rangle$, whereas $\Delta\omega_a$ decreases with the intensity [see (VI.5.20)].

γ2) In the "adiabatic" region but well above threshold the following expression is found for homogeneous solid state and inhomogeneous solid state, gas, and semiconductor lasers (for vanishing detuning)[11]

$$K(\tau) = (n_{sp} + n_{th})\,\frac{\varkappa}{\beta}\, e^{-2\beta\langle n\rangle\tau} \quad \text{(VI.2.11)}$$

which has the form

$$K(\tau) = K(0)\, e^{-\tau/\tau_0}. \quad \text{(VI.2.12)}$$

The quantities n_{sp}, n_{th}, $\varkappa$ are explained on p. 102, β is defined in (VI.2.8).

The relaxation time

$$\tau_0 = \frac{1}{2\beta\langle n\rangle} \quad \text{(VI.2.13)}$$

thus decreases with increasing mean photon number $\langle n\rangle$.

The result that $K(0)$ does not change with the photon number follows from the fact[12] that well above threshold the laser amplitude has the form $r_0 + \varrho(t)$. r_0 is a stable amplitude, whereas $\varrho(t)$ describes small residual fluctuations, which decrease with increasing photon numbers, $\langle n\rangle$, i.e. the average $\langle\varrho^2(t)\rangle$ varies with $1/\langle n\rangle$ and $\langle\varrho(t)\rangle = 0$.

γ3) In the close vicinity of threshold several relaxation times occur, so that $K(\tau)$ reads[13]

$$K(\tau) = K(0) \sum_{m=1}^{\infty} M_m\, e^{-\lambda_{0m}\tau}. \quad \text{(VI.2.14)}$$

Numerical values for the constants M's and λ's as functions of the pump parameter a (VI.2.7), are given on p. 160. The λ's are real throughout. The effect of the different relaxation times may well be taken care of by a single, effective relaxation time[13] so that (VI.2.14) has again the form (VI.2.12) where

$$\tau_{0\,eff} = \frac{1}{\sqrt{\beta Q}} \sum_{m=1}^{\infty} \frac{M_m}{\lambda_{0m}}. \quad \text{(VI.2.14a)}$$

A comparison between (VI.2.10) and (VI.2.11) shows that even qualitatively laser light behaves in completely different ways below and above threshold. In their pioneering work, ARMSTRONG and SMITH[14] could fully substantiate this

[10] L. MANDEL, and E. WOLF: Phys. Rev. **124**, 1696 (1961).

[11] H. HAKEN: Z. Physik **181**, 96 (1964), and the papers quoted under 1, 3, 4, and 5.

[12] H. HAKEN: Phys. Rev. Letters **13**, 329 (1964), and the references given in [11].

[13] H. RISKEN and H. D. VOLLMER: Z. Physik **201**, 323 (1967).

[14] J. A. ARMSTRONG and A. W. SMITH: Phys. Rev. Letters **14**, 68 (1965); — Phys. Rev. **140**, A 155 (1965). — C. FREED and H. A. HAUS: Appl. Phys. Letters **6**, 85 (1965); — Phys. Rev. **141**, 287 (1966). — F. T. ARECCHI, A. BERNE, and P. BURLAMACCHI: Phys. Rev. Letters **16**, 32 (1966); — For reviews of the experimental work see J. A. ARMSTRONG and A. W. SMITH, in: Progr. Optics (ed. E. WOLF). Amsterdam: North Holland Publ. Co. 1967, as well as the articles of F. T. ARECCHI, H. HAUS and E. R. PIKE in the Proceedings of the Varenna Summer School 1967, editor R. J. GLAUBER, to be published.

change predicted by HAKEN[15], both as regards the change in $K(0)$ and the change in τ_0. The narrow threshold region however, was not covered by these experiments, nor by the expressions (VI.2.10), (VI.2.11). The behaviour of $\lambda_{eff} = 1/\tau_{0\,eff}$ in this region is exhibited in Fig. 20 which shows an experimental check of RISKEN'S[16] theory by ARECCHI et al.[17]

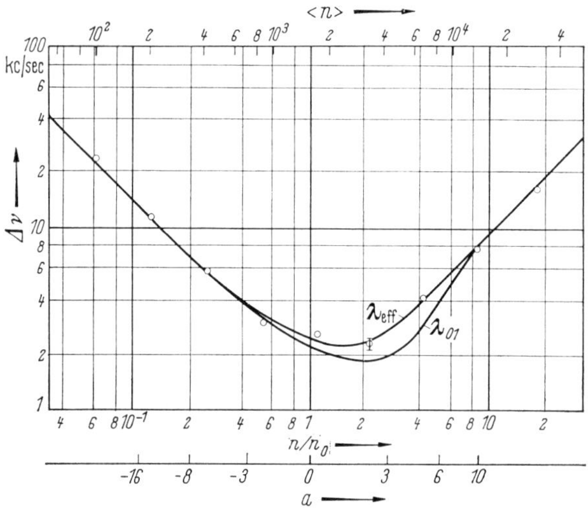

Fig. 20. Measured linewidth of the intensity fluctuations near threshold as a function of the normalized photon number, real photon number $\langle n\rangle$ and the pump parameter a, compared to the theoretical linewidth λ_{eff} and the width λ_{01} (ARECCHI, l. c. [17] on p. 106) n/n_0 is the ratio between actual mean photon number and mean photon number at laser threshold. The λ's are explained in detail in Chap. VI.13.

γ4) High above threshold both sidebands and several relaxation times occur: For a *homogeneously broadened* line $K(\tau)$ has the form[18]

$$K(\tau) = \sum_{m=1}^{m_0} k_m \, e^{(i\omega_m - \varkappa_m)\tau}. \tag{VI.2.15}$$

ω_m and $\varkappa_m$ are real constants, the k_m's are complex constants. m_0 is given by the number of atomic levels, which are connected by pump and relaxation processes, plus one. The Fourier transform of $K(\tau)$ (VI.2.15) shows resonances at $\omega = \omega_m + i\varkappa_m$. Some of these are connected with undamped spiking; others give rise to "holeburning" of a homogeneously broadened line. For a discussion see VI.7, VI.8.

For *homogeneously and inhomogeneously broadened* atomic lines numerical examples[19] are represented in Figs. 20—23 in which the spectral density $S_I(\omega)$

[15] H. HAKEN, Z. Physik **181**, 96 (1964).
[16] l.c. [13] on p. 105.
[17] F. T. ARECCHI: Proceedings of the Varenna Summer School 1967, ed. R. J. GLAUBER (to be published), and Phys. Letters **25 A**, 59 (1967).
[18] H. HAKEN: Z. Physik **190**, 327 (1966) [three-level atoms, (IV.2.15) holds quite generally, however]. — H. RISKEN, C. SCHMID, and W. WEIDLICH: l.c. [1] on p. 103.
[19] V. ARZT, C. SCHMID, and H. HAKEN: Unpublished.

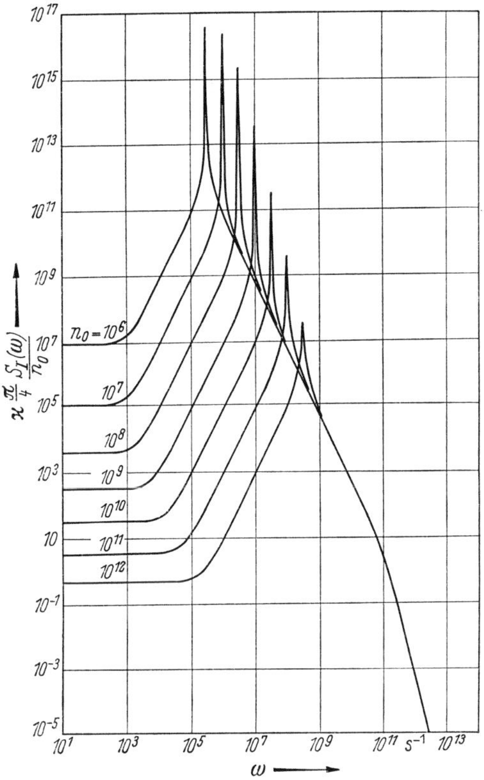

Fig. 21. Spectral density of intensity fluctuations in a solid state laser (homogeneously broadened). Two-level system. The curves are obtained by a computer solution of the quantum mechanical Langevin equations (after V. ARZT: Diplom-Thesis. Stuttgart 1967 and ref. [19]). The parameters are:

cavity half-width	$\varkappa = 10^9$ sec^{-1}
atomic half-width	$\gamma = 10^{11}$ sec^{-1}
inverse relaxation time of population inversion	$1/T_p = 10^3$ sec^{-1}
number of atoms	$N = 10^{19}$
coupling constant	$g^2 = 10^6$ sec^{-2}
threshold inversion	$d_{thr} = 10^{-5}$

n_0: mean number of photons. The line shows a dip in the center, which has been called holeburning of a homogeneously broadened line [HAKEN, Z. Physik **181**, 96 (1964)]. The resonances lie at $\omega_r = 2g \cdot \sqrt{2 n_0 \varkappa/\gamma}$, provided $1/T_p \ll \varkappa \ll \gamma$ and $4 g^2 n_0 \ll \gamma^2$. Here, these conditions are well fulfilled.

is plotted. $S_I(\omega)$ is the Fourier transform of $K(\tau)$:

$$S_I(\omega) = \frac{1}{\pi} \int_{-\infty}^{+\infty} e^{-i\omega\tau} K(\tau) \, d\tau .$$

δ) *Photon statistics.* Because the photon numbers are of the order of several thousand even at threshold, the photon distribution within a mode may be treated as a continuous function. In the adiabatic region the probability distri-

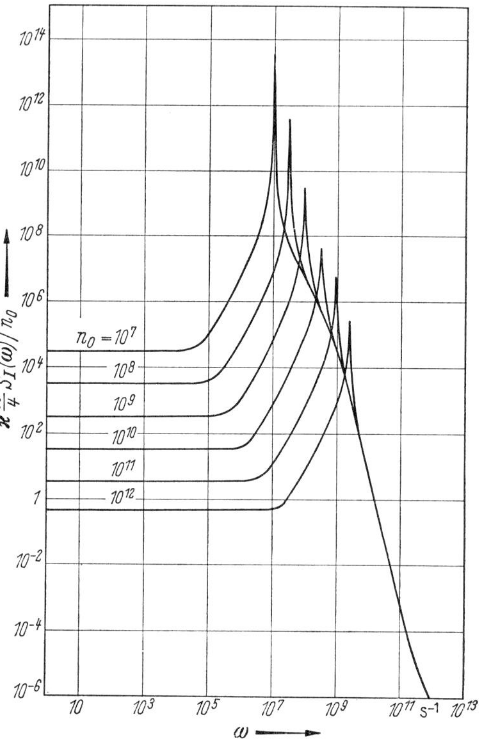

Fig. 22. Spectral density of the intensity fluctuations of a solid state laser (homogeneously broadened line). Computer solution, after V. ARZT, Diplom Thesis, Stuttgart 1967 and ref. [19]. The parameters are

$$\varkappa = 10^7\ \text{sec}^{-1}, \quad \gamma = 10^7\ \text{sec}^{-1}, \quad 1/T_p = 10^3\ \text{sec}^{-1}, \quad d_{thr} = 10^{-9}, \quad g^2 = 10^6\ \text{sec}^{-2}.$$

The resonances lie at $\omega_r = 2g\sqrt{n_0}$ [compare Eg. (VI.8.18)], provided $4g^2 n_0 \gg \varkappa^2, \gamma^2, (1/T_p)^2$.

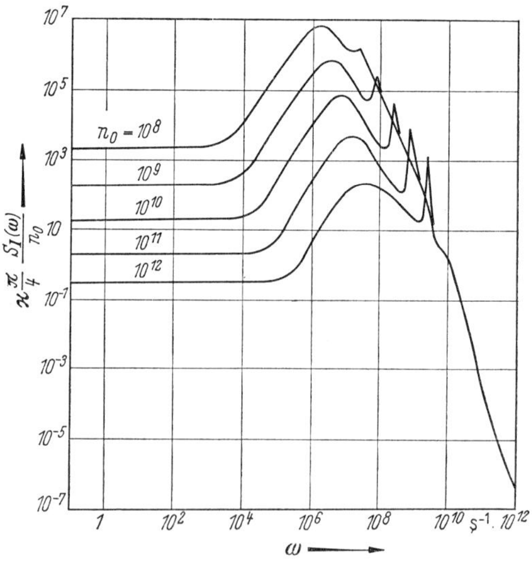

Fig. 23.

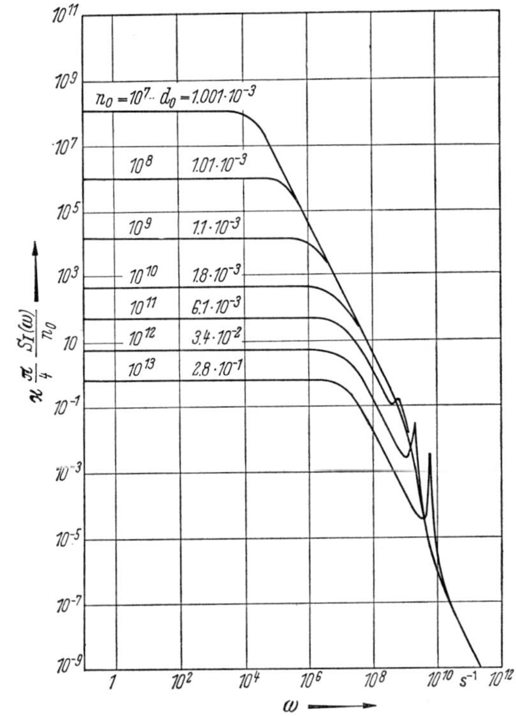

Fig. 24. Spectral density of intensity fluctuations in a gas laser (inhomogeneously broadened). Two-level system. (Computer solution, after ARZT: Diplom-Thesis, 1967 and ref. [19]).

$\varkappa = 10^7\ \text{sec}^{-1}$, $\gamma = 10^8\ \text{sec}^{-1}$, $1/T_p = 2 \cdot 10^8\ \text{sec}^{-1}$,

Doppler-halfwidth $\alpha = 10^9\ \text{sec}^{-1}$, $N = 10^{13}$, $g^2 = 10^6\ \text{sec}^{-2}$, $d_{thr} = 10^{-3}$, n_0: photon number.

At low photon numbers and moderate frequencies the curves exhibit a Lorentzian form, which corresponds to the van der Pol approximation. It is remarkable, that the plot of a *homogeneously* broadened laser for the same physical situation shows nearly no difference to the one for the inhomogeneously broadened line presented here.

bution of the normalized photon number* $\hat{n}$ is given by[20]

$$W(\hat{n}) = N\, e^{-\frac{\hat{n}^2}{4} + a\frac{\hat{n}}{2}} \tag{VI.2.16}$$

* $\hat{n}$ is considered here as a continuous variable so that we should call it "intensity" rather than "photon number". For more details see VI.13.

[20] H. RISKEN: Z. Physik **186**, 85 (1965).

Fig. 23. Spectral density of the intensity fluctuations in a solid state laser, inhomogeneously broadened line. Computer solution. After ARZT, Diplom-Thesis, Stuttgart 1967 and ref. [19]. The shown parameters may be those of ruby at 10 °K.

$\varkappa = 10^7\ \text{sec}^{-1}$, $\gamma = 10^7\ \text{sec}^{-1}$, $1/T_p = 10^3\ \text{sec}^{-1}$, $g^2 = 10^6\ \text{sec}^{-2}$,

inhomogeneous width: $10^9\ \text{sec}^{-1}$, $d_{thr} = 10^{-7}$.

There are now two resonances. At high photon numbers the dominant one lies at $\omega_r = 2g\, r_0$. If the line were homogeneously broadened, a completely different behaviour would result, see Fig. 22.

or, written in a different form

$$W(\hat{n}) = N\, e^{\frac{a^2}{4}}\, e^{-\frac{1}{4}(\hat{n}-a)^2} \tag{VI.2.17}$$

a is the pump parameter, the normalization constant is given by

$$\frac{1}{N} = \int_0^\infty e^{-\frac{\hat{n}^2}{4}+a\frac{\hat{n}}{2}}\, d\hat{n}. \tag{VI.2.18}$$

The derivation of (VI.2.16) is presented in (VI.13). The first experimental check of RISKEN's distribution (VI.2.16) performed by SMITH and ARMSTRONG[21] is shown in Fig. 25.

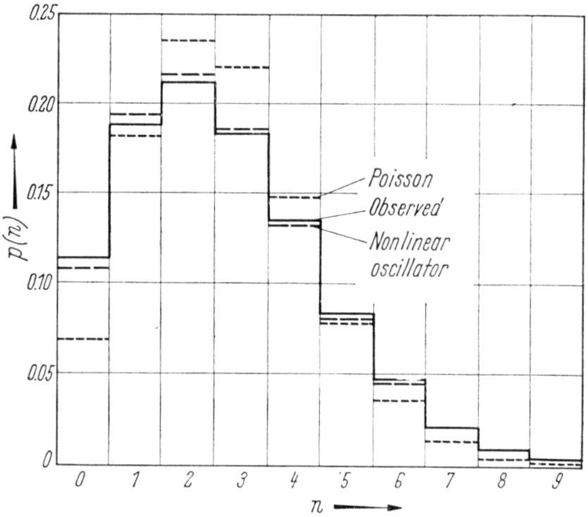

Fig. 25. Counting distribution just above threshold (solid line), the Poisson distribution for the same n (dotted lines) and the nonlinear oscillator distribution give the best fit (dashed line). For $n=7$, 8 and 9 the nonlinear oscillator and observed distributions are coincident. (Reproduced from SMITH and ARMSTRONG, l.c. [21] on p. 110.)

The experiments to check (VI.2.10) and (VI.2.16) are based on photon counting techniques. This means that the functions (VI.2.16) or their moments of are not directly measured, but must be corrected by the photon-counting statistics. If the intervals of measurement T are small compared to the relaxation times of the laser light, the distribution of the *discrete* photon numbers n found within T is connected with the function $W(\hat{n})$ of the *continuous* variable $\hat{n}$ by[22]

$$p(n, T) = \int_0^\infty \frac{(\nu\,\hat{n})^n}{n!}\, e^{-\nu\hat{n}}\, W(\hat{n})\, d\hat{n} \tag{VI.2.19}$$

where ν is proportional to T and contains the detector sensitivity etc. [compare (V.4.13)] In the adiabatic region $W(\hat{n})$ is given by (VI.2.16). A similar correction must be applied to the correlation function (VI.2.9).

The distribution of the average photon number for different pump parameters a is plotted in Fig. 26. The statistics changes at threshold from Einstein statistics (below threshold) to a Poisson distribution (well above threshold) which holds for classical particles. This smooth transition is best seen by considering the mean

[21] A. W. SMITH and J. A. ARMSTRONG: Phys. Rev. Letters **16**, 1169 (1966).

[22] L. MANDEL: Proc. Phys. Soc. (London) **72**, 1037 (1958), and in Progr. Optics (ed. E. WOLF). Amsterdam: North Holland Publ. Co. 1963.

square deviation $\langle(\Delta n)^2\rangle \equiv \langle n^2\rangle - \langle n\rangle^2$ which may be expressed in the form*

$$\langle n^2\rangle - \langle n\rangle^2 = \langle n\rangle (1 + H_2 \langle n\rangle) \tag{VI.2.20}$$

Resolving Eq. (VI.2.20) with respect to H_2, and using the abbreviation $M_2 = \langle n^2\rangle$, $M_1 = \langle n\rangle$ we obtain

$$H_2 = \frac{M_2 - M_1}{M_1^2} - 1 . \tag{VI.2.21}$$

Thus H_2 is expressed by directly measurable quantities. Fig. 27 shows a comparison between RISKEN's theory and an experimental check by ARECCHI[23]. The pump parameter is replaced by a normalized output, M_1/M_{10}.

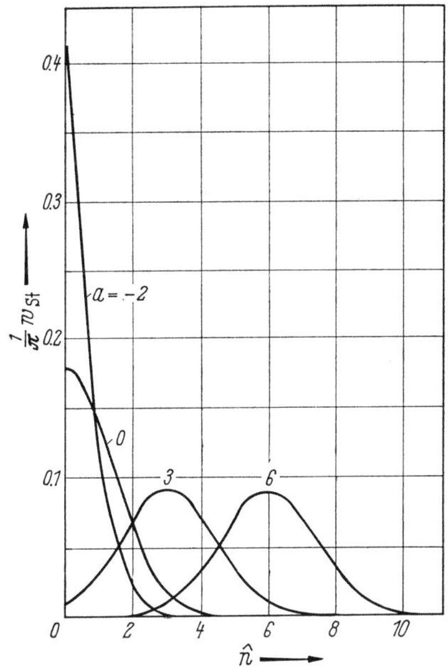

Fig. 26. The stationary distribution (VI.2.16) as a function of the normalized "intensity" $\hat{n}$. (After RISKEN.)

Thus both theoretically and experimentally the following change is found: below threshold (small pumping, $a \ll 1$) one obtains $H_2 = 1$ and thus

$$\langle(\Delta n)^2\rangle = \langle n\rangle (\langle n\rangle + 1) \tag{VI.2.22}$$

and is thus a law typical of Bose-Einstein statistics. Above threshold for large pumping ($a \gg 1$), H_2 tends to zero, so that

$$\langle(\Delta n)^2\rangle = \langle n\rangle \tag{VI.2.23}$$

corresponding to a Poisson distribution.

In Fig. 27 we exhibit a comparison between the experimental and theoretical points found for H_2. A similarly good agreement is found for H_3 and H_4 where

* In RISKEN's treatment H_2 is called $\delta_1(a)$ and treated as a function of the pump parameter a. The pump parameter can be further expressed by the intensity of the laser output (i.e. the first moment, $M_1 = \langle n\rangle$).

[23] F. T. ARECCHI, G. S. RODARI and A. SONA: Phys. Letters **25** A, 59 (1967).

moments of $W(n)$ up to the fourth order are invol ed (R. F. CHANG, R. W. DETENBECK, V. KORENMAN, C. O. ALLEY jr., U. HO ·HULI and others[24]).

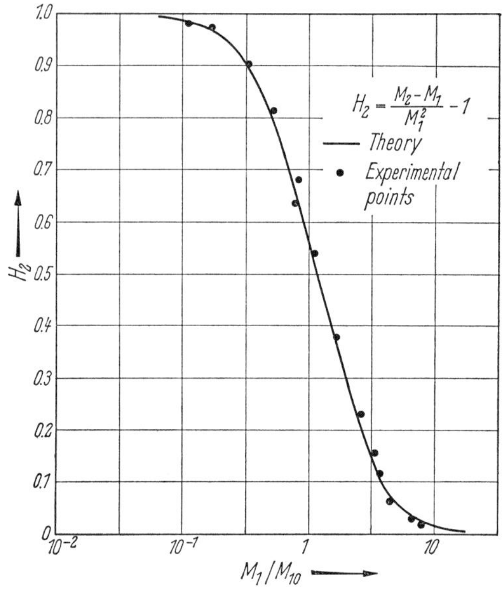

Fig. 27. Theoretical and experimental results for H_2 (for the explanation see text). The theoretical curve is taken from RISKEN's theory. The experimental points are those found by ARECCHI et al. (after ARECCHI et al.: Phys. Letters **25** A, 59 [1967]).

VI.3. The quantum mechanical Langevin equations for the solid state laser. The results of Chap. IV, in which the laser equations have been derived, can be summarized as follows:

We decompose the vector potential, $\boldsymbol{A}$, of the lightfield into a set of cavity modes $\boldsymbol{u}_\lambda(\boldsymbol{x})$:

$$\boldsymbol{A}(\boldsymbol{x}) = \sum_\lambda \boldsymbol{u}_\lambda(\boldsymbol{x})\,(b_\lambda^+ + b_\lambda)\sqrt{\frac{2\pi c^2 \hbar}{\omega_\lambda}}\,. \tag{VI.3.1}$$

The spatial dependence of the modes is assumed to be known (comp. II.1 to II.5). When standing waves are in the resonator, the $\boldsymbol{u}_\lambda$'s can be represented in a good approximation by

$$\boldsymbol{u}_\lambda(\boldsymbol{x}) = \boldsymbol{u}_{0,\lambda} \sin(k_{\lambda,x}\,x)\sin(k_{\lambda,y}\,y)\sin(k_{\lambda,z}\,z)\,. \tag{VI.3.2}$$

For purely axial modes (the nodes being in the x direction), we have instead

$$\boldsymbol{u}_\lambda(\boldsymbol{x}) = \boldsymbol{u}_{0,\lambda} \sin k_\lambda x\,. \tag{VI.3.3}$$

In a classical description of the electromagnetic field the coefficients b_λ^+, b_λ are functions of time, whose time dependence is essentially given by $e^{i\omega_\lambda t}$ and $e^{-i\omega_\lambda t}$ respectively. Quantum mechanically, they are creation and annihilation operators, respectively.

---

[24] R. F. CHANG, R. W. DETENBFCK, V. KORENMAN, C. O. ALLEY jr., and U. HOCHULI: Phys. Letters **25** A, 272 (1967). R. F. CHANG, V. KORENMAN, C. O. ALLEY, and R. DETENBECK: Phys. Rev. **178**, 612 (1969). — F. T. ARECCHI, M. GIGLIO, and A. SONA: Phys. Letters **25** A, 341 (1967). — F. DAVIDSON and L. MANDEL: Phys. Letters **25** A, 700 (1967).

They obey the commutation relations:

$$b_\lambda b_{\lambda'}^+ - b_{\lambda'}^+ b_\lambda = \delta_{\lambda\lambda'}, \quad b_\lambda b_{\lambda'} - b_{\lambda'} b_\lambda = 0, \quad b_\lambda^+ b_{\lambda'}^+ - b_{\lambda'}^+ b_\lambda^+ = 0. \tag{VI.3.4}$$

Although the sum (VI.3.1) runs over a complete set of modes, in general only a few modes will be strongly supported during laser action. In our analysis we therefore divide the modes into the "lasing" ones (which can also be those near laser threshold) and the "nonlasing" ones. Such a division can be done in a self-consistent way. In our analysis we keep explicitly only the "lasing" modes. The others act merely as noise sources in a manner which is explained in detail in IV.3.—IV.5. In the following, the index λ refers only to the lasing modes.

We now write down the equation for the laser modes and the atoms which support laser action for the following situation: The atoms may be placed either

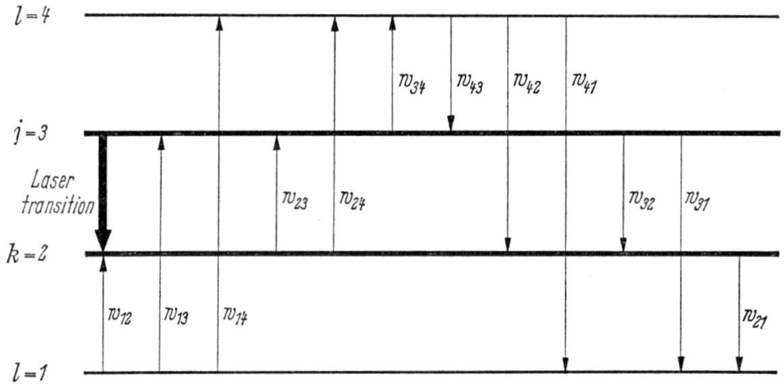

Fig. 28. Example of a four-level system, showing all transition rates.

at random or periodic lattice sites. We admit that they have a homogeneously or an inhomogeneously broadened line, but that they are otherwise identical. Each may possess L levels. We assume that laser action occurs only between two definite levels, j and k. The other levels serve for the pump and loss. An example is represented in Fig. 28. The initial level j can also be the uppermost level, as k can be the ground level. We use the (excellent) rotating wave approximation (compare III.4). The most general equations are presented in IV.9 and IV.10. The basic equations for the present case can be divided into two groups:

a) the field Eq. (VI.3.5) and

b) the matter Eqs. (VI.3.13) to (VI.3.19).

a) Field equations

$$\dot{b}_\lambda^+ = (i\omega_\lambda - \varkappa_\lambda)\, b_\lambda^+ + i \sum_\mu g_{\mu\lambda}^* (a_j^+ a_k)_\mu + F_\lambda^+ \tag{VI.3.5*}$$

ω_λ is the frequency in the "unloaded" cavity, i.e. without the laser atoms (a passive material which can change the velocity of light is, however, admitted). $\frac{1}{2\varkappa_\lambda}$ is equal to the lifetime of the mode. Thus the first term would give rise to a damped oscillation. The sum describes the coupling to the atoms at random lattice sites, $\boldsymbol{x}_\mu$. We assume a laser transition from the atomic state, j, to another, k. In a classical description, $(a_j^+ a_k)_\mu$ is proportional to the dipole moment

* The dot · indicates the differentiation with respect to the time t.

of the atom μ, and describes just its time dependence. Thus this term represents the generation of the field by oscillating dipoles, where the oscillation frequency is given as usual by the transition frequency between levels j and k. In a quantum mechanical description $a_{j,\mu}^{+}$ and $a_{k,\mu}$ are creation and annihilation operators for an electron of atom μ in the respective states.

Their commutation relations are

$$\left.\begin{aligned} a_{l,\mu}^{+} a_{l',\mu'} + a_{l',\mu'} a_{l,\mu}^{+} = \delta_{l l'}\, \delta_{\mu\mu'}; \quad & a_{l,\mu} a_{l',\mu'} + a_{l',\mu'} a_{l,\mu} = 0; \\ & a_{l,\mu}^{+} a_{l',\mu'}^{+} + a_{l',\mu'}^{+} a_{l,\mu}^{+} = 0 \end{aligned}\right\} \tag{VI.3.6}$$

provided there is no overlap between wavefunctions of different atoms.

The coupling coefficient $g_{\mu\lambda}$ describes the strength of interaction between the light mode λ and the atom μ.

It is proportional to the atomic dipole matrix element*

$$\int \varphi_j^*(\boldsymbol{x})\, \boldsymbol{e}\, \boldsymbol{p}\, \varphi_k(\boldsymbol{x})\, dV \tag{VI.3.7}$$

and to the field amplitude $u(\boldsymbol{x}_\mu)$ at this lattice point. Its exact form is

$$g_{\mu,\lambda} = e^{-i\boldsymbol{k}_\lambda \boldsymbol{x}_\mu} \cdot g \tag{VI.3.8}$$

where

$$g = \left(-\frac{e}{m}\sqrt{\frac{2\pi}{\hbar\omega V}}\right) \int \varphi_k^*\, \boldsymbol{e}\, \boldsymbol{p}\, \varphi_j\, dV \tag{VI.3.9}$$

for a running wave and

$$g_{\mu,\lambda} = g\, 2\sqrt{2}\, \sin(k_{\lambda,x} x_\mu)\, \sin(k_{\lambda,y} y_\mu)\, \sin(k_{\lambda,z} z_\mu) \tag{VI.3.10}$$

for a standing wave.

e and m are the electronic charge and mass, respectively. φ_j and φ_k are the corresponding electronic wave functions of one atom. V represents the volume of the cavity, $\boldsymbol{p}$ is the momentum operator. $\boldsymbol{k}_\lambda$ is the wave vector of mode λ, $\boldsymbol{e}$ is the polarization vector. F_λ^+ represents a fluctuating force, exerted on the light-mode by the thermally fluctuating currents in the cavity mirrors (and other passive noise sources as well). It is known from the Brownian motion in classical physics that fluctuating forces (or random collisions) on a particle lead to a damping (by a friction force), and residual fluctuating forces. Whereas the damping through the heatbaths is taken care of by the damping constant $\varkappa_\lambda$, F_λ^+ is just the residual fluctuating force. In a quantum theoretical treatment, F_λ^+ becomes an operator which is responsible for the quantum theoretical consistency, i.e. through its action the b's obey the commutation relations for all times. The F_λ's and F_λ^+'s have the following properties:

$$\langle F_\lambda^+ \rangle = \langle F_\lambda \rangle = 0, \tag{VI.3.11}$$

$$\left.\begin{aligned} \langle F_\lambda^+(t)\, F_{\lambda'}^+(t') \rangle &= \langle F_\lambda(t)\, F_{\lambda'}(t') \rangle = 0, \\ \langle F_\lambda^+(t)\, F_{\lambda'}(t') \rangle &= n_{th,\lambda}(T)\, 2\varkappa_\lambda\, \delta(t-t')\, \delta_{\lambda\lambda'}, \\ \langle F_\lambda(t)\, F_{\lambda'}^+(t') \rangle &= \left(n_{th,\lambda}(T)+1\right) 2\varkappa_\lambda\, \delta(t-t')\, \delta_{\lambda\lambda}\,. \end{aligned}\right\} \tag{VI.3.12}$$

where $n_{th,\lambda}$ is the number of thermal quanta.

$\langle \cdots \rangle$ means the thermal average over the states of the heatbaths. In our following analysis we need only the properties (VI.3.11) and (VI.3.12), but we never have to do the average explicitly.

---

* Compare footnote * on p. 28.

b) Matter equations.

1. The light field acts, of course, back on the atoms. This has two effects:

α) *The motion of the atomic dipole moment* is influenced:

1. *Dipole moment between levels j and k*

$$(a_j^+ a_k)^{\cdot}_\mu = (i\nu_{jk,\mu} - \gamma_{jk})\,(a_j^+ a_k)_\mu - i\sum_\lambda g_{\mu\lambda}\, b_\lambda^+ (N_{j,\mu} - N_{k,\mu}) + \Gamma_{jk,\mu}\,. \tag{VI.3.13}$$

$\nu_{jk,\mu}$ is the transition frequency between levels j and k at atom μ, with which the atomic dipole moment precesses. Due to the interaction with lattice vibrations, the continuum of nonlasing modes etc., this motion is steadily disturbed, so that the *phase* of the dipole moments fluctuates. After a time $1/2\gamma$ the phase memory is essentially lost. 2γ is the full optical linewidth of each atom. The second term describes how the field amplitude b_λ^+ acts as driving force on the atomic dipole. The interaction is again proportional to $g_{\mu\lambda}$. Most important for the whole laser theory is the factor $(N_j - N_k)_\mu \equiv (a_j^+ a_j - a_k^+ a_k)_\mu$ which represents the difference of occupation numbers N of the levels j and k. It determines whether the atomic dipole is in phase or in opposite phase to the driving field and thus, whether (induced) emission or absorption takes place. $\Gamma_{jk,\mu}$ finally is a fluctuating force, which is connected with the Brownian motion of the atomic dipole moment and maintains quantum mechanical consistency.

2. *Dipole moment between levels j and $l \neq k, j$ and between levels k and $l = j, k$.* Besides the two laser-active levels, j and k, the occupation numbers and the relative dipole moments of all other atomic levels, which are involved in the pumping and loss processes, change, of course, too.

$$(a_j^+ a_l)^{\cdot}_\mu = (i\nu_{jl,\mu} - \gamma_{jl})\,(a_j^+ a_l)_\mu + i\sum_\lambda g_{\mu\lambda}\, b_\lambda^+ (a_k^+ a_l)_\mu + \Gamma_{jl,\mu}, \tag{VI.3.14}$$

$$(a_k^+ a_l)^{\cdot}_\mu = (i\nu_{kl,\mu} - \gamma_{kl})\,(a_k^+ a_l)_\mu + i\sum_\lambda g_{\mu\lambda}^*\, b_\lambda (a_j^+ a_l)_\mu + \Gamma_{kl,\mu}\,. \tag{VI.3.15}$$

The equations for $a_l^+ a_j$ and $a_l^+ a_k$ can be obtained by taking the Hermitian conjugate equation.

3. *Dipole moment between levels $i \neq k, j$ and $l \neq k, j$*

$$(a_i^+ a_l)^{\cdot}_\mu = (i\nu_{il,\mu} - \gamma_{il})\,(a_i^+ a_l)_\mu + \Gamma_{il,\mu}\,. \tag{VI.3.16}$$

The quantities occurring in (VI.3.14)—(VI.3.16) were explained above.

The Eqs. (VI.3.14)—(VI.3.16) have no influence on the laser process, and can be dropped. They become influential, however, in laser cascades etc. (see Chap. VIII.3).

β) *The occupation numbers change.*

1. *For the laser levels j and k.* This change is brought about not only by the lightfield, but also by the pump and loss systems.

$$\left.\begin{aligned}(a_j^+ a_j)^{\cdot}_\mu = {} & \sum_{i\neq j} (a_i^+ a_i)_\mu\, w_{ij} - (a_j^+ a_j)_\mu \sum_{i\neq j} w_{ji} + i\,(a_k^+ a_j)_\mu \sum_\lambda g_{\mu\lambda}\, b_\lambda^+ - \\ & - i\,(a_j^+ a_k)_\mu \sum_\lambda g_{\mu\lambda}^*\, b_\lambda + \Gamma_{jj,\mu},\end{aligned}\right\} \tag{VI.3.17}$$

$$\left.\begin{aligned}(a_k^+ a_k)^{\cdot}_\mu = {} & \sum_{i\neq k} (a_i^+ a_i)_\mu\, w_{ik} - (a_k^+ a_k)_\mu \sum_{i\neq k} w_{ki} - i\,(a_k^+ a_j)_\mu \sum_\lambda g_{\mu\lambda}\, b_\lambda^+ + \\ & + i\,(a_j^+ a_k)_\mu \sum_\lambda g_{\mu\lambda}^*\, b_\lambda + \Gamma_{kk,\mu}\,.\end{aligned}\right\} \tag{VI.3.18}$$

The effect of the pump and loss systems can be described by transition rates w_{il} and again by residual fluctuating forces, $\Gamma_{ll,\mu}$ (w_{il} = number of transitions from i to l per second, provided the level i is initially occupied). The rate of change of the occupation number caused by the field is proportional to both the dipole moment and the field amplitude.

2. *For the non-laser levels*

$$(a_l^+ a_l)_\mu^{\cdot} = \sum_{i \neq l} (a_i^+ a_i)_\mu w_{il} - (a_l^+ a_l)_\mu \sum_{i \neq l} w_{li} + \Gamma_{ll,\mu}. \tag{VI.3.19}$$

The *fluctuating forces* Γ_{il} (where i, l includes also j and k) have the following properties:

$$\langle \Gamma_{il}(t) \rangle = 0, \tag{VI.3.20}$$

$$\langle \Gamma_{i_1,i_2}(t)\, \Gamma_{j_1,j_2}(t') \rangle = G_{i_1,i_2;j_1,j_2}\, \delta(t-t') \tag{VI.3.21}$$

where $\langle \cdots \rangle$ means average over all "heatbaths" (i.e. pump and loss systems etc.). The general formula for the G's is given in (IV.5.26). For the following analysis it is of importance, that the a's commute with the b's.

Although the above equations referring to field and atomic operators are fully quantum mechanical, their meaning can be visualized in terms of classical quantities like field amplitude, dipole moment and occupation number. In the following it will turn out that it is even possible to treat these quantities in a way very similar to classical variables. The deeper justification of this procedure can be expressed in the following two ways:

a) In the Heisenberg picture our further treatment will show that above laser threshold the operators can be split into a c-number and a residual operator. This residual operator only describes small fluctuations, so that the above equations are very close to c-number equations besides small quantum fluctuations compare e.g. footnote * on p. 129).

b) In the Schrödinger picture it can be shown that the quantum mechanical laser system is approximately in an eigenstate of the field operator b and the atomic operators. The latter property comes from the fact that the laser atoms form a macroscopic system. The order of magnitude by which the true eigenstate differs from the above-mentioned state can be determined, e.g. in the framework of the density matrix equation (see e.g. WEIDLICH and HAAKE, ref. 1 on p. 51), and turns out to be very small above laser threshold.

VI.4. Qualitative discussion of single mode operation[1]. A series of implications of the above equations can be explained by the following specialization:

We treat a system of two-level atoms with a single transition frequency ν, which are in resonance with a single running cavity mode. The equations of motion then read:

$$\dot{b}^+ = (i\omega - \varkappa)\, b^+ + i \sum_\mu g\, e^{-ikx_\mu} (a_2^+ a_1)_\mu + F^+, \tag{VI.4.1}$$

$$(a_2^+ a_1)_\mu^{\cdot} = (i\omega - \gamma)\, (a_2^+ a_1)_\mu - i\, g\, e^{ikx_\mu}\, b^+ (a_2^+ a_2 - a_1^+ a_1)_\mu + \Gamma_{21,\mu} \tag{VI.4.2}$$

[1] H. HAKEN: Z. Physik **181**, 96 (1964), and in: Dynamical Processes in Solid State Optics 1966, Tokyo Summer Lectures in Theoretical Physics (ed. R. KUBO and H. KAMIMURA). Tokyo-New York: Syokabō and W. A. Benjamin, Inc. 1967.

and the complex conjugate equation, and further

$$\left.\begin{aligned}(a_2^+ a_2)^{\boldsymbol{\cdot}}_\mu = (a_1^+ a_1)_\mu\, w_{12} - (a_2^+ a_2)_\mu\, w_{21} + i\,(a_1^+ a_2)_\mu\, g_\mu\, b^+ -\\ - i\,(a_2^+ a_1)_\mu\, g_\mu^*\, b + \Gamma_{22,\mu},\end{aligned}\right\} \tag{VI.4.3}$$

$$\left.\begin{aligned}(a_1^+ a_1)^{\boldsymbol{\cdot}}_\mu = (a_2^+ a_2)_\mu\, w_{21} - (a_1^+ a_1)_\mu\, w_{12} - i\,(a_1^+ a_2)_\mu\, g_\mu\, b^+ +\\ + i\,(a_2^+ a_1)_\mu\, g_\mu^*\, b + \Gamma_{11,\mu}.\end{aligned}\right\} \tag{VI.4.4}$$

We eliminate ω and the factors e^{ikx_μ} and e^{-ikx_μ} from the above equations by the transformations

$$b^+ \to b^+ e^{i\omega t}, \qquad b \to b\, e^{-i\omega t}, \tag{VI.4.5}$$

$$(a_2^+ a_1)_\mu \to e^{ikx_\mu} e^{i\omega t} (a_2^+ a_1)_\mu; \qquad (a_1^+ a_2)_\mu \to e^{-ikx_\mu} e^{-i\omega t} (a_1^+ a_2)_\mu. \tag{VI.4.6}$$

Because $a_1^+ a_1 + a_2^+ a_2 = 1$ which holds exactly, we introduce as new variables $(a_1^+ a_1 + a_2^+ a_2)$ and

$$(a_2^+ a_2 - a_1^+ a_1)_\mu = \sigma_\mu. \tag{VI.4.7}$$

The Eqs. (VI.4.1)—(VI.4.4) then take the very simple form:

$$\dot{b}^+ = -\varkappa\, b^+ + i\, g \sum_\mu (a_2^+ a_1)_\mu + F^+, \tag{VI.4.8}$$

$$(a_2^+ a_1)^{\boldsymbol{\cdot}} = -\gamma\, (a_2^+ a_1)_\mu - i\, g\, b^+ \sigma_\mu + \Gamma_{21,\mu}, \tag{VI.4.9}$$

$$\dot{\sigma}_\mu = \frac{d_0 - \sigma_\mu}{T} + 2\, i\, g\, (a_1^+ a_2\, b^+ - a_2^+ a_1\, b)_\mu + (\Gamma_{22} - \Gamma_{11})_\mu \tag{VI.4.10}$$

where

$$d_0 = \frac{w_{12} - w_{21}}{w_{12} + w_{21}} \tag{VI.4.11}$$

and

$$\frac{1}{T} = w_{12} + w_{21}. \tag{VI.4.12}$$

Because the mode interacts with the atoms through a sum, one is led to introduce $\sum_\mu (a_2^+ a_1)_\mu$ as a new variable S^+ and correspondingly to put

$$S = \sum_\mu \sigma_\mu. \tag{VI.4.13}$$

Eqs. (VI.4.8)—(VI.4.10) thus simplify [by summation over μ in (VI.4.9) and (VI.4.10)] to

$$\dot{b}^+ = -\varkappa\, b^+ + i\, g\, S^+ + F^+, \tag{VI.4.14}$$

$$\dot{S}^+ = -\gamma\, S^+ - i\, g\, b^+ S + \Gamma_{21}, \tag{VI.4.15}$$

where

$$\Gamma_{21} = \sum_\mu \Gamma_{21,\mu},$$

$$\dot{S} = \frac{S^0 - S}{T} + 2\, i\, g\, (S^- b^+ - b\, S^+) + \Gamma_{22} - \Gamma_{11}, \tag{VI.4.16}$$

where

$$\Gamma_{jj} = \sum_\mu \Gamma_{jj,\mu},$$

$$S^0 = N\, d_0 \tag{VI.4.17}$$

and N: total number of atoms.

S^+ can be visualized as the macroscopic total dipole moment of all atoms, while $S = N_2 - N_1$ is the total occupation difference (total inversion). As is shown in III.6, S^+, S^- and S can also be interpreted as operators for the total spin in a formal manner

$$S=(S_x, S_y, S_z) \quad \text{where} \quad S^+=S_x+i\,S_y, \qquad S^-=S_x-i\,S_y, \qquad S=2\,S_z.$$

We eliminate S^+ from the first two equations and we substitute S^+ in Eq. (VI.4.16) by means of Eq. (VI.4.14) and S^- by means of an equation which is conjugate complex to Eq. (VI.4.14). This leaves us with

$$\left.\begin{array}{l}(\text{VI.4.14})\\(\text{VI.4.15})\end{array}\right\}: \ddot{b}^+ + (\varkappa+\gamma)\,\dot{b}^+ + (\varkappa\gamma - g^2 S)\,b^+ = \underbrace{i\,g\,\Gamma_{21} + \dot{F}^+ + \gamma F^+}_{\tilde{F}^+_{tot}(t)}, \qquad (\text{VI.4.18})$$

$$\left.\begin{array}{l}(\text{VI.4.16})\\(\text{VI.4.14})\end{array}\right\}: \dot{S} = \frac{S^0-S}{T} - 2\left((b^+b)^{\cdot} + 2\varkappa\,b^+b - (b^+F + F^+b)\right) + \Gamma_{22} - \Gamma_{11}. \quad (\text{VI.4.19})$$

a) The linear range (subthreshold region). For an interpretation of Eqs. (VI.4.18) and (VI.4.19) we treat b^+ as a classical complex time-dependent variable, which we decompose into real and imaginary parts:

$$b^+(t) = x(t) + i\,y(t). \qquad (\text{VI.4.20})$$

Eq. (VI.4.18) can then also be decomposed into a real and an imaginary part. The resulting equations then have the same appearance as those of a classical particle, moving in two dimensions under a force

$$\hat{K}_x = -\left(\varkappa\,\gamma - g^2 S(x, y)\right) x; \quad \hat{K}_y = -\left(\varkappa\,\gamma - g^2 S(x, y)\right) y.$$

Although the motion of the light mode has, of course, nothing to do with particle motion, this formal analogy is most useful for an understanding of the nonlinear Eq. (VI.4.18), (VI.4.19). Consider first the case where the inversion $S=N_2-N_1$ is kept fixed, which is actually done in the linear theories of laser noise, or in the amplifier range. Let us start with a small inversion, so that

$$\varkappa\,\gamma - g^2 S > 0. \qquad (\text{VI.4.21})$$

Eq. (VI.4.18) is then that of a damped harmonic oscillator, being subject to a random force $K(t)$. This random force can be interpreted in such a way that it exerts pushes in a random time sequence. After each such excitation the "particle falls down the potential curve" (compare Fig. 29). When we increase the inversion S, $(\varkappa\gamma - g^2 S)$ becomes smaller (compare Fig. 29) so that the particle comes down the potential curve more slowly. Translating this statement into the language of waves, we may say that the modulation frequency, caused by the fluctuations, becomes smaller, so that a line narrowing results. The pushes in $\tilde{F}^+_{tot}(t)$ stem from statistically independent events. Consider for instance Γ_{21} which is equal to the sum of the atomic noise operators $\Gamma_{21,\mu}$. Each operator is responsible for the fluctuations of the atomic dipole moment, which are indeed statistically independent. As we will see later explicitly, Γ_{21} is responsible for the spontaneous emission noise, while F accounts for the thermal noise of the walls and the vacuum fluctuations. Each emission process creates a contribution to b^+. Due to the flatter curve, the amplitude falls more slowly or, in other words, the losses are partly compensated by induced emission. Because the total amplitude b^+ is a sum of many statistical contributions, the probability P of finding a definite value of b^+ is a Gaussian:

$$P(b) = \text{const.}\ e^{-\frac{|b|^2}{|b_0|^2}}. \qquad (\text{VI.4.22})$$

When S becomes so big, that $\varkappa\gamma - g^2 S < 0$, the system *seems* to become unstable (compare Fig. 29). In the linear theories, S can never become bigger than $S_c = \frac{\varkappa\gamma}{g^2}$, because otherwise the light output $\sim |b|^2$ would become infinite. These latter considerations are, however, incorrect, because they neglect the dependence of S on b^+ completely.

b) The nonlinear range (at threshold and somewhat above). Our discussion is based again on Eqs. (VI.4.18) and (VI.4.19) where we investigate now the consequence of the dependence of S on b. On account of the line narrowing we may expect the frequencies inherent in (b^+b) to be smaller than $\varkappa$, so that we can

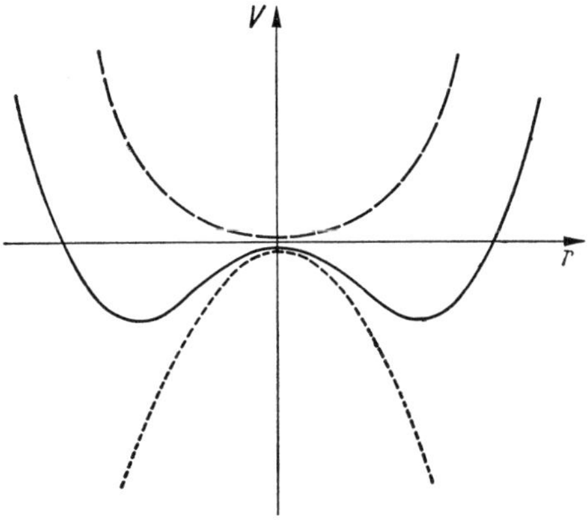

Fig. 29. The fictive potential V. Dashed line below threshold, dotted line above threshold in the linear case, solid line above threshold in the nonlinear case. (After H. HAKEN, l.c.[1] on p. 116.)

neglect $(b^+b)^{\cdot}$ compared with $2\varkappa b^+b$. Because the atomic relaxation time T is also small compared to times inherent in (b^+b) (again on account of the line narrowing), S can follow the motion of b^+b adiabatically. Neglecting the fluctuating terms in (VI.4.19) we thus obtain:

$$S \approx S_0 - 4\varkappa T b^+ b$$

so that (VI.4.18) takes now the form:

$$\ddot{b}^+ + (\varkappa + \gamma)\,\dot{b}^+ + (\varkappa\gamma - g^2 S_0)\,b^+ + 4g^2\varkappa T b^+ b b^+ = \widetilde{F}^+_{tot}(t) \qquad \text{(VI.4.23)}$$

which can be understood as the equation of a particle moving in the potential

$$V(b) = (\varkappa\gamma - g^2 S_0)\,\frac{|b|^2}{2} + g^2\varkappa T |b|^4 . \qquad \text{(VI.4.24)}$$

For $\varkappa\gamma - g^2 S > 0$ i.e. small inversion, the potential curve has qualitatively the same appearance as discussed under a).

For $\varkappa\gamma - g^2 S < 0$, the amplitude oscillates around a new stable value $r_0 \neq 0$. Because the potential has rotation symmetry, there is no restoring force in tangential direction. Thus the phase undergoes some kind of diffusion under the action of the random force $\widetilde{F}^+_{tot}(t)$.

We write

$$b^+ = e^{i\varphi(t)}\,(r_0 + \varrho(t)) \qquad \text{(VI.4.25)}$$

and discuss the phase diffusion and the amplitude fluctuation ϱ in more detail.

α) *Phase diffusion.* With increasing S the stable value r_0 increases. Consequently the path of the particle, and thus its migration time, increases. Because in a diffusion process $\Delta x^2 \sim t$, and for a tangential motion $\Delta x = r_0 \Delta\varphi$, we have

$$\overline{(\Delta\varphi)^2} \sim \frac{t}{r_0^2}. \tag{VI.4.26}$$

The factor of t however, is just the linewidth: With increasing inversion the linewidth $\Delta\omega$ due to phase diffusion drops down with $1/r_0^2$ or, because the coherent output power, P, is $\hbar\omega\, 2\varkappa\, r_0^2$:

$$\Delta\omega \sim \frac{1}{P}. \tag{VI.4.27}$$

β) *Amplitude (intensity) fluctuations.* With increasing r_0, the slope of the potential curve becomes steeper in the vicinity of r_0. Therefore:

1. The oscillation frequency increases, so that the relaxation time of the amplitude fluctuation decreases

$$\tau \sim \frac{1}{P}, \tag{VI.4.28}$$

2. The fluctuating part, ϱ, of the amplitude decreases,

$$\langle\varrho\rangle^2 \sim \frac{1}{P}. \tag{VI.4.29}$$

c) *The nonlinear range at high inversion.* With increasing inversion, the oscillation frequency of the amplitude fluctuations finally becomes bigger than $2\varkappa$, so that the dominant terms in (VI.4.19) are given by:

$$\dot{S} \approx \frac{S^0 - S}{T} - 2(b^+ b)^{\boldsymbol{\cdot}} \tag{VI.4.30}$$

(the terms with F and $\Gamma_{22} - \Gamma_{11}$ are not completely negligible, however), which, for rapid oscillation has the solution:

$$S = \text{Const} - 2(b^+ b). \tag{VI.4.31}$$

Thus the behaviour is qualitatively similar to that of the region b).

d) *Exact elimination of all atomic coordinates.* For the qualitative discussion we have expressed S only approximately by means of b^+, b. This can be done exactly, however, by solving Eq. (VI.4.19).

$$\left.\begin{aligned} S = d_0 - 2\int^t e^{-\frac{1}{T}(t-\tau)} \big((b^+ b)^{\boldsymbol{\cdot}} + 2\varkappa (b^+ b) - \\ - (b^+ F + F^+ b) - \tfrac{1}{2}(\Gamma_{22} - \Gamma_{11})\big)\, d\tau. \end{aligned}\right\} \tag{VI.4.32}$$

Inserting (VI.4.32) into (VI.4.18) yields an exact equation for the b-field alone.

VI.5. Quantitative treatment of a homogeneously broadened transition: emission below threshold (intensity, linewidth, amplification of signals).

a) *No external signals.* For our analysis we use the equations of VI.3 which were written down for a system of atoms each with two optically active levels and an arbitrary number of auxiliary (pumping) levels. In order to simplify the notation somewhat, we put $j=3$ and $k=2$, although these optically active levels may still have any position within the atomic level system. We confine ourselves to a *homogeneous broadening*. We start from the field Eq. (VI.3.5):

$$\dot{b}_\lambda^+ = (i\omega_\lambda - \varkappa_\lambda)\, b_\lambda^+ + i \sum_\mu g_{\mu\lambda}^*\, a_{3,\mu}^+ a_{2,\mu} + F_\lambda^+ \tag{VI.5.1}$$

and the matter Eq. (VI.3.13)

$$(a_{3,\mu}^{+} a_{2,\mu})^{\cdot} = (i\nu_{32} - \gamma_{32})(a_{3,\mu}^{+} a_{2,\mu}) - i \sum_{\lambda} g_{\mu\lambda} b_{\lambda}^{+} (N_3 - N_2)_{\mu} + \Gamma_{32,\mu}. \qquad \text{(VI.5.2)}$$

In (VI.5.2) we have abbreviated $(a_3^+ a_3 - a_2^+ a_2)_\mu$ by $(N_3 - N_2)_\mu$. In the following, we assume that the inversion $(N_3 - N_2)_\mu$ is determined merely by the pumping process, by spontaneous emission and by radiationless transition, but it is not changed by any stimulation process. Therefore we replace $(N_3 - N_2)_\mu$ by the "unsaturated" inversion $(N_3 - N_2)_\mu^0$. We assume further, that $(N_3 - N_2)_\mu$ does not depend on μ. In accordance with our later notation, we write

$$(S_{33} - S_{22})^0 = \sum_{\mu} (N_3 - N_2)_\mu^0. \qquad \text{(VI.5.3)}$$

From (VI.5.1) and (VI.5.2) we can now easily eliminate $a_{3,\mu}^+ a_{2,\mu}$ so that we find:

$$\left.\begin{aligned} &\ddot{b}_\lambda^+ + (-i(\omega_\lambda + \nu_{32}) + \varkappa_\lambda + \gamma_{32})\, \dot{b}_\lambda^+ + [(i\nu_{32} - \gamma_{32})(i\omega_\lambda - \varkappa_\lambda) \\ &\quad - g^2 (S_{33} - S_{22})^0]\, b_\lambda^+ = \underbrace{-(i\nu_{32} - \gamma_{32}) F_\lambda^+ + i g\, \Gamma_{32}^\lambda + \dot{F}_\lambda^+}_{F_{tot}^+(t)} \end{aligned}\right\} \qquad \text{(VI.5.4)}$$

where the $g_{\mu\lambda}$'s were assumed orthogonal:

$$\frac{1}{N} \sum_{\mu} g_{\mu\lambda}^* g_{\mu\lambda'} = \delta_{\lambda'\lambda}\, g^2. \qquad \text{(VI.5.5)}$$

This equation defines g^2 (for $\lambda = \lambda'$). In particular, g is real.

We have further abbreviated:

$$g\, \Gamma_{32}^\lambda = \sum_{\mu} g_{\mu\lambda}^* \Gamma_{32,\mu}. \qquad \text{(VI.5.5a)}$$

We drop the indeces λ and 32 and treat first *a single mode.*

The main time-dependence of b^+ in Eq. (VI.5.4) can be done away by putting

$$b^+ = e^{i\Omega t}\, \tilde{b}^+ \qquad \text{(VI.5.6)}$$

where

$$\Omega = \frac{\nu\varkappa + \gamma\omega}{\varkappa + \gamma} \qquad \text{(VI.5.7)}$$

so that the following equation for b^+ results:

$$\left.\begin{aligned} &\ddot{\tilde{b}}^+ + \dot{\tilde{b}}^+ \left(\varkappa + \gamma + i\,\frac{\delta(\varkappa - \gamma)}{\varkappa + \gamma}\right) + \\ &\quad + \left[\varkappa\gamma\left(1 + \frac{\delta^2}{(\varkappa + \gamma)^2}\right) - g^2 (S_{33} - S_{22})^0\right] \tilde{b}^+ = \tilde{F}_{tot}^+(t) \end{aligned}\right\} \qquad \text{(VI.5.8)}$$

where

$$\tilde{F}_{tot}^+ = e^{-i\Omega t} F_{tot}^+ \qquad \text{(VI.5.9)}$$

and

$$\delta = \nu - \omega. \qquad \text{(VI.5.10)}$$

(VI.5.7) describes a frequency-pulling effect, which will be discussed in more detail in VII.3.

Provided the mode frequency ω is not too far apart from the atomic line center, $\ddot{\tilde{b}}$ can be safely neglected (see below). The solution of (VI.5.8) then takes the simple form

$$\tilde{b}^+ = \int^t e^{-\hat{\varkappa}(t-\tau)}\, \tilde{F}_{tot}^+(\tau) \left\{\varkappa + \gamma + i\,\frac{(\nu - \omega)(\varkappa - \gamma)}{\varkappa + \gamma}\right\}^{-1} d\tau \qquad \text{(VI.5.11)}$$

where

$$\hat{\varkappa} = \frac{\varkappa\gamma\left(1+\delta^2/(\varkappa+\gamma)^2\right) - g^2(S_{33}-S_{22})^0}{\varkappa+\gamma+i(\nu-\omega)(\varkappa-\gamma)/(\varkappa+\gamma)} \tag{VI.5.12}$$

($\hat{\varkappa} > 0$ below threshold).

In (VI.5.11) we have already dropped the solution of the homogeneous equation, $b_h^+ = b^+(0)\, e^{-\hat{\varkappa} t}$, because it represents only a switching-on effect. The steady state solution is thus obtained by the noisy driving force $\tilde{F}_{tot}^+(t)$. In the subthreshold region it is therefore correct to say that the laser (maser) is driven by its noise. The stimulation process acts as a frequency filter, which does not change the statistical properties of the light. On account of the properties (VI.3.11) and (VI.3.20) of the noise operators, we find immediately $\langle b^+(t)\rangle = 0$. In order to find the linewidth, we insert (VI.5.11) into the coherence function (V.1.32): $\langle b^+(t)\, b(t')\rangle$. In order to calculate $\langle F_{tot}^+(\tau)\, F_{tot}(\tau')\rangle$ which occurs then in (V.1.32), we use (VI.3.11), (VI.3.12), (VI.3.20) and (VI.3.21). We find finally

$$\langle b^+(t)\, b(t')\rangle = \frac{R\cdot e^{i\Omega(t-t')}}{\left\{(\varkappa+\gamma)^2 + \dfrac{(\nu-\omega)^2(\varkappa-\gamma)^2}{(\varkappa+\gamma)^2}\right\}(\hat{\varkappa}+\hat{\varkappa}^*)} \cdot \begin{cases} e^{-\hat{\varkappa}(t-t')} & \text{for } t>t' \\ e^{-\hat{\varkappa}^*(t'-t)} & \text{for } t<t' \end{cases} \tag{VI.5.13}$$

where

$$R = g^2(2\gamma_{32}N_3 + P_{coh}) + 2\varkappa\,\gamma_{32}^2\, n_{th}\left(1+\delta^2/(\varkappa+\gamma)^2\right) \tag{VI.5.14}$$

with

$$N_3 = \sum_\mu N_{3,\mu}.$$

P_{coh} denotes the coherent output-power which can be safely neglected below threshold, so that we use in the following

$$R = g^2\, 2\gamma_{32}\, N_3 + 2\varkappa\,\gamma_{32}^2\, n_{th}\left(1+\delta^2/(\varkappa+\gamma)^2\right). \tag{VI.5.15}$$

From the form of Eq. (VI.5.13) it follows clearly, that the linewidth $\Delta\nu$ is given by Re $\hat{\varkappa}$.

The essential features of the line narrowing can be seen especially clearly in the resonant case ($\omega=\nu$) in which

$$\Delta\omega = \hat{\varkappa} = \frac{\varkappa\gamma - g^2(S_{33}-S_{22})^0}{\varkappa+\gamma}. \tag{VI.5.16}$$

When the inversion $(S_{33}-S_{22})^0$ is very small,

$$\hat{\varkappa} \approx \frac{\varkappa\gamma}{\varkappa+\gamma}. \tag{VI.5.17}$$

For $\gamma \gg \varkappa$ we find $\hat{\varkappa}\approx\varkappa$ while for $\varkappa\gg\gamma$ we have $\hat{\varkappa}\approx\gamma$.

The single mode linewidth is always smaller than the smaller one of the widths $\varkappa, \gamma$. With growing inversion $(S_{33}-S_{22})^\circ$ the line narrows, as can be seen immediately from (VI.5.16). The linewidth $\Delta\omega = Re\,\hat{\varkappa}$ can be expressed by means of the output power

$$P = 2\varkappa\,\hbar\omega\,\langle b^+(t)\, b(t)\rangle \tag{VI.5.18}$$

$$= \varkappa \frac{\hbar\omega R}{\left\{(\varkappa+\gamma)^2 + \dfrac{(\nu-\omega)^2(\varkappa-\gamma)^2}{(\varkappa+\gamma)^2}\right\}\Delta\omega} \tag{VI.5.19}$$

so that

$$\Delta\omega = \frac{\hbar\omega R\cdot\varkappa}{P\left\{(\varkappa+\gamma)^2 + \dfrac{(\nu-\omega)^2(\varkappa-\gamma)^2}{(\varkappa+\gamma)^2}\right\}}. \tag{VI.5.20}$$

The coupling constant g^2 which occurs in (VI.5.15) can be eliminated by means of the threshold condition $\hat{\varkappa}_{th}=0$:

$$g^2=\frac{\varkappa\gamma\left(1+\delta^2/(\varkappa+\gamma)^2\right)}{(N_3-N_2)_{thr}} \tag{VI.5.21}$$

where N_3 and N_2 are the total numbers of atoms in the upper and lower levels, respectively. *thr* means threshold.

With (VI.5.14) and (VI.5.21) we obtain the expression for the

α) *Single-mode linewidth below threshold* (halfwidth at half power)

$$\Delta\omega=\frac{\hbar\omega}{P}\,\frac{\varkappa^2\gamma^2\,2\left\{\frac{N_3}{(N_3-N_2)_{thr}}+n_{th}\right\}\left(1+\delta^2/(\varkappa+\gamma)^2\right)}{\{(\varkappa+\gamma)^2+\Delta^2\}}$$

where

$$\Delta^2=\frac{(\nu-\omega)^2(\varkappa-\gamma)^2}{(\varkappa+\gamma)^2}\,. \tag{VI.5.22}$$

This formula holds for atoms with an arbitrary number of "pump" levels. The number of thermal light quanta, n_{th} dominates in the case for a maser, whereas for a laser ($\hbar\omega/kT\gg 1$) it can be safely neglected compared to $\frac{N_3}{(N_3-N_2)_{thr}}$.

Because the noise sources inherent in the driving force of Eq. (VI.5.4), $F_{tot}(t)$ consist of very many statistically independent contributions, the distribution of b is a "Gaussian". Some care has to be exercised compared to the classical Gaussian distribution, because the b's are operators. One establishes, however, readily that the higher order correlation functions of the type

$$\langle b^+(t_1)\dots b^+(t_n)\,b(t'_m)\dots b(t'_1)\rangle \tag{VI.5.23}$$

can be expressed by those of type $\langle b^+(t_j)\,b(t_k)\rangle$ through the usual classical connections.

We assume merely, that

$$\langle b(t)\,b(t')\rangle=\langle b^+(t)\,b^+(t')\rangle=0\,.$$

It then follows

$$(\text{VI.5.23})=0\quad\text{for}\quad m\neq n \tag{VI.5.24}$$

and

$$(\text{VI.5.23})=\sum_p\langle b^+(t_1)\,b(t'_{P_1})\rangle\langle b^+(t_2)\,b(t'_{P_2})\rangle\cdots\langle b^+(t_n)\,b(t'_{P_n})\rangle \tag{VI.5.25}$$

where the sum runs over all permutations $(P_1, P_2\dots)$ of the numbers $(1, 2, \dots, n)$. If all times are equal, we find especially

$$\langle(b^+(t))^n((b(t))^n\rangle=n!\ \langle b^+(t)\,b(t)\rangle^n\,. \tag{VI.5.26}$$

β) *Many modes below threshold.* Because the resonator is highly overmoded, we have to consider all modes, at least at low inversion.

Because the equations for the single modes are decoupled as long as we neglect nonlinear terms and as long as the $g_{\mu\lambda}$'s are orthogonal to each other, we can immediately use the above solution (VI.5.13), where we have merely to supplement Ω, ω, and $\varkappa$ with the index λ. We calculate

$$\langle \boldsymbol{E}^{(-)}(t)\,\boldsymbol{E}^{(+)}(t+\tau)\rangle \tag{VI.5.27}$$

where we take $\boldsymbol{E}^{(-)}(t)$ and $\boldsymbol{E}^{(+)}(t)$ at the same space point $\boldsymbol{x}$. After averaging over a space region the $\boldsymbol{x}$-dependence drops out, and due to the orthogonality of the expansion functions in $\boldsymbol{E}$ [compare for instance (VI.3.2)] we obtain

$$(\text{VI.5.27}) = \text{const.} \sum_\lambda \langle b_\lambda^+(t)\, b_\lambda(t+\tau)\rangle. \tag{VI.5.28}$$

In order to find the spectral distribution $I(\omega)$ we insert (VI.5.13) into (VI.5.28) and take the Fourier transform, which yields

$$I(\omega) = \text{const.} \cdot \sum_\lambda \frac{R_\lambda}{\{(\varkappa_\lambda+\gamma)^2 + \Delta_\lambda^2\}\{|\hat{\varkappa}_\lambda|^2 + (\omega-\Omega_\lambda)^2\}}. \tag{VI.5.29}$$

It is interesting to note that $\varkappa$ and γ enter (VI.5.29) in a completely symmetric way. The sum λ runs, of ocurse, over all modes. (VI.5.29) contains the inversion $N_3 - N_2$ and the number of atoms N_3 in the upper state as parameters. They must be determined in such a way, that the total power output by spontaneous and induced emission equals the power input minus losses by other loss mechanisms of the atomic system. In general, for the evaluation of (VI.4.29) a machine integration is required. A lucid discussion of the properties of an expression of the type (VI.5.29) has been given by WAGNER and BIRNBAUM[1]. They have shown that the output characteristics depend critically on the relative loss rates of the various modes.

The physical meaning of (VI.5.29) can be seen in the following way. Let us consider the special case that the atomic linewidth γ is much bigger than the resonator linewidth $\varkappa$ which holds in almost all cases of the laser region. In this case the first factor $\approx \frac{1}{\gamma^2 + (\omega_\lambda - \nu)^2}$ represents the relative intensity emitted by spontaneous emission within the atomic linewidth into a mode at a frequency ω_λ. The second factor $\frac{1}{|\hat{\varkappa}_\lambda|^2 + (\omega - \Omega_\lambda)^2}$ describes the influence of the stimulated emission on the amplification of that mode. The size of $\hat{\varkappa}_\lambda$ is responsible for the gain (the smaller $\hat{\varkappa}_\lambda$ the higher the gain). If there is a set of modes with high gain which can be achieved for instance by having the losses small, these modes give the essential contribution to the sum, whereas all the other modes contribute to a background.

b) External signals. We now consider the case that the modes λ are generated by external, fixed sources (see also V.1, p. 81), and that these modes are amplified by an active material. The right hand side of the field equations is then to be supplemented by these external sources, so that we have to replace F_λ^+ by $F_\lambda^+ + F_{\lambda,\,signal}^+$ in Eq. (VI.5.1), as well as in Eq. (VI.5.4). Because (VI.5.4) is a *linear equation* the total solution $b^+(t) = \tilde{b}^+ e^{i\Omega t}$ is a linear combination of the noisy contribution (VI.5.11) and

$$\left.\begin{aligned} b_{sign}^+(t) = {} & \left\{\varkappa + \gamma + i\,\frac{(\nu-\omega)(\varkappa-\gamma)}{\varkappa+\gamma}\right\}^{-1} \int^t e^{(i\Omega - \hat{\varkappa})(t-\tau)} \times \\ & \times \{\dot{F}_{\lambda,\,sign}^+ - (i\nu - \gamma)\, F_{\lambda,\,sign}^+\}_\tau \, d\tau. \end{aligned}\right\} \tag{VI.5.30}$$

(This solution holds, if the signal frequencies are lying close to the atomic line center, otherwise the full Eq. (VI.5.8) with its second derivative $\ddot{\tilde{b}}^+$ has to be solved).

In the subthreshold region, *noise and signal are additive* in the field amplitudes. They are, however, also additive in the field intensities, if those are averaged over the fluctuations, because $\langle F_{tot}^+(\tau)\rangle = 0$ (see also p. 122).

[1] W. G. WAGNER and G. BIRNBAUM: J. Appl. Phys. 32, 1185 (1961).

$F^+_{\lambda, sign}$ in (VI.5.30) can be decomposed into a Fourier series, so that it suffices to consider a single component, e.g. $a e^{i\omega_s t}$. (VI.5.30) then yields:

$$b^+_{\lambda, sign}(t) = \frac{\gamma + i(\omega_s - \nu)}{\tilde{\varkappa}_\lambda + i(\omega_s - \Omega_\lambda)} \frac{1}{\{\varkappa_\lambda + \gamma + i(\nu - \omega_\lambda)(\varkappa_\lambda - \gamma)/(\varkappa_\lambda + \gamma)\}} a e^{i\omega_s t}. \quad \text{(VI.5.31)}$$

The original signal suffers a phase shift due to detuning.

A comparison between (VI.5.30) and (VI.5.11) shows that signal and noise are amplified in exactly the same manner.

VI.6. Exact elimination of atomic variables in the case of a homogeneously broadened line[1]. Running or standing waves. We base our equations on the laser equations of VI.3. Though the method described in the following is capable of treating the general case of atoms with an arbitrary number of levels, we confine ourselves to a typical case of a 3-level atom and assume the following pumping scheme: The atom is pumped from its ground level, 1, to its third level. From there it can decay into a middle level 2 by laser action, or any other transition (radiative or radiationless). Finally it decays through an arbitrary, incoherent process to level 1*.

In the laser Eqs. (VI.3.5), (VI.3.13), (VI.3.17)—(VI.3.19)** we put therefore $j=3$, $k=2$, $l=1$. Because the line is assumed to be homogeneously broadened, the ν_μ's do not depend on μ. We are ultimately interested in the "motion" of the light field alone, so that it is desirable to eliminate the atomic coordinates. An example of such an elimination procedure was treated in VI.4.

Using essentially the steps as before, we first treat

α) *Standing waves*: We define quite generally[2] for $i \neq j$

$$S_{ij;\lambda} = \frac{1}{g} \sum_\mu g_{\mu\lambda} (a_i^+ a_j)_\mu . \quad \text{(VI.6.1)}$$

The field Eqs. (VI.3.5) then take the form

$$\dot{b}^+_\lambda = (i\omega_\lambda - \varkappa_\lambda)\, b^+_\lambda + i g S_{32;\lambda} + F^+_\lambda . \quad \text{(VI.6.2)}$$

We introduce further

$$S_{jj;\lambda\lambda'} = \frac{1}{g^2} \sum_\mu g_{\mu\lambda}\, g_{\mu\lambda'} (a_j^+ a_j)_\mu . \quad \text{(VI.6.3)}$$

Multiplying Eq. (VI.3.13) by $\frac{1}{g} g_{\mu\lambda}$ and summing up over μ yields

$$\dot{S}_{32,\lambda} = (i\nu_{32} - \gamma_{32})\, S_{32,\lambda} - i g \sum_{\lambda'} b^+_{\lambda'} (S_{33,\lambda\lambda'} - S_{22,\lambda\lambda'}) + \Gamma^\lambda_{32} \quad \text{(VI.6.4)}$$

where

$$\Gamma^\lambda_{32} = \frac{1}{g} \sum_\mu g_{\mu\lambda} \Gamma_{32,\mu} . \quad \text{(VI.6.5)}$$

Finally, we multiply the Eqs. (VI.3.17) and (VI.3.18) for the occupation numbers by $\frac{1}{g^2} g_{\mu\lambda}\, g_{\mu\lambda'}$ and sum up over μ. Using the orthogonality between the g's and

---

* The corresponding laser equations can, however, also be interpreted in such a way that laser action takes place between the two lower levels. Level 2 then serves as ground level. From there the atom is pumped to the upper state 1, from where it falls down to 3. Finally the laser transition occurs between levels 3 and 2.

** The Eqs. (VI.3.14)—(VI.3.16) are not needed for the laser process.

[1] H. HAKEN: Z. Physik **190**, 327 (1966).

[2] These collective modes were introduced by H. HAKEN: Z. Physik **181**, 96 (1964). For the definition of g see (VI.5.5).

that $a_1^+ a_1 + a_2^+ a_2 + a_3^+ a_3 = 1$ for each atom we then obtain

$$\dot{S}_{33,\lambda\lambda'} = w_{13}\,\delta_{\lambda\lambda'}\,N - w_{13}\,S_{22,\lambda\lambda'} - (w_{32}+w_{13})\,S_{33,\lambda\lambda'} + A_{\lambda\lambda'}, \tag{VI.6.6}$$

$$\dot{S}_{22,\lambda\lambda'} = w_{32}\,S_{33,\lambda\lambda'} - w_{21}\,S_{22,\lambda\lambda'} + B_{\lambda\lambda'}. \tag{VI.6.7}$$

(N is the total number of atoms.)

A is given by

$$A_{\lambda\lambda'} = \left\{\frac{i}{g^2}\sum_{\mu}\sum_{\lambda''} g_{\mu\lambda}\, g_{\mu\lambda'}\, g_{\mu\lambda''}\, b_{\lambda''}^+ (a_2^+ a_3)_\mu - \text{conj. compl.}\right\} + \Gamma_{33}^{\lambda\lambda'} \tag{VI.6.8}$$

and B by

$$B_{\lambda\lambda'} = -\{\cdots\} + \Gamma_{22}^{\lambda\lambda'} \tag{VI.6.9}$$

where the curly brackets are the same. We have further introduced the abbreviation

$$\Gamma_{jj}^{\lambda\lambda'} = \frac{1}{g^2}\sum_{\mu} g_{\mu\lambda}\, g_{\mu\lambda'}\, \Gamma_{jj,\mu}. \tag{VI.6.10}$$

Instead of the old equations of VI.3 we now have the Eqs. (VI.6.2), (VI.6.4), (VI.6.6) and (VI.6.7). While (VI.6.2) and (VI.6.4) contain only the new collective atomic modes (VI.6.1) and (VI.6.3), Eqs. (VI.6.6) and (VI.6.7) still contain via A and B the single atomic operators $(a_2^+ a_3)_\mu$ and $(a_3^+ a_2)_\mu$. In order to express the operators by the collective mode operators (VI.6.1), we now consider as a formal trick a complete set of cavity modes (whose wavelength is cut off at the mean distance between laser atoms). For this set we may assume

$$\sum_{\lambda} g_{\mu\lambda}\, g_{\mu'\lambda} = N g^2 \delta_{\mu\mu'}. \tag{VI.6.11}$$

Using (VI.6.11) we obtain readily from (VI.6.1) for $i \neq j$

$$(a_i^+ a_j)_\mu = \frac{1}{Ng}\sum_{\lambda} g_{\mu\lambda}\, S_{ij,\lambda}. \tag{VI.6.12}$$

By this procedure we introduce of course more modes than the modes which ultimately show laser action and which we are interested in. All the modes, which are artificially added and which will then occur in (VI.6.8), (VI.6.9) (see below), will have (due to their high loss or low gain) such a small amplitude that they can safely be neglected in the final equations. Thus it will become obvious at the end that the lasing and nonlasing modes can be indeed separated in a self-consistent way.

Expressing the atomic operators in A and B by the collective mode operators by means of (VI.6.12) yields for the curly bracket in (VI.6.8) and (VI.6.9)

$$\{\cdots\} = i g \left(\sum_{\lambda'',\lambda'''} C(\lambda, \lambda', \lambda'', \lambda''')\, b_{\lambda''}^+ S_{23,\lambda'''} - \text{hermitian conj.}\right) \tag{VI.6.13}$$*

where

$$C(\lambda, \lambda', \lambda'', \lambda''') = \frac{1}{N g^4}\sum_{\mu} g_{\mu\lambda}\, g_{\mu\lambda'}\, g_{\mu\lambda''}\, g_{\mu\lambda'''}. \tag{VI.6.14}$$

---

* It is important to note that C vanishes for most combinations of the λ's, so that also for this reason the modes, which have been introduced in addition, do not spoil the whole procedure.

We eliminate now $S_{23,\lambda}$ and $S_{32,\lambda}$ from Eq. (VI.6.13) by means of (VI.6.2) and its conjugate equations, and insert the curly bracket into (VI.6.8), (VI.6.9):

$$A_{\lambda\lambda'} = -\left[\sum_{\lambda'',\lambda'''} C(\lambda,\lambda',\lambda'',\lambda''')\, L(\lambda'',\lambda''') + \text{h.c.}\right] + \Gamma_{33}^{\lambda\lambda'}, \tag{VI.6.15}$$

$$B_{\lambda\lambda'} = [\cdots] + \Gamma_{22}^{\lambda\lambda'} \tag{VI.6.16}$$

where

$$L(\lambda'',\lambda''') = b_{\lambda''}^{+}\left(\dot{b}_{\lambda'''} + (i\,\omega_{\lambda'''} + \varkappa_{\lambda'''})\, b_{\lambda'''} - F_{\lambda'''}\right). \tag{VI.6.17}$$

The further procedure consists in steps which are completely analogous to those of Chap. VI.4: We eliminate $S_{32,\lambda}$ from Eqs. (VI.6.4) and (VI.6.2), so that a second order differential equation for b_λ^+ results, in which $(S_{33,\lambda\lambda'} - S_{22,\lambda\lambda'})$ still occurs. We now integrate Eqs. (VI.6.6) and (VI.6.7) [with A and B given by (VI.6.15), (VI.6.16)] and insert the resulting expression for $(S_{33,\lambda\lambda'} - S_{22,\lambda\lambda'})$ into the equation for b_λ^+. Thus we obtain the following new *basic set of equations, which contains only the light modes* but no longer the atomic operators:

$$\left.\begin{aligned}
&\ddot{b}_\lambda^+ - \left(i(\omega_\lambda + \nu_{32}) - \varkappa_\lambda - \gamma_{32}\right)\dot{b}_\lambda^+ + (i\nu_{32} - \gamma_{32})(i\omega_\lambda - \varkappa_\lambda)\, b_\lambda^+ - \\
&\quad - g^2 b_\lambda^+ (S_{33} - S_{22})^0 - g^2 \sum_{\lambda'} b_{\lambda'}^+ \int^t K_1(t,\tau)\, d\tau \times \\
&\quad \times \Big\{\sum_{\lambda'',\lambda'''} C(\lambda,\lambda',\lambda'',\lambda''')\, b_{\lambda''}^+ \times \\
&\quad \times \left(\dot{b}_{\lambda'''} + (i\omega_{\lambda'''} + \varkappa_{\lambda'''})\, b_{\lambda'''} - F_{\lambda'''}\right) + \text{conj. compl.}\Big\}_\tau - \\
&\quad - g^2 \sum_{\lambda'} b_{\lambda'}^+ \left(\int^t K_3(t,\tau)\, \Gamma_{33}^{\lambda\lambda'} d\tau + \int^t K_4(t,\tau)\, \Gamma_{22}^{\lambda\lambda'} d\tau\right) \\
&= -(i\nu_{32} - \gamma_{32})\, F_\lambda^+ + i g\, \Gamma_{32}^{\lambda} + \dot{F}_\lambda^+ .
\end{aligned}\right\} \tag{VI.6.18}$$

In accordance with (VI.6.4) we abbreviate the right hand side of Eq. (VI.6.18) by $F_{tot}^+(t)$ [see (VI.7.2)].

We have used the following abbreviations

$$(S_{33} - S_{22})^0 = \frac{w_{13} N}{D}(w_{21} - w_{32}) \tag{VI.6.19}$$

where

$$D = (w_{32} + w_{13})\, w_{21} + w_{13} w_{32}, \tag{VI.6.20}$$

$$\left.\begin{aligned}
K_1 &= \frac{1}{\Lambda_2 - \Lambda_1}\{e^{-\Lambda_1(t-\tau)}(2\Lambda_1 - 2w_{13} - w_{21}) - \\
&\quad - e^{-\Lambda_2(t-\tau)}(2\Lambda_2 - 2w_{13} - w_{21})\}, \\
K_3 &= \frac{1}{\Lambda_2 - \Lambda_1}\{e^{-\Lambda_1(t-\tau)}(\Lambda_2 - w_{13} - 2w_{32}) - e^{-\Lambda_2(t-\tau)}(\Lambda_1 - w_{13} - 2w_{32})\}, \\
K_4 &= \frac{1}{\Lambda_2 - \Lambda_1}\{e^{-\Lambda_1(t-\tau)}(\Lambda_1 - w_{32} - 2w_{13}) - e^{-\Lambda_2(t-\tau)}(\Lambda_2 - w_{32} - 2w_{13})\}
\end{aligned}\right\} \tag{VI.6.21}$$

where

$$\Lambda_{1,2} = \tfrac{1}{2}(w_{21} + w_{32} + w_{13}) \pm \tfrac{1}{2}\sqrt{(w_{32} + w_{13} - w_{21})^2 - 4 w_{13} w_{32}}. \tag{VI.6.22}$$

The Eqs. (VI.6.18) represent the required result: The atomic operators are explicitly eliminated, and one obtains a set of coupled, nonlinear, quantum mechanical integro-differential equations for the laser field alone*. Although they have a rather complicated appearance, we will show that they can be considerably simplified and solved in excellent approximation. We mention briefly that a system with an arbitrary number of pump levels leads to exactly the same structure of equations. One has merely to replace the K_j's by different expressions.

β) In the case of *running waves* we have merely to put

$$C(\lambda, \lambda', \lambda'', \lambda''') = \frac{1}{N} \sum_{\mu} \exp\{i\,\boldsymbol{x}_\mu(\boldsymbol{k}_\lambda - \boldsymbol{k}_{\lambda'} + \boldsymbol{k}_{\lambda''} - \boldsymbol{k}_{\lambda'''})\}. \quad \text{(VI.6.23)}$$

The Eqs. (VI.6.18) can now be considered as the basic equations for single as well as multimode laser action. The simplification reached is enormous compared to the original set of Eqs. (VI.3.5), (VI.3.13)—(VI.3.19). If there are N atoms and M high gain modes, then the system (VI.3.5), (VI.3.13)—(VI.3.19) is equivalent to a (nonlinear) differential equation of order $5N+M$, whereas the new system is, in very good approximation, equivalent to a differential equation of order $4M$.

The quantum noise sources Γ still depend implicitly on the atomic density matrices which can be treated, however, in the averaged equations in the sense of the self-consistent field method (compare also IV.5). This will become evident in VI.7.

VI.7. Single mode operation above threshold[1], homogeneously broadened line. We continue the treatment of single-mode operation, which was begun qualitatively in VI.4, in a quantitative way. We assume as before that only one mode has high gain, so that the amplitudes of all other modes are negligibly small and can be neglected in the nonlinear term. Thus we obtain from (VI.6.18)

$$\left.\begin{aligned} &\ddot{b}^+ - \left(i(\omega+\nu_{32}) - \varkappa - \gamma_{32}\right)\dot{b}^+ + (i\nu_{32} - \gamma_{32})(i\omega - \varkappa)\, b^+ - \\ &- g^2 b^+ \Big\{(S_{33} - S_{22})^0 + \\ &+ C \int^t K_1(t,\tau)\, d\tau \left((b^+ b)^{\cdot} + 2\varkappa b^+ b - b^+ F - F^+ b\right)_\tau + \\ &+ \left(\int^t K_3(t,\tau)\, \Gamma_{33}^{00}\, d\tau + \int^t K_4(t,\tau)\, \Gamma_{22}^{00}\, d\tau\right)\Big\} = F_{tot}^+ \end{aligned}\right\} \quad \text{(VI.7.1)}$$

where

$$F_{tot}^+ = -(i\nu_{32} - \gamma_{32})\, F^+ + i g\, \Gamma_{32}^{0} + \dot{F}^+. \quad \text{(VI.7.2)}$$

One finds $C = \frac{3}{2}$ for a standing wave and $C = 1$ for a running wave.

(VI.7.1) is an equation for the photon creation operator b^+. The qualitative discussion of (VI.7.1) indicates that after a phasefactor is split off from b^+, $|b^+|$ oscillates around a stable value, so that one is led to the decomposition:

$$b^+ = e^{i\varphi}(r_0 + \varrho)\, e^{i\Omega t}, \quad \text{(VI.7.3)}$$

* Eqs. (VI.6.18) permit us to substantiate our assumption, that we can split the system of cavity modes into "lasing" and "nonlasing" modes. As pointed out in the discussion following Eq. (VI.6.12) the index λ now refers to all cavity modes. From the structure of Eqs. (VI.6.18) it follows, however, immediately that modes with high losses (or those far away from the atomic resonance) retain a very small amplitude, so that they can be neglected in (VI.6.18). One is thus left again with the lasing modes (or those close to threshold) alone. It might appear that this treatment holds only if there exists a strong enough discrimination among modes. However, Graham and Haken [R. Graham and H. Haken: Z. Physik **213**, 420 (1968)] have shown that even for a continuum of modes with equal losses a pronounced discrimination of modes occurs, favouring the resonant mode. In this analysis running modes are assumed.

[1] We follow the paper of H. Haken: Z. Physik **190**, 327 (1966); — The method of solution is that of "quantum mechanical quasi-linearization", introduced in Z. Physik **181**, 96 (1964).

at least in a classical theory. It can be shown, by methods of quantum field theory, that (VI.7.3) can be used too as a decomposition of the operator b^+ with the following interpretations: φ is an operator, which can still refer to the noise source. r_0 is a c-number, whereas ϱ is still an operator*. When we neglect the (slowly varying) phase factor, the decomposition (VI.7.3) or (VI.7.3a) is just the one investigated in V.1, p. 81, where it was shown that the corresponding wave function represents a *coherent state*.

On account of our qualitative discussion in VI.4, we expect ϱ and φ to become considerably smaller than r_0 with increasing inversion above threshold. Thus we can solve the problem by an iteration procedure with respect to increasing powers of ϱ and φ. It is assumed that also the noise sources F and Γ are an order of magnitude smaller than r_0 (multiplied by a suitable quantity).

a) Lowest order: terms independent of φ, ϱ, F and Γ. Inserting (VI.7.3) into (VI.7.1) yields:

$$\left.\begin{aligned}&-\Omega^2 r_0 - \left(i(\omega+\nu_{32}) - \varkappa - \gamma_{32}\right) i\Omega r_0 + (i\nu_{32} - \gamma_{32})(i\omega - \varkappa)\, r_0 - \\ &\quad - g^2 r_0 \left\{(S_{33} - S_{22})^0 - C\, 2\varkappa r_0^2 \frac{2w_{13} + w_{21}}{w_{13}(w_{32} + w_{21}) + w_{32}\, w_{21}}\right\} = 0.\end{aligned}\right\} \tag{VI.7.4}$$

Because Ω and r_0 must be real, we find readily from (VI.7.4) by splitting it into a real and an imaginary part:

$$\Omega = \frac{\nu_{32}\,\varkappa + \omega\,\gamma_{32}}{\varkappa + \gamma_{32}} \tag{VI.7.5}$$

and

$$2\varkappa r_0^2 = \frac{D}{(2w_{13} + w_{21})\, C} \left\{\frac{N w_{13}(w_{21} - w_{32})}{D} - \frac{\varkappa\gamma}{g^2}\left(1 + \frac{(\nu-\omega)^2}{(\varkappa+\gamma)^2}\right)\right\} \tag{VI.7.6}$$

where

$$D = w_{13}(w_{32} + w_{21}) + w_{32}\, w_{21}.$$

N: total number of laser atoms.

(VI.7.5) describes a frequency shift of the laser mode, from that in the unloaded (laser-inactive) cavity towards the center of the atomic resonance (*frequency pulling*), which is *pump-power independent for the homogeneously broadened line*[2].

After multiplying both sides of (VI.7.6) with $\hbar\omega$ we obtain for the left hand side

$$P = 2\varkappa\hbar\omega r_0^2 \tag{VI.7.7}$$

which is the coherent output power. (VI.7.6) thus describes the dependence of this quantity on the pump power $\sim w_{13}$. In VIII.1 this connection is treated in more detail.

---

* As had been shown in V.1, the operator b^+ can be decomposed into a new $\tilde{b}^+$ and a c-number (with respect to the field):

$$b^+ = \tilde{b}^+ + f \tag{VI.7.3a}$$

(where f can still be an operator with respect to the atoms). The replacement $b^+ \to \tilde{b}^+$ can be achieved by a unitary transformation (compare p. 83). One can further check that a replacement of b^+ by $e^{i\varphi} b^+$, where φ is independent of the b-field, does not violate the commutation relations (III.1.10–12). Therefore the decomposition (VI.7.3) is completely justified, as long as φ does not depend on the b's. But even if it does, (VI.7.3) represents an excellent approximation on account of the high photon number involved (compare V.2). — H. SAUERMANN: Z. Physik **189**, 312 (1966), has shown that HEITLER's commutation relations between amplitude and phase (which are good for high photon numbers) are fulfilled by our procedure.

[2] C. H. TOWNES, in: Quantum Electronics, ed. by J. R. SINGER. New York: Columbia University Press 1961.

b) First order: terms linear in ϱ, φ or in the noise-sources. For our explicit considerations we assume complete resonance between the laser mode and the atoms, $\nu=\omega$. If $\nu \neq \omega$, the deviations are small and are given in VI.2, p. 103, for the linewidth. The first order equation can be written

$$e^{i\varphi}(E_2+E_3)=M \tag{VI.7.8}$$

where E_2 contains the expressions linear in ϱ:

$$\left.\begin{aligned} E_2 = \ddot{\varrho} + (\varkappa+\gamma_{32})\dot{\varrho} + \underbrace{\left\{\varkappa\gamma_{32}\varrho - g^2\left[(S_{33}-S_{22})^0 + 2\varkappa r_0^2 \int^t K_1(t,\tau)\,d\tau\right]\varrho\right\}}_{=0} \\ -g^2 r_0\left[\int^t (2r_0\dot{\varrho}+4\varkappa r_0\varrho)K_1(t,\tau)\,d\tau\right]. \end{aligned}\right\} \tag{VI.7.9}$$

The curly bracket vanishes due to (VI.7.6).

E_3 contains terms linear in φ:

$$E_3 = \left(i r_0\ddot{\varphi} + i r_0(\varkappa+\gamma_{32})\dot{\varphi}\right). \tag{VI.7.10}$$

M contains the noise sources:

$$\left.\begin{aligned} M = e^{i\varphi} g^2 r_0 \left\{ r_0 \int^t (\tilde{F}+\tilde{F}^+) K_1(t,\tau)\,d\tau - \int^t \Gamma_{33}^0 K_3\,d\tau - \int^t \Gamma_{22}^0 K_4\,d\tau \right\} \\ \underbrace{+\gamma_{32}\tilde{F}^+ + \dot{\tilde{F}}^+ + ig\,\Gamma_{32}^0}_{\tilde{F}_{tot}^+(t)}. \end{aligned}\right\} \tag{VI.7.11}$$

In it we have defined

$$\tilde{F}^+ = e^{-i\omega t}F^+, \quad \text{and} \quad \tilde{F}_{tot}^+ = e^{-i\omega t}F_{tot}^+. \tag{VI.7.12}$$

We multiply both sides of (VI.7.8) with $e^{-i\varphi}$ and add or subtract from this new equation a hermitian conjugate equation. More loosely speaking, we take the real and the imaginary part, which are responsible for the amplitude fluctuations (real part) and the phase fluctuations (imaginary part).

c) Phase noise. Linewidth formula. The equation for the imaginary part reads

$$\ddot{\varphi} + (\varkappa+\gamma_{32})\dot{\varphi} = \frac{1}{r_0}\,\mathrm{Im}\,\{e^{-i\varphi}\tilde{F}_{tot}^+(t)\}. \tag{VI.7.13}$$

The factor $e^{-i\varphi}$ represents a strong nonlinearity. It can be absorbed into the fluctuating forces for the following reason: As can be shown[1], the fluctuating forces can be represented by random pushes with random phases. Thus $e^{i\varphi}$ can be combined to a new random phase factor. Eq. (VI.7.13) is readily integrated. As we will show at the end of this paragraph, the linewidth (half-width at half power) due to phase noise is given by

$$\Delta\omega = \frac{1}{2t}\langle(\varphi(t)-\varphi(0))^2\rangle \qquad \left(t > \frac{1}{\gamma+\varkappa}\right) \tag{VI.7.14}$$

where the average refers to all coordinates of the noise sources. Using the properties of the noise sources F, Γ (VI.3.11), (VI.3.12), (VI.3.20), (VI.3.21), one obtains

$$\Delta\omega = \frac{1}{2r_0^2(\varkappa+\gamma_{32})^2}\left\{\frac{1}{2}g^2\gamma_{32}(N_2+N_3) + \varkappa\gamma_{32}^2\left(n_{Th}+\frac{1}{2}\right)\right\} \tag{VI.7.15}$$

where $N_2 = N\langle\varrho_{22}\rangle$, $N_3 = N\langle\varrho_{33}\rangle$ is the mean occupation number of level 2 and 3, respectively, of all atoms.

A simple analysis shows that the coupling constant g is connected with the saturated inversion by

$$g^2 = \frac{\gamma_{32}\varkappa}{N_3 - N_2} \tag{VI.7.16}$$

[note, that N_j is the actual occupation number under the joint action of "heat-baths" and laser field, whereas $(S_{33} - S_{22})^0 = (N_3 - N_2)^0$ is the inversion under the action of the "heatbaths" alone].

Thus we find finally for the *line-width (half-width at half-power)*:

$$\Delta\omega = \frac{\gamma_{32}^2 \varkappa^2}{(\varkappa + \gamma_{32})^2} \frac{\hbar\omega}{P} \left\{ \frac{1}{2} \frac{(N_3 + N_2)}{N_3 - N_2} + n_{th} + \frac{1}{2} \right\} \tag{VI.7.17}$$

where $P = 2\varkappa r_0^2 \hbar\omega$ is the output power.

ω is the mode frequency, γ_{32} and $\varkappa$ are the atomic and the cavity half-width, respectively. For a detailed discussion on which γ is to be used, the reader is referred to IV.7, p. 58. (For instance in a two-level system γ depends both on the spontaneous emission rate and on the pump rate.) n_{th} is the number of thermal quanta. In the maser $n_{th} \gg 1$ because usually $kT \gg h\nu$, so that the thermal noise of the resonator determines the linewidth. In the laser, this contribution is entirely negligible. Because $\frac{(N_3 + N_1)}{(N_3 - N_1)} > 1$ the first term is here dominant.

In a two-level system it is many orders of magnitude bigger than the other two terms.

A comparison of the formula (VI.7.17) for the linewidth above threshold with formula (VI.5.22) (valid below threshold) shows, that at resonance, $\nu = \omega$, both expressions differ merely by a factor $\frac{1}{2}$ which is known from the classical theory. The expressions in the curly brackets agree with each other after a rearrangement of terms. [Note that in a homogeneously broadened laser $(N_3 - N_2) = (N_3 - N_2)_{thr.}$ above threshold. Somewhat below threshold the difference between $N_3 - N_2$ and $(N_3 - N_2)^0$ is negligible anyway.]

The line shape is a Lorentzian, which is cut off at frequencies $\approx \varkappa + \gamma$. In order to derive the *line shape* as well as the relation (VI.7.14), we consider the coherence function*

$$\langle b^+(t)\, b(0) \rangle \tag{VI.7.18}$$

$e^{i\Omega t}$ is supposed to be split off. We insert

$$b^+ \approx r_0\, e^{i\varphi(t)} + \cdots \tag{VI.7.19**}$$

into (VI.7.18) and use the fact that both $\Gamma = \Sigma\, \Gamma_\mu$ and F^+ consist of many statisti cally (or quantum mechanically) independent systems, so that

$$\varphi(t) - \varphi(0) = \sum_\mu \varphi_\mu \tag{VI.7.20}$$

where the φ's are independent of each other.

* The following investigation can be done equally well with the expectation value of $b^+(t)$ alone, provided one assumes, that the initial phase at $t = 0$ is measured. The expectation value $\langle b^+(t) \rangle$ is then not vanishing and behaves like the amplitude of a classical coherent field (this statement is, of course, equivalent with the one, that the laser light-field is in a coherent state (compare footnote * on p. 129).

** In using the form (VI.7.19) for $b^+(t)$ we anticipate that somewhat above laser threshold the amplitude-fluctuations ϱ are much smaller than r_0 in the decomposition (VI.7.3) of $b^+(t)$.

We write

$$\langle b^+(t)\, b(0)\rangle \approx r_0^2 \prod_\mu \langle e^{i\varphi_\mu(t)}\rangle \approx r_0^2 \prod_\mu \{1 + i\,\langle\varphi_\mu(t)\rangle - \tfrac{1}{2}\,\langle\varphi_\mu(t)^2\rangle\}. \tag{VI.7.21}$$

Because

$$\langle\varphi_\mu\rangle = 0, \qquad \langle\varphi_\mu\,\varphi_\nu\rangle = 0 \quad (\mu \neq \nu) \tag{VI.7.22}$$

and thus

$$\sum_\mu \langle\varphi_\mu^2\rangle = \langle(\varphi - \varphi_0)^2\rangle \tag{VI.7.23}$$

we have finally

$$\langle b^+(t)\, b(0)\rangle \approx r_0\, e^{-\langle(\varphi-\varphi_0)^2\rangle \frac{1}{2}}. \tag{VI.7.24}$$

The explicit evaluation of (VI.7.23), which we performed above, has shown that the quantity $\langle(\varphi - \varphi_0)^2\rangle$ is proportional to the time, t. On the other hand we would expect for a classical field with a Lorentzian lineshape $E(t) \sim e^{-\Delta\omega t}$ where $\Delta\omega$ is the half-width at half power. Thus the laser line shape is indeed a Lorentzian. The above linewidth formula holds well above threshold, i.e. for

$$\langle\varrho^2\rangle \ll r_0^2. \tag{VI.7.25}$$

d) Amplitude fluctuations. We start again with Eq. (VI.7.8) where we now compare the "real" parts of both sides (again after multiplication with $e^{-i\varphi(t)}$). This yields the following equation:

$$\left.\begin{aligned} &\ddot\varrho + (\varkappa + \gamma_{32})\,\dot\varrho - 2\,g^2 r_0^2 \int^t (\dot\varrho + 2\varkappa\,\varrho)\,K_1(t,\tau)\,d\tau \\ &= g^2 r_0 \left[r_0 \int^t (-1)\,(\tilde F + \tilde F^+)\,K_1(t,\tau)\,d\tau - \int^t \Gamma_{33}^0\,K_3\,d\tau - \int^t \Gamma_{44}^0\,K_4\,d\tau\right] + \\ &\quad + \mathrm{Re}\,\{e^{-i\varphi}\,\tilde F_{tot}^+(t)\}. \end{aligned}\right\} \tag{VI.7.26}$$

For the solution we perform a Fourier transformation:

$$\varrho(t) = \int_{-\infty}^{+\infty} e^{i\omega t}\,\varrho(\omega)\,d\omega. \tag{VI.7.27}$$

We then obtain readily for the Fourier transform of Eq. (VI.7.26)

$$\varrho(\omega)\,O(\omega) = M(\omega) \tag{VI.7.28}$$

where

$$O(\omega) = \{-\omega^2 + i\omega\,(\varkappa + \gamma_{32}) - 2\,g^2 r_0^2 (i\omega + 2\varkappa)\,K_1(\omega)\} \tag{VI.7.29}$$

with

$$K_1(\omega) = -\frac{2w_{13} + w_{21} + 2i\omega}{(\Lambda_1 + i\omega)\,(\Lambda_2 + i\omega)}. \tag{VI.7.30}$$

$M(\omega)$ is of a complicated structure, so that we drop this intermediate result. If $O(\omega)$ possesses one or several *real* roots ω_j, the modulation ϱ of $|b| = r_0 + \varrho$ shows an undamped oscillation ("spiking"). These phenomena are treated in detail in VI.8, where instabilities are also discussed. Because the calculation is elementary, but lengthy, we quote the final result for the correlation functions of the amplitude fluctuations:

$$\langle\varrho(t)\,\varrho(t+\tau)\rangle = \frac{1}{2\pi} \int_{-\infty}^{+\infty} \frac{G(\omega)\,e^{i\omega\tau}}{h(\omega)\,h(-\omega)}\,d\omega \tag{VI.7.31}$$

where

$$G(\omega) = b_0\,\omega^6 + b_1\,\omega^4 + b_2\,\omega^2 + b_3. \tag{VI.7.32}$$

The b's are rather complicated expressions of the parameters of the 3-level-system, so that the reader is referred to the original paper.

The function $h(\omega)$ is defined as

$$h(\omega) = \omega^4 - i\omega^3 A - \omega^2 B + i\omega C + D \tag{VI.7.33}$$

where

$$A = K + W, \tag{VI.7.34}$$

$$B = KW + \Lambda_1\Lambda_2 + 4g^2 r_0^2, \tag{VI.7.35}$$

$$C = K\Lambda_1\Lambda_2 + 2g^2 r_0^2 (2w_{13} + w_{21}) + 8g^2 r_0^2 \varkappa, \tag{VI.7.36}$$

$$D = 4g^2 r_0^2 \varkappa (2w_{12} + w_{21}). \tag{VI.7.37}$$

The Λ's are given in (VI.6.22), whereas K and W have the following meaning:

$$K = (\varkappa + \gamma_{32}), \tag{VI.7.38}$$

$$W = \Lambda_1 + \Lambda_2. \tag{VI.7.39}$$

Whereas the noise intensity is calculated explicitly and quite generally in the original paper[1], we quote here only two special cases:

α) *The special case of a moderate photon number.* Provided $r_0 g < \varkappa, w_{ij}, \gamma_{32}$ the expression for the total noise intensity can be simplified to:

$$\langle \varrho^2(t) \rangle \approx \frac{1}{2} \frac{\left\{\varkappa\left(n_{th} + \frac{1}{2}\right)\gamma_{32}^2 + N\frac{g^2}{2}\gamma_{32}(\varrho_{22} + \varrho_{33})\right\}\left((w_{13} + w_{21})w_{32} + w_{13}w_{21}\right)}{4g^2 r_0^2 \varkappa (2w_{13} + w_{21})(\varkappa + \gamma_{32})}. \tag{VI.7.40}$$

Apparently, the noise intensity $\langle \varrho^2(t) \rangle$ decreases with the inverse of the output power

$$P = \hbar\omega\, 2\varkappa r_0^2 \tag{VI.7.41}$$

in accordance with experiments (compare VI.2).

A closer inspection of the atomic noise contribution $\sim N g^2 \gamma_{32}(\varrho_{22} + \varrho_{33})$ reveals that it comes from the nondiagonal elements Γ_{32}, Γ_{23} of the noise sources, which enter the linewidth. In this region of r_0 it does not stem from the diagonal elements, which represent shot noise. In (VI.7.40) the coupling constant g can be eliminated by means of (VI.7.16). The curly bracket in (VI.7.40) is identical with that occurring in the linewidth formula (VI.7.15). We obtain therefore the following relation:

$$\langle \varrho^2(t) \rangle = \frac{1}{4} \Delta\omega (N_3 - N_2) \frac{\varkappa + \gamma_{32}}{\varkappa^2 \gamma_{32}} \cdot \frac{(w_{13} + w_{21})w_{32} + w_{13}w_{21}}{(2w_{13} + w_{21})}. \tag{VI.7.42}$$

The results for a two-level system can be obtained from (VI.7.42) by letting $w_{13} \to \infty$.

We discuss now the *frequency dependence* of (VI.7.31) for small r_0^2. According to (VI.7.31) the characteristic frequencies of $\langle \varrho(t)\, \varrho(t+\tau) \rangle$ are determined by the roots of $h(\omega)$, which can best be seen if (VI.7.31) is evaluated by means of Cauchy's theorem. For our present discussion it is more convenient, however, to study the roots, ω_r of $O(\omega)$ [Eq. (VI.7.29)] instead.

Let us assume that

$$|\omega_r| < \varkappa, w_{ik} \quad \text{and thus} \quad |\omega_r| < \Lambda_1, \Lambda_2. \tag{VI.7.43}$$

Then we can neglect ω completely in the factor of $2g^2 r_0^2$ so that

$$O(\omega) \approx -\omega^2 + i\omega(\varkappa + \gamma_{32}) + \underbrace{2g^2 r_0^2 \varkappa \frac{2w_{13} + w_{21}}{(w_{13} + w_{32})w_{21} + w_{13}w_{32}}}_{S} \tag{VI.7.44}$$

which has the roots

$$\omega_r = \frac{i(\varkappa + \gamma_{32})}{2} \pm \frac{1}{2}\sqrt{-(\varkappa + \gamma_{32})^2 + 4S}. \tag{VI.7.45}$$

One root lies close to $i(\varkappa+\gamma_{32})$. For $S<\varkappa+\gamma_{32}$ the other root can be simplified to

$$\omega_r \approx i\frac{S}{(\varkappa+\gamma_{32})} = i\frac{4g^2 r_0^2 \varkappa(2w_{13}+w_{21})}{(\varkappa+\gamma_{32})\left((w_{13}+w_{32})w_{21}+w_{13}w_{32}\right)} \tag{VI.7.46}$$

or, using (VI.7.41) and (VI.7.16)

$$\omega_r = i\frac{2P}{\hbar\omega}\frac{\gamma_{32}\varkappa}{(\varkappa+\gamma_{32})}\frac{1}{N_3-N_2}\frac{(2w_{13}+w_{21})}{(w_{13}+w_{32})w_{21}+w_{13}w_{32}}. \tag{VI.7.47}$$

As is evident from (VI.7.47), the frequency (or better decay constant) $|\omega_r|$ increases linearly with the photon flux, in agreement with experiments.

To be sure, there are three other decay constants in the system, because $O(\omega)$ (VI.7.29) possesses four roots. All these are of the magnitude $\varkappa+\gamma_{32}$ or Λ_1 or Λ_2 and are connected with intensities which are much smaller than the contribution connected with (VI.7.40). This latter statement can be verified simply by inserting these roots ω_j into $\frac{O(\omega)}{\omega-\omega_j}$ and $O^*(\omega)$. Our results for the amplitude-fluctuations can be summarized as follows: for

$$r_0^2\cdot g^2<\gamma_{32}^2,\, w_{ik}^2 \tag{VI.7.48}$$

the correlation function takes the form

$$\langle\varrho(t)\,\varrho(t+\tau)\rangle=\langle\varrho^2\rangle\, e^{-|\omega_r||\tau|} \tag{VI.7.49}$$

where $\langle\varrho^2\rangle$ and $|\omega_r|$ are given by (VI.7.40) and (VI.7.47), respectively. The above result (VI.7.49) with (VI.7.40) and (VI.7.47) was obtained for a *homogeneously broadened line*. For an *inhomogeneously broadened line* and also for the *gas-laser (inhomogeneously broadened line)* see (VI.10).

β) The special case of a big photon number. For $gr_0\gg w_{ik},\gamma_{32}$, the total noise intensity can be considerably simplified. Keeping only the leading term, we obtain

$$\langle\varrho^2\rangle=\left(n_{th}+\frac{1}{2}\right)\frac{\varkappa}{(\varkappa+w_{13}+w_{21}/2)}. \tag{VI.7.50}$$

For $\varkappa>w_{13}, w_{21}$ it represents just half the thermal noise plus half the vacuum fluctuations. This result is to be expected, because the two degrees of freedom of a complex amplitude have been decomposed into that of "radial" and that of "tangential" (or phase) motion. On the other hand, for $\varkappa\ll w_{13}, w_{21}$, these fluctuations are largely suppressed.

VI.8. Stability of amplitude. Spiking and damped oscillations. Single-mode operation, homogeneously broadened line. As we have shown above, the phase is unstable, and undergoes a diffusion process. After splitting off the phase-factor, $e^{i\varphi(t)}$, we now investigate the stability of the amplitude r_0 [compare (VI.7.3)]. The standard method consists in adding to r_0 a small term ϱ and studying its behaviour in time[1]. This is just the procedure of VI.7, where an equation for ϱ was obtained.

[1] It should be noted that so far stability criteria have been derived only for *classical* nonlinear equations. General treatments of such equations, which belong mainly to the field of nonlinear mechanics or nonlinear network-theory, can be found e.g. in the monographs M. M. Bogoliubov and Y. A. Mitropolsky: Nonlinear oscillations. Delhi: Hindustan Publ. Corp. 1961. — M. Minorsky: Nonlinear oscillations. New York: D. van Nostrand Co. 1962; — The stability criterion, which is based on the linearization, can also be applied to a quantum mechanical equation, e.g. of the type (VI.7.1), because the time-dependence of the solutions of a classical linear equation and the corresponding quantum mechanical equation is the same (the only difference consists in the coefficients, which are operators in the latter case).

For the present purpose it suffices to investigate the homogeneous Eq. (VI.7.28) (with $M(\omega)\equiv 0$), of the Fourier transform of ϱ. We denote the roots of $O(\omega)=0$ by $\omega_{r,\sigma}\cdot \varrho(t)$ then has the form

$$\varrho(t)=A_1\, e^{i\omega_{r,1}}+A_2\, e^{i\omega_{r,2}}+\cdots. \qquad \text{(VI.8.1)}$$

The amplitude is stable, if the imaginary part of all ω's is positive, i.e. if all roots are lying in the upper complex plane. Instead of $O(\omega)$ we consider

$$h(\omega)=O(\omega)\,(i\omega+\Lambda_1)\,(i\omega+\Lambda_2) \qquad \text{(VI.8.2)}$$

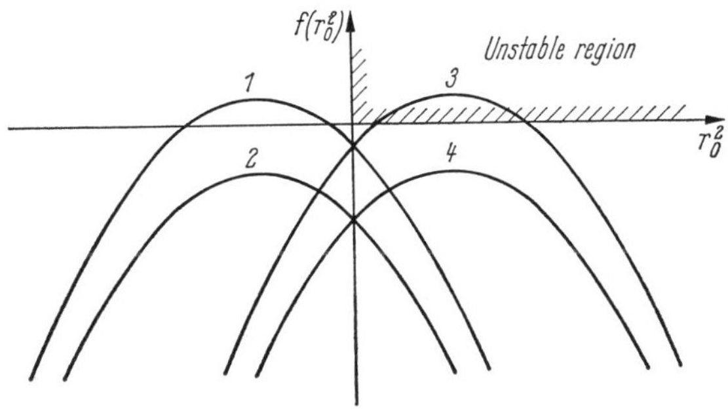

Fig. 30. The function $f(r_0^2)$ [Eq. (VI.8.3)] for $\gamma_{32}+w_{32}+\frac{1}{2}w_{21}-\varkappa>O$ or equivalently, $V<O$.
1. $U<O$; $U^2+V>O$.
2. $U<O$; $U^2+V<O$.
3. $U>O$; $U^2+V>O$.
4. $U>O$; $U^2+V<O$.

The curves 1, 2 and 4 present a stable situation for all amplitudes r_0. Curve 3 describes a situation where an unstability occurs for a certain finite region of r_0^2. This case is excluded for a two-level system; and in the three-level system no combination of the γ's and w's is known so far which would produce 3.

which is allowed as long as $\omega=\Lambda_1, \Lambda_2$ is not among the roots ω_r. The condition that all roots are lying in the upper complex plane is

$$f(r_0^2)\equiv A^2 D+C^2-C\,A\,B<0 \qquad \text{(VI.8.3)}$$

(and $C/A>0$, which is always fulfilled in the present context).

The quantities A, B, C, D, are explicitly given in (VI.7.33)—(VI.7.37) and depend on the photon number, r_0^2.

When we insert these expressions into (VI.8.3) we find that $f(r_0^2)$ is a quadratic function of r_0^2, which is negative for $r_0^2=0$. Thus, according to (VI.8.3), the amplitude is always stable for small r_0^2. With increasing r_0^2, $f(r_0^2)$ may become zero and then change its sign, so that the amplitude becomes unstable. The zeros of $f(r_0^2)$ are given by

$$4g^2 r_0^2=U\pm\sqrt{U^2+V} \qquad \text{(VI.8.4)}$$

where U and V are given below [see (VI.8.7)—(VI.8.9)]. According to the position of the zeros, different physical situations arise, which we now discuss:

a) Qualitative discussion: $f(r_0^2)$ is plotted in Figs. 30 and 31. Fig. 30 refers to the case

$$\gamma_{32}+w_{32}+\tfrac{1}{2}\,w_{21}-\varkappa>0 \qquad \text{(VI.8.5)}$$

which neccessitates $V<0$ [see below, (VI.8.8) and (VI.8.9)]. In the cases 1, 2 and 4 the unstable domain is never reached. The case 3 describes a situation, where the instability occurs within a certain finite range of values of r_0^2.

In the two-level laser such a situation can never occur (see below). For the three-level laser the general case has not been discussed completely, because U and V are given by rather lengthy expressions. So far, it has not been possible to find combinations of these parameters, so that the case 3 would hold.

Fig. 31 refers to the case

$$\gamma_{32}+w_{32}+\tfrac{1}{2}w_{21}-\varkappa<0 \tag{VI.8.6}$$

which neccessitates $V>0$ [see below, (VI.8.8) and (VI.8.9)]. Both curves 1 and 2 cut the r_0^2-axis at a positive value, so that for sufficiently high photon numbers the amplitude becomes unstable.

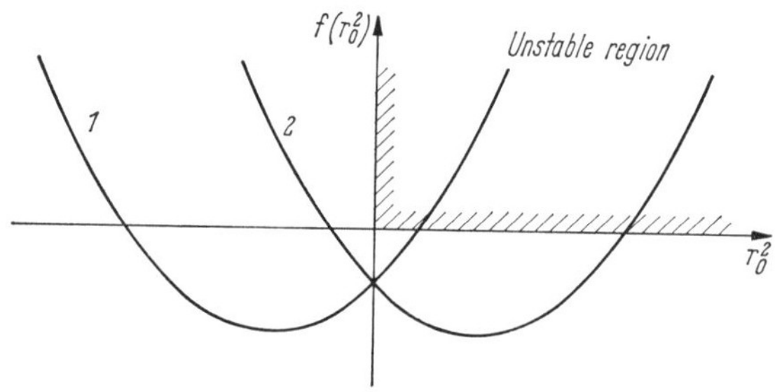

Fig. 31. The function $f(r_0^2)$ [Eq. (VI.8.3)] for $\gamma_{32}+w_{32}+\frac{1}{2}w_{21}-\varkappa<0$ (or, equivalently, $V>0$)
1. $U<0$; 2. $U>0$.
In both cases at sufficiently high amplitudes the unstable region is reached.

In the above and following discussion, r_0^2 is treated as an independent variable. It should be noted that r_0^2 is connected with the γ's and w's by Eq. (VI.7.6), but because the photon number depends in addition on the *concentration* N/V of the impurity atoms, it can be considered for the mathematical treatment as an independent variable.

Because in most laser systems $\varkappa<\gamma_{32}$ so that (VI.8.5) holds, the amplitude is stable and no undamped spiking should occur. Thus undamped spiking, which is observed in some systems, for instance in ruby, may be caused by other mechanisms. A discussion of other possible mechanisms is given in VI.9.d.

b) Quantitative treatment. The coefficients U and V occurring in (VI.8.4) are given by

$$U=\frac{\tilde{A}^2 d+2\tilde{\varkappa}c-\tilde{A}(\tilde{\varkappa}b+c)}{2(\tilde{A}\tilde{\varkappa}-\tilde{\varkappa}^2)}\gtreqless 0, \tag{VI.8.7}$$

$$V=\frac{\tilde{A}bc-c^2}{\tilde{\varkappa}^2-\tilde{A}\tilde{\varkappa}} \tag{VI.8.8}$$

where

$$\left.\begin{aligned}
\tilde{\varkappa}&=2\varkappa+\tfrac{1}{2}(2w_{13}+w_{21}),\\
\tilde{A}-\tilde{\varkappa}&=\gamma_{32}+w_{32}+\tfrac{1}{2}w_{21}-\varkappa\gtreqless 0,\\
(\tilde{A}b-c)\,c&=K^2W(K+W)\Lambda_1\Lambda_2>0,\\
b&=KW+\Lambda_1\Lambda_2,\\
c&=K\Lambda_1\Lambda_2,\\
d&=K(2w_{13}+w_{21}).
\end{aligned}\right\} \tag{VI.8.9}$$

K, W, Λ_1, Λ_2 are defined in the Eqs. (VI.7.38), (VI.7.39), (VI.6.22) respectively. Inserting numerical values one can check readily whether $f(r_0^2) \gtrless 0$.

c) The special case $w_{13} \to \infty$ ("two level system")[2]: In this limit both U and V diverge, and thus also one root of $f(r_0^2) = 0$, whereas the other one remains finite:

$$4g^2 r_0^2 = -\frac{V}{2U}. \qquad \text{(VI.8.10)}$$

Because $g^2 r_0^2$ must be positive the sign of U must be opposite to that of V:

a) We find

$$V < 0 \quad \text{if} \quad \gamma_{32} + w_{32} + \tfrac{1}{2} w_{21} - \varkappa > 0 \qquad \text{(VI.8.11)}$$

and

$$U > 0 \quad \text{if} \quad \gamma_{32} + w_{32} + w_{21} - \varkappa < 0. \qquad \text{(VI.8.12)}$$

(VI.8.11) and (VI.8.12) contradict each other, i.e. no cross-section with the positive r_0^2-axis exists and no instability is possible.

b) On the other hand,

$$V > 0 \quad \text{for} \quad \gamma_{32} + w_{32} + \tfrac{1}{2} w_{21} - \varkappa < 0 \qquad \text{(VI.8.13)}$$

and

$$U < 0 \quad \text{for} \quad \gamma_{32} + w_{32} + w_{21} + \varkappa > 0. \qquad \text{(VI.8.14)}$$

(VI.8.13) and (VI.8.14) can be fulfilled simultaneously for a certain range of $\varkappa$, so that the amplitude becomes unstable. In this case, r_0^2 must be chosen so that

$$4g^2 r_0^2 \geqq \left| \frac{V}{2U} \right|. \qquad \text{(VI.8.15)}$$

Let us discuss this instability more closely. If there is a slight perturbation, $r_0 + \varrho$ will increase at first exponentially. Due to induced emission more atoms fall into the ground state, so that the inversion is depressed. Consequently, the light-amplitude must drop again. This shows that in the case of instability an undamped spiking can be expected. For adequate treatment nonlinear terms of ϱ must also be taken into account (compare[3]).

On the other hand, *damped oscillation occurs*[4], if the roots, ω_r, of $h(\omega)$ possess a real part. Because $x = (-i\omega)$ obeys an algebraic equation of fourth order (in the general case of a three-level laser) with real coefficients, there must be pairs of complex conjugate roots $x = a \pm ib$, where a, b are real. Thus, the four roots ω_r must be of the form:

$$\omega_{r,1} = ia_1 + b_1; \quad \omega_{r,2} = ia_1 - b_1; \quad \omega_{r,3} = ia_2 + b_2; \quad \omega_{r,4} = ia_2 - b_2 \qquad \text{(VI.8.16)}$$

where in the case of stability (VI.8.3) both a_1 and a_2 are positive.

Because $h(\omega)$ is of fourth order, in general the roots ω_r cannot be determined explicitly.

The limiting case

$$w_{13} \to \infty \qquad \text{(two-level system)}$$

has been treated, however.

---

[2] The stability of the two-level system has been investigated by A. S. GRASJUK and A. N. ORAEWSKIJ, in: Quantum Electronics and Coherent Light (ed. P. A. MILES), p. 192. New York-London: Acad. Press 1964. We present here a somewhat different treatment.

[3] A detailed treatment of undamped spiking has been given by V. V. KOROBKIN and A. V. USPENSKI: Soviet Phys. JETP **45**, 1003 (1963), by analytical methods and by E. R. BULEY and F. W. CUMMINGS: Phys. Rev. **134**, A 1454 (1964), using a computer calculation.

[4] For a treatment of damped oscillations by means of rate equations compare Chap. VIII.4.

In order to do the limit properly, one has to start with $O(\omega)$, then take $w_{13}\to\infty$, and then to form $h(\omega)=(i\omega+w_{21}+w_{32})\,O(\omega)$ where

$$\left.\begin{aligned} h(\omega)=&-i\omega^3-\omega^2(w_{21}+w_{32}+\varkappa+\gamma_{32})+\\ &+i\omega\left((\varkappa+\gamma_{32})(w_{21}+w_{32})+4g^2r_0^2\right)+8\varkappa g^2r_0^2. \end{aligned}\right\} \quad \text{(VI.8.17)}$$

The roots can be calculated by the standard formula for an equation of the 3rd order. The expressions are, however, formidable, so that we quote only the limiting cases:

1. small photon numbers: $(gr_0<w_{ik})$:
only the root occurring in the expression (VI.7.49) for the amplitude fluctuation is of importance [compare (VI.7.47)].

2. high photon numbers $(gr_0>w_{ik},\varkappa)$:

$$\left.\begin{aligned} \omega_{r,1}&=2\varkappa i,\\ \omega_{r,2}&=2g\,r_0+i\tfrac{1}{2}(w_{32}+w_{21}+\gamma_{32}-\varkappa),\\ \omega_{r,3}&=-2g\,r_0+i\tfrac{1}{2}(w_{32}+w_{21}+\gamma_{32}-\varkappa). \end{aligned}\right\} \quad \text{(VI.8.18)}$$

From (VI.8.18) one sees again, that the amplitude becomes unstable, if

$$\varkappa>(w_{32}+w_{21}+\gamma_{32}). \quad \text{(VI.8.19)}$$

Two damping constants appear, $2\varkappa$ and $\frac{1}{2}(w_{32}+w_{21}+\gamma_{32}-\varkappa)$, the latter usually being bigger. Superimposed on the decay there are oscillations with frequency $2gr_0$. This behaviour is clearly exhibited by Fig. 22 on p. 108.

VI.9. Qualitative discussion of two-mode operation[1].

a) Some transformations. For the qualitative discussion it is convenient to proceed in close analogy to the single mode case and to use equations, which *explicitly* contain the *inversion*. It suffices to write down the equation for mode 1. The other one can then be obtained simply by exchanging the indices 1 and 2. We start from the field Eq. (VI.6.2) and the matter Eq. (VI.6.4) from which we can easily eliminate $S_{32,\lambda}$. Choosing $\lambda=1$, we are then left with

$$\left.\begin{aligned} &\ddot{b}_1^+-\left(i(\omega_1+\nu_{32})-\varkappa_1-\gamma_{32}\right)\dot{b}_1^++(i\nu_{32}-\gamma_{32})(i\omega_1-\varkappa_1)\,b_1^+-\\ &\quad-g^2b_1^+S^{11}-g^2b_2^+S^{12}=\underbrace{-(i\nu_{32}-\gamma_{32})F_1^++ig\,\Gamma_{32}^1+\dot{F}_1^+}_{F^+_{tot,1}(t)}. \end{aligned}\right\} \quad \text{(VI.9.1)}$$

$S^{\lambda\lambda'}$ is an abbreviation for

$$S^{\lambda\lambda'}=S_{33,\lambda\lambda'}-S_{22,\lambda\lambda'} \quad \text{(VI.9.2)}$$

[compare (VI.6.3)] and depends accordingly on the inversion $(N_3-N_2)_\mu$ of the atom μ in the following way (for standing waves, where g_μ is real)

$$S^{11}=\frac{1}{g^2}\sum_\mu g_{\mu 1}^2(N_3-N_2)_\mu, \quad \text{(VI.9.3)}$$

$$S^{12}=\frac{1}{g^2}\sum_\mu g_{\mu 1}\,g_{\mu 2}(N_3-N_2)_\mu. \quad \text{(VI.9.4)}$$

We have now to establish equations for the variables $S^{\lambda\lambda'}(\lambda,\lambda'=1,2)$. Because the equations become formidable, we confine ourselves to the simplest possible case, namely to atoms having only two levels, which are connected by a pump rate w_{23} and a decay rate w_{32}.

---

[1] We follow essentially H. HAKEN, in: Dynamical processes in Solid State Physics (ed. R. KUBO and H. KAMIMURA). Tokyo-New York: Syokabo, and W. A. Benjamin, Inc. 1967.

Instead of the Eqs. (VI.6.6) and (VI.6.7) we then obtain

$$\dot{S}_{33,\lambda\lambda'} = w_{23} S_{22,\lambda\lambda'} - w_{32} S_{33,\lambda\lambda'} + A_{\lambda\lambda'}, \tag{VI.9.5}$$

$$\dot{S}_{22,\lambda\lambda'} = w_{32} S_{33,\lambda\lambda'} - w_{23} S_{22,\lambda\lambda'} + B_{\lambda\lambda'} \tag{VI.9.6}$$

where A and B are functions of the light-mode amplitudes b^+, b and of the fluctuating forces F^+ and Γ.

Because in a two-level atom, $(N_2+N_3)_\mu = 1$ still holds, we can easily introduce $S^{\lambda\lambda'}$ [compare (VI.9.2)], and

$$\left\{\frac{1}{g^2}\sum_\mu g_{\mu\lambda} g_{\mu\lambda'} (N_2+N_3)_\mu\right\} \equiv \delta_{\lambda\lambda'} \cdot N \tag{VI.9.7}$$

as new variables, replacing $S_{33,\lambda\lambda'}$ and $S_{22,\lambda\lambda'}$. In analogy to (VI.4.10) we thus obtain from (VI.9.5), (VI.9.6) an equation for $S^{\lambda\lambda'}$:

$$\dot{S}^{\lambda\lambda'} = \{w_{23} - w_{32}\}\, \delta_{\lambda\lambda'} N - (w_{23} + w_{32})\, S^{\lambda\lambda'} + 2A_{\lambda\lambda'} \tag{VI.9.8}$$

or, after insertion of $A_{\lambda\lambda'}$ according to (VI.6.15):

$$\left.\begin{aligned} \dot{S}^{11} = {} & \frac{N d_0 - S^{11}}{T} - 2\cdot\frac{3}{2}\{(b_1^+ b_1)^{\cdot} + 2\varkappa_1 b_1^+ b_1\} - \\ & - 2\{(b_2^+ b_2)^{\cdot} + 2\varkappa_2 (b_2^+ b_2)\} + \text{fluctuations}, \end{aligned}\right\} \tag{VI.9.9}$$

$$\left.\begin{aligned} \dot{S}^{12} = {} & -\frac{S^{12}}{T} - 2\{(b_1^+ b_2)^{\cdot} + (b_2^+ b_1)^{\cdot} + (\varkappa_1+\varkappa_2)(b_1^+ b_2 + b_2^+ b_1)\} + \\ & + \text{fluctuations} \end{aligned}\right\} \tag{VI.9.10}$$

where

$$\frac{1}{T} = w_{23} + w_{32} \tag{VI.9.11}$$

and

$$d_0 = \frac{w_{23} - w_{32}}{w_{23} + w_{32}}. \tag{VI.9.12}$$

As in the single-mode case, the fluctuations in (VI.9.9), (VI.9.10) play a minor role (except perhaps at very high photon numbers) and are neglected in our present discussion.

In the following we eliminate the main time dependence by substituting

$$b_j^+ \to e^{i\nu_{32}t} b_j^+, \quad a_j^+ \to e^{i\nu_{32}t} a_j^+. \tag{VI.9.13}$$

As can be followed up in the original equations, this has the effect that ν_{32} vanishes in (VI.3.13), and that ω_j is to be replaced by $\omega_j - \nu_{32} \equiv \delta_j$. The corresponding substitutions have to be made in (VI.9.1). We do not assume complete resonance, because the frequency difference $\omega_1 - \omega_2$ plays a major role in our subsequent discussion.

b) Both modes well below threshold. Because both modes have a small amplitude in this region, expressions quadratic in b^+, b can be neglected. The inversion S^{11} is then time-independent:

$$S^{11} = N d_0 \tag{VI.9.14}$$

whereas the "mixed" inversion

$$S^{12} = 0, \tag{VI.9.15}$$

in the steady state.

Eq. (VI.9.1) then reduces to an equation which is completely equivalent to (VI.5.4), if the transformation (VI.9.13) is taken into account. The further discussion has been performed in VI.5, for the multimode case also and need not be repeated here.

c) Modes somewhat above or somewhat below threshold. As we have seen in the treatment of a single mode, the laser line narrows with increasing inversion as we approach the threshold. Because below threshold the phase and amplitude fluctuations contribute in a similar way, we may safely neglect $(b_j^+ b_j)^{\cdot}$ compared to $2\varkappa_j b_j^+ b_j$ in the neighbourhood of the threshold.

Putting

$$b_j^+ = e^{i(\omega_j - \nu_{32})t}\, \tilde{b}^+(t) \tag{VI.9.16}$$

where $\tilde{b}_j^+(t)$ changes again slowly, we see that $(b_1^+ b_2)^{\cdot} + (b_2^+ b_1)^{\cdot}$ is negligibly small compared to $(\varkappa_1 + \varkappa_2)\,(b_1^+ b_2 + b_2^+ b_1)$.

Provided the changes of $\tilde{b}^+$ are small in times $1/T$, S^{11} follows the "motion" of $b_j^+ b_j$ adiabatically, so that

$$S^{11} \approx N d_0 - 2\left(\tfrac{3}{2}\, T\, 2\varkappa_1\, b_1^+ b_1 + T\, 2\varkappa_2\, b_2^+ b_2\right). \tag{VI.9.17}$$

S^{11} is called the *time-independent inversion,* because above threshold $b_j^+ b_j$ is time-independent, if the small intensity fluctuations are neglected. As in the single-mode case, S^{11} is responsible for the gain of mode 1. In S^{11} the inversion $N d_0$ which is caused by pumping and incoherent decay processes, is lowered by the intensity of both modes, so that S^{11} describes the gain saturation.

Consider now (VI.9.10). Because $(b_1^+ b_2)$ contains still a factor $\sim e^{i(\omega_1-\omega_2)t}$ the "adiabatic condition" is not necessarily fulfilled, so that we solve (VI.9.10) more exactly:

$$\left.\begin{aligned} S^{12} = -2 \int^{t} e^{-\frac{1}{T}(t-\tau)} \{ & i(\omega_1 - \omega_2)\, b_1^+ b_2 - i(\omega_1 - \omega_2)\, b_2^+ b_1 + \\ & + (\varkappa_1 + \varkappa_2)\,(b_1^+ b_2 + b_2^+ b_1)\}\, d\tau, \end{aligned}\right\} \tag{VI.9.18}$$

$$\approx -2\left\{\frac{i(\omega_1 - \omega_2) + (\varkappa_1 + \varkappa_2)}{\frac{1}{T} + i(\omega_1 - \omega_2)}\, b_1^+ b_2 + \text{c.c.}\right\}. \tag{VI.9.19}$$

S^{12} pulsates above threshold with (approximately) a frequency $\omega_1 - \omega_2$ and thus describes the *population pulsation.* It is also called the *time-dependent (atomic) response.*

S^{12} gives rise to a change in the gain of mode 1 and, as we will show below, also to a frequency shift. In order to see under what conditions this term becomes influential on the gain, we compare the b-dependent part of S^{11} with

$$2 b_1^+ b_2 \,\mathrm{Re}\left\{\frac{i(\omega_1 - \omega_2) + (\varkappa_1 + \varkappa_2)}{\frac{1}{T} + i(\omega_1 - \omega_2)}\right\}. \tag{VI.9.20}$$

[The reason why we choose the part (VI.9.20) of S^{12} as representative of the change in gain will become clearer below.] We assume that both modes are above threshold and have comparable amplitudes. A comparison between (VI.9.20) and

$$2\left(\tfrac{3}{2}\, T\, 2\varkappa_1\, b_1^+ b_1 + T\, 2\varkappa_2\, b_2^+ b_2\right) \tag{VI.9.21}$$

then shows that (VI.9.20) can be neglected if

$$|\omega_1 - \omega_2| \gg \frac{1}{T} \quad \text{and} \quad \varkappa T \gg 1 \tag{VI.9.22}$$

and that (VI.9.20) is comparable with (VI.9.21) if

$$|\omega_1 - \omega_2| \ll \frac{1}{T}. \tag{VI.9.23}$$

In our following analysis, we have therefore to distinguish between these two cases.

For a further discussion, we have to insert S^{11} and S^{12} into (VI.9.1) [after the substitution (VI.9.13) has been made]:

$$\left.\begin{aligned} &\ddot{b}_1^+ + (\varkappa_1 + \gamma_{32} - i\delta_1)\,\dot{b}_1^+ + \gamma_{32}(\varkappa_1 - i\delta_1)\,b_1^+ - \\ &\quad - g^2 b_1^+ (N d_0 - 6T\varkappa_1 b_1^+ b_1 - 4\varkappa_2 T b_2^+ b_2) + \\ &\quad + g^2 b_2^+ \, 2\left(\frac{\varkappa_1 + \varkappa_2 + i(\omega_1 - \omega_2)}{\frac{1}{T} + i(\omega_1 - \omega_2)}\right)(b_1^+ b_2) + \\ &\quad + g^2 b_2^+ \, 2\,\frac{\varkappa_1 + \varkappa_2 - i(\omega_1 - \omega_2)}{\frac{1}{T} - i(\omega_1 - \omega_2)}\,(b_2^+ b_1) = \tilde{F}^+_{tot,1}(t). \end{aligned}\right\} \tag{VI.9.24}$$

The equation for the second mode can be obtained from (VI.9.24) simply by interchanging all indices 1 and 2.

Under the assumption that the mode frequencies are not much influenced by the mode coupling (an assumption which has to be checked later),

$$b_1^+ \sim e^{i(\omega_1 - \nu_{32})t}, \tag{VI.9.25}$$

$$b_2^+ \sim e^{i(\omega_2 - \nu_{32})t} \tag{VI.9.26}$$

(if the small fluctuations of the b's are neglected).

We then see from (VI.9.24) that all expressions containing b_1^+ are $\sim e^{i(\omega_1 - \nu_{32})t}$, whereas the last term on the left hand side, which contains b_1, is proportional to

$$e^{i(2\omega_2 - \omega_1 - \nu_{32})t} \tag{VI.9.27}$$

so that it differs from the first terms by a factor

$$e^{i(2\omega_2 - 2\omega_1)t}. \tag{VI.9.28}$$

When the average over a time $t' \gg \frac{1}{2|\omega_2 - \omega_1|}$ is performed, this last term vanishes. As we will show below, this reasoning breaks down if $|\omega_2 - \omega_1|$ becomes small $\left(\approx \frac{1}{T}\right)$, because then frequency locking occurs, so that $\Omega_2 - \Omega_1 = 0$.

d) Both modes above threshold[2]. In analogy to the single mode we put

$$b_j^+ = e^{i\tilde{\Omega}_j t}\, e^{i\varphi_j(t)}\,(r_{0j} + \varrho_j). \tag{VI.9.29}$$

[Note that the total mode frequency is $\Omega_j = \tilde{\Omega}_j + \nu_{32}$, because the frequency dependence with $e^{i\nu_{32}t}$ had been removed in (VI.9.13)]. $\tilde{\Omega}_j$, $\varphi_j(t)$, r_{0j} and ϱ_j are quantities still to be determined, which is done in analogy to the single-mode case.

---

[2] For a quantitative treatment see: H. Haken and H. Sauermann: Z. Physik **173**, 261 (1963), where the case $|\omega_1 - \omega_2| \gg 1/T$ is treated, and N. G. Basov, V. N. Morosov, and A. N. Oraewskij: J. Exptl. Theoret. Phys. USSR **49**, 895 (1965). — Ostrowskij: J. Exptl. Theoret. Phys. USSR **48**, 1087 (1965); **49**, 1534 (1965), who also treat the case $|\omega_1 - \omega_2| < 1/T$. The latter author neglects, however, the spatial difference between the modes. All papers quoted under[2] neglect noise.

α) $|\omega_1-\omega_2|\gg 1/T$. We neglect in Eq. (VI.9.24) the term with the "wrong" time-dependence (VI.9.27), and the change of gain caused by the time-dependent atomic response S^{12}, but we keep that term of S^{12}, which causes a frequency shift, and which is one order of magnitude bigger than the other (gain) term.

From (VI.9.24) then follows:

$$\left.\begin{aligned}&\ddot{b}_1^+ + (\varkappa_1+\gamma_{32}-i\delta)\,\dot{b}_1^+ + b_1^+\{\gamma_{32}\varkappa_1 -\\ &\quad - g^2(N d_0 - 6T\varkappa_1 b_1^+ b_1 - 4T\varkappa_2 b_2^+ b_2)\} -\\ &\quad - i b_1^+\left[\delta_1\gamma_{32} + \frac{\omega_1-\omega_2}{\frac{1}{T^2}+(\omega_1-\omega_2)^2}\, 2(\varkappa_1+\varkappa_2)\, g^2 b_2^+ b_2\right] = \tilde{F}^+_{tot,1}(t)\,.\end{aligned}\right\} \tag{VI.9.30}$$

Because b_2^+ and b_2 occur in the combination $b_2^+\cdot b_2$, the phase-fluctuations of mode 2 do not influence those of mode 1.

Consequently, the phase noise is that of single mode laser action (if r_{01} is the same for single and two-mode laser action), see also below, p. 143. In order to discuss (VI.9.30) further, we must first treat the

lowest order of approximation

where we neglect all fluctuations ($\tilde{F}^+_{tot}=\varphi=\varrho\equiv 0$). Inserting (VI.9.29) into (VI.9.30) and splitting (VI.9.30) into real and imaginary parts yields

imaginary part: (responsible for the frequency shift)

$$\tilde{\Omega}_1 = \frac{\gamma_{32}\delta_1}{\varkappa_1+\gamma_{32}} + \frac{(\omega_1-\omega_2)}{\varkappa_1+\gamma_{32}}\cdot\frac{\varkappa_1+\varkappa_2}{\frac{1}{T^2}+(\omega_1-\omega_2)^2}\, g^2 b_2^+ b_2\,. \tag{VI.9.31}$$

Besides the frequency pulling described by the first expression, which is known from single-mode operation, an additional frequency shift occurs, which depends on the intensity of the other mode, 2.

real part:

When we use

$$\tilde{\Omega}_1 \approx \frac{\gamma_{32}\delta_1}{\varkappa_1+\gamma_{32}} \tag{VI.9.32}$$

we find immediately

$$\gamma_{32}\varkappa_1\left(1+\frac{\delta_1^2}{(\varkappa_1+\gamma_{32})^2}\right) r_{01} - g^2\{N d_0 - 6T\varkappa_1 r_{0,1}^2 - 4\varkappa_2 T r_{0,1}^2\}\, r_{01} = 0, \tag{VI.9.33}$$

$$\gamma_{32}\varkappa_2\left(1+\frac{\delta_2^2}{(\varkappa_2+\gamma_{32})^2}\right) r_{02} - g^2\{N d_0 - 6T\varkappa_2 r_{0,2}^2 - 4\varkappa_1 T r_{0,2}^2\}\, r_{02} = 0 \tag{VI.9.34}$$

which are two algebraic equations for r_{01} and r_{02}.

Because above threshold $r_{0j}\neq 0$, we can divide these equations by r_{0j}. After a slight rearrangement we find with $r_{0j}^2 = n_j$

$$3\varkappa_1 n_1 + 2\varkappa_2 n_2 = \frac{N d_0}{2T} - \frac{\gamma_{32}\varkappa_1}{2g^2 T}\left(1+\frac{\delta_1^2}{(\varkappa_1+\gamma_{32})^2}\right), \tag{VI.9.35}$$

$$2\varkappa_1 n_1 + 3\varkappa_2 n_2 = \frac{N d_0}{2T} - \frac{\gamma_{32}\varkappa_2}{2g^2 T}\left(1+\frac{\delta_2^2}{(\varkappa_1+\gamma_{32})^2}\right). \tag{VI.9.36}$$

These equations allow us to determine n_1 and n_2 uniquely. If n_1 or n_2 turns out to be <0, the corresponding equation must be dropped, since this mode is still below threshold. For high enough pumping, i.e. for high enough d_0, two modes can coexist. This result can be generalized to an arbitrary number of modes. Because this can be done in the framework of rate equations, the reader is referred to Chap. (VIII.2).

This coexistence is only possible for modes connected with standing, but not with running waves.[3]

In the latter case, the factor "3" must be replaced by "2", so that the determinant on the left hand side vanishes and the two coupled equations allow no solution. A consistent solution can only be obtained if only one $n_j \neq 0$ (the one with the highest gain).

First order, linear in the fluctuations.

We proceed in complete analogy to the single-mode case [compare (VI.7)] and find two equations for ϱ_1 and φ_1:

$$\ddot{\varphi}_1 + (\varkappa_1 + \gamma_{32})\dot{\varphi}_1 + \frac{1}{r_{01}}(2\tilde{\Omega}_1 - \delta_1)\dot{\varrho}_1 = \frac{1}{r_{01}} \operatorname{Im}\left(e^{-i\tilde{\Omega}_1 t - i\varphi_1(t)} F^{+}_{tot,1}(t)\right), \tag{VI.9.37}$$

$$\left.\begin{aligned} \ddot{\varrho}_1 + (\varkappa_1 + \gamma_{32})\dot{\varrho}_1 + 12 g^2 T \varkappa_1 r_{01}^2 \varrho_1 + 8 g^2 T \varkappa_2 r_{01} r_{02} \varrho_2 + \delta_1 \dot{\varphi}_1 r_{01} \\ = \operatorname{Re}\left(e^{-i\tilde{\Omega}_1 t - i\varphi_1(t)} F^{+}_{tot,1}(t)\right). \end{aligned}\right\} \tag{VI.9.38}$$

These equations must be supplemented by two others with the indices 1 and 2 interchanged.

Eq. (VI.9.37) describes the phase-fluctuations, as in the single-mode case, and shows explicitly that there is a small coupling to the amplitude fluctuations in the case of detuning ($\delta_1 \tilde{\Omega} \neq 0$). This coupling leads to a small increase of the linewidth with detuning. The Eq. (VI.9.38) and the corresponding one for mode 2 describe a dynamic coupling of the amplitude fluctuations. From (VI.9.38) it follows that the gain of the amplitude fluctuation ϱ_1 increases if ϱ_2 decreases and vice versa. This explains why the detailed mathematical treatment yields[4] an *anticorrelation* of the *amplitudes*. If we do not go too far beyond threshold, we may safely neglect $\ddot{\varrho}$ compared to $(\varkappa_1 + \gamma_{32})\dot{\varrho}$.

β) $|\omega_1 - \omega_2| \lesssim 1/T$. Whereas the time-dependent response S^{12} now causes only a very small frequency shift in (VI.9.24), it changes the factor of $b_2^+ b_2$ essentially. The relative size of the factors of $b_1^+ b_1$ and $b_2^+ b_2$ plays a decisive role, however, for the coexistence of modes, as shown above, and thus also for their stability. The coexistence of modes of the type considered above certainly breaks down if the coefficients become equal, i.e. if

$$3\varkappa_1 = 2\varkappa_2 + \frac{(\varkappa_1 + \varkappa_2) + T(\omega_1 - \omega_2)^2}{1 + T^2(\omega_1 - \omega_2)^2}. \tag{VI.9.39}$$

If the losses are equal and we assume $\varkappa_1 + \varkappa_2 > T(\omega_1 - \omega_2)^2$, (VI.9.39) reduces to

$$|\omega_1 - \omega_2| = \frac{1}{T}. \tag{VI.9.40}$$

It can be shown that the solution $b_1^+ \neq 0$ $b_2^+ = 0$ or vice versa, is in general not stable. In order to demonstrate some of the pecularities which may happen here, we consider a special case: the cavity losses are assumed equal: $\varkappa_1 = \varkappa_2 = \varkappa$, and the frequencies in the unloaded cavity have a symmetric position with respect to the atomic line center: $\delta_2 = -\delta_1$.

---

[3] H. Haken and H. Sauermann: l.c. [2] on p. 141. — C. L. Tang, H. Statz, and E. A. de Mars: J. Appl. Phys. **34**, 2289 (1963).

[4] M. Helm and H. Haken: Unpublished manuscript.

We now make an ansatz for a *time-independent* solution: (neglecting all fluctuations):

$$b_1^+ = r_1\, e^{i\varphi_1}, \quad b_2^+ = r_2\, e^{i\varphi_2} \tag{VI.9.41}$$

(r and φ are assumed real), and insert it into Eq. (VI.9.24) and the corresponding equation for mode 2. This leaves us with two algebraic equations for $r_1, r_2, \varphi_1, \varphi_2$. As can be shown in detail, there are regions of the inversion, where (VI.9.41) is stable. Taking into account that the time-dependence $e^{i\nu_{32}t}$ had been removed from b^+ by the transformation (VI.9.13), we find that both modes oscillate with the frequency of the atomic line-center. Because φ_1 and φ_2 are constants, their relative phase is fixed. Thus we obtain phase locking*.

There remains a region depending on the inversion and the frequency difference $(\omega_1 - \omega_2)$ where neither (VI.9.41) (single-frequency operation) nor (VI.9.29) (two-frequency operation) are stable. Basov et al. and Ostrowskij[5] produced evidence that in this region undamped spiking occurs.

VI.10. Gas laser and solid-state laser with an inhomogeneously broadened line. The van der Pol equation, single-mode operation[1].

a) Solid-state laser with an inhomogeneously broadened line and an arbitrary number of levels. The Langevin equations are given by (VI.3.5) and (VI.3.13). We allow for a spread of atomic frequencies ν_{ik} for the lasing transition $i \to k$. For simplicity of notation we choose $i=2$; $k=1$, although these levels may be arbitrarily grouped within the level system.

The atomic variables are eliminated by an iteration procedure** which holds for not too high photon numbers, i.e.

$$n|g|^2 \ll \gamma^2, \quad \text{and} \ll \left(\frac{1}{T_p}\right)^2, \tag{VI.10.1}$$

where $1/T_p$ is an effective pump time, according to which the two levels i, j equilibrate.

We put in lowest approximation

$$(a_2^+ a_2 - a_1^+ a_1)_\mu^{(0)} = d_0 \tag{VI.10.2}$$

where d_0 is the unsaturated population inversion (i.e. the inversion under the action of the external pump source, but without laser action).

We then solve (VI.3.13) for $(a_2^+ a_1)_\mu$:

$$\left.\begin{aligned} (a_2^+ a_1)_\mu^{(1)} &= \frac{1}{\gamma - i\nu_\mu + \dfrac{d}{dt}}\left(-d_0\, i g_\mu\, b^+ + \Gamma_{21,\mu}\right) \qquad (\gamma \equiv \gamma_{21}), \\ &= \frac{-i g_\mu\, d_0}{i(\Omega - \nu_\mu) + \gamma}\, b^+ + \frac{1}{\gamma - i\nu_\mu + \dfrac{d}{dt}}\,\Gamma_{21,\mu}. \end{aligned}\right\} \tag{VI.10.3}$$

For the derivation of the last result use has been made of $b^+(t) = \tilde{b}^+(t)\, e^{i\Omega t}$ (where Ω is the mode frequency in the loaded cavity) and of the assumption that the frequencies inherent in $\tilde{b}^+$ are small compared to γ.

* For the influence of noise on phase-locking see VI.14c.

** For an exact elimination technique see Chap. IX.6.

[5] N. G. Basov, V. N. Morosov, and A. N. Oraewskij: l.c. [2] on p. 141. — Ostrowskij: l.c. [2] on p. 141.

[1] V. Arzt, H. Haken, H. Risken, H. Sauermann, C. Schmid, and W. Weidlich: Z. Physik **197**, 207 (1966); — For a *classical*, nonlinear treatment of the phase and amplitude fluctuations of a gas laser, see J. A. Fleck: J. Appl. Phys. **37**, 188 (1966).

Introducing this quantity and its complex conjugate into Eqs. (VI.3.17) to (VI.3.19) leads to a set of equations for the diagonal elements $(a_k^+ a_k)$ alone:

$$\frac{d}{dt}(a_j^+ a_j)_\mu - \sum_k w_{kj}(a_k^+ a_k)_\mu + \sum_k (a_j^+ a_j)_\mu w_{jk} = O_\mu(\delta_{1j} - \delta_{2j}) + \Gamma_{jj,\mu} \tag{VI.10.4}$$

where

$$O_\mu = b^+ b |g_\mu|^2 \frac{2\gamma d_0}{\gamma^2 + (\Omega - \nu_\mu)^2} + \text{noise terms}. \tag{VI.10.5}$$

A detailed analysis shows that the diagonal noise terms $\Gamma_{jj,\mu}$ and the noise terms occurring in (VI.10.5) can be neglected in the threshold region and somewhat above.

Provided the effective relaxation time of the inversion $(a_2^+ a_2 - a_1^+ a_1)$ is short compared to the fluctuation times of $b^+ b$ we may put $\frac{d}{dt}(a_j a_j) = 0$ in Eq. (VI.10.4). Thus the Eqs. (VI.10.4) reduce to purely algebraic, linear equations. Their solution (for the inversion) can be written in the form*

$$(a_2^+ a_2 - a_1^+ a_1)_\mu = d_0 - Z b^+ b \tag{VI.10.6}$$

where the constant Z_μ is given by

$$Z_\mu = \frac{d_0 2\gamma |g_u|^2}{\gamma^2 + (\Omega - \nu_\mu)^2} Z'. \tag{VI.10.7}$$

Z' only depends on the transition rates w_{ik} and is defined as follows:

The equations

$$\sum_k w_{kj} z_k - z_j \sum w_{jk} = -\delta_{j1} + \delta_{j2}, \tag{VI.10.8}$$

$$\sum z_j = 0 \tag{VI.10.9}$$

are to be solved for the unknown quantities z_j. Then $Z' = z_2 - z_1$.

Inserting (VI.10.6) into (VI.3.13) and integrating yields an improved expression

$$\left.\begin{aligned}(a_2^+ a_1)_\mu = {} & \frac{-i g_\mu d_0 b^+}{i(\Omega - \nu_\mu) + \gamma} + \frac{i g_u |g_u|^2 d_0 2\gamma}{(\gamma^2 + (\Omega - \nu_\mu)^2)(i(\Omega - \nu_\mu) + \gamma)} Z' b^+ b b^+ + \\ & + \frac{1}{\gamma - i\nu_\mu + \frac{d}{dt}} \Gamma_{21,\mu}.\end{aligned}\right\} \tag{VI.10.10}$$

Again the adiabatic hypothesis has been made. Evidently the dipole moment operators are now explicit functions of the field, b, and of the fluctuating forces. When we insert the expression (VI.10.10) into the field equation (VI.3.5), we obtain an equation for the field operators b^+, b alone:

$$\left\{\frac{d}{dt} + i(\Omega - \omega) + \varkappa - C\right\} \tilde{b}^+ + \beta \tilde{b}^+ \tilde{b} \tilde{b}^+ = \hat{F}_{\text{tot}}^+. \tag{VI.10.11}$$

This equation has the same form as the classical van der Pol equation. There are, however, two essential differences: It is now an operator equation, and it possesses a noisy driving term $(\hat{F}_{\text{tot}}^+)$.

---

* The proof follows immediately, if the atomic occupation numbers $N_{j,\mu}$ are decomposed into $N_{j,\mu} = N_j^0 + z_j \cdot O_\mu$, where the N_j^0's are the *unsaturated* occupation numbers.

The coefficients of Eq. (VI.10.11) are given by

$$C = \sum_\mu |g_\mu|^2 d_0 \frac{1}{\gamma - i(\nu_\mu - \Omega)}, \tag{VI.10.12}$$

$$\beta = d_0 Z' \sum \frac{2|g_\mu|^4 \gamma}{(\gamma^2 + (\Omega - \nu_\mu)^2)(i(\Omega - \nu_\mu) + \gamma)} \tag{VI.10.13}$$

$$e^{i\Omega t} \hat{F}_{tot}^+ = F^+ + i \sum_\mu g_\mu^* \frac{1}{\gamma - i\nu_\mu + \frac{d}{dt}} \Gamma_{21,\mu}. \tag{VI.10.14}$$

On the right hand side of Eq. (VI.10.13) we can replace the unsaturated inversion d_0 by its threshold value d_{thr}, because β is multiplied by the photon number $(\tilde{b}^+ \tilde{b})$ which is small in the threshold region and not too far above it.

The correlation functions of $\hat{F}_{tot}^+$, $\hat{F}_{tot}$ can be evaluated by means of those of the atomic and field noise sources (VI.3.11), (VI.3.12), and (VI.3.20), (VI.3.21), (IV.5.26) respectively. This yields

$$\left.\begin{aligned}
&\langle \hat{F}_{tot}^+(t) \rangle = \langle \hat{F}_{tot}^+(t) \rangle = 0, \\
&\langle \hat{F}_{tot}^+(t)\, \hat{F}_{tot}^+(t') \rangle = \langle \hat{F}_{tot}(t)\, \hat{F}_{tot}(t') \rangle = 0, \\
&\langle \hat{F}_{tot}(t')\, \hat{F}_{tot}^+(t) \rangle + \langle \hat{F}_{tot}^+(t)\, \hat{F}_{tot}(t') \rangle \\
&\qquad = \delta(t-t')\, 2 \left\{ \sum_\mu |g_\mu|^2 \frac{(N_2+N_1)_{\mu,s}\gamma}{\gamma^2 + (\nu_\mu - \Omega)^2} + \varkappa(2n_{th}+1) \right\}
\end{aligned}\right\} \tag{VI.10.15}$$

where the index "s" means saturated. We define

$$\overline{(N_2+N_1)_s} = \left\{ \sum_\mu \frac{|g_\mu|^2 \gamma}{\gamma^2 + (\nu_\mu - \Omega)^2} \right\}^{-1} \left[\sum_\mu |g_\mu|^2 \frac{(N_2+N_1)_{\mu,s}\gamma}{\gamma^2 + (\nu_\mu - \Omega)^2} \right]. \tag{VI.10.16}$$

The curly bracket equals $\frac{\varkappa}{(N_2 - N_1)_{thr}}$.

Eq. (VI.10.11) may be solved below threshold by linearization, i.e. by neglecting the term $\sim b^+ b\, b^+$, and above threshold by quantum mechanical quasi-linearization. The details of the calculation are very similar to those of VI.5 and VI.7, so that we merely quote the final results above threshold:

linewidth [half width at half-power (no detuning)].

$$\Delta\omega = \frac{\hbar\omega\varkappa^2}{2P} \left(\frac{\overline{(N_2+N_1)_s}}{(N_2-N_1)_{thr}} + 2n_{th} + 1 \right) \tag{VI.10.17}$$

and *correlation function for the intensity fluctuations:*

$$K(\tau) = \left(\frac{\overline{(N_2+N_1)_s}}{(N_2-N_1)_{thr}} + 2n_{th} + 1 \right) \frac{\varkappa}{\beta} e^{-2\beta n \tau}. \tag{VI.10.18}$$

Within the framework of our approximation $\overline{(N_2+N_1)_s}$ may be replaced by the unsaturated value $(N_2+N_1)^0$. n is the photon number and $P = 2\hbar\omega\varkappa n$ the light intensity.

b) Gas laser. The treatment of a gas laser is completely analogous, and yields the same results (VI.10.17), (VI.10.18).

VI.11. Direct solution of the density matrix equation[1]. We specialize the density matrix equation, which was derived in VI.7, to a single mode of infinite

[1] We follow essentially the paper of W. Weidlich, H. Risken, and H. Haken: Z. Physik **201**, 396 (1967); — A completely different method of solving the master equation (including gas lasers) has been used by C. R. Willis: Phys. Rev. **147**, 406 (1966); **156**, 320 (1967); **165**, 420 (1968).

wavelength [or a running wave with finite wavelength, see (VI.4)], and two-level atoms with a homogeneously broadened line.

We assume in the following complete resonance, i.e. $\omega = \nu$

$$\frac{d\varrho}{dt} = \left(\frac{\partial\varrho}{\partial t}\right)_{coherent} + \left(\frac{\partial\varrho}{\partial t}\right)_{incoherent} \tag{VI.11.1}$$

$\left(\frac{\partial\varrho}{\partial t}\right)_{coherent}$ describes the interaction between the mode and the atomic system without the action of external heatbaths like the pumping etc. [compare (IV.7.68)]. In the interaction representation [compare (III.4)], $\left(\frac{\partial\varrho}{\partial t}\right)_{coherent}$ is given by

$$\left.\begin{aligned}\left(\frac{\partial\varrho}{\partial t}\right)_{coherent} &= -\frac{i}{\hbar}[\tilde{H}_{I_1}, \varrho]\\ &\equiv -ig\left[\left(\sum_\mu (a_2^+ a_1)_\mu b + (a_1^+ a_2)_\mu b^+\right), \varrho\right].\end{aligned}\right\} \tag{VI.11.2}$$

H_{I1} is introduced in (III.3), its explicit form occurs in the last row of Eq. (VI.11.2). $[A, B]$ means the commutator between the operators A and B. $a_{j,\mu}^+$, $a_{j,\mu}$ are the creation and annihilation operators, respectively, of an electron at the atom μ in the level j (see III.2 or VI.3). b^+ and b are the creation and annihilation operators, respectively, of the field mode (see III.1 or VI.3). g is the coupling constant between the atoms and the field mode (compare VI.3.9). $\left(\frac{\partial\varrho}{\partial t}\right)_{incoherent}$ describes the incoherent decay of the field mode $\left(\frac{\partial\varrho}{\partial t}\right)_L$ and that of the atoms $\left(\frac{\partial\varrho}{\partial t}\right)_{A,\mu}$ so that

$$\left(\frac{\partial\varrho}{\partial t}\right)_{incoh.} = \left(\frac{\partial\varrho}{\partial t}\right)_L + \sum_\mu \left(\frac{\partial\varrho}{\partial t}\right)_{A,\mu}. \tag{VI.11.3}$$

These quantities are defined as follows:

$$\left(\frac{\partial\varrho}{\partial t}\right)_L = \xi\left([b, \varrho b^+] + [b\varrho, b^+]\right) + \delta\left([b^+, \varrho b] + [b^+\varrho, b]\right). \tag{VI.11.4}$$

The right hand side of Eq. (VI.11.4) can be rearranged in the form

$$= \varkappa\{[b\varrho, b^+] + [b, \varrho b^+]\} + 2\varkappa\, n_{th}[[b, \varrho], b^+] \tag{VI.11.5}$$

where the constants $\varkappa = \xi - \delta$ and $n_{th} = \frac{\delta}{\xi - \delta}$ have the following physical meaning [compare (IV.7.α)]: $\varkappa$ is the cavity half width, n_{th} is the number of thermal light quanta.

$$\left.\begin{aligned}\left(\frac{\partial\varrho}{\partial t}\right)_{A,\mu} &= \frac{w_{21}}{2}\{[(a_1^+ a_2)_\mu, \varrho(a_2^+ a_1)_\mu] + [(a_1^+ a_2)_\mu \varrho, (a_2^+ a_1)_\mu]\} +\\ &+ \frac{w_{12}}{2}\{[(a_2^+ a_1)_\mu, \varrho(a_1^+ a_2)_\mu] + [(a_2^+ a_1)_\mu \varrho, (a_1^+ a_2)_\mu]\}\end{aligned}\right\} \tag{VI.11.6}$$

[compare (IV.7.β)].

In (VI.11.6), we took into account only real transitions [compare (IV.7.β)]. w_{21} is the incoherent decay rate from level 2 to level 1, w_{12} is the pump rate from level 1 to level 2. For further use we introduce the longitudinal and transverse transition rates

$$\gamma_\parallel \equiv \frac{1}{T} \qquad \text{[notation of Chap. (VI.4)]}$$

$$\gamma_\perp \equiv \gamma \qquad \text{[notation of Chap. (VI.4)]}$$

and the unsaturated inversion σ_0 of a single atom.

$$\gamma_{\|} = w_{21} + w_{12}; \quad \gamma_{\perp} = \frac{1}{2}(w_{21} + w_{12}); \quad d_0 = \sigma_0 = \frac{w_{12} - w_{21}}{w_{12} + w_{21}}. \tag{VI.11.7}$$

We now turn to the direct solution of Eq. (VI.11.1). With respect to that part of ϱ which refers to the field, two different methods have been applied:

a) it is represented in the coherent state representation in diagonal form [compare (IV.9a), (V.1.6.γ)] [2].

b) it is represented in the photon number representation [3].

At least in the latter treatments, all authors had to eliminate the electron variables by an adiabatic approximation. In our article we represent the method a) and express ϱ in the following form:

$$\varrho = \int P(\beta, \beta^*, \ldots) \, |\beta\rangle \langle\beta| \, d^2\beta \tag{VI.11.8}$$

where the transformed density matrix P does not only depend on the classical quantities β, β^* (which refer to the mode), but is still an operator acting on the electronic ("atomic") variables.

The next step in the solution of (VI.11.1) consists in specifying P. Two approaches are presently known:

a) Following WEIDLICH, RISKEN and HAKEN we put

$$P = \prod_{\mu=1}^{N} \varrho_\mu \tag{VI.11.9}$$

where

$$\left.\begin{aligned} \varrho_\mu = \big(C_E(r)\, 1_\mu + C_\sigma(r)\,(a_2^+ a_2 - a_1^+ a_1)_\mu + C_\pi(r)\,(a_2^+ a_1 + a_1^+ a_2)_\mu + \\ + C_\tau(r)\, i\,(a_2^+ a_1 - a_1^+ a_2)_\mu\big) \end{aligned}\right\} \tag{VI.11.10}$$

1_μ is the unity matrix of the atom μ.

The hypothesis (VI.11.10) is coined for the stationary solution of (VI.11.1), where P depends only on $|\beta| = r$. The C's are arbitrary functions of the field amplitude r, and still have to be determined (see below). An extension of (VI.11.10) to the unstationary solution has been given by GNUTZMANN and WEIDLICH [4].

b) Following J. P. GORDON we put

$$P = \int P_1(\beta, \beta^*, \chi_j, t)\, \varrho_a(\beta, \beta^*, \chi_j)\, d^4\chi \tag{VI.11.11}$$

where

$$\varrho_a(\beta, \beta^*, \chi_j) = \prod_{\mu=1}^{N} \sigma_\mu(\chi_1, \ldots, \chi_4) \tag{VI.11.12}$$

with

$$\sigma_\mu = \chi_1 \cdot (a_1^+ a_1)_\mu + \chi_2 \cdot (a_2^+ a_2)_\mu + \chi_3 \cdot (a_2^+ a_1)_\mu + \chi_4 \cdot (a_1^+ a_2)_\mu \tag{VI.11.13}$$

P_1 is a still arbitrary function.

In both cases, a) and b), it is assumed without proof that the density matrix contains the electronic (atomic) variables in a completely symmetric way. This

---

[2] W. WEIDLICH, H. RISKEN, and H. HAKEN: l.c. [1] on p. 146. — J. P. GORDON: Phys. Rev. **161**, 367 (1967). — M. LAX: Phys. Rev. **157**, 213 (1967).

[3] M. SCULLY and W. E. LAMB JR.: Phys. Rev. Letters **16**, 853 (1966); — Phys. Rev. **159**, 208 (1967); **166**, 246 (1968). — J. A. FLECK: Phys. Rev. **149**, 309, 322 (1966); **152**, 278 (1966); — FLECK's papers include more complicated cases.

[4] U. GNUTZMANN and W. WEIDLICH: Phys. Letters A **27**, 179 (1968). — U. GNUTZMANN: Thesis Stuttgart 1968; — Z. Physik **222**, 283; **225**, 416 (1969).

assumption can be justified by the use of macroscopic variables [see (e.g. p. 65)]. The function (VI.11.11) is in principle exact. GORDON obtains for P_1 an equation of a generalized Fokker-Planck type, which, however, has some strange properties (negative eigenvalues) so that no solutions are known so far. GORDON has solved a different equation, which he obtained for the field alone after adiabatic elimination of the atomic variables*.

The function (VI.11.9) is somewhat restricted, but general enough to allow for a nearly exact solution where no difficulties occur, so that we represent this approach. In order to determine the coefficients C, we insert (VI.11.8) into Eq. (VI.11.1). Using the properties of the coherent state representation, one easily finds that

$$[b, \varrho] \to \frac{\partial P}{\partial \beta^*}; \quad [b^+, \varrho] \to -\frac{\partial P}{\partial \beta}. \tag{VI.11.14}$$

Now making the substitution

$$\beta = \tfrac{1}{2}(v_1 - i v_2), \quad \beta^* = \tfrac{1}{2}(v_1 + i v_2) \tag{VI.11.15}$$

we thus obtain from (VI.11.1) the modified but completely equivalent master equation for P.

$$\left.\begin{aligned} \frac{\partial P}{\partial t} &= (-i)\, g \left\{\frac{v_1}{2}[(S^+ + S^-), P] - \frac{v_2}{2}[i(S^+ - S^-), P] + \right.\\ &\quad + \left(\frac{\partial P}{\partial v_1} S^+ - S^- \frac{\partial P}{\partial v_1}\right) - i\left.\left(\frac{\partial P}{\partial v_2} S^+ + S^- \frac{\partial P}{\partial v_2}\right)\right\} + \\ &\quad + \sum_\mu \left(\frac{\partial P}{\partial t}\right)_{A_\mu} + \left(\frac{\partial P}{\partial t}\right)_L \end{aligned}\right\} \tag{VI.11.16}$$

where

$$S^+ = \sum_\mu (a_2^+ a_1)_\mu; \quad S^- = \sum_\mu (a_1^+ a_2)_\mu. \tag{VI.11.17}$$

Here $\left(\frac{\partial P}{\partial t}\right)_{A_\mu}$ has the same form as (VI.11.6), since no light field operators appear in this expression, and from (VI.11.5) there follows immediately, using (VI.11.14)**

$$\left(\frac{\partial P}{\partial t}\right)_L = \varkappa\left(\frac{\partial(v_1 P)}{\partial v_1} + \frac{\partial(v_2 P)}{\partial v_2}\right) + 2\varkappa\, n_{th}\left(\frac{\partial^2 P}{\partial v_1^2} + \frac{\partial^2 P}{\partial v_2^2}\right). \tag{VI.11.18}$$

The ansatz (VI.11.9), (VI.11.10) for P as an operator acting on the electrons now rests on the fact that the operators

$$\left.\begin{aligned} &1_\mu, \sigma_\mu = (a_2^+ a_2 - a_1^+ a_1)_\mu, \quad \pi_\mu = (a_2^+ a_1 + a_1^+ a_2)_\mu, \\ &\tau_\mu = i(a_2^+ a_1 - a_1^+ a_2)_\mu \end{aligned}\right\} \tag{VI.11.19}$$

span a complete set of operators in the two level system of one atom. Therefore

$$\varrho_\mu = (C_E(v)\, 1_\mu + C_\sigma(v)\, \sigma_\mu + C_\pi(v)\, \pi_\mu + C_\tau(v)\, \tau_\mu) \tag{VI.11.20}$$

is the most general form of the statistical operator of one two-level atom coupled to one light field mode. If we now assume that the N atoms of the laser are coupled to the light field symmetrically in the same way, we may write for

---

* Recently F. HAAKE has developed a method to eliminate the atomic variables exactly. See Chap. (IX.8).

** For the proof we observe that (VI.11.8) is equivalent to (IV.9.10a) as far as it concerns the field. The further analysis runs as in (IV.10.26), p. 69 and (IV.10.34), p. 71.

$P(v, v^*, \ldots)$ as in (VI.11.9).

$$P = \prod_{\mu=1}^{N} \varrho_\mu \tag{VI.11.21}$$

with ϱ_μ given by (VI.11.20).

Comparing this ansatz with the most general symmetrical form of P

$$\left.\begin{aligned} P = {} & f_E(v)\, E + f_\sigma(v) \sum_\mu \sigma_\mu + f_\pi(v) \sum_\mu \pi_\mu + f_\tau(v) \sum_\mu \tau_\mu + \\ & + f_{\sigma\sigma}(v) \sum_{\mu \neq \nu} \sigma_\mu\, \sigma_\nu + f_{\sigma\pi}(v) \sum_{\mu \neq \nu} \sigma_\mu\, \pi_\nu + f_{\sigma\tau}(v) \sum_{\mu \neq \nu} \sigma_\mu\, \tau_\nu + \\ & + f_{\pi\pi}(v) \sum_{\mu \neq \nu} \pi_\mu\, \pi_\nu + \cdots \end{aligned}\right\} \tag{VI.11.22}$$

we obtain for the functions $f_E(v)$, $f_\alpha(v)$, $f_{\alpha\beta}(v), \ldots$ $(\alpha, \beta = \sigma, \pi, \tau)$ the following relations:

$$\left.\begin{aligned} & f_E(v) = C_E^N(v); \quad f_\alpha(v) = C_E^{N-1}(v)\, C_\alpha(v); \\ & f_{\alpha\beta}(v) = C_E^{N-2}(v)\, C_\alpha(v)\, C_\beta(v) = \frac{f_\alpha(v)\, f_\beta(v)}{f_E(v)}. \end{aligned}\right\} \tag{VI.11.23}$$

The higher expansion functions $f_{\alpha\beta}(v), f_{\alpha\beta\gamma}(v) \ldots$ are thus expressed by the lower ones.

In order to get coupled equations of motion for the four functions $f_E(v)$, $f_\alpha(v)$ $(\alpha = \sigma, \pi, \tau)$, we introduce the traces over the atomic variables

$$\left.\begin{aligned} Tr_A\, P &= 2^N f_E(v) \equiv F_E(v); \\ Tr\Big(P \sum_\mu \sigma_\mu\Big) &= N\, 2^N f_\sigma(v) \equiv N F_\sigma(v); \\ Tr\Big(P \sum_\mu \pi_\mu\Big) &= N\, 2^N f_\pi(v) \equiv N F_\pi(v); \\ Tr\Big(P \sum_\mu \tau_\mu\Big) &= N\, 2^N f_\tau(v) \equiv N F_\tau(v). \end{aligned}\right\} \tag{VI.11.24}$$

and insert the density matrix equation (VI.11.16) in dF_E/dt, dF_α/dt. Thus we obtain by straightforward calculation, using the commutation relations of the atomic operators, the following set of equations:

$$\frac{dF_E(v)}{dt} = \left(\frac{\partial F_E(v)}{\partial t}\right)_L - gN\left(\frac{\partial F_\tau(v)}{\partial v_1} + \frac{\partial F_\pi(v)}{\partial v_2}\right), \tag{VI.11.25}$$

$$\left.\begin{aligned} \frac{dF_\sigma(v)}{dt} = {} & \left(\frac{\partial F_\sigma(v)}{\partial t}\right)_L + \sigma_0\, \gamma_\parallel\, F_E(v) - \gamma_\parallel\, F_\sigma(v) - \\ & - g\left\{v_1\, F_\tau(v) - \frac{\partial F_\tau(v)}{\partial v_1} + v_2\, F_\pi(v) - \frac{\partial F_\pi(v)}{\partial v_2}\right\} - \\ & - (N-1)\, g\left\{\frac{\partial}{\partial v_1}\left(\frac{F_\sigma(v)\, F_\tau(v)}{F_E(v)}\right) + \frac{\partial}{\partial v_2}\left(\frac{F_\sigma(v)\, F_\pi(v)}{F_E(v)}\right)\right\}, \end{aligned}\right\} \tag{VI.11.26}$$

$$\left.\begin{aligned} \frac{dF_\pi(v)}{dt} = {} & \left(\frac{\partial F_\pi(v)}{\partial t}\right)_L - \gamma_\perp\, F_\pi(v) + g\left\{v_2\, F_\sigma(v) - \frac{\partial F_\sigma(v)}{\partial v_2} - \frac{\partial F_E(v)}{\partial v_2}\right\} - \\ & - (N-1)\, g\left\{\frac{\partial}{\partial v_1}\left(\frac{F_\pi(v)\, F_\tau(v)}{F_E(v)}\right) + \frac{\partial}{\partial v_2}\left(\frac{F_\pi(v)\, F_\pi(v)}{F_E(v)}\right)\right\} \end{aligned}\right\} \tag{VI.11.27}$$

$$\left.\begin{aligned} \frac{dF_\tau(v)}{dt} = {} & \left(\frac{\partial F_\tau(v)}{\partial t}\right)_L - \gamma_\perp\, F_\tau(v) + g\left\{v_1\, F_\sigma(v) - \frac{\partial F_\sigma(v)}{\partial v_1} - \frac{\partial F_E(v)}{\partial v_1}\right\} - \\ & - (N-1)\, g\left\{\frac{\partial}{\partial v_1}\left(\frac{F_\tau(v)\, F_\tau(v)}{F_E(v)}\right) + \frac{\partial}{\partial v_2}\left(\frac{F_\tau(v)\, F_\pi(v)}{F_E(v)}\right)\right\}. \end{aligned}\right\} \tag{VI.11.28}$$

The following form of the functions $F_E(v)$, $F_\alpha(v)$ is compatible with the above equations:

$$\left.\begin{aligned} F_E(v) &= G_E(r); \quad F_\sigma(v) = G_\sigma(r); \\ F_\pi(v) &= v_2\, G(r); \quad F_\tau(v) = v_1\, G(r) \end{aligned}\right\} \tag{VI.11.29}$$

with

$$4r^2 = v_1^2 + v_2^2 = 4\beta^*\beta. \tag{VI.11.30}$$

The system (VI.11.25)—(VI.11.28) now reduces to the form $\big($where $G_j'(r) \equiv \partial G_j(t)/\partial r\big)$

$$\left.\begin{aligned} \frac{dG_E(r)}{dt} &= \varkappa\big(2G_E(r) + r\,G_E'(r)\big) + \frac{1}{2}\varkappa\, n_{th}\Big(G_E''(r) + \frac{1}{r}\,G_E'(r)\Big) - \\ &\quad - g\,N\big(2G(r) + r\,G'(r)\big), \end{aligned}\right\} \tag{VI.11.31}$$

$$\left.\begin{aligned} \frac{dG_\sigma(r)}{dt} &= \varkappa\big(2G_\sigma(r) + rG_\sigma'(r)\big) + \frac{1}{2}\varkappa\, n_{th}\Big(G_\sigma''(r) + \frac{1}{r}\,G_\sigma'(r)\Big) + \\ &\quad + \sigma_0\,\gamma_\parallel G_E(r) - \gamma_\parallel G_\sigma(r) + g\big(rG'(r) - (4r^2-2)\,G(r)\big) - \\ &\quad - (N-1)\,g\left\{2\left(\frac{G_\sigma(r)\,G(r)}{G_E(r)}\right) + r\left(\frac{G_\sigma(r)\,G(r)}{G_E(r)}\right)'\right\}, \end{aligned}\right\} \tag{VI.11.32}$$

$$\left.\begin{aligned} \frac{dG(r)}{dt} &= \varkappa\big(3G(r) + rG'(r)\big) + \frac{1}{2}\varkappa\, n_{th}\Big(G''(r) + \frac{3}{r}\,G'(r)\Big) - \\ &\quad - \gamma_\perp G(r) + g\Big(G_\sigma(r) - \frac{1}{4}\,\frac{1}{r}\,G_\sigma'(r) - \frac{1}{4}\,\frac{1}{r}\,G_E'(r)\Big) - \\ &\quad - (N-1)\,g\left\{3\left(\frac{G^2(r)}{G_E(r)}\right) + r\left(\frac{G^2(r)}{G_E(r)}\right)'\right\}. \end{aligned}\right\} \tag{VI.11.33}$$

Though these equations still have a rather complicated structure, they may be solved exactly, except for very small corrections, in the stationary case, i.e. for $\dfrac{dG_E}{dt} = \dfrac{dG_\sigma}{dt} = \dfrac{dG}{dt} = 0$.

Because of the extremely small value of n_{th} for optical frequencies, we may neglect in (VI.11.31)—(VI.11.33) all terms with $\varkappa\, n_{th}$ in a very good approximation. (VI.11.31) then has the exact solution

$$G(r) = \frac{\varkappa}{gN}\,G_E(r) \tag{VI.11.34}$$

which may be inserted in (VI.11.32), (VI.11.33) to give

$$\frac{\varkappa}{N}\,r\big(G_E'(r) + G_\sigma'(r)\big) = \Big(\gamma_\parallel - \frac{2\varkappa}{N}\Big)\,G_\sigma(r) + \Big(\frac{\varkappa}{N}\,(4r^2-2) - \sigma_0\,\gamma_\parallel\Big)\,G_E(r) \tag{VI.11.35}$$

or with $\varkappa/N \ll \gamma_\parallel,\ \sigma_0\,\gamma_\parallel$

$$\frac{1}{2}\,\big(G_E'(r) + G_\sigma'(r)\big) = \Big(2r - \frac{2\sigma_0 Z}{r}\Big)\,G_E(r) + \frac{2Z}{r}\,G_\sigma(r) \tag{VI.11.36}$$

or in the variable $z = r^2$

$$z\,(G_{E/z} + G_{\sigma/z}) = 2\,(z - \sigma_0\,Z)\;G_E + 2Z\,G_\sigma \tag{VI.11.37}$$

with

$$Z = \frac{N\gamma_\parallel}{4\varkappa}; \quad G_{/z} \equiv \frac{dG}{dz} \tag{VI.11.38}$$

and

$$\frac{1}{2}\,\big(G_E'(r) + G_\sigma'(r)\big) = 2r\,G_\sigma(r) - \bar{\sigma}\,2r\,G_E(r) + \frac{\varkappa^2}{g^2N^2}\,\big(6r\,G_E(r) + 2r^2G_E'(r)\big) \tag{VI.11.39}$$

with

$$\bar{\sigma} = \frac{\gamma_{\perp} \varkappa}{g^2 N}. \tag{VI.11.40}$$

Because of

$$\frac{\varkappa^2}{g^2 N^2} \ll \bar{\sigma} \tag{VI.11.41}$$

and

$$\frac{\varkappa^2 r^2}{g^2 N^2} \ll 1 \tag{VI.11.42}$$

for all possible values of $r^2 = z \leqq Z_m$ [comparing the form of the solution (VI.11.49)], we may neglect the terms with $\varkappa^2/g^2 N^2$ and get in z:

$$(G_{E/z} + G_{\sigma/z}) = 2G_\sigma - 2\bar{\sigma} G_E. \tag{VI.11.43}$$

The elimination of $(G_{E/z} + G_{\sigma/z})$ in (VI.11.37), (VI.11.43) leads to a relation between G_E and G_σ:

$$G_\sigma (Z - z) = (\sigma_0 Z - z(1 + \bar{\sigma}))\, G_E, \tag{VI.11.44}$$

$$G_{\sigma/z}(Z - z) - G_\sigma = (\sigma_0 Z - z(1 + \bar{\sigma}))\, G_{E/z} - (1 + \sigma)\, G_E. \tag{VI.11.45}$$

Inserting in (VI.11.45) $G_{\sigma/z}$ from (VI.11.43) and G_σ from (VI.11.44) we arrive at a first order differential equation for $G_E(z)$ alone:

$$(Z(1 + \sigma_0) - z(2 + \bar{\sigma}))\, G_{E/z} = 2(Z(\sigma_0 - \bar{\sigma}) - z)\, G_E \tag{VI.11.46}$$

or, because of $\bar{\sigma} \ll 2$

$$\frac{G_{E/z}}{G_E} = \frac{(\hat{z} - z)}{(z_m - z)} \tag{VI.11.47}$$

with

$$\hat{z} = Z(\sigma_0 - \bar{\sigma}); \qquad z_m = \frac{(1 + \sigma_0)}{2} Z = \frac{w_{12} N}{4\varkappa}. \tag{VI.11.48}$$

Eq. (VI.11.47) can easily be solved and gives

$$G_E(z) \begin{cases} = c_E\, e^z (z_m - z)^{(z_m - \hat{z})}; & 0 \leqq z \leqq z_m, \\ = 0; & z > z_m. \end{cases} \tag{VI.11.49}$$

The constant c_E is determined by the trace condition $\operatorname{Tr} P = 1$. The meaning of the parameters $\hat{z}$, z_m is clear; because of $G_{E/z} = 0$ for $z = \hat{z}$, $\hat{z}$ gives the number $\widehat{\beta^* \beta}$ where the distribution function (VI.11.49) has its maximum, while z_m represents the absolute maximum of possible photon numbers, which can be present in the lasing mode for given constants σ_0, $\gamma_\|$, $\varkappa$, N. The formula (VI.11.48) for z_m may be interpreted as follows: As level 1 is occupied by $\frac{N}{2}$ $\left(\text{more exactly } N \frac{(1 - \bar{\sigma})}{2}\right)$ atoms in the stationary state, $w_{12} \frac{N}{2}$ atoms per second make transitions into level 2 by pumping. If all these atoms fall to level 1 by induced emission of a photon (and not by other processes) the rate equation for produced and absorbed photons reads in this optimal case $w_{12} \frac{N}{2} = 2\varkappa z_m$ which agrees with (VI.11.48). It is useful to expand the distribution function (VI.11.49)

$$G_E(z) = c_E \exp[z + (z_m - \hat{z}) \ln (z_m - z)] \tag{VI.11.50}$$

about its maximum $z = \hat{z}$ (provided $\hat{z} > 0$, which holds above threshold. We obtain in very good approximation for all values of σ_0

$$G_E(z) \approx \tilde{c}_E\, e^{-\frac{(z - \hat{z})^2}{q^2}} \tag{VI.11.51}$$

with

$$q = \sqrt{2(z_m - \hat{z})}. \qquad \text{(VI.11.52)}$$

The remarkable result which follows in the frame of our ansatz for the statistical operator is the dependence of the width q of the distribution function $G_E(z)$ on pumping. Starting at threshold ($\hat{z}=0$; $\sigma_0=\bar{\sigma}\ll 1$) we have

$$q^2(\sigma_0=\bar{\sigma}) = 2z_m \approx z = \frac{N\gamma_\parallel}{4\varkappa} \qquad \text{(VI.11.53)}$$

and for very high pumping $\sigma_0=1$ we obtain

$$q^2(\sigma_0=1) = 2Z\bar{\sigma} = \frac{\gamma_\parallel \gamma_\perp}{2g^2}; \qquad \hat{z} = \hat{z}_m = Z(1-\bar{\sigma}). \qquad \text{(VI.11.54)}$$

For reasonable values of $N, \gamma_\parallel, \gamma_\perp, \varkappa, g$ there is

$$1 \ll q^2(\sigma_0=1) \ll q^2(\sigma_0=\bar{\sigma}).$$

Of course we must keep in mind that $G_E(z)$ is not the photon number distribution function $P(n)$. Using (VI.11.24) and the formula

$$|\langle n|\beta\rangle|^2 = e^{-z}\frac{z^n}{n!} = p_z(n); \qquad z = |\beta|^2 \qquad \text{(VI.11.55)}$$

we get for $P(n)$:

$$P(n) = Tr_{1\ldots N,F}\left(|n\rangle\langle n| \int d^2\beta\, |\beta\rangle P(\beta,\beta^*)\langle\beta|\right) = \int_0^{z_m} p_z(n)\, G_E(z)\, dz. \qquad \text{(VI.11.56)}$$

The width $q_p = \sqrt{z}$ of the Poisson distribution $p_z(n)$ for the relevant values of z is much smaller than q for low pumping, but much broader than q for very high pumping. Therefore we expect that $P(n)$ approaches the Poisson distribution for very high pumping, while it practically agrees with $G_E(n)$ for lower pumping.

VI.12. Reduction of the generalized Fokker-Planck equation for single-mode action*. In this chapter we show by means of an example (2-level atoms, and single-mode operation) how the exact generalized Fokker-Planck equation (whose derivation was given in IV.10) can be reduced step by step. In particular, the adiabatic approximation is used which assumes that the relaxation times of the atomic system are much shorter than all other relevant frequencies. This approximation holds to an excellent degree in the threshold region and above up to rather high photon numbers. The resulting Fokker-Planck equation is valid for solid-state, gas and semiconductor lasers. It is solved in Chap. VI.13 for both the steady and transient state. In VI.14 we will represent the Fokker-Planck equation for multimode action near threshold and its exact or nearly exact stationary solution. Finally in Chap. VI.15 we will show how the Fokker-Planck equation can be solved outside the threshold region by linearization and quasi-linerization.

In IV.10 we derived an equation for a c-number "distribution" function f for classical quantities which correspond to the complex field amplitude

$$u, u^* \leftrightarrow b, b^+$$

to the complex total dipole moment of the atoms

$$v \leftrightarrow S^- = \sum_\mu (a_1^+ a_2)_\mu,$$

$$v^* \leftrightarrow S^+ = \sum_\mu (a_2^+ a_1)_\mu$$

* For a different reduction scheme see Chaps. IX.8, IX.9.

and to the total inversion

$$D \leftrightarrow 2S_z = \sum_\mu (a_2^+ a_2 - a_1^+ a_1)_\mu .$$

We have further shown how we may calculate all quantum-mechanical expectation values for the field by pure c-number procedures using f.

The equation for f reads

$$\dot{f} = Lf, \tag{VI.12.1}$$

where

$$\left.\begin{aligned}
L &= L_A + L_{AL} + L_L,\\
L_A &= \frac{w_{12}}{2}\left\{N\left(e^{-2\frac{\partial}{\partial D}} - 1\right) + N e^{2\frac{\partial}{\partial D}} \frac{\partial^4}{\partial v^2\,\partial v^{*2}} + 2N \frac{\partial^2}{\partial v\,\partial v^*} + \right.\\
&\quad + \frac{\partial}{\partial v}\left[2\frac{\partial^2}{\partial v\,\partial v^*} + 2e^{-2\frac{\partial}{\partial D}} - 1\right] v +\\
&\quad + \frac{\partial}{\partial v^*}\left[2\frac{\partial^2}{\partial v\,\partial v^*} + 2e^{-2\frac{\partial}{\partial D}} - 1\right] v^* -\\
&\quad \left. - 2\left[\left(e^{-2\frac{\partial}{\partial D}} - 1\right) - e^{2\frac{\partial}{\partial D}} \frac{\partial^4}{\partial^2 v\,\partial^2 v^*}\right]\frac{D}{2}\right\} +\\
&\quad + \frac{w_{21}}{2}\left\{N\left(e^{2\frac{\partial}{\partial D}} - 1\right) + \frac{\partial}{\partial v} v + \frac{\partial}{\partial v^*} v^* + \right.\\
&\quad \left. + 2\left(e^{2\frac{\partial}{\partial D}} - 1\right)\frac{D}{2}\right\},
\end{aligned}\right\} \tag{VI.12.2)*}$$

$$\left.\begin{aligned}
L_{AL} = -ig\left\{\left[e^{-2\frac{\partial}{\partial D}} v^* - \frac{\partial^2}{\partial v^2} v + \frac{\partial}{\partial v} D\right] u - \right.\\
- \left[-\frac{\partial^2}{\partial v^{*2}} v^* + e^{-2\frac{\partial}{\partial D}} v + \frac{\partial}{\partial v^*} D\right] u^* +\\
\left. + \left[-\frac{\partial}{\partial u} + u^*\right] v - \left[-\frac{\partial}{\partial u^*} + u\right] v^*\right\},
\end{aligned}\right\} \tag{VI.12.3}$$

$$L_L = \varkappa\left[\frac{\partial}{\partial u} u + \frac{\partial}{\partial u^*} u^*\right] + 2\varkappa\, n_{th} \frac{\partial^2}{\partial u\,\partial u^*} . \tag{VI.12.4}$$

a) Expansion in powers of $N^{-\frac{1}{2}}$ (N: number of atoms)[1]. For the first reduction we expand the equation in terms of powers of $N^{-\frac{1}{2}}$. We introduce reduced units

$$\tilde{u} = \frac{u}{\sqrt{n_{thr}}}, \qquad \tilde{v} = \frac{v}{v_{thr}}, \qquad \tilde{D} = \frac{D}{D_{thr}}, \tag{VI.12.5}$$

where

$$n_{thr} = \delta_1 N^{\frac{1}{2}} \tag{VI.12.6}$$

with

$$\delta_1 = \frac{1}{4}\sqrt{\frac{\gamma_\perp}{\varkappa + \gamma_\perp}\,\frac{\gamma_\parallel}{\varkappa}} \tag{VI.12.6a}$$

---

* If the phases of the atomic dipole moments are destroyed not only by real, but also by virtual transitions, the following expression must be added to the r.h.s. of Eq. (VI.12.2):

$$\frac{\eta}{2}\left\{\frac{\partial}{\partial v} v + \frac{\partial}{\partial v^*} v^* + 2\frac{\partial^2}{\partial v\,\partial v^*} e^{2\frac{\partial}{\partial D}} \frac{D}{2} + N + \frac{\partial^2}{\partial v\,\partial v^*} e^{2\frac{\partial}{\partial D}}\right\}.$$

[1] H. Haken, H. Risken, H. D. Vollmer, and W. Weidlich: Unpublished manuscript.

and

$$\left.\begin{aligned} \gamma_\parallel &= w_{12} + w_{21}, \\ 2\gamma_\perp &= w_{12} + w_{21} (+\eta)^*, \end{aligned}\right\} \tag{VI.12.6b}$$

$$v_{thr}^2 = \frac{\varkappa^2}{g^2} n_{thr} = \frac{\delta_2}{K} N^{\frac{3}{2}} \tag{VI.12.7}$$

with

$$K = \frac{g^2 N}{\varkappa \gamma_\perp} = \frac{1}{d_{thr}} \tag{VI.12.7a}$$

and

$$\delta_2 = \frac{1}{4} \sqrt{\frac{\varkappa \gamma_\parallel}{(\varkappa + \gamma_\perp)\gamma_\perp}}, \tag{VI.12.7b}$$

$$D_{thr} = \frac{N}{K}. \tag{VI.12.8}$$

We consider K as fixed, but N very large.

In the present context, n_{thr}, v_{thr}^2 and D_{thr} merely serve as abbreviations of the corresponding right hand sides of Eqs. (VI.12.6)—(VI.12.8). Later it will be shown, however, that these quantities are just the values of the photon number, of the absolute square of the total atomic dipole moment, and of the total inversion *at threshold*. d_{thr} is the threshold inversion of the single atom. Taking into account terms of the order one and $N^{-\frac{1}{2}}$, we find

$$\left.\begin{aligned} L_A = {} & (w_{21} - w_{12}) K \frac{\partial}{\partial \tilde{D}} + \gamma_\perp \left(\frac{\partial}{\partial \tilde{v}} \tilde{v} + \frac{\partial}{\partial \tilde{v}^*} \tilde{v}^* \right) + \gamma_\parallel \frac{\partial}{\partial \tilde{D}} \tilde{D} + \\ & + N^{-\frac{1}{2}} \frac{w_{12} K}{\delta_2} \frac{\partial^2}{\partial \tilde{v}\, \partial \tilde{v}^*}, \end{aligned}\right\} \tag{VI.12.9}$$ **

$$\left.\begin{aligned} L_{A\text{-}L} = -ig \Big\{ & -2 \frac{\partial}{\partial \tilde{D}} \tilde{v}^* \tilde{u} \sqrt{\delta_1 \delta_2} \sqrt{K} + \\ & + \frac{\partial}{\partial \tilde{v}} \tilde{D} \tilde{u} \sqrt{\delta_1/\delta_2} \frac{1}{\sqrt{K}} N^{\frac{1}{2}} - \\ & - \frac{\partial}{\partial \tilde{u}} \tilde{v} \sqrt{\delta_2/\delta_1} \frac{1}{\sqrt{K}} N^{\frac{1}{2}} - \text{c.c.} \Big\}. \end{aligned}\right\} \tag{VI.12.10}$$

Because for fixed K, $g \sim N^{-\frac{1}{2}}$ [see (VI.12.7a)], we have neglected all terms in the curly bracket which are proportional to $N^{-\frac{1}{2}}, \ldots$

In order to obtain a concise formula, we replace the complex variables u and v by two-dimensional vectors:

$$\vec{v} = \begin{pmatrix} v_1 \\ v_2 \end{pmatrix}; \quad \vec{u} = \begin{pmatrix} u_1 \\ u_2 \end{pmatrix};$$

where

$$u_1 = \frac{1}{2}(u + u^*); \quad u_2 = \frac{1}{2i}(u - u^*);$$

$$v_1 = \frac{1}{2i}(v - v^*); \quad v_2 = -\frac{1}{2}(v + v^*).$$

We further put

$$D_0 = N \frac{w_{12} - w_{21}}{w_{12} + w_{21}} = N(N_2 - N_1) \tag{VI.12.11}$$

---

* The additional term η accounts for the processes described in footnote * on p. 154.

** If additional phase-destroying processes are included, the factor w_{12} of $\frac{\partial^2}{\partial \tilde{v}\, \partial \tilde{v}^*}$ must be replaced by $w_{12} + \frac{1}{2}\eta(1 + \tilde{D}/K)$.

where

$$\left.\begin{aligned} N_2 &= \frac{w_{12}}{w_{12}+w_{21}}; \\ N_1 &= \frac{w_{21}}{w_{12}+w_{21}}. \end{aligned}\right\} \tag{VI.12.12}$$

D_0 is the unsaturated inversion of all atoms, whereas N_2 and N_1 are the unsaturated occupation numbers of the upper and lower state, respectively, of a single atom. After a short transformation we obtain

$$\left.\begin{aligned} &\frac{\partial f}{\partial t} + \nabla_{\vec{u}}\{(-\varkappa\vec{u}+g\vec{v})\,f\} + \nabla_{\vec{v}}\{(-\gamma_\perp\vec{v}+gD\vec{u})\,f\} + \\ &\quad + \frac{\partial}{\partial D}\{[\gamma_\parallel(D_0-D)-4g\vec{u}\vec{v}]\,f\} \\ &= \frac{\varkappa}{2}n_{th}\Delta_{\vec{u}}f + \frac{1}{4}Nw_{12}\Delta_{\vec{v}}f + \gamma_\parallel N\frac{\partial^2 f}{\partial D^2}. \end{aligned}\right\} \tag{VI.12.13}$$ * **

An equation of this form was first represented by RISKEN, SCHMID and WEIDLICH by the method described on p. 66. There is only a slight difference between their equation and Eq. (VI.12.13). They use symmetrized diffusion coefficients and find $n_{th}+\frac{1}{2}$ instead of n_{th}, and $\frac{w_{21}+w_{12}}{2}$ instead of w_{12} in the first two terms.

A detailed analysis shows that both forms lead to the same results, as long as higher derivatives are neglected.

b) Adiabatic elimination of the atomic variables. With respect to the statistical properties of the laser field, the distribution function f contains more information than is needed, because it also contains the atomic variables. Thus it is desirable to eliminate these variables, so that a function $W(u, u^*)$ of the field variables alone is to be determined.

If the relaxation times of the atomic system are short compared to all other times of the systems, the atomic variables (i.e. the dipole moment and the inversion) follow the motion of the field adiabatically. Under this assumption the atomic variable can be easily eliminated:

Two ways are known for this:

1. The direct elimination within the Fokker-Planck equation[2].
2. An elimination via the Langevin equations[3].

Because the second procedure is more transparent, we present the latter[3]:

The Fokker-Planck equation (VI.12.13) possesses the following associate Langevin equations for the complex light amplitude u, the total complex dipole

* If additional phase-destroying processes (connected with virtual transitions are included, the expression Nw_{12} must be replaced by

$$N\,w_{12} + \frac{N}{2}\eta + D\frac{\eta}{2}$$

(compare the footnotes on p. 71, 154 and 155).

** The last term on the right hand side of Eq. (VI.12.13) is of the order N^{-1}. It is contained in the systematic expansion of L_A and has been included in (VI.12.13) in order to allow a comparison with the equation of RISKEN, SCHMID and WEIDLICH.

[2] J. P. GORDON: Phys. Rev. **161**, 367 (1967).

[3] H. RISKEN, C. SCHMID, and W. WEIDLICH: Z. Physik **194**, 337 (1966), Appendix.

moment v, and the inversion D:

$$\left(\frac{d}{dt}+\varkappa\right)u+ig\,v=\Gamma_u, \tag{VI.12.14}$$

$$\left(\frac{d}{dt}+\gamma_\perp\right)v-ig\,uD=i\Gamma_v, \tag{VI.12.15}$$

$$\left(\frac{d}{dt}+\gamma_\parallel\right)(D-D_0)+2ig(v^*u-vu^*)=\Gamma_D. \tag{VI.12.16}$$

Note that these equations are now *classical* with *classical* random forces Γ.

The diffusion constants are given by

$$Q_j=\frac{1}{4T}\int_0^T\int_0^T\langle\Gamma_j(t_1)\,\Gamma_j^*(t_2)\rangle\,dt_1dt_2,\quad j=u,v,D. \tag{VI.12.17}$$

[compare (IV.2.7)].

A comparison with the RSW-Fokker-Planck equation[3] (in the symmetrized form) yields

$$Q_u=\frac{\varkappa}{2}\left(n_{th}+\frac{1}{2}\right); \tag{VI.12.18}$$

$$Q_v=\frac{1}{4}N\gamma_\perp=\frac{1}{4}N(N_1+N_2)\,\gamma_\perp=\frac{1}{4}N\frac{(w_{21}+w_{12})}{2}. \tag{VI.12.19}$$

Had we used the Fokker-Planck equation (VI.12.13) instead, the result would read

$$Q_u=\frac{\varkappa}{2}n_{th}, \tag{VI.12.20}$$

$$Q_v=\frac{1}{4}Nw_{12}. \tag{VI.12.21}$$

We will see below that the results are not affected by the difference between (VI.12.18)—(VI.12.21) so that the corresponding Fokker-Planck equations are equivalent.

We now eliminate the atomic variables: In accordance with the adiabatic assumption we neglect d/dt compared to $\gamma_\perp$ and $\gamma_\parallel$. Neglecting Γ_D, and in the first step also Γ_v, we obtain from Eqs. (VI.12.15) and (VI.12.16)

$$D=D_0\left(1+\frac{4g^2}{\gamma_\parallel\gamma_\perp}u^*u\right)^{-1},$$

or, for small photon numbers:

$$D=D_0-\frac{4g^2}{\gamma_\parallel\gamma_\perp}D_0uu^*. \tag{VI.12.22}$$

Inserting (VI.12.22) into (VI.12.15) we obtain an equation for the dipole moment as a function of the field.

We now apply the operator $\left(\frac{d}{dt}+\gamma_\perp\right)$ to Eq. (VI.12.14), so that we can eliminate v from this equations by means of (VI.12.15). When we finally neglect $\frac{d^2u}{dt^2}$ and $\frac{d}{dt}\Gamma_u$, we find

$$\frac{d}{dt}u-\beta(\bar{n}-u^*u)\,u=\Gamma \tag{VI.12.23}$$

where

$$\beta = 4\frac{\varkappa^2\gamma_\perp}{\gamma_\parallel(\varkappa+\gamma_\perp)D_{thr}} \approx \frac{4\varkappa^2}{\gamma_\parallel D_{thr}}, \tag{VI.12.24}*$$

$$\bar{n} \equiv \tilde{d} = \frac{\gamma_\parallel}{4\varkappa}(D_0 - D_{thr}), \tag{VI.12.25}$$

$$D_{thr} \equiv N(N_2 - N_1)_{thr} = \varkappa\gamma_\perp/g^2 \tag{VI.12.26}$$

$$\Gamma = \frac{\gamma_\perp}{\varkappa+\gamma_\perp}\Gamma_u + \frac{g}{\varkappa+\gamma_\perp}\Gamma_v. \tag{VI.12.27}$$

Eq. (VI.12.23) is the so called van der Pol equation in the rotating wave approximation with a noisy driving force Γ. If $\Gamma=0$, the stationary solution for u is given by $|u|^2=\bar{n}$, where $\bar{n}$ can be interpreted as the average value of the photon number, at least if the right hand side of Eq. (VI.12.25) is positive. Because, below laser threshold, the right-hand side of Eq. (VI.12.25) is negative, it is better to denote this expression by $\tilde{d}$ and to call it a pump parameter.

An equation of the form (VI.12.23) occurs also in Chap. VI.10, where Eq. (VI.10.13) refers to Bose-*operators* b, b^+ and *quantum mechanical* Langevin forces $\hat{F}^+_{tot}$ instead of the classical quantities u, u^* and Γ in Eq. (VI.12.23). From this comparison we see that Eq. (VI.12.23) holds equally well for the inhomogeneously broadened solid state laser (and the gas laser). The nonlinearity factor β is then given by (VI.10.13) (we assume exact resonance, $\Omega=\omega=\nu$), and $\tilde{d}=\frac{C-\varkappa}{\beta}$, where C is defined by (VI.10.12). For a two-level system with a homogeneously broadened line, the corresponding expressions agree with each other*.

c) The Fokker-Planck equation. We write the complex field amplitude u in polar coordinates: $u=r\,e^{-i\varphi}$. The Fokker-Planck equation belonging to the *classical van der Pol equation* (in the rotating wave approximation) (VI.12.23) is easily derived using (IV.2.5) and reads[4]

$$\frac{\partial W}{\partial t} + \beta\frac{1}{r}\frac{\partial}{\partial r}\{(\tilde{d}-r^2)\,r^2 W\} = Q\left\{\frac{1}{r}\frac{\partial}{\partial r}\left(r\frac{\partial W}{\partial r}\right) + \frac{1}{r^2}\frac{\partial^2 W}{\partial\varphi^2}\right\}. \tag{VI.12.28}$$

The diffusion constant Q is given by

$$Q = \frac{1}{4}\frac{1}{T}\int_0^T\!\!\int_0^T \langle\Gamma(t_1)\Gamma^*(t_2)\rangle\,dt_1\,dt_2. \tag{VI.12.29}$$

Using the form (VI.12.27) of Γ and the fact that Γ_u and Γ_v are uncorrelated we obtain

$$Q = \left(\frac{\gamma_\perp}{\varkappa+\gamma_\perp}\right)^2 Q_u + \frac{g^2}{(\varkappa+\gamma_\perp)^2}Q_u. \tag{VI.12.30}$$

Inserting (VI.12.18), (VI.12.19) into (VI.12.30) we find

$$Q = \left(\frac{\gamma_\perp}{\varkappa+\gamma_\perp}\right)^2\frac{\varkappa}{2}\left(n_{th}+\frac{1}{2}\right) + \frac{g^2}{(\varkappa+\gamma_\perp)^2}\,\frac{1}{4}N\gamma_\perp. \tag{VI.12.31}$$

---

* Note that the adiabatic approximation implies $\varkappa \ll \gamma_\perp$, so that $(\gamma_\perp+\varkappa)$ may be replaced by $\gamma_\perp$ in (VI.12.24), (VI.12.27), and in the following equations of this chapter.

[4] See e.g. H. Risken: Z. Physik **186**, 85 (1965); for the general method of deriving Fokker-Planck equations see e.g. R. L. Stratonovich: Topics in the Theory of Random Noise, Vol. 1. New York: Gordon and Breach 1963.

We now eliminate the coupling constant using the threshold condition

$$N(N_2 - N_1)_{thr} = \varkappa \gamma_\perp / g^2 . \tag{VI.12.32}$$

After rearranging (VI.12.31) we end up with

$$\left.\begin{aligned} Q &= \frac{\varkappa}{2}\left[\frac{\gamma_\perp}{\varkappa+\gamma_\perp}\right]^2 \left[n_{th} + \frac{1}{2} + \frac{1}{2}\,\frac{(N_2+N_1)_{thr}}{(N_2-N_1)_{thr}}\right] \\ &= \frac{\varkappa}{2}\left[\frac{\gamma_\perp}{\varkappa+\gamma_\perp}\right]^2 \left[n_{th} + \frac{N_{2\,thr}}{(N_2-N_1)_{thr}}\right] \\ &\approx \frac{\varkappa}{2}\,(n_{th}+n_{sp})\,. \end{aligned}\right\} \tag{VI.12.33}$$

[Note that $1 = N_2 + N_1 = (N_2 + N_1)_{thr}$].

Let us now discuss the diffusion coefficient using (VI.12.20) and (VI.12.21):

$$Q = \left(\frac{\gamma_\perp}{\varkappa+\gamma_\perp}\right)^2 \frac{\varkappa}{2}\, n_{th} + \frac{\varkappa\gamma_\perp}{(\varkappa+\gamma_\perp)^2}\,\frac{1}{(N_2-N_1)_{thr}}\cdot\frac{1}{4}\,w_{12} \tag{VI.12.34}$$

where again use of (VI.12.32) has been made.

Because $N_2 = \dfrac{w_{12}}{w_{12}+w_{21}} \approx N_{2,thr}$, (VI.12.34) and (VI.12.33) are indeed equivalent (for not too high photon numbers).

VI.13. Solution of the reduced Fokker-Planck equation.

a) Steady state solution[1]. First we introduce normalized coordinates for the real field amplitude r, the time t, and the pump parameter $\tilde{d}$ [see Eq. (VI.12.25)]. Using the nonlinearity constant β (VI.12.24), and the diffusion constant (VI.12.29), (VI.12.33) or (VI.12.34) we define:

$$\hat{r} = \sqrt[4]{\beta/Q}\, r; \quad \hat{t} = \sqrt{\beta Q}\, t; \quad a = \sqrt{\beta/Q}\, \tilde{d}\,. \tag{VI.13.1}$$

The Fokker-Planck equation (VI.12.28) then takes the form

$$\frac{\partial W}{\partial \hat{t}} + \frac{1}{\hat{r}}\frac{\partial}{\partial \hat{r}}\{(a-\hat{r}^2)\,\hat{r}^2 W\} = \frac{1}{\hat{r}}\frac{\partial}{\partial \hat{r}}\left(\hat{r}\,\frac{\partial W}{\partial \hat{r}}\right) + \frac{1}{\hat{r}^2}\,\frac{\partial^2 W}{\partial \varphi^2}\,. \tag{VI.13.2}$$

Its stationary solution is given by (N: normalization constant)

$$W(\hat{r}) = \frac{N}{2\pi}\, e^{-\frac{\hat{r}^4}{4} + a\frac{\hat{r}^2}{2}}; \quad \frac{1}{N} = \int_0^\infty \hat{r}\, e^{-\frac{\hat{r}^4}{4} + a\frac{\hat{r}^2}{2}}\, d\hat{r}\,. \tag{VI.13.3}$$

The distribution function (VI.13.3) is a continuous function of $\hat{r}$ (or r). A plot of W as a function of $\hat{n} = \hat{r}^2$ is given on p. 111. In order to compare it with the distribution, $p(n, T)$, of the discrete photon numbers n which are measured in the time interval T outside the cavity in a photo detector, a correction, derived in (V.4b), must be added*.

If T is small compared to the relaxation times of the van der Pol oscillator, $p(n, T)$ is given by (V.4.13).

* Note that the *discrete* photon number n, which is measured by counting experiments, has another meaning than r^2 which represents the *averaged* unnormalized photon number.

[1] H. Risken: Z. Physik **186**, 85 (1965). This solution is a special case of a class of solutions ("potential case") of a general Fokker-Planck equation, described in the book of Stratonovich, l.c. [4] of Chap. VI.12 on p. 158. This class of solutions is contained in a more general class, given in Chap. VI.14.

Using the normalized coordinates (VI.13.1), we transform (V.4.13) into

$$p(n, T) = \int_0^\infty \frac{(\hat{\alpha}\,\hat{I})^n}{n!}\, e^{-\hat{\alpha}\hat{I}}\, \hat{W}(\hat{I})\, d\hat{I} \tag{VI.13.4}$$

where $\hat{I} = \hat{r}^2$; $\hat{\alpha} = \alpha\sqrt{Q/\beta}\,T$ [for the definition of α see (V.4.9)].

On account of $\hat{W}(\hat{I}) = \pi W(\hat{r})$ and using the explicit form of $W(\hat{r})$, (VI.13.3), we find

$$p(n, T) = \frac{\hat{\alpha}^n}{n!}\, \frac{F_n(a - 2\hat{\alpha})}{F_0(a)} \tag{VI.13.5}$$

where $F_n(a)$ is given by

$$F_n(a) = \int_0^\infty \hat{I}^n\, e^{-\frac{\hat{I}^2}{4} + a\frac{\hat{I}}{2}}\, d\hat{I}\,. \tag{VI.13.6}$$

For a large average number of photons, i.e. for large $\hat{\alpha}$, we obtain a (nearly) continuous distribution

$$p(n, T) = \frac{1}{\hat{\alpha}\, F_0(a)} \exp\left\{-\frac{1}{4}\left(\frac{n}{\hat{\alpha}}\right)^2 + \frac{1}{2}\, a\, \frac{n}{\hat{\alpha}}\right\}. \tag{VI.13.7}$$

In order to obtain *correlation functions*, e.g. of the type (V.4.8), the *non-stationary solutions* of the Fokker-Planck equation must be used. Since general analytical expressions are not known, either approximation methods (like a variational method) or machine calculations were performed. The latter were performed with great accuracy, so that we represent the relevant results[2]:

First the Eq. (VI.13.2) is reduced to a *one-dimensional Schrödinger equation.*

The hypothesis

$$W(\hat{r}, \varphi, \hat{t}) = \sum_{m=0}^{\infty} \sum_{n=-\infty}^{\infty} A_{nm}\left[\frac{1}{\sqrt{\hat{r}}}\, e^{-\frac{\hat{r}^4}{8} + a\frac{\hat{r}^2}{4}}\, \Psi_{nm}(\hat{r})\right] e^{in\varphi}\, e^{-\lambda_{nm}\hat{t}} \tag{VI.13.8}$$

leads to

$$\Psi_{nm}'' + (\lambda_{nm} - V_n(\hat{r}))\, \Psi_{nm} = 0 \qquad \left(\text{where } \Psi'' \equiv \frac{d^2}{d\hat{r}^2}\Psi\right) \tag{VI.13.9}$$

for the eigenfunctions $\Psi_{nm} = \Psi_{-nm}$ and the eigenvalues $\lambda_{nm} = \lambda_{-nm}$.

a	λ_{01}	M_1	λ_{02}	M_2	λ_{03}	M_3	λ_{04}	M_4
10	19.1142	0.4614	19.1237	0.4885	34.5184	0.0212	35.3947	0.0226
8	14.6507	0.4423	14.9666	0.4622	23.6664	0.0492	28.3894	0.0344
7	12.0787	0.4085	13.0891	0.4569	20.0129	0.0895	26.4382	0.0354
6	9.4499	0.4061	11.5823	0.4459	18.0587	0.1132	25.6136	0.0287
5	7.2368	0.4717	10.6059	0.4095	17.3876	0.0980	25.8079	0.0179
4	5.6976	0.5925	10.2361	0.3344	17.6572	0.0634	26.9004	0.0086
3	4.8564	0.7246	10.4763	0.2387	18.6918	0.0330	28.7914	0.0033
2	4.6358	0.8284	11.2857	0.1553	20.3871	0.0151	31.3963	0.0011
0	5.6266	0.9370	14.3628	0.0601	25.4522	0.0028	38.4621	0.0001

[2] H. Risken and H. D. Vollmer: Z. Physik **201**, 323 (1967). For a more complete list of eigenvalues and matrix elements see H. Risken, Fortschr. Physik **16**, 261 (1968). This article contains also a detailed review of laser light statistics. For an equivalent treatment see R. D. Hempsteadt and M. Lax: Phys. Rev. **161**, 350 (1967).

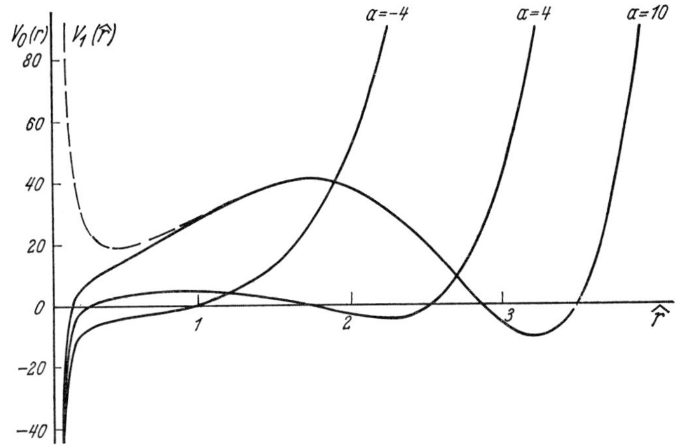

Fig. 32. The potential $V_0(\hat{r})$ of the Schrödinger equation (2.2) for three pump parameters (solid line) and $V_1(\hat{r})$ for $a = 10$ (broken line). (From H. RISKEN and H. D. VOLLMER, l.c. [2] on p. 160.)

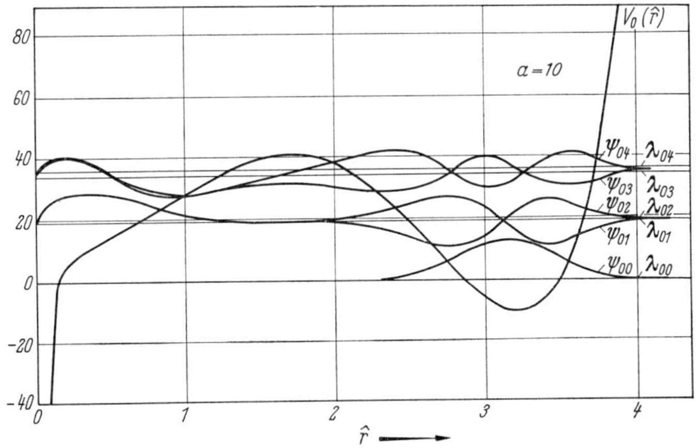

Fig. 33. The potential V_0 of the Schrödinger equation (3.12) and the first five eigenvalues and eigenfunctions for the pump parameter $a = 10$. (From H. RISKEN and H. D. VOLLMER: l.c. [2] on p. 160.)

The potential is given by (see Figs. 32 and 33)

$$\left.\begin{aligned} V_n(\hat{r}) &= \frac{n^2}{\hat{r}^2} + \frac{\Psi_{00}''}{\Psi_{00}} \\ &= \left(n^2 - \frac{1}{4}\right)\frac{1}{\hat{r}^2} + a + \left(\frac{a^2}{4} - 2\right)\hat{r}^2 - \frac{a}{2}\hat{r}^4 + \frac{\hat{r}^6}{4}. \end{aligned}\right\} \tag{VI.13.10}$$

The eigenfunction Ψ_{00} belonging to the stationary eigenvalue $\lambda_{00} = 0$ is, according to (VI.13.3) and (VI.13.8)

$$\Psi_{00} = \sqrt{N\hat{r}}\exp\left[-\hat{r}^4/8 + a\,\hat{r}^2/4\right]. \tag{VI.13.11}$$

It follows from Eq. (VI.13.9) that the Ψ_{nm}'s are orthogonal for different m. If they are normalized to unity, we have

$$\int_0^\infty \Psi_{nm}(\hat{r})\,\Psi_{nm'}(\hat{r})\,d\hat{r} = \delta_{mm'}. \tag{VI.13.12}$$

The completeness relation

$$\delta(\hat{r}-\hat{r}') = \sum_{m=0}^{\infty} \Psi_{nm}(\hat{r})\,\Psi_{nm}(\hat{r}') \tag{VI.13.13}$$

leads immediately to the Green's function of the Fokker-Planck Eq. (VI.13.2). It is obtained from the general solution (VI.13.8) by putting

$$A_{nm} = \frac{1}{2\pi\sqrt{\hat{r}'}}\, e^{\frac{\hat{r}'^4}{8} - a\frac{\hat{r}'^2}{4}}\, \Psi_{nm}(\hat{r}')\, e^{-in\varphi'}.$$

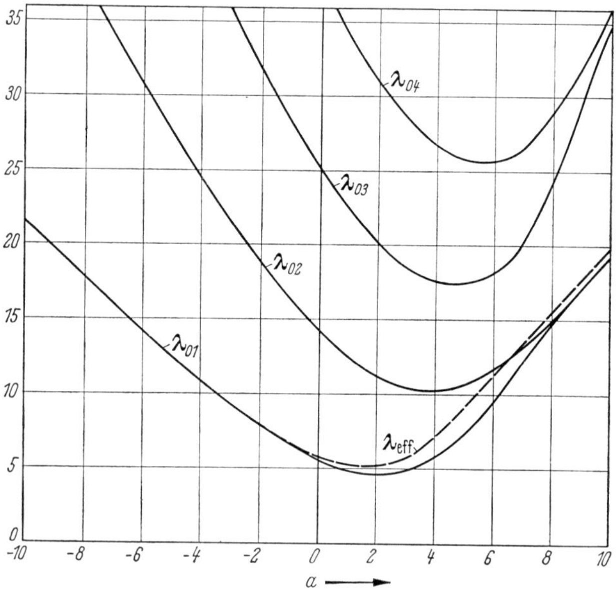

Fig. 34. The first four non-zero eigenvalues λ_{0m} and the effective eigenvalue λ_{eff} (see VI.13.25) as functions of the pump parameter a. (From RISKEN and H. D. VOLLMER: l.c. [2] on p. 160.)

Thus the Green's function reads

$$\left.\begin{aligned} G(\hat{r},\varphi;\hat{r}',\varphi',\hat{\tau}) &= \frac{1}{2\pi\sqrt{\hat{r}\,\hat{r}'}}\, e^{-\frac{\hat{r}^4}{8}+a\frac{\hat{r}^2}{4}}\, e^{+\frac{\hat{r}'^4}{8}-a\frac{\hat{r}'^2}{4}} \times \\ &\times \sum_{m=0}^{\infty}\sum_{n=-\infty}^{\infty} \Psi_{nm}(\hat{r})\,\Psi_{nm}(\hat{r}')\, e^{in(\varphi-\varphi')}\, e^{-\lambda_{nm}\hat{\tau}}. \end{aligned}\right\} \tag{VI.13.14}$$

For the calculation of stationary two-time correlation functions the joint distribution function $F(\hat{r},\varphi;\hat{r}',\varphi';\hat{\tau})$ is needed. $F(\hat{r},\varphi;\hat{r}',\varphi';\tau)\,\hat{r}\,d\hat{r}\,d\varphi\,\hat{r}'\,d\hat{r}'\,d\varphi'$ is the probability that $\hat{r}(t+\tau)$, $\hat{\varphi}(t+\tau)$ lie in the interval $\hat{r},..,\hat{r}+d\hat{r}$; $\varphi,..,\varphi+d\varphi$ and that $\hat{r}'(t)$, $\varphi'(t)$ lie in the interval $\hat{r}',\ldots,\hat{r}'+d\hat{r}'$; $\varphi',..,\varphi'+d\varphi'$.

F can be expressed by means of the Green's function $G(\hat{r},\varphi;\hat{r}';\varphi';\hat{\tau})$ and of $W(\hat{r}',\varphi')$, which describes the distribution at the initial time:

$$F(\hat{r},\varphi;\hat{r}',\varphi';\hat{\tau}) = G(\hat{r},\varphi;\hat{r}',\varphi',\hat{\tau})\, W(\hat{r}',\varphi'). \tag{VI.13.15}$$

The correlation function of the intensity fluctuations is obtained by

$$\left.\begin{aligned} K(a,\hat{\tau}) &\equiv \langle (\hat{r}^2(\hat{t}+\hat{\tau}) - \langle \hat{r}^2\rangle)\,(\hat{r}^2(\hat{t}) - \langle \hat{r}^2\rangle)\rangle \\ &\equiv \iiiint \hat{r}\,d\hat{r}\,\hat{r}'\,d\hat{r}'\,d\varphi\,d\varphi'\,(\hat{r}^2 - \langle \hat{r}^2\rangle)\,(\hat{r}'^2 - \langle \hat{r}'^2\rangle)\,F(\hat{r},\varphi;\hat{r}',\varphi',\hat{\tau}) \\ &= K(a,0)\sum_{m=1}^{\infty} \mathrm{M}_m\, e^{-\lambda_{0m}\hat{\tau}}, \end{aligned}\right\} \tag{VI.13.16}$$

$$\mathrm{M}_m = \frac{N}{K(a,0)}\left[\int_0^\infty \sqrt{\hat{r}}\,\hat{r}^2\, e^{-\frac{\hat{r}^4}{8}+a\frac{\hat{r}^2}{4}}\,\Psi_{0m}(\hat{r})\,d\hat{r}\right]^2. \tag{VI.13.17}$$

(The use of the classical correlation function (VI.13.16) instead of the corresponding quantum mechanical expression has been justified in IV.11).

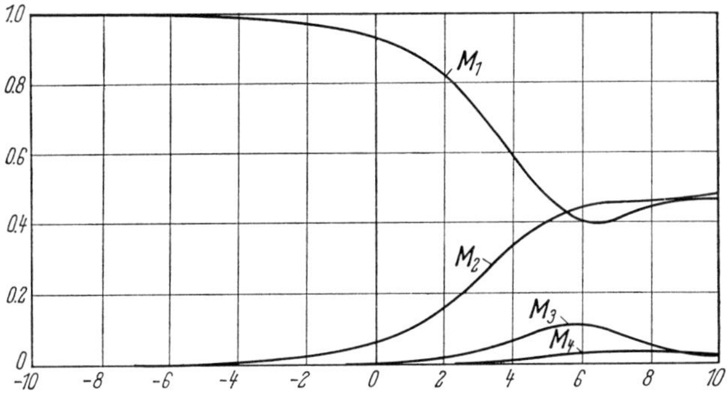

Fig. 35. The first four matrix elements M_m as functions of the pump parameter a. (From H. RISKEN and H. D. VOLLMER: l.c. [2] on p. 160.)

K(a, 0) is given by

$$\langle \hat{r}^4\rangle - \langle \hat{r}^2\rangle^2 = 2\pi\int_0^\infty \hat{r}^5\, W(\hat{r})\,d\hat{r} - \left[2\pi\int_0^\infty \hat{r}^3\, W(\hat{r})\,d\hat{r}\right]^2 \tag{VI.13.18}$$

where $W(\hat{r})$ is defined in (VI.13.3)

Both N and $K(a, 0)$ may be reduced to the error integral

$$\Phi(y) = \frac{2}{\sqrt{\pi}}\int_0^y e^{-x^2}\,dx. \tag{VI.13.19}$$

Introducing the new variable $v = \hat{r}^2$ we define

$$I_n(a) = \int_0^\infty v^n\, e^{-\frac{v^2}{4}+a\frac{v}{2}}\,dv. \tag{VI.13.20}$$

Then the following recurrence relations hold:

$$I_0(a) = \sqrt{\pi}\exp\{a^2/4\}\,[1+\Phi(a/2)], \tag{VI.13.21}$$

$$I_1(a) = 2 + a\, I_0(a), \tag{VI.13.22}$$

$$I_n(a) = 2(n-1)\, I_{n-2}(a) + a\, I_{n-1}(a) \quad \text{for} \quad n \geqq 2. \tag{VI.13.23}$$

The spectrum $S(a, \hat{\omega})$ of the intensity fluctuations is given by the FOURIER transform of the correlation function (VI.13.16)*

$$S(a, \hat{\omega}) = K(a, 0) \sum_{m=1}^{\infty} M_m \frac{\lambda_{0m}}{\hat{\omega}^2 + \lambda_{0m}^2}. \qquad \text{(VI.13.24)}$$

Although it is a sum of Lorentzian lines of widths λ_{0m}, $S(a, \hat{\omega})$ may be well approximated by an "effective" Lorentzian curve

$$S_{eff}(a, \hat{\omega}) = K(a, 0) \frac{\lambda_{eff}}{\hat{\omega}^2 + \lambda_{eff}^2} \quad \text{with} \quad \frac{1}{\lambda_{eff}} = \sum_{m=1}^{\infty} \frac{M_m}{\lambda_{0m}} \qquad \text{(VI.13.25)}$$

which has the same area and the same maximum value (see Fig. 36).

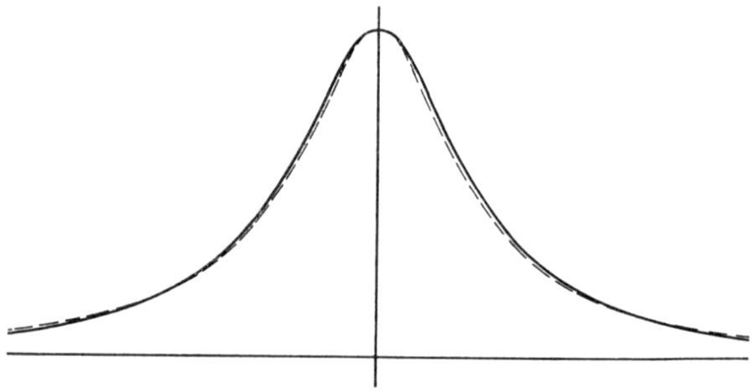

Fig. 36. A comparison between the exact noise spectrum and the effective Lorentzian line for $a = 5$. (From H. RISKEN and H. D. VOLLMER: l.c. [2] on p. 160.)

This effective width λ_{eff} is, however, about 25% larger than λ_{01} for $a \approx 5$. The eigenvalues and matrixelements were calculated numerically by RISKEN and VOLLMER[3], in particular for the threshold region $-10 \leq a \leq 10$. For a comparison with experimental results the reader is referred to Chap. VI.2.

Similar calculations for the correlation function of the amplitude yield

$$\left. \begin{aligned} g(a, \hat{\tau}) &= \langle \hat{r}(\hat{t} + \hat{\tau}) e^{i\varphi(\hat{t}+\hat{\tau})} \hat{r}(\hat{t}) e^{-i\varphi(\hat{t})} \rangle \\ &= \iiiint \hat{r}\, d\hat{r}\, \hat{r}'\, d\hat{r}'\, d\varphi\, d\varphi'\, \hat{r}\, \hat{r}'\, e^{i\varphi} e^{-i\varphi'} F(\hat{r}, \varphi; \hat{r}', \varphi'; \hat{\tau}) \\ &= g(a, 0) \sum_{m=0}^{\infty} V_m e^{-\lambda_{1m}\hat{\tau}}, \end{aligned} \right\} \qquad \text{(VI.13.26)}$$

$$V_m = \frac{N}{g(a, 0)} \left[\int_0^{\infty} \sqrt{\hat{r}}\, \hat{r}\, e^{-\frac{\hat{r}^4}{8} + a\frac{\hat{r}^2}{4}} d\hat{r} \right]^2 \qquad \text{(VI.13.27)}$$

$g(a, 0)$ is given by

$$g(a, 0) = \langle \hat{r}^2 \rangle = 2\pi \int_0^{\infty} \hat{r}^3 W(\hat{r})\, d\hat{r}$$

and can be reduced to the error integral (VI.13.19) by the same substitution leading to (VI.13.20). A calculation of V_0 shows that $1 - V_0 = \sum_{m=1}^{\infty} V_m$ is of the

* The use of $\hat{\omega}$ instead of ω indicates that we are using reduced units defined by $\hat{\omega} = \omega/\sqrt{\beta Q}$.

[3] H. RISKEN, and H. D. VOLLMER: l.c. [2] on p. 160.

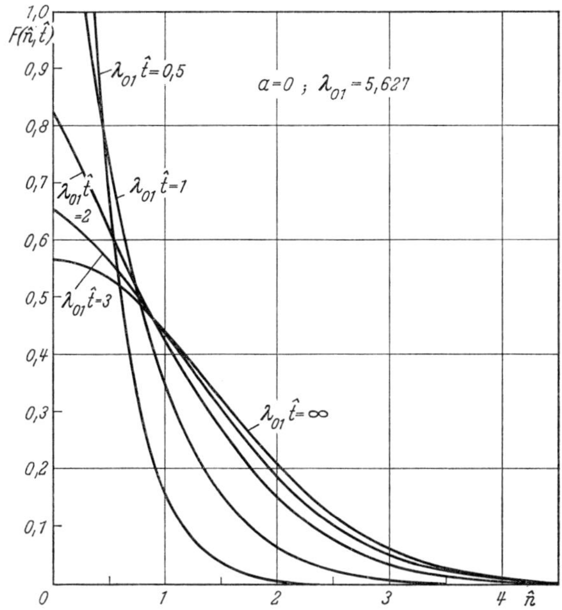

a

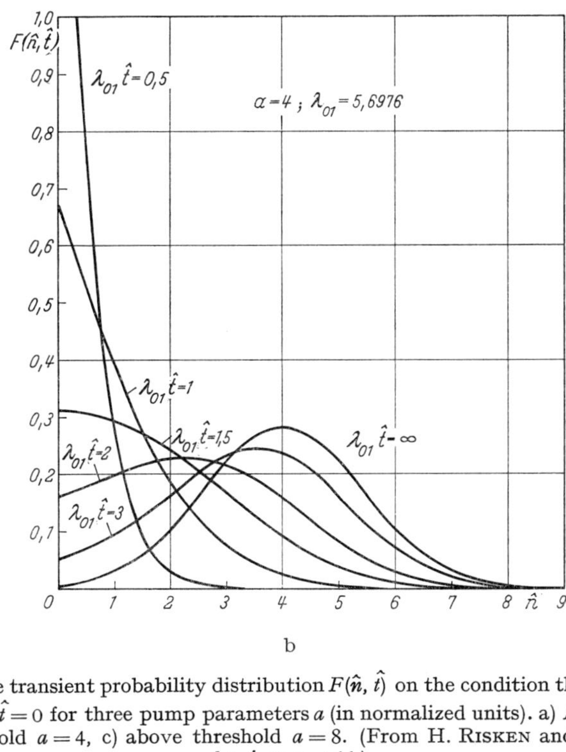

b

Fig. 37 a—c. The transient probability distribution $F(\hat{n}, \hat{t})$ on the condition that no photons are present at time $\hat{t}=0$ for three pump parameters a (in normalized units). a) At threshold $a=0$, b) above threshold $a=4$, c) above threshold $a=8$. (From H. Risken and H. D. Vollmer: l.c. [4] on p. 166.)

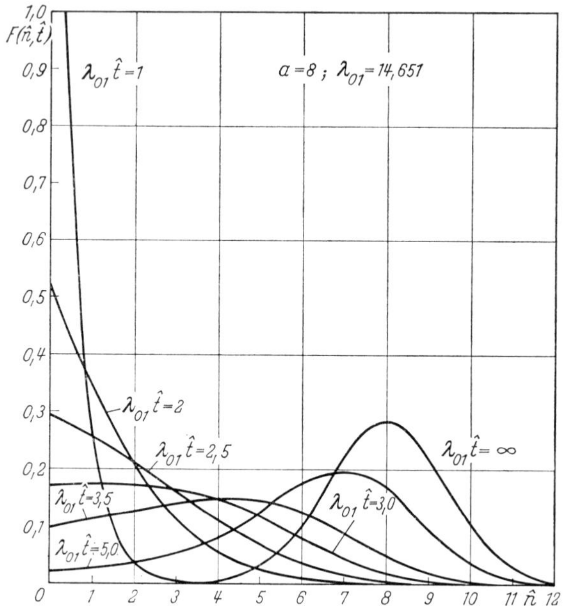

Fig. 37c.

order of 2% near threshold and smaller outside. Therefore the spectral profile is nearly a Lorentzian one with a line width (in unnormalized units)

$$\left.\begin{aligned} \Delta\omega = \sqrt{\beta Q}\,\lambda_{10} = \alpha(a)\,Q/\langle n\rangle, \\ \alpha = \lambda_{10}(a)\,\langle \hat{r}^2(a)\rangle. \end{aligned}\right\} \tag{VI.13.28}$$

The factor α is plotted in Fig. 19.

b) Transient solution[4]. The general transient solution of Eq. (VI.13.2) is given by

$$W(\hat{r},\varphi,\hat{t}) = \int_0^\infty \int_0^{2\pi} G(\hat{r},\varphi;\hat{r}',\varphi';\hat{t})\,W(\hat{r}',\varphi',0)\,\hat{r}'\,d\hat{r}'\,d\varphi' \tag{VI.13.29}$$

where G is the Green's function (VI.13.14) and $W(\hat{r}',\varphi',0)$ is the initial distribution. Risken and Vollmer have evaluated (VI.13.26) for an initial distribution, where no photons are present, i.e.

$$W(\hat{r},\varphi,0) = \lim_{\varepsilon\to 0} \frac{1}{2\pi\hat{r}}\,\delta(\hat{r}-\varepsilon) \tag{VI.13.30}$$

using eigenfunctions Ψ_{nm} up to order 10.

To facilitate a comparison with experiments, we introduce the photon reduced number $\hat{n} = \hat{r}^2$ instead of the amplitude $\hat{r}$, and show the plot of $F(\hat{n},\hat{t})$ instead of $W(\hat{r},\varphi,\hat{t})$. $F(\hat{n},\hat{t})\,d\hat{n}$ is the probability that $\hat{n}(\hat{t})$ lies in the interval $\hat{n} \leqq \hat{n}(\hat{t}) \leqq \hat{n} + d\hat{n}$.

[4] H. Risken and H. D. Vollmer: Z. Physik **204**, 240 (1967).

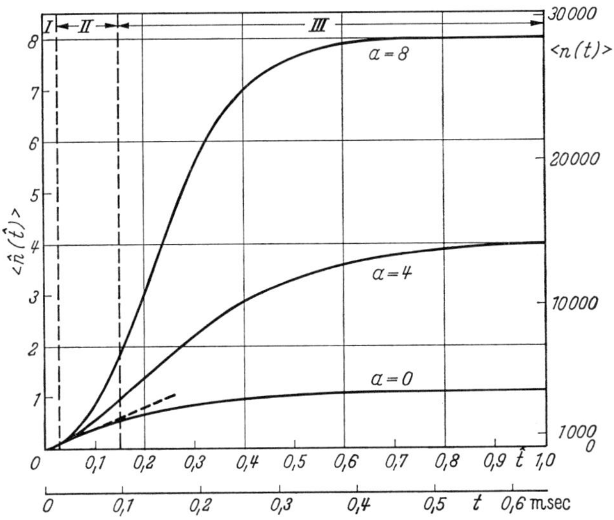

Fig. 38. The transient mean photon number $\langle\hat{n}(\hat{t})\rangle$ for the condition that no photons are present at time $\hat{t}=0$ in the laser for three pump parameters (in normalized units). The values given for t and $\langle n(t)\rangle$ hold for a laser with a threshold photon number of $\langle n\rangle=4000$ and a linewidth of the intensity fluctuations at threshold of 1400 c/sec (as found by ARECCHI et al.) The three indicated regions are: I spontaneous emission region, II amplification of spontaneous emission by induced emission, III saturated region. (From H. RISKEN and H. D. VOLLMER: l.c. [4] on p. 166.)

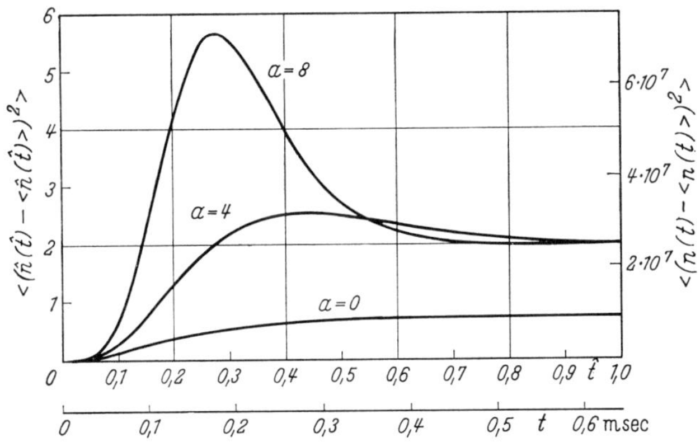

Fig. 39. The transient mean squared deviation of the photon number from its average, when no photons are present at time $\hat{t}=0$. The unnormalized units hold for the same laser as Fig. 38. (From H. RISKEN and H. D. VOLLMER: l.c. [4] on p. 166.)

The experiments of ARECCHI[5] et al. are only in qualitative agreement with (VI.13.29). According to ARECCHI this is due to the fact that the Fokker-Planck equation (VI.13.2) does not hold well below threshold. Another mathematical treatment by SCULLY and LAMB[6] is limited in so far as only photon numbers of the order of 100 are treated and again only qualitative agreement could be found.

[5] F. T. ARECCHI, V. DEGIORGIO, and B. QUERZOLA: Phys. Rev. Letters **19**, 20, 1168 (1967).

[6] M. SCULLY and W. E. LAMB: Proc. of the Fourth Internat. Quantum Electr. Conference, Phoenix, Arizona, 1966, and Proc. of the Internat. School on Quantum Optics "Enrico Fermi", Varenna, Italy, 1967 ed. R. GLAUBER.

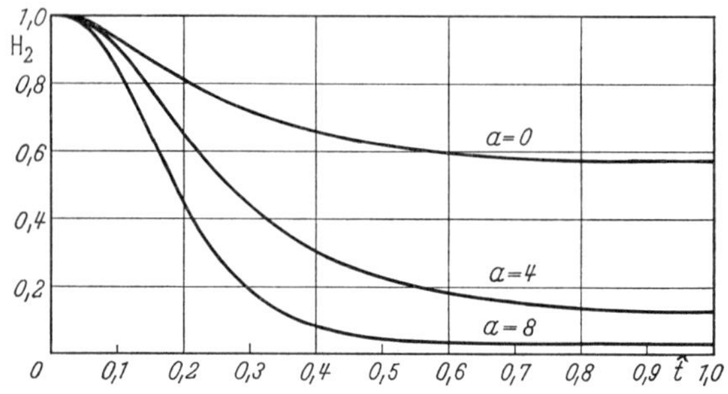

Fig. 40. The parameter $H_2 = \langle(\hat{n}(\hat{t}) - \langle\hat{n}(\hat{t})\rangle)^2\rangle / \langle\hat{n}(\hat{t})\rangle^2$ of the photon variance as a function of the normalized time $\hat{t}$ for three pump parameters a. (From H. RISKEN and H. D. VOLLMER: l.c. [4] on p. 166.)

VI.14. The Fokker-Planck equation for multimode action near threshold. Exact or nearly exact stationary solution.

a) The explicit form of the Fokker-Planck equation. In Chap. IV.10 we have derived an exact generalized Fokker-Planck equation for a single mode and a system of two-level atoms. In a region not too far above threshold we have reduced that equation to an ordinary Fokker-Planck equation for the field variables alone. The whole procedure can be extended to multimode laser action with multi-level atoms. We do not intend to repeat all the steps in detail, because the equations are very lengthy. We want rather to give a prescription for how to derive this final equation by merely looking at some of the intermediate results of our single-mode analysis:

The first important result was that the Fokker-Planck equation after the first reduction (p. 156) can be interpreted as being connected with a set of classical Langevin equations. If we anticipate the later results of Chap. VII.1 on the semiclassical treatment of the laser equations, we find by comparison, that the just mentioned classical Langevin equations coincide with the *semiclassical* laser equations (if the fluctuating forces are dropped). Because the drift coefficients are identical to the right hand sides of these Langevin equations (without fluctuating forces), we can immediately use the corresponding expressions of the semiclassical analysis. Furthermore we have seen that the subsequent adiabatic elimination can best be done in the framework of the Langevin equations. This elimination procedure will be performed for multimode action in Chap. VII.2, so that we can use the results of these chapters for the explicit form of the drift coefficients. Finally it is obvious from our above analysis that the diffusion coefficients can be found from fluctuating forces of the quantum mechanical Langevin equations. It can be shown that the diffusion coefficients Q_{ij} [compare (IV.2.5)] vanish for different modes, i, j, (the proof simply uses the fact that the thermal heatbaths of the modes are uncorrelated and that the modes are orthogonal with respect to their spatial dependence). The only difficulty in the evaluation of the Q's according to (IV.2.5) or (VI.12.17) consists in the noncommutivity of the quantum mechanical, fluctuating forces F, F^+ [see e.g. (VI.10.15)]. It can be overcome, however, by evaluating $\langle\Gamma(t)\Gamma(t')\rangle$ in Eq. (VI.12.17) or (VI.12.29) by using a certain order of F, F^+:

$$\alpha)\quad \langle F_{tot}^+(t)\, F_{tot}(t')\rangle$$

or

$$\beta)\quad \tfrac{1}{2}\left(\langle F_{tot}(t')\, F_{tot}^+(t)\rangle + \langle F_{tot}^+(t)\, F_{tot}(t')\rangle\right).$$

$\langle\ldots\rangle$ is a quantum mechanical average over the system coordinates including the heatbath variables. The form α) is suggested from the treatment by means of the generalized Fokker-Planck equations (see e.g. Chap. VI.12) whereas β) is the prescription given by RISKEN, SCHMID and WEIDLICH. We have shown above by means of our example, that at least in the threshold region the difference between α) and β) is negligible. In Eq. (VI.14.3) we choose β).

After these remarks it is quite simple to write down the Fokker-Planck equation for multi-mode laser action in a homogeneously or inhomogeneously broadened laser. Denoting the complex field amplitudes of the modes λ by u_λ, u_λ^*, we obtain

$$\left.\begin{aligned}\frac{d}{dt}W(u_1,\ldots,u_M)=\Big\{-\sum_\lambda\frac{\partial}{\partial u_\lambda}A_\lambda(u_1,\ldots,u_M)-\\ -\sum_\lambda\frac{\partial}{\partial u_\lambda^*}A_\lambda^*(u_1,\ldots,u_M)+\sum_\lambda Q_\lambda\frac{\partial^2}{\partial u_\lambda\,\partial u_\lambda^*}\Big\}W(u_1,\ldots,u_M)\end{aligned}\right\}\quad\text{(VI.14.1)}$$

where

$$\left.\begin{aligned}A_\lambda^*(u_1,\ldots,u_M)=(i\omega_\lambda-\varkappa_\lambda)\,u_\lambda^*-u_\lambda^*\sum_\mu d_{0\mu}|g_{\mu\lambda}|^2\frac{i}{(\Omega_\lambda-\nu_\mu-i\gamma)}+\\ +2\sum_{\mu,\lambda_1,\lambda_2,\lambda_3}d_{0\mu}\,g_{\lambda\mu}^*\,g_{\mu\lambda_1}^*\,g_{\mu\lambda_2}\,g_{\mu\lambda_3}\,u_{\lambda_1}\,u_{\lambda_2}^*\,u_{\lambda_3}^*\times\\ \times\Big\{\Big(\frac{1}{T}+i(\Omega_{\lambda_2}-\Omega_{\lambda_1})\Big)(\Omega_{\lambda_2}+\Omega_{\lambda_3}-\Omega_{\lambda_1}-\nu_\mu-i\gamma)\Big\}^{-1}\times\\ \times\frac{\Omega_{\lambda_1}-\Omega_{\lambda_2}+2i\gamma}{(\Omega_{\lambda_2}-\nu_\mu-i\gamma)(\Omega_{\lambda_1}-\nu_\mu+i\gamma)}\,.\end{aligned}\right\}\quad\text{(VI.14.2)}$$

ω_λ and $\varkappa_\lambda$ are mode frequency and half-width in the unloaded cavity, Ω_λ is the mode frequency in the loaded cavity, $d_{0\mu}$ is the unsaturated inversion of atom μ, $g_{\mu\lambda}$ is the coupling constant (VI.3.8), ν_μ is the frequency at line center of atom μ, γ is the homogeneous atomic width, T is the longitudinal relaxation time. We further have for the diffusion coefficients [compare (VI.10.15)]

$$Q_\lambda=\frac{1}{8}\Big\{\sum_\mu\frac{2\gamma(N_2+N_1)_\mu}{\gamma^2+(\nu_\mu-\Omega_\lambda)^2}\,|g_{\mu\lambda}|^2+2\varkappa(2n_{th}+1)\Big\}.\qquad\text{(VI.14.3)}$$

Before giving the (nearly exact) solution of (VI.14.1) with (VI.14.2), (VI.14.3) we present a

b) Theorem on the exact stationary solution of a Fokker-Planck equation[1]. We assume that the drift- and diffusion coefficients of the Fokker-Planck equation

$$\frac{\partial W}{\partial t}=\Big\{\sum_j\Big(\frac{\partial}{\partial u_j}C_j+\frac{\partial}{\partial u_j^*}\tilde{C}_j\Big)+\sum_{kj}Q_{kj}\frac{\partial^2}{\partial u_k\,\partial u_j^*}\Big\}W\qquad\text{(VI.14.4)}$$

have the following form

$$C_j=\frac{\partial B}{\partial u_j^*}+I_j^{(1)},\qquad\text{(VI.14.5)}$$

$$\tilde{C}_j=\frac{\partial B}{\partial u_j}+I_j^{(2)},\qquad\text{(VI.14.6)}$$

$$Q_{kj}=\delta_{kj}\,Q.\qquad\text{(VI.14.7)}$$

We further assume

$$\sum_j\Big(\frac{\partial B}{\partial u_j}I_j^{(1)}+\frac{\partial B}{\partial u_j^*}I_j^{(2)}\Big)=0,\qquad\text{(VI.14.8)}$$

$$\sum_j\Big(\frac{\partial I_j^{(1)}}{\partial u_j}+\frac{\partial I_j^{(2)}}{\partial u_j^*}\Big)=0.\qquad\text{(VI.14.9)}$$

[1] H. HAKEN: Z. Physik **219**, 246 (1969).

The theorem states that the stationary solution of the Fokker-Planck Eq. (VI.14.4) is given by

$$W = N e^{-2B/Q} \tag{VI.14.10}$$

where N is the normalization constant. It is implicitly assumed that W is normalizable which is always the case in laser systems (due to the gain saturation).

c) Nearly exact solution of (VI.14.1)[2]. Because modes near the atomic line center have the highest gain, we confine our analysis to modes for which Q_λ is nearly independent of λ, so that the condition (VI.14.7) is fulfilled.

Whether A^* und A (VI.14.2) fulfill the conditions (VI.14.5), (VI.14.6), (VI.14.8), (VI.14.9) requires a detailed and lengthy analysis, so that we present only the results.

α) *Normal multimode action.* For the definition see (VII.5a).

We assume that the frequency spacing $\Delta\Omega = \Omega_{\lambda_2} - \Omega_{\lambda_1}$ is large compared to $1/T$. The important contributions to (VI.14.2) then stem from those terms in the sum where $\lambda_1 = \lambda_2$. If we further neglect combination tones [compare (VII.5.6)], the sum over λ_3 reduces to a single term: $\lambda_3 = \lambda$.

Under these assumptions Eq. (VI.14.1) possesses the following *exact solution*

$$W = N \exp\left\{-\frac{2}{Q}\left(B^{(1)} + B^{(2)} + B^{(3)}\right)\right\} \tag{VI.14.11}$$

where

$$B^{(1)} = \sum_\lambda \varkappa_\lambda u_\lambda^* u_\lambda \tag{VI.14.12}$$

is connected with the cavity losses.

$$B^{(2)} = \sum_\lambda \Delta\varkappa_\lambda u_\lambda^* u_\lambda \tag{VI.14.13}$$

with

$$\Delta\varkappa_\lambda = -\sum_\mu d_{0\mu} |g_{\mu\lambda}|^2 \frac{\gamma}{(\Omega_\lambda - \nu_\mu)^2 + \gamma^2} \tag{VI.14.14}$$

is connected with the unsaturated gain.

$$B^{(3)} = 2\gamma^2 T \sum_{\lambda\lambda_1\mu} |u_\lambda|^2 |u_{\lambda_1}|^2 \cdot d_{0\mu} |g_{\mu\lambda}|^2 |g_{\mu\lambda_1}|^2 \cdot \frac{1}{(\Omega_\lambda - \nu_\mu)^2 + \gamma^2} \cdot \frac{1}{(\Omega_{\lambda_1} - \nu_\mu)^2 + \gamma^2} \tag{VI.14.15}$$

is connected with the gain saturation (including mode interaction). If (VI.14.11) with (VI.14.12)—(VI.14.15) is specialized to a single mode, the result (VI.13.3) is reobtained.

β) *Phase locking of many modes.* [For a treatment of phase locking without noise see (VII.5c) and (VII.9.36—43)].

We assume

$$Q_\lambda \text{ independent of } \lambda: \tag{VI.14.16}$$

$$\text{axial, running modes in one direction} \tag{VI.14.17}$$

$$\Delta\Omega < \gamma \tag{VI.14.18}$$

and

$$\Delta\Omega \lesssim \frac{1}{T}. \tag{VI.14.19}$$

From (VI.14.17) it follows that under the triple sum in (VI.14.2) the following relation for the mode vectors k_λ holds:

$$k_\lambda + k_{\lambda_1} - k_{\lambda_2} - k_{\lambda_3} = 0. \tag{VI.14.20}$$

[2] H. Haken: l.c. [1] on p. 169.

If we assume that the frequencies Ω_λ in the loaded cavity depend linearly on k_λ (i.e. linear pulling or pushing), we obtain from (VI.14.21)

$$\Omega_\lambda + \Omega_{\lambda_1} - \Omega_{\lambda_2} - \Omega_{\lambda_3} = 0. \tag{VI.14.21}$$

Under slight approximations we obtain the following stationary solution

$$W = N \exp\left\{-\frac{2}{Q}\left(B^{(1)} + B^{(2)} + B^{(4)}\right)\right\} \tag{VI.14.22}$$

where $B^{(1)}$ and $B^{(2)}$ are given by (VI.14.12) and (VI.14.13) respectively. The function $B^{(4)}$ is, however, new:

$$B^{(4)} = 2\gamma^2 \frac{1}{T} \sum_{\lambda_0 \lambda_1 \lambda_2 \lambda_3, \mu} u_{\lambda_0} u_{\lambda_1} u^*_{\lambda_2} u^*_{\lambda_3} d_0 g^*_{\mu\lambda_0} g^*_{\mu\lambda_1} g_{\mu\lambda_2} g_{\mu\lambda_3} p_\mu(\lambda_0, \lambda_1, \lambda_2, \lambda_3) \tag{VI.14.23}$$

where

$$p_\mu(\lambda_0, \lambda_1, \lambda_2, \lambda_3) = \tfrac{1}{2}\{m(\lambda_1, \lambda_2; \lambda_0, \lambda_3) + m(\lambda_1, \lambda_3; \lambda_0, \lambda_2)\} l_\mu(\lambda_0, \lambda_1, \lambda_2, \lambda_3) \tag{VI.14.24}$$

with

$$m(\lambda_1, \lambda_2; \lambda_0, \lambda_3) = \left[\left(\left(\frac{1}{T}\right)^2 + (\Omega_{\lambda_2} - \Omega_{\lambda_1})^2\right)\left(\left(\frac{1}{T}\right)^2 + (\Omega_{\lambda_0} - \Omega_{\lambda_3})^2\right)\right]^{-\frac{1}{2}} \tag{VI.14.25}$$

and

$$l_\mu(\lambda_0, \lambda_1, \lambda_2, \lambda_3) = \left[\prod_{j=0}^{3} \left((\Omega_{\lambda_j} - \nu_\mu)^2 + \gamma^2\right)\right]^{-\frac{1}{2}}. \tag{VI.14.26}$$

The distribution function W has its local extrema at those configurations of the u's where

$$\frac{\partial(B^{(1)} + B^{(2)} + B^{(4)})}{\partial u_\lambda} = 0; \qquad \frac{\partial(B^{(1)} + B^{(2)} + B^{(4)})}{\partial u^*_\lambda} = 0. \tag{VI.14.27}$$

This condition coincides with the requirement that the real parts of the forces on the right side of the averaged equations of motion (VII.2.6) vanish.

γ) *A qualitative discussion of phase locking (example of three modes*).* We now want to show that the solution (VI.14.22) with (VI.14.23) describes phase-locking phenomena. This would mean that the probability of finding a prescribed configuration of the u's depends on their relative phases φ. For this end we write $u_\lambda = r_\lambda e^{i\varphi_\lambda}$ (r_λ real) and investigate in which manner $B = B^{(1)} + B^{(2)} + B^{(4)}$ depends on the φ_λ's. Evidently $B^{(1)}$ and $B^{(2)}$ do not depend on phases so that we investigate $B^{(4)}$. $B^{(4)}$ contains a number of terms without phase locking, namely those for which

$$\left.\begin{aligned} &\lambda_0 = \lambda_3, \quad \lambda_1 = \lambda_2 \\ \text{and} \quad &\lambda_0 = \lambda_2, \quad \lambda_1 = \lambda_3 \end{aligned}\right\} \tag{VI.14.28}$$

holds. There occur, however, other terms, which depend on the phase difference of different modes. In order to show this most clearly we consider the

example of three modes: $\lambda = 1, 2, 3$, where

$$k_1 + k_3 - 2k_2 = 0. \tag{VI.14.29}$$

The table lists all different terms which depend on the phase difference

$$\Delta\varphi = \varphi_1 + \varphi_3 - 2\varphi_2.$$

* For a detailed treatment of phase locking of three modes see [3].

[3] H. Haken, H. Sauermann, C. Schmid, and H. D. Vollmer: Z. Physik **206**, 369 (1967); The classical theory of phase-locking is represented in the books: F. M. Gardner: Phase-lock techniques, New York-London-Sidney: John Wiley 1966, and A. J. Viterbi: Principles of coherent communication, New York: McGraw Hill 1966, where many further references are given.

Table. *Combinations of* β's.

λ_0	λ_1	λ_2	λ_3	Phase combinations
1	3	2	2	$\varphi_0+\varphi_3-2\varphi_2=\Delta\varphi$
3	1	2	2	$\varphi_0+\varphi_3-2\varphi_2=\Delta\varphi$
2	2	1	3	$2\varphi_2-\varphi_0-\varphi_3=-\Delta\varphi$
2	2	3	1	$2\varphi_2-\varphi_0-\varphi_3=-\Delta\varphi$

Selecting only those terms of $B^{(3)}$, which are listed in the table, we find that the corresponding sum has the following form:

$$\sum_{\lambda_0\lambda_1\lambda_2\lambda_3}\cdots\sim\text{const.}\ r_2^2\, r_1\, r_3\cos(\Delta\varphi) \tag{VI.14.30}$$

where the constant is proportional to

$$\sum_{\mu} d_{0\mu}\, g^*_{\mu\lambda_0}\, g^*_{\mu\lambda_1}\, g_{\mu\lambda_2}\, g_{\mu\lambda_3}. \tag{VI.14.31}$$

We obtain a maximum of the photon number, if W (VI.14.22) has a maximum. If (VI.14.31) is positive, such a mode configuration has the highest probability of realization for which

$$\cos\Delta\varphi=-1. \tag{VI.14.32}$$

The sign of (VI.14.31) may change if the active material fills only a certain part of the cavity, i.e. if the unsaturated inversion $d_{0\mu}$ is nonvanishing only within that part[4].

VI.15. The linear and quasi-linear solution of the general Fokker-Planck equation[1]. Far below and far above threshold the special "adiabatic" Fokker-Planck Eq. (VI.12.28) or (VI.13.2) or that for multimode action (VI.14.1) cease to be valid, so that in these regions a Fokker-Planck equation of the type (VI.12.13), which still contains the atomic variables, must be used. Fortunately, this equation can be solved in the regions far below threshold by linearization and far above threshold by quasi-linearization.

a) Far below threshold. In the term $gD\vec{u}$ of (VI.12.13) we put $D=D_0$ and neglect the term with $\vec{u}\,\vec{v}$. The drift coefficients of Eq. (VI.12.13) thus become linear, while the diffusion coefficients are constants. Thus we obtain a Fokker-Planck equation of the form

$$\frac{\partial W}{\partial t}+\sum_{ij=1} C_{ij}\frac{\partial}{\partial x_i}(x_j W)=\frac{1}{2}\sum_{ij=1}^{n} Q_{ij}\frac{\partial^2 W}{\partial x_i\,\partial x_j} \tag{VI.15.1}$$

with constant coefficients C, Q. This equation which determines a linear Gaussian process, can be solved by standard methods (see e.g. WANG and UHLENBECK[2]).

b) Well above threshold. The coefficients of the Fokker-Planck Eq. (VI.12.13) can be linearized if we use polar coordinates for u and v, decompose the amplitudes of u and v into a stable value and a residual coordinate, and neglect higher powers of the residual coordinates (which include the phases). Again an equation of the type (VI.15.1) results. RISKEN, SCHMID and WEIDLICH[3] have demonstrated, that the correlation functions calculated by this procedure agree with those obtained by the quantum mechanical Langevin equations [see (VI.5), (VI.7) or [4]].

[4] C. L. TANG and H. STATZ: J. Appl. Phys. **38**, 2963 (1967).

[1] For more details see: H. RISKEN, C. SCHMID, and W. WEIDLICH: Phys. Letters **20**, 489 (1966); — Z. Physik **194**, 337 (1966).

[2] M. C. WANG and G. E. UHLENBECK: Rev. Mod. Phys. **17**, 323 (1945).

[3] See l.c.[1] above.

[4] H. HAKEN: Z. Physik **190**, 327 (1966).

We briefly indicate the method of solution: We denote the set $x_1, \dots x_n$ by x. The Green's function $G(x, x', t)$ of Eq. (VI.14.1), which obeys the initial condition

$$G(x, x', 0) = \prod_j \delta(x_j - x'_j) \tag{VI.15.2}$$

is given by $(t \geqq 0)$

$$\left.\begin{aligned} G(x, x', t) = {} & \left(\pi^n \operatorname{Det} \boldsymbol{\sigma}(t)\right)^{-\frac{1}{2}} \times \\ & \times \exp\left\{-\sum_{ij} (\sigma^{-1})_{ij} \left(x_i - \sum_k b_{ik}(t)\, x'_k\right)\left(x_j - \sum_l b_{jl}(t)\, x'_l\right)\right\} \end{aligned}\right\} \tag{VI.15.3}$$

with

$$\begin{aligned} \boldsymbol{\sigma} &= (\sigma_{ij}), \\ \sigma_{ij}(t) &= \sum_{sr} \left[\delta_{is}\, \delta_{jr} - b_{is}(t)\, b_{jr}(t)\right] \sigma_{sr}(\infty). \end{aligned} \tag{VI.15.4}$$

The matrix $b_{ij}(t)$ obeys the following equation

$$\dot{b}_{is} = \sum_j C_{ij}\, b_{js} \tag{VI.15.5}$$

with the initial condition

$$b_{js}(0) = \delta_{js}. \tag{VI.15.6}$$

$\boldsymbol{\sigma}(\infty)$ is determined by the equation

$$\boldsymbol{C}\boldsymbol{\sigma}(\infty) + \boldsymbol{\sigma}(\infty)\, \boldsymbol{C}^T = -2\boldsymbol{Q} \tag{VI.15.7}$$

where the index T indicates the transposed matrix, and $\boldsymbol{C}$ and $\boldsymbol{Q}$ stand for the matrices C_{ij} and Q_{ij}, respectively. The stationary solution of (VI.14.1) is given by

$$W(x) = G(x, x', \infty) = \left(\pi^n \operatorname{Det} \boldsymbol{\sigma}(\infty)\right) \exp\left\{-\sum_{ij} (\sigma^{-1})_{ij}(\infty)\, x_i\, x_j\right\}. \tag{VI.15.8}$$

The stationary joint distribution reads

$$F(x, x', t) = G(x, x', t)\, W(x'). \tag{VI.15.9}$$

VII. The semiclassical approach and its applications.

VII.1. Spirit of the semiclassical approach[1]. The equations for the solid-state laser. In the foregoing Chap. VI, we have shown that well above laser threshold (i.e. for photon numbers which are bigger than about 10 times the photon number at threshold) the quantum mechanical operator for the electric field strength

[1] The semiclassical approach has been used by a number of authors, who introduced it more or less independently of each other. Clearly the semiclassical approach draws heavily on the analogy between the problems of spin-resonance and the maser (consisting of two-level atoms) [for the mathematical equivalence of spins and two-level atoms see Chap. III.6].

The magnetic and electromagnetic fields, respectively, are treated as classical quantities obeying Maxwell's equation, whereas the matter, i.e. the spins and electrons (or atoms), respectively, are treated by the density matrix equations, which include the effects of pumping, incoherent decays etc. through the addition of phenomenological relaxation terms of the type introduced by Bloch [F. Bloch: Phys. Rev. **70**, 460 (1946)] for magnetic resonance problems.

For a derivation of these equations see the fundamental paper of R. K. Wangsness and F. Bloch: Phys. Rev. **89**, 728 (1953) (who treated spins); — The relevance of the Bloch equations to a description of the maser has been recognized by R. P. Feynman, F. L. Vernon, and R. W. Hellwarth: J. Appl. Phys. **28**, 49 (1957); — Later treatments may be found e.g. in the papers of L. W. Davis: Proc. IEE (London) **51**, 76 (1963). — W. E. Lamb Jr.: Phys. Rev. **134**, 1429 (1964). — H. Haken and H. Sauermann: Z. Physik **173**, 261 (1963); **176**, 47 (1963) and others.

can be decomposed into a classical quantity and a small additional quantity which describes quantum fluctuations. In addition, the "phase" of the total field changes slowly due to quantum fluctuations. If we are interested only in the gross features of laser light, we may discard these quantum fluctuations by averaging the fully quantum mechanical laser equations of Chap. IV over the quantum fluctuations. This procedure has been performed in Chap. IV.6; the resulting *semiclassical* laser equations are listed there in detail.

According to our promise to write each chapter in such a way that it can be understood without a knowledge of the other ones, we rederive these equations in a more elementary and somewhat heuristic way. It is important to note that all occurring quantities are classical, i.e. given by c-numbers.

a) The field equations. From Maxwell's equation

$$\operatorname{curl} \boldsymbol{H} = \frac{4\pi}{c}\boldsymbol{j} + \frac{1}{c}(\dot{\boldsymbol{E}} + 4\pi\dot{\boldsymbol{P}}) \tag{VII.1.1}$$

we find with the relations

$$\boldsymbol{j} = \sigma\boldsymbol{E}, \quad \boldsymbol{H} = \operatorname{curl}\boldsymbol{A}, \quad \boldsymbol{E} = -\frac{1}{c}\dot{\boldsymbol{A}} \tag{VII.1.2)*$$

and the Coulomb gauge

$$\operatorname{div}\boldsymbol{A} = 0 \tag{VII.1.3}$$

a differential equation for the vector potential $\boldsymbol{A}$:

$$-\Delta\boldsymbol{A} + \frac{1}{c^2}\ddot{\boldsymbol{A}} + \frac{4\pi\sigma}{c^2}\dot{\boldsymbol{A}} = \frac{4\pi}{c}\dot{\boldsymbol{P}} \tag{VII.1.4)*$$

and, by differentiation with respect to the time, a differential equation for the electric field strength

$$-\Delta\boldsymbol{E} + \frac{1}{c^2}\ddot{\boldsymbol{E}} + \frac{4\pi\sigma}{c^2}\dot{\boldsymbol{E}} = -\frac{4\pi}{c^2}\ddot{\boldsymbol{P}}. \tag{VII.1.5}$$

Usually, $\boldsymbol{E}(\boldsymbol{x}, t)$ is decomposed into the normalized eigenfunctions $\boldsymbol{u}_\lambda(\boldsymbol{x})$ of the unloaded interferometer ("cavity") (see Chap. II)

$$\boldsymbol{E}(\boldsymbol{x}, t) = \sum_\lambda E_\lambda(t)\,\boldsymbol{u}_\lambda(\boldsymbol{x}). \tag{VII.1.6}$$

The $\boldsymbol{u}$'s obey the homogeneous wave equation, so that

$$\Delta\boldsymbol{u}_\lambda = -k_\lambda^2\boldsymbol{u}_\lambda = -\frac{\omega_\lambda^2}{c^2}\boldsymbol{u}_\lambda. \tag{VII.1.7}$$

Because the $\boldsymbol{u}$'s are defined within an open cavity, they are not exactly orthogonal, although for practical purposes it is sufficient to assume their orthogonality. The open cavity causes diffraction losses, which may be taken care of by an "effective" conductivity σ in Eq. (VII.1.5).

Inserting (VII.1.6) into (VII.1.5), multiplying the resulting equation with $\boldsymbol{u}_\lambda(\boldsymbol{x})$ and integrating over the volume of the "cavity" we find

$$\omega_\lambda^2 E_\lambda + \ddot{E}_\lambda + \sum_{\lambda'} 4\pi\sigma_{\lambda\lambda'}\dot{E}_{\lambda'} = -4\pi\ddot{P}_\lambda \tag{VII.1.8}$$

where

$$\sigma_{\lambda\lambda'} = \int \boldsymbol{u}_\lambda\,\sigma\,\boldsymbol{u}_{\lambda'}\,dV, \tag{VII.1.9}$$

$$\ddot{P}_\lambda = \int \boldsymbol{u}_\lambda\,\ddot{\boldsymbol{P}}\,dV. \tag{VII.1.10}$$

* Note, that σ may be a tensor.

We abbreviate: $2\pi\sigma_{\lambda\lambda}=\varkappa_\lambda$, where $\varkappa_\lambda$ is the cavity half-width of mode λ and $2\pi\sigma_{\lambda\lambda'}=\varkappa_{\lambda\lambda'}$, $(\lambda\neq\lambda')$. In the following we put $\sigma_{\lambda\lambda'}=0$ for $\lambda\neq\lambda'$ except in VII.11, where we treat loss modulation.

b) The material equations. The polarization $\boldsymbol{P}$ in Eq. (VII.1.5) [or (VII.1.8)] is not a fixed source, but is caused in turn by the electric field $\boldsymbol{E}$. We adopt in this chapter the following laser model: The material consists of a solid state matrix, e.g. a crystalline lattice, in which the laser-active atoms labeled with an index μ, are embedded at random lattice sites $\boldsymbol{x}_\mu$.

We assume that the solid-state matrix causes a polarization, which is proportional to $\boldsymbol{E}$. The effect of the inactive solid-state matrix may be taken care of by using an effective light velocity. In the following we therefore consider only that part of $\boldsymbol{P}$ which stems from the laser-active atoms. We decompose $\boldsymbol{P}$ into the single-atom contributions

$$\boldsymbol{P}(\boldsymbol{x})=\sum_\mu \delta(\boldsymbol{x}-\boldsymbol{x}_\mu)\,\boldsymbol{p}_\mu \tag{VII.1.11}$$

where δ is Dirac's δ-function and the sum runs over the laser-active atoms.

Each $\boldsymbol{p}_\mu$ can be visualized as decomposed in $e\boldsymbol{\xi}$, where e is the electronic charge and $\boldsymbol{\xi}$ is the displacement of a local oscillator. In a *purely classical* treatment, the equation for $\boldsymbol{p}_\mu$ would read (as in the usual dispersion theory):

$$\ddot{\boldsymbol{p}}_\mu+2\gamma\,\dot{\boldsymbol{p}}_\mu+(\nu_\mu)^2\,\boldsymbol{p}_\mu=\frac{e^2}{m}\,\boldsymbol{E}(\boldsymbol{x}_\mu) \tag{VII.1.12}$$

(ν_μ is the frequency of the oscillator).

The damping constant γ can be identified with the atomic half-width. This equation is, however, insufficient, because it does not account for the effect of saturation. The quantum mechanical calculation shows that the strength and the sign (i.e. the relative phase) between $\boldsymbol{p}$ and the driving field depends on the atomic inversion, which is defined as the difference $d_\mu=(N_2-N_1)_\mu$ between the occupation numbers of the upper and lower level of the lasing atom, μ. Anticipating further a quantum mechanical replacement of the factor e^2/m, we may write the quantum mechanical analogue of (VII.1.12) in the form

$$\ddot{\boldsymbol{p}}_\mu+2\gamma\,\dot{\boldsymbol{p}}_\mu+(\nu_\mu)^2\boldsymbol{p}_\mu=-\,2\omega_0\,\frac{|\vartheta_{12}^2|}{\hbar}\,\boldsymbol{E}(\boldsymbol{x}_\mu)\,d_\mu \tag{VII.1.13}$$

ϑ_{12} is the atomic dipole moment matrix element [compare (V.3.13)]. For simplicity we have assumed that the vectors ϑ_{12}, $\boldsymbol{p}_\mu$ and $\boldsymbol{E}$ are parallel, so that ϑ_{12} enters Eq. (VII.1.13) in the given way*. ω_0 is the circular frequency at the atomic line center. The occurrence of the inversion d_μ in Eq. (VII.1.13) means that the atom absorbs energy if it is in its lower state and emits energy if it is in its upper state. This factor d_μ introduces into the equations of motion a nonlinearity which is of basic importance for the whole laser theory.

The inversion (population difference) $d_\mu=(N_2-N_1)_\mu$ itself may change due to radiative emission and due to relaxation processes (which include the pumping of the atom). Accordingly we write for the temporal changing of the atomic energy $\hbar\nu d_\mu$:

$$\frac{d}{dt}(\hbar\nu d_\mu)=\left(\frac{d(\hbar\nu d_\mu)}{dt}\right)_{\textit{field}}+\left(\frac{d(\hbar\nu d_\mu)}{dt}\right)_{\textit{relaxation}} \tag{VII.1.14}$$

---

* If the dipole moments point in random directions and an average is performed over all directions, a factor of $\frac{1}{3}$ must be added to $|\vartheta_{12}|^2$ on the right-hand side of Eq. (VII.1.13).

where

$$\left(\frac{d(\hbar\nu d_\mu)}{dt}\right)_{field} = 2\boldsymbol{E}(\boldsymbol{x}_\mu)\,\dot{\boldsymbol{p}}_\mu \tag{VII.1.15}$$

is the usual classical expression.

$\left(\frac{d(\hbar\nu d_\mu)}{dt}\right)_{relaxation}$ obeys a single equation, if we confine ourselves *to a two-level atom*:

$$\frac{d(\hbar\nu d_\mu)}{dt} = \frac{\hbar\nu(d_0 - d)_\mu}{T} \tag{VII.1.16}$$

where T is the effective relaxation time to restore the equilibrium value d_0.

Putting (VII.1.14)—(VII.1.16) together we find

$$\frac{d}{dt}\,d_\mu = \frac{(d_0 - d)_\mu}{T} + \frac{2}{\hbar\nu}\,\boldsymbol{E}(\boldsymbol{x}_\mu)\,\dot{\boldsymbol{p}}_\mu \tag{VII.1.17}$$

Eqs. (VII.1.5) or (VII.1.8), (VII.1.13) and (VII.1.17) are the basic equations for a set of laser modes interacting with a set of two-level atoms. In the case of multi-level atoms the Eqs. (VII.1.5), (VII.1.13) remain valid, but (VII.1.17) acquires a more complicated structure (see p. 180).

We now introduce the *rotating wave approximation* (compare also Chap. III.4).

We decompose E_λ and $\boldsymbol{p}_\mu$ [see (VII.1.6)] into a positive and negative frequency part:

$$E_\lambda(t) = e^{-i\omega_\lambda t}\tilde{E}_\lambda^{(+)}(t) + e^{i\omega_\lambda t}\tilde{E}_\lambda^{(-)}(t) \equiv E_\lambda^{(+)} + E_\lambda^{(-)}(t), \tag{VII.1.18}$$

$$\boldsymbol{p}_\mu(t) = e^{-i\nu_\mu t}\tilde{\boldsymbol{p}}_\mu^{(+)}(t) + e^{i\nu_\mu t}\tilde{\boldsymbol{p}}_\mu^{(-)}(t) \equiv \boldsymbol{p}_\mu^{(+)}(t) + \boldsymbol{p}_\mu^{(-)}(t) \tag{VII.1.19}$$

where ω_λ is the frequency in the unloaded cavity.

A decomposition corresponding to (VII.1.19) holds also for the macroscopic polarization $\boldsymbol{P}_\lambda$.

It will become evident below that in a laser $\omega \approx \nu$, and that the remaining amplitudes $E^{(\pm)}$, $\boldsymbol{p}^{(\pm)}$ vary much more slowly than $e^{i\omega t}$. We therefore neglect $\ddot{\tilde{E}}_\lambda$ compared to $\omega_\lambda \dot{\tilde{E}}_\lambda$, and $\ddot{\tilde{\boldsymbol{p}}}_\mu$ compared to $\nu_\mu \dot{\tilde{\boldsymbol{p}}}_\mu$. Comparing the coefficients of terms $\sim e^{i\omega_\lambda t}$, $e^{-i\omega_\lambda t}$ in Eqs. (VII.1.8) and (VII.1.11) separately, we find

$$\dot{\tilde{E}}_\lambda^{(-)} + \varkappa_\lambda \tilde{E}_\lambda^{(-)} = -i2\pi\,\omega_0 \sum_\mu \boldsymbol{u}_\lambda(\boldsymbol{x}_\mu)\,\boldsymbol{p}_\mu^{(-)}\,e^{i\nu_\mu t - i\omega_\lambda t} \tag{VII.1.20}$$

or, with

$$E_\lambda^{(\mp)} = e^{\pm i\omega_\lambda t}\tilde{E}_\lambda^{(\mp)}; \qquad \boldsymbol{p}_\mu^{(\mp)} = \tilde{\boldsymbol{p}}_\mu^{(\mp)}\,e^{\pm i\nu_\mu t} \tag{VII.1.21}$$

and (VII.1.11) we end up with

$$\dot{E}_\lambda^{(-)} = (i\omega_\lambda - \varkappa_\lambda)\,E_\lambda^{(-)} - i\omega_0\,2\pi \sum_\mu \boldsymbol{p}_\mu^{(-)}\,\boldsymbol{u}_\lambda(\boldsymbol{x}_\mu)\,. \tag{VII.1.22}$$

Similarly we find

$$\dot{\boldsymbol{p}}_\mu^{(-)} = (i\nu_\mu - \gamma)\,\boldsymbol{p}_\mu^{(-)} + i\,\frac{|\vartheta_{12}|^2}{\hbar}\sum_\lambda \boldsymbol{u}_\lambda(\boldsymbol{x}_\mu)\,E_\lambda^{(-)}(N_2 - N_1)_\mu, \tag{VII.1.23}$$

$$\frac{d\,d_\mu}{dt} = \frac{d_{0\mu} - d_\mu}{T} + \frac{2i}{\hbar}\left(\sum_\lambda E_\lambda^{(+)}\,\boldsymbol{u}_\lambda(\boldsymbol{x}_\mu)\,\boldsymbol{p}_\mu^{(-)} - \sum_\lambda E_\lambda^{(-)}\,\boldsymbol{u}_\lambda(\boldsymbol{x}_\mu)\,\boldsymbol{p}_\mu^{(+)}\right) \tag{VII.1.24}$$

where we have dropped terms $\sim e^{\pm i\omega t}$ (Rotating-wave approximation). We have further put $\omega_\lambda = \nu_\mu = \omega_0$ in all factors on the right hand side of Eqs. (VII.1.20), (VII.1.22), (VII.1.23), (VII.1.24). The Eqs. (VII.1.22)—(VII.1.24) again represent a complete set of laser equations.

In our article, we use a notation which directly makes contact with the quantum theoretical treatment:

We put [see also formula (III.1.15)]

$$E_\lambda^{(-)} = -i\sqrt{2\pi\hbar\omega_\lambda}\, b_\lambda^*, \tag{VII.1.25}$$

$$E_\lambda^{(+)} = i\sqrt{2\pi\hbar\omega_\lambda}\, b_\lambda. \tag{VII.1.26}$$

In a fully quantum mechanical treatment, b_λ^*, b_λ are creation and annihilation operators. Here, they are merely time dependent, complex *dimensionless* amplitudes.

We decompose the atomic polarization $\boldsymbol{p}^{(+)}$ into a quantity of the dimension of a dipole moment $(e x_{21})$, and a remaining dimensionless factor ϱ_{21}.

$$\boldsymbol{p}_\mu^{(+)} = \boldsymbol{\vartheta}_{21} \cdot \varrho_{21,\mu} \tag{VII.1.27}$$

and correspondingly

$$\boldsymbol{p}_\mu^{(-)} = \boldsymbol{\vartheta}_{12}\, \varrho_{12,\mu}. \tag{VII.1.28}$$

We further write

$$N_{2,\mu} = \varrho_{22,\mu}, \tag{VII.1.29}$$

$$N_{1,\mu} = \varrho_{11,\mu}. \tag{VII.1.30}$$

In the quantum theoretical treatment, ϱ_{21}, ϱ_{22} etc. are the elements of the density matrix of the two-level atom. Both there and here they are time dependent classical functions. Finally we introduce the coupling constant

$$g_{\mu\lambda} = i\boldsymbol{\vartheta}_{21}\sqrt{\frac{2\pi\omega_\lambda}{\hbar}}\,\boldsymbol{u}_\lambda(\boldsymbol{x}_\mu). \tag{VII.1.31}$$

By use of (VII.1.25)—(VII.1.30), the original Eqs. (VII.1.22)—(VII.1.24) acquire the following form which we will use in all subsequent chapters.

The complete set of the semiclassical laser equations (two-level atoms)

$$\dot{b}_\lambda^*(t) = (i\omega_\lambda - \varkappa_\lambda)\, b_\lambda^*(t) + i\sum_\mu g_{\mu,\lambda}^*\, \varrho_{12,\mu}(t), \tag{VII.1.32}$$ * **

$$\dot{\varrho}_{12,\mu}(t) = (i\nu_{21,\mu} - \gamma_{21})\, \varrho_{12,\mu} - i\sum_\lambda g_{\mu\lambda}\, b_\lambda^*\,(\varrho_{22} - \varrho_{11})_\mu, \tag{VII.1.33}$$

$$\dot{d}_\mu = \frac{d_{0,\mu} - d_\mu}{T} + 2i\,\varrho_{21,\mu}\sum_\lambda g_{\mu\lambda}\, b_\lambda^* - 2i\,\varrho_{21,\mu}^*\sum_\lambda g_{\mu\lambda}^*\, b_\lambda \tag{VII.1.34}$$

where

$$d_{0,\mu} = \frac{w_{12} - w_{21}}{w_{12} + w_{21}} \tag{VII.1.34a}$$

is the unsaturated inversion of a single atom [compare e.g. (VI.4.11)] and

$$\frac{1}{T} = w_{12} + w_{21} \tag{VII.1.34b}$$

* If we retain the mode coupling caused by the conductivity σ [see (VII.1.8) (VII.1.9), we have to add to the right hand side of Eq. (VII.1.32) the term:

$$\sum_{\lambda'} \sigma_{\lambda\lambda'}\sqrt{\frac{\omega_{\lambda'}}{\omega_\lambda}}\, b_{\lambda'}^*. \tag{VII.1.32a}$$

** These equations can also be obtained directly from a Hamiltonian (i.e. an energy expression) in the framework of second quantization, if damping terms are added. See H. Haken and H. Sauermann: l.c.[1] on p. 173 or Chap. IV.6b. This procedure can be readily applied to the semiconductor laser, where it has certain advantages. See H. Haken and E. Haken: Z. Physik **176**, 421 (1964). — H. Haug: Z. Physik **194**, 482 (1966); **195**, 74 (1966). — T. L. Paoli and J. E. Ripper: Phys. Rev. Letters **22**, 1085 (1969).

is the relaxation time of the inversion [compare e.g. (VI.4.12)]. $d_\mu = (\varrho_{22} - \varrho_{11})_\mu$ is the saturated inversion of atom μ.

The complete set of corresponding equations for multilevel atoms is given in (IV.6.5) *and* (IV.6.6).

In most of the following chapters we will use Eqs. (VII.1.32)—(VII.1.34).

c) Macroscopic treatment.*

α) *Wave picture, inhomogeneous atomic line.* Instead of the single dipole-moments, we may use the macroscopic polarization $\boldsymbol{P}$ [see Eq. (VII.1.11)] instead. We split the atomic index μ into one part, which refers to the atomic position, μ' and a second one, which refers to the transition frequency ν, i.e. $\mu = (\mu', \nu)$. We assume that the transition frequencies and the positions are uncorrelated. We define a macroscopic polarization for each transition frequency by

$$\boldsymbol{P}_\nu(\boldsymbol{x}, t) = \sum_{\mu'} \delta(\boldsymbol{x} - \boldsymbol{x}_{\mu'})\, \boldsymbol{p}_{(\mu',\nu)}(t). \tag{VII.1.35}$$

The total polarization is given by the sum

$$\boldsymbol{P}(\boldsymbol{x}, t) = \sum_\nu \boldsymbol{P}_\nu(\boldsymbol{x}, t), \tag{VII.1.36}$$

which more properly must be replaced by an integral over the inhomogeneous frequency distribution.

The *field equation* follows directly from (VII.1.5)

$$-\Delta \boldsymbol{E}(\boldsymbol{x}, t) + \frac{1}{c^2}\ddot{\boldsymbol{E}}(\boldsymbol{x}, t) + \frac{4\pi\sigma}{c^2}\dot{\boldsymbol{E}}(\boldsymbol{x}, t) = -\frac{4\pi}{c^2}\sum_\nu \boldsymbol{P}_\nu(\boldsymbol{x}, t). \tag{VII.1.37}$$

The *material equations* are obtained from (VII.1.12) and (VII.1.17) by multiplying those equations with $\delta(\boldsymbol{x} - \boldsymbol{x}_{\mu'})$ and summing up over μ':

Using (VII.1.35) and

$$D_\nu(\boldsymbol{x}, t) = \sum_{\mu'} \delta(\boldsymbol{x} - \boldsymbol{x}_{\mu'})\, D_\mu(t) \tag{VII.1.38}$$

we obtain

$$\ddot{\boldsymbol{P}}_\nu(\boldsymbol{x}, t) + 2\gamma\dot{\boldsymbol{P}}_\nu(\boldsymbol{x}, t) + \nu^2 \boldsymbol{P}_\nu(\boldsymbol{x}, t) = -2\omega_0 \frac{|\vartheta_{12}^2|}{\hbar}\boldsymbol{E}(\boldsymbol{x}, t)\, D_\nu(\boldsymbol{x}, t) \tag{VII.1.39}$$

and

$$\dot{D}_\nu(\boldsymbol{x}, t) = \frac{D_{0,\nu} - D_\nu(\boldsymbol{x}, t)}{T} + \frac{2}{\hbar\nu}\boldsymbol{E}(\boldsymbol{x}, t)\,\dot{\boldsymbol{P}}_\nu(\boldsymbol{x}, t). \tag{VII.1.40}$$

β) *Wave picture, homogeneous atomic line.*

Field equation. We replace in Eq. (VII.1.37) $\boldsymbol{P}_\nu(\boldsymbol{x}, t)$ by $\boldsymbol{P}(\boldsymbol{x}, t)$ and drop the sum over ν

$$-\Delta \boldsymbol{E}(\boldsymbol{x}, t) + \frac{1}{c^2}\ddot{\boldsymbol{E}}(\boldsymbol{x}, t) + \frac{4\pi\sigma}{c^2}\dot{\boldsymbol{E}}(\boldsymbol{x}, t) = -\frac{4\pi}{c^2}\boldsymbol{P}(\boldsymbol{x}, t). \tag{VII.1.41}$$

Material equation. In Eqs. (VII.1.39) and (VII.1.40) we replace $\boldsymbol{P}_\nu(\boldsymbol{x}, t)$ and $D_\nu(\boldsymbol{x}, t)$ by $\boldsymbol{P}(\boldsymbol{x}, t)$ and $D(\boldsymbol{x}, t)$ respectively, and replace ν by ν_0:

$$\ddot{\boldsymbol{P}}(\boldsymbol{x}, t) + 2\gamma\dot{\boldsymbol{P}}(\boldsymbol{x}, t) + \nu_0^2 \boldsymbol{P}(\boldsymbol{x}, t) = -2\omega_0 \frac{|\vartheta_{12}|^2}{\hbar}\boldsymbol{E}(\boldsymbol{x}, t)\, D(\boldsymbol{x}, t), \tag{VII.1.42}$$

$$\dot{D}(\boldsymbol{x}, t) = \frac{D_0 - D(\boldsymbol{x}, t)}{T} + \frac{2}{\hbar\nu_0}\boldsymbol{E}(\boldsymbol{x}, t)\,\dot{\boldsymbol{P}}(x, t). \tag{VII.1.43}$$

---

* For the extension to a fully quantum mechanical treatment see R. GRAHAM and H. HAKEN: Z. Physik **213**, 420 (1968).

γ) *Wave picture, homogeneous atomic line, rotating wave approximation, slowly varying amplitude approximation.* The steps on p. 176 can be repeated and lead from the Eqs. (VII.1.41), (VII.1.42), (VII.1.43) to the corresponding equations in the rotating wave approximation: We introduce the decomposition (with $\omega_0 = \nu_0$)

$$\boldsymbol{E}(\boldsymbol{x}, t) = e^{-i\omega_0 t}\tilde{\boldsymbol{E}}^{(+)}(\boldsymbol{x}, t) + e^{i\omega_0 t}\tilde{\boldsymbol{E}}^{(-)}(\boldsymbol{x}, t), \tag{VII.1.44}$$

$$\boldsymbol{P}(\boldsymbol{x}, t) = e^{-i\omega_0 t}\tilde{\boldsymbol{P}}^{(+)}(\boldsymbol{x}, t) + e^{i\omega_0 t}\tilde{\boldsymbol{P}}^{(-)}(\boldsymbol{x}, t) \tag{VII.1.45}$$

into the Eqs. (VII.1.41)—(VII.1.43) and obtain for the leading terms

$$\left.\begin{aligned}\dot{\tilde{\boldsymbol{E}}}^{(-)}(\boldsymbol{x}, t) + i\frac{\omega_0}{2}\tilde{\boldsymbol{E}}^{(-)}(\boldsymbol{x}, t) + i\frac{c^2}{2\omega_0}\Delta\tilde{\boldsymbol{E}}^{(-)}(\boldsymbol{x}, t) + 2\pi\sigma\tilde{\boldsymbol{E}}^{(-)}(\boldsymbol{x}, t)\\ = -i2\pi\,\omega_0\tilde{\boldsymbol{P}}^{(-)}(\boldsymbol{x}, t),\end{aligned}\right\} \tag{VII.1.46}*$$

$$\dot{\tilde{\boldsymbol{P}}}^{(-)}(\boldsymbol{x}, t) + \gamma\tilde{\boldsymbol{P}}^{(-)}(\boldsymbol{x}, t) = i\frac{|\vartheta_{12}|^2}{\hbar}\cdot\tilde{\boldsymbol{E}}^{(-)}(\boldsymbol{x}, t)\,D(\boldsymbol{x}, t), \tag{VII.1.47}$$

$$\dot{D}(\boldsymbol{x}, t) = \frac{D_0 - D(\boldsymbol{x}, t)}{T} + \frac{2i}{\hbar}\left(\tilde{\boldsymbol{E}}^{(+)}(\boldsymbol{x}, t)\tilde{\boldsymbol{P}}^{(-)}(\boldsymbol{x}, t) - \tilde{\boldsymbol{E}}^{(-)}(\boldsymbol{x}, t)\boldsymbol{P}^{(+)}(\boldsymbol{x}, t)\right). \tag{VII.1.48}$$

For wave propagation in *one direction*, e.g. the z-direction, Eq. (VII.1.46) can be further simplified by putting

$$\tilde{\boldsymbol{E}}^{(-)}(\boldsymbol{x}, t) = e^{-ikz}\hat{\boldsymbol{E}}^{(-)}(z, t), \tag{VII.1.49}$$

$$\tilde{\boldsymbol{P}}^{(-)}(\boldsymbol{x}, t) = e^{-ikz}\hat{\boldsymbol{P}}^{(-)}(z, t) \tag{VII.1.50}$$

where $k = \frac{\omega_0}{c}$. Again neglecting terms $\frac{\partial^2\hat{E}}{\partial z^2}$ compared to $k\frac{\partial\hat{E}}{\partial z}$ we obtain from Eq. (VII.1.46)

$$\dot{\hat{\boldsymbol{E}}}^{(-)}(z, t) + c\frac{\partial\hat{\boldsymbol{E}}^{(-)}}{\partial z}(z, t) + 2\pi\sigma\hat{\boldsymbol{E}}^{(-)}(z, t) = -2\pi i\omega_0\hat{\boldsymbol{P}}^{(-)}(z, t). \tag{VII.1.51}$$

Because the factor e^{-ikz} drops out of Eqs. (VII.1.47) and (VII.1.48) we have merely to replace ~ by ^ and $\boldsymbol{x}$ by z:

$$\dot{\hat{\boldsymbol{P}}}^{(-)}(z, t) + \gamma\hat{\boldsymbol{P}}^{(-)}(z, t) = i\frac{|\vartheta_{12}|^2}{\hbar}\hat{\boldsymbol{E}}^{(-)}(z, t)\,D(z, t), \tag{VII.1.52}$$

$$\dot{D}(z, t) = \frac{D_0 - D(z, t)}{T} + \frac{2i}{\hbar}\left(\hat{\boldsymbol{E}}^{(+)}(z, t)\hat{\boldsymbol{P}}^{(-)}(z, t) - \hat{\boldsymbol{E}}^{(-)}(z, t)\hat{\boldsymbol{P}}^{(+)}(z, t)\right). \tag{VII.1.53}$$

δ) *Mode picture, polarization waves.* In order to avoid too many indices we confine our analysis to a *homogeneously broadened* atomic line, although the whole procedure can easily be extended to an inhomogeneous atomic line (compare Chap. VII.1.*c*.α). We decompose the macroscopic polarization $\boldsymbol{P}(\boldsymbol{x}, t)$ into the cavity modes $U_\lambda(\boldsymbol{x})$**:

$$P(\boldsymbol{x}, t) = \sum_\lambda P_\lambda(t)\,U_\lambda(\boldsymbol{x}). \tag{VII.1.54}$$

From (VII.1.11) we find

$$P_\lambda(t) = \sum_\mu U_\lambda(\boldsymbol{x}_\mu)\,p_\mu(t). \tag{VII.1.55}$$

* The equation for $\tilde{E}^{(+)}$, $\tilde{P}^{(+)}$ are conjugate complex to those for $\tilde{E}^{(-)}$, $\tilde{P}^{(-)}$.

** For simplicity we assume that all vectors are parallel, so that we use scalar quantities. We put $\boldsymbol{u}_\lambda(\boldsymbol{x}) = \boldsymbol{e}_\lambda U_\lambda(\boldsymbol{x})$, where $\boldsymbol{e}_\lambda$: vector of polarization.

Multiplying Eq. (VII.1.13) with $U_\lambda(\boldsymbol{x}_\mu)$ and summing up over μ we obtain

$$\ddot{P}_\lambda + 2\gamma \dot{P}_\lambda + \nu^2 P_\lambda = -2\omega_0 \frac{|\vartheta_{12}|^2}{\hbar} \sum_{\lambda'} E_{\lambda'} D_{\lambda\lambda'} \tag{VII.1.56}$$

where

$$D_{\lambda\lambda'} = (N_2 - N_1)_{\lambda\lambda'} \equiv \sum_\mu (N_2 - N_1)_\mu \cdot U_\lambda(\boldsymbol{x}_\mu)\, U_{\lambda'}(\boldsymbol{x}_\mu). \tag{VII.1.57}$$

In order to obtain an equation of motion for $D_{\lambda\lambda'}$ we multiply Eq. (VII.1.17) with $U_\lambda(\boldsymbol{x}_\mu)\, U_{\lambda'}(\boldsymbol{x}_\mu)$ and sum up over $\boldsymbol{x}_\mu$.

After a slight transformation which expresses P_μ by P_λ the resulting equation reads

$$\frac{d}{dt} D_{\lambda\lambda'} = \frac{D^0_{\lambda\lambda'} - D_{\lambda\lambda'}}{T} + \frac{2}{\hbar\nu} \sum_{\lambda''\lambda'''} A(\lambda, \lambda', \lambda'', \lambda''') \cdot E_{\lambda''}(t)\, \dot{P}_{\lambda'''}(t) \tag{VII.1.58}$$

where

$$A(\lambda, \lambda', \lambda'', \lambda''') = \int U_\lambda(\boldsymbol{x})\, U_{\lambda'}(\boldsymbol{x})\, U_{\lambda''}(\boldsymbol{x})\, U_{\lambda'''}(\boldsymbol{x})\, dV. \tag{VII.1.59}$$

If the unsaturated inversion $d_0 = (N_2 - N_1)^0_\mu$ is independent of the atomic index μ, $D^0_{\lambda\lambda'}$ reduces to $d_0\,\delta_{\lambda\lambda'}$.

d) Extension to multilevel atoms. So far we have been considering two-level atoms. The extension to multilevel atoms is, however, straightforward. We assume, that only a single atomic transition, say from level 2 to level 1, is in resonance with the laser light. Levels 1 and 2 may have any position in the atomic level scheme. The required changes of the field and matter equations are given in the following scheme:

Field equations

(VII.1.32)	(VII.1.37)	(VII.1.41)	(VII.1.46)	(VII.1.51)

Remain completely unchanged

Matter equations for the polarization

(VII.1.38)	(VII.1.39)	(VII.1.42)	(VII.1.47)	(VII.1.52)
Unchanged	Replace on the right-hand side $D(\boldsymbol{x}, t)$ by $N_2(\boldsymbol{x}, t) - N_1(\boldsymbol{x}, t)$			

Matter equations for the inversion

(VII.1.34)	(VII.1.40)	(VII.1.43)	(VII.1.48)	(VII.1.53)

Replace the single equation for D by a set of equations for the

Occupation numbers $\varrho_{jj} = N_j$	Occupation number densities $N_{j,\nu}(\boldsymbol{x}) = \sum_{\mu'} \delta(\boldsymbol{x} - \boldsymbol{x}_{\mu'})\, N_{j,\nu,\mu'}$ (j: level index, ν: line broadening index, μ': atomic position index)

This set of equations reads*:

$$\frac{dN_1}{dt} = \left(\frac{\partial N_1}{\partial t}\right)_{inc.} + \left(\frac{\partial N_1}{\partial t}\right)_{coh.},$$

$$\frac{dN_2}{dt} = \left(\frac{\partial N_2}{\partial t}\right)_{inc.} + \left(\frac{\partial N_2}{\partial t}\right)_{coh.},$$

$$\frac{dN_l}{dt} = \left(\frac{\partial N_l}{\partial t}\right)_{inc.}; \quad l \neq 1, 2.$$

* All quantities N_j bear additional indices ν or μ and/or arguments $(\boldsymbol{x})$ depending on what picture is used.

$\left(\frac{\partial N_j}{\partial t}\right)_{inc}$ is given by the incoherent changes of N_j (compare also e.g. Sect. IV.7) with transition rates w_{jk} from level j to level k:

$$\left(\frac{\partial N_j}{\partial t}\right)_{inc} = \sum_k N_k w_{kj} - \sum_k N_j w_{jk}.$$

The recipe for obtaining $\left(\frac{\partial N_2}{\partial t}\right)_{coh} = -\left(\frac{\partial N_1}{\partial t}\right)_{coh}$ is:
on the right-hand sides of Eqs.

(VII.1.34) | (VII.1.40) | (VII.1.43) | (VII.1.48) | (VII.1.53)

put $1/T=0$, and divide the rest by two. The resulting expression is identical with $\left(\frac{\partial N_2}{\partial t}\right)_{coh}$.

e) Systematics of the semiclassical approach. The light field is treated as a classical quantity which obeys Eq. (VII.1.5) [or Eq. (VII.1.22)]. The polarization $\boldsymbol{P}$ may be decomposed into the contribution of the single atoms, $\boldsymbol{p}_\mu$. $\boldsymbol{p}_\mu$ must be calculated as the quantum statistical average of the dipole moment operator $e\boldsymbol{x}$ of the single atom. This is achieved by taking the trace over the product of $e\boldsymbol{x}$ and the density matrix of the single atom*:

$$\boldsymbol{p} = \mathrm{tr}\,(e\boldsymbol{x}\,\varrho_{e.c}), \tag{VII.1.60}$$

where $\varrho_{e.c}$ is a function of the electron coordinate. When the integration over the electron coordinate is performed, $\boldsymbol{p}$ takes the form:

$$\boldsymbol{p} = \boldsymbol{\vartheta}_{12}\,\varrho_{12} + \boldsymbol{\vartheta}_{21}\,\varrho_{21} \tag{VII.1.61}$$

(note that $\boldsymbol{\vartheta}_{12} = e\boldsymbol{x}_{12}$), where ϱ_{ik} is the density matrix in the occupation number representation, and $\boldsymbol{x}_{12} = \int \varphi_1^*(\boldsymbol{x})\,\boldsymbol{x}\,\varphi_2(\boldsymbol{x})\,dV$; φ_i are the wave functions of the levels $i=1, 2$, which are connected by the optical (laser) transition. ϱ obeys an equation of the form

$$\frac{d\varrho}{dt} = \left(\frac{\partial\varrho}{\partial t}\right)_{coh} + \left(\frac{\partial\varrho}{\partial t}\right)_{incoh} \tag{VII.1.62}$$

where $\left(\frac{\partial\varrho}{\partial t}\right)_{coh}$ describes the coherent change of ϱ

$$\left(\frac{\partial\varrho}{\partial t}\right)_{coh} = -\frac{i}{\hbar}\,[H, \varrho]. \tag{VII.1.63}$$

The electron Hamiltonian H comprises the kinetic energy, the potential energy which stems from the atomic field, energy contributions due to the vector potential (or the electric field strength) of the light field, energy contributions due to externally applied fields.

An explicit example for H (in the coordinate representation) would be

$$H = -\frac{\hbar^2}{2m}\,\Delta - \frac{e^2 Z}{|\boldsymbol{x}|} + e\boldsymbol{x}\boldsymbol{E}(t) \tag{VII.1.64}$$

m: electronic mass, Z: charge of the nucleus, $\boldsymbol{E}(t)$ is assumed independent of $\boldsymbol{x}$.

$\left(\frac{\partial\varrho}{\partial t}\right)_{incoh}$ stems from all kinds of incoherent processes like the pumping, radiationless transition, spontaneous emission etc. The general theory and explicit examples were given in IV.7. Since H is a function of the vector potential, $\boldsymbol{A}$, or of the electric field strength, $\boldsymbol{E}$, ϱ itself becomes a functional of $\boldsymbol{A}$ or $\boldsymbol{E}$, e.g.

* At present, we drop the index μ of the single atom.

$\varrho = \varrho(\boldsymbol{E})$. For not too high field strengths, ϱ may be expanded into a power series of E. Inserting this expansion into (VII.1.61), the cartesian components of $\boldsymbol{p}$ acquire the form

$$\left.\begin{aligned} p_j(t) &= \alpha_j^{(0)} + \sum_k \int^t \alpha_{jk}^{(1)}(t,\tau)\, E_k(\tau)\, d\tau + \\ &\quad + \sum_{kl} \int^t\!\!\int^t \alpha_{jkl}^{(2)}(t,\tau_1,\tau_2)\, E_k(\tau_1)\, E_l(\tau_2)\, d\tau_1\, d\tau_2 + \cdots . \end{aligned}\right\} \tag{VII.1.65}$$

The values of the tensor components $\alpha^{(i)}$ are restricted by symmetry conditions, which are discussed in detail in articles about nonlinear optics[2]. The time integral takes care of retardation effects.

In all cases the task is to express $\boldsymbol{p}$ as a functional of $\boldsymbol{E}$, so that an equation is obtained for $\boldsymbol{E}$ alone, which replaces (VII.1.5). The resulting equation is in general nonlinear.

VII.2. Method of solution for the stationary state[1]. We solve the system (VII.1.32)—(VII.1.34) of the nonlinear coupled equations by an iteration procedure. To this end we assume that by the pumping as well as by relaxation processes an inversion

$$d_\mu^{(0)} = d_{\mu,0} \tag{VII.2.1}$$

is established without laser action in zero'th approximation. We assume further that due to the laser process coherent fields of the form

$$b_\lambda^* = B_\lambda^*\, e^{i\Omega_\lambda t} \tag{VII.2.2}$$

are built up. The complex amplitude B_λ^* and the frequency Ω_λ are unknown parameters which are to be determined in a self-consistent way. Ω_λ is assumed to be real, because we confine ourselves in the present paragraph to the stationary state. Under certain circumstances B_λ^* may not be completely time-independent but may vary very slowly. The atomic dipole moments are changed under the influence of a given light field according to Eq. (VII.1.33) whose integration yields[2]

$$\varrho_{12,\mu}^{(1)}(t) = -d_{\mu,0} \sum_\lambda B_\lambda^*\, g_{\mu\lambda} \frac{e^{i\Omega_\lambda t}}{(\Omega_\lambda - \nu_\mu - i\gamma)} . \tag{VII.2.3}$$

Inserting this result into (VII.1.34) and integrating this equation we obtain as an improved value for the inversion d_μ

$$d_\mu^{(1)} = d_{\mu,0} + \left\{ 2i d_{0,\mu} \sum_{\lambda,\lambda'} B_\lambda B_{\lambda'}^* g_{\mu\lambda}^* g_{\mu\lambda'} \frac{e^{i(\Omega_{\lambda'}-\Omega_\lambda)t}}{(\Omega_{\lambda'} - \nu_\mu - i\gamma)\left(\frac{1}{T} + i[\Omega_{\lambda'} - \Omega_\lambda]\right)} + \text{c.c.} \right\} . \tag{VII.2.4}$$

---

[2] S. A. Akhmanov and R. V. Khokhlov: Usp. Fiz. Nauk **88**, 439 (1966); translated as Soviet Phys.-Usp. **9**, 210 (1966); — Usp. Fiz. Nauk **95**, 231 (1968). — N. Bloembergen: Nonlinear Optics. New York: W. A. Benjamin, Inc. (1965). — P. Pershan, in: Progr. in Optics **5**, 85 (1966) (E. Wolf, ed.). Amsterdam: North Holland Publ. Co. 1966.

[1] H. Haken and H. Sauermann: Z. Physik **176**, 47 (1963); also in the Proc. Internat. School of Physics, "Enrico Fermi", Course XXXI: Quantum Electronics and Coherent Light (ed. P. A. Miles), New York and London: Academic Press 1964.

[2] The iteration steps (VII.2.3)—(VII.2.5) are equivalent to a perturbation theoretical treatment with respect to the coupling constants g. Similar procedures are applied in nonlinear optics (see e.g. the book of Bloembergen, l.c. [2] above, where further references are given). The consistency requirement (VII.2.6) yields the fundamental laser equations of ref. [1]. This iteration procedure is not restricted to two-level atoms but can also be applied to multilevel atoms. See e.g. Chap. VI.10, where single mode action (including fluctuating forces) is treated.

On account of the improved inversion we can determine further improved atomic dipole moment. For this purpose we insert (VII.2.4) into (VII.1.33) and obtain after the integration

$$\left.\begin{aligned}\varrho^{(2)}_{12,\mu} &= -d_{\mu,0}\sum_{\lambda} B^*_\lambda g_{\mu\lambda}\frac{e^{i\Omega_\lambda t}}{(\Omega_\lambda-\nu_\mu-i\gamma)}+\\ &+2d_{\mu,0}\sum_{\lambda,\lambda',\lambda''} B_\lambda B^*_{\lambda'} B^*_{\lambda''} g^*_{\mu\lambda} g_{\mu\lambda'} g_{\mu\lambda''}\times\\ &\times\frac{\exp\left[i(\Omega_{\lambda'}+\Omega_{\lambda''}-\Omega_\lambda)t\right]}{i\left(\frac{1}{T}+i\left[\Omega_{\lambda'}-\Omega_\lambda\right]\right)(\Omega_{\lambda'}-\Omega_\lambda+\Omega_{\lambda''}-\nu_\mu-i\gamma)}\times\\ &\times\left(\frac{1}{\Omega_{\lambda'}-\nu_\mu-i\gamma}-\frac{1}{\Omega_\lambda-\nu_\mu+i\gamma}\right).\end{aligned}\right\}\qquad\text{(VII.2.5)}$$

Finally we have to take into account that the dipole moments react on the light field. Therefore we insert the $\varrho^{(2)}_{12,\mu}$'s, which have been obtained above, in the field equations instead of $\varrho^{(1)}_{12,\mu}$. This yields

$$\left.\begin{aligned}\dot b^*_\lambda &= (i\omega_\lambda-\varkappa)\,b^*_\lambda - i\sum_\mu d_{0,\mu}|g_{\mu\lambda}|^2\frac{B^*_\lambda e^{i\Omega_\lambda t}}{(\Omega_\lambda-\nu_\mu-i\gamma)}+\\ &+2\sum_{\mu,\lambda_1\lambda_2\lambda_3} d_{0,\mu} B_{\lambda_1} B^*_{\lambda_2} B^*_{\lambda_3} g^*_{\mu\lambda} g^*_{\mu\lambda_1} g_{\mu\lambda_2} g_{\mu\lambda_3}\times\\ &\times\frac{e^{i(\Omega_{\lambda_2}+\Omega_{\lambda_3}-\Omega_{\lambda_1})t}}{\left[\frac{1}{T}+i(\Omega_{\lambda_2}-\Omega_{\lambda_1})\right]\left[\Omega_{\lambda_2}+\Omega_{\lambda_3}-\Omega_{\lambda_1}-\nu_\mu-i\gamma\right]}\times\\ &\times\left(\frac{1}{\Omega_{\lambda_2}-\nu_\mu-i\gamma}-\frac{1}{\Omega_{\lambda_1}-\nu_\mu+i\gamma}\right).\end{aligned}\right\}\qquad\text{(VII.2.6)}$$

As can be shown in more detail the above iteration procedure is well justified if the atomic linewidth γ is larger than the other frequencies inherent in the problem. The Eqs. (VII.2.6) represent the basis of our investigations in this chapter. In fact the whole further analysis consists in performing the summations or integrations over energies and atomic indices and in solving the set of Eqs. (VII.2.6). These equations describe the oscillation of the modes taking into account their interaction, which is brought about by the response of the atomic system.

According to different combinations $\Omega_{\lambda_2}+\Omega_{\lambda_3}-\Omega_{\lambda_1}$ on the right hand side of (VII.2.6), new frequencies may occur. Thus by the combination of two or three fields, additional field modes may be created. An important example is the creation of $\Omega=2\Omega_2-\Omega_1$ which we will treat below in detail. This kind of frequency mixing is related to that in inactive materials.

In the following we will mostly solve the above equations in detail for one and two modes. To this end we write down the equations explicitly for these two cases:

a) Single-mode operation, general features

In Eqs. (VII.2.6) we drop the indices λ, λ_1, λ_2, λ_3 (which are all equal) and drop accordingly the sum over the λ's. We then make the ansatz (VII.2.2) and insert it into (VII.2.6). After division of the resulting equation by B^* we obtain:

$$\varkappa+i(\Omega-\omega) = -i\sum_\mu\frac{|g_\mu|^2 d_{0,\mu}}{(\Omega-\nu_\mu-i\gamma)}+\sum_\mu\frac{B^*B\,|g_\mu|^4\,d_{0\mu}\cdot 4i\gamma T}{[\Omega-\nu_\mu-i\gamma]\,[(\Omega-\nu_\mu)^2+\gamma^2]}\,. \qquad\text{(VII.2.7)}$$

We discuss Eq. (VII.2.7) briefly in order to give a deeper insight into our treatment. To this end we split Eq. (VII.2.7) into its real and imaginary parts:

real part (multiplied by 2):

$$2\varkappa = \sum_\mu \frac{d_{0\mu}|g_\mu|^2 2\gamma}{(\Omega-\nu_\mu)^2+\gamma^2} - \sum_\mu \frac{n|g_\mu|^4 d_{0,\mu} 8\gamma^2 T}{[(\Omega-\nu_\mu)^2+\gamma^2]^2}. \tag{VII.2.8}$$

After multiplication with the number of photons, $n = B^*B$, on both sides of (VII.2.8), the left hand side represents the loss rate of photons, whereas the first sum on the right hand side gives the increase of photon number due to induced emission from atoms with a given inversion.

If γ is small compared to the spread of atomic energies, $\frac{\gamma}{(\Omega-\nu_\mu)^2+\gamma^2}$ can be replaced by the δ-function $\delta(\Omega-\nu_\mu)$ representing energy conservation. Only those atoms whose transition energy coincides with the mode frequency participate in the process of stimulated emission. This expression can also be derived by usual first order perturbation theory (replacing ω by Ω). The second sum takes into account that the inversion is decreased by the laser process. For an inhomogeneously broadened line this effect leads to the so-called "holeburning into the gain curve" and will be treated in Chap. VII.4, VII.5, VII.7.

The *imaginary part* of (VII.2.7) reads:

$$(\Omega-\omega) = -\sum_\mu \frac{|g_\mu|^2(\Omega-\nu_\mu)\, d_{0\mu}}{(\Omega-\nu)^2+\gamma^2} + 4\gamma T \sum_\mu \frac{n|g_\mu|^4 d_{0\mu}(\Omega-\nu_\mu)}{[(\Omega-\nu)^2+\gamma^2]^2}. \tag{VII.2.9}$$

According to this, the active laser atoms change the frequency ω of the light mode in the unloaded cavity. The first sum on the right hand side has the same form as the usual dispersion formula with a given occupation of the atomic system.

This expression could have also been obtained by second order perturbation theory. The only difference is that we have already used a renormalized frequency Ω instead of ω as occurring in the usual perturbation procedure. What is essentially new, however, is represented by the occurrence of the second sum. Here the square of the light amplitude or correspondingly the number of photons occurs. This term therefore describes the change of the dispersion due to the adjustment of the atomic occupation to the intensity of the light field. It may be noted that this response of the atomic system is time independent. In the steady-state case the photon number adjusts in such a way that gain and loss compensate each other, which is described by Eq. (VII.2.8). With the photon number fixed in this way we can calculate the dispersion part of the right hand side of Eq. (VII.2.9), which leads immediately to an expression for the frequency shift in the inverted system as compared to the unloaded cavity.

b) Two-mode operation, general features. Specialization of (VII.2.6) yields in analogy to the above treatment of single mode action

$$\left.\begin{aligned} \varkappa + i(\Omega_\lambda-\omega_\lambda) = &\sum_\mu \frac{d_{0\mu}|g_{\mu\lambda}|^2}{\gamma+i(\Omega_\lambda-\nu_\mu)} - \\ &-4\gamma T \sum_{\mu\lambda'} \frac{d_{0\mu} B^*_{\lambda'} B_{\lambda'}|g_{\mu\lambda}|^2|g_{\mu\lambda'}|^2}{[\gamma+i(\Omega_\lambda-\nu_\mu)]\,[\gamma^2+(\Omega_{\lambda'}-\nu_\mu)^2]} + \\ &+2\sum_{\substack{\mu\\(\lambda'\neq\lambda)}} \frac{d_{0,\mu} B^*_{\lambda'} B_{\lambda'}|g_{\mu\lambda}|^2|g_{\mu\lambda'}|^2}{\left[\frac{1}{T}+i(\Omega_\lambda-\Omega_{\lambda'})\right][\Omega_\lambda-\nu_\mu-i\gamma]} \times \\ &\times\left(\frac{1}{\Omega_\lambda-\nu_\mu-i\gamma} - \frac{1}{\Omega_{\lambda'}-\nu_\mu-i\gamma}\right) \end{aligned}\right\} \tag{VII.2.10}$$

for $\lambda = 1$ and 2.

Whereas the left-hand side and the first sum on the right hand side are the same as for single-mode operation, new features occur due to the second and third sum:

α) *Time-independent atomic response.* The second sum gives rise to a change both of gain and dispersion in the first mode caused by the photon numbers of the first and the second modes. Whereas the change of the gain of mode λ is composed of contributions proportional to

$$\frac{n_{\lambda'}\gamma}{[\gamma^2+(\Omega_{\lambda'}-\nu_\mu)^2]} \tag{VII.2.11}$$

that of the dispersion is composed of terms proportional to

$$\frac{n_{\lambda'}(\Omega_{\lambda'}-\nu_\mu)}{[\gamma^2+(\Omega_{\lambda'}-\nu_\mu)^2]}\,. \tag{VII.2.12}$$

If we take further into account that the main contributions of the sums over μ stem from $\nu_\mu\approx\Omega_\lambda$, we see that the mutual influence of both modes has a longer range for the dispersion than for the gain. In fact, if $|\Omega_{\lambda'}-\Omega_\lambda|$ is bigger than γ there is practically no influence of one photon number, n_λ, on the other, $n_{\lambda'}$. On the other hand, the "holeburning" changes the dispersion. As we will see in VII.4 explicitly, this effect leads to a frequency pushing, if two modes show laser action. Whereas the holeburning stems from the time-independent atomic response, the

β) *Time-dependent atomic response* gives rise to the third sum in (VII.2.10). This can be seen by following up the iteration steps performed above. Under certain circumstances (being discussed in VI.9 and VII.4) it leads also to an appreciable frequency pushing of two modes.

VII.3. The solid-state laser with a homogeneously broadened line. Single- and multimode laser action[1]. The Eqs. (VII.1.32)—(VII.1.34) or the Eqs. (VII.2.6) allow us to treat mainly two problems:

1. *The determination of the mean number of photons in the laser modes.*
2. *The determination of the actual laser frequencies.*

Although in general the solid-state laser will have an inhomogeneously broadened line, we treat first the simpler case of a homogeneously broadened line, which allows us, incidentally, to elucidate the general procedure for the solution of the Eqs. (VII.2.6).

a) Single-mode operation. We start with the Eq. (VII.2.7) where we have to perform the sums (or integrals) over μ. Because in the present case the atom transition frequency ν does not depend on the atomic index μ, no integration over such frequencies is required. On the other hand we still have to sum up over the random atomic positions $\boldsymbol{x}_\mu$. This sum can be performed in a very good approximation by an integration over $\boldsymbol{x}$:

$$\sum_\mu \ldots = \varrho \int \ldots dV \tag{VII.3.1}$$

where ϱ is the density of the laser active atoms. In the following we assume that the unsaturated inversion $d_{0\mu}=d_0$ [compare Eq. (VII.1.34a)] does not depend on the position $\boldsymbol{x}_\mu$, i.e. we assume a spatially homogeneous pumping (which is also time-independent, of course).

---

[1] We follow closely the procedure of H. Haken and H. Sauermann: Z. Physik **173**, 261 (1963), and l.c.[1] on p. 182; — The single-mode operation can also be treated exactly, see p. 201.

In order to perform the integration (VII.3.1), we adopt, for the coupling constant $g_{\mu,\lambda}$ between the light mode λ and the atom μ the form [compare (VI.3.9), (VI.3.10) and also (VII.1.31), where we put $\nu\approx\omega_\lambda\approx\omega_0$]

$$g_{\mu\lambda}=i\sqrt{\frac{4\pi\nu^2}{\hbar\omega_\lambda V}}\,\boldsymbol{e}_\lambda\,\boldsymbol{\vartheta}_{21}\sin(k_\lambda\cdot x_\mu) \qquad \text{(VII.3.2)}$$

i.e. purely axial modes. The integration for modes with an arbitrary $\boldsymbol{k}$-direction can be done in an analogous way. $\boldsymbol{\vartheta}_{21}$ is the atomic dipole moment, V is the volume of the active material. The other quantities are explained in Eq. (VI.3.10). We multiply (VII.2.7) on both sides with $\gamma+i(\Omega-\nu)$. The real part of the new equation then reads*:

$$\gamma\varkappa-(\Omega-\omega)(\Omega-\nu)=\frac{2\pi\nu^2|\vartheta_{12}|^2\varrho d_0}{\hbar\omega}\left\{1-\frac{12\pi\nu^2|\vartheta_{12}|^2\gamma T\bar{n}}{\hbar\omega[\gamma^2+(\Omega-\nu)^2]}\right\} \qquad \text{(VII.3.3)}$$

whereas the imaginary part is given by

$$\Omega=\frac{\gamma\omega+\varkappa\nu}{\gamma+\varkappa}. \qquad \text{(VII.3.4)}$$

As we will see in VIII.1, Eq. (VII.3.3) is identical with a rate equation, which determines the photon density $\bar{n}$ as a function of the pumping (via d_0). $\bar{n}$ is defined as $\frac{1}{V}B^*B$. Solving (VII.3.3) for $\bar{n}$ yields indeed:

$$\bar{n}=\frac{d_{thr}(\Delta)\,\varrho}{6T\varkappa}\left(1-\frac{d_{thr}(\Delta)}{d_0}\right) \qquad \text{(VII.3.5)}$$

where we abbreviated**

$$d_{thr}(\Delta)=\frac{2\hbar\gamma\varkappa\omega}{4\pi\varrho\nu^2|\vartheta_{12}|^2}\left(1+\frac{\Delta^2}{(\gamma+\varkappa)^2}\right) \qquad \text{(VII.3.6)}$$

and $\Delta=\omega-\nu$.

The laser condition[2] follows from (VII.3.5): Laser action starts if $\bar{n}$ becomes positive or, according to (VII.3.6), if the atomic inversion d_0, which is fixed by the pumping and the incoherent decay processes, becomes bigger than d_{thr}:

$$d_0\geqq d_{thr} \qquad \text{(VII.3.7)}$$

or explicitly

$$\frac{D_0}{V}\equiv\frac{N(N_2-N_1)_0}{V}\geqq\frac{2\hbar\gamma\varkappa\omega}{4\pi\nu^2|\vartheta_{12}|^2}\left(1+\frac{\Delta^2}{(\gamma+\varkappa)^2}\right). \qquad \text{(VII.3.8)}^2$$

[N total number of atoms; $(N_2-N_1)_0$: inversion of a single atom (without laser action).] D_0: Total (unsaturated) inversion.

The laser frequency Ω given by Eq. (VII.3.4) is shifted from the frequency ω in the unloaded cavity to the center of the line[3] ("frequency pulling"). For a homogeneously broadened line, this shift does not depend on pump power.

b) Multiple-mode operation. A detailed discussion of two-mode cases was given in VI.9, and will not be repeated here. The general discussion of the Eqs. (VII.2.6)

* We assume in VII.3.3 and VII.3.6, that due to a crystal symmetry the atomic dipole moments point in only one direction. If the atomic dipole moments point in random directions, the formulas must be slightly changed. The corresponding results are given in (VII.7.12).

** The index "thr" of d_{thr} is intended to remind the reader that d_{thr} is just the inversion (of a single atom) at laser threshold.

[2] This laser-condition was first derived (for the special case of exact resonance, $\omega=\nu$) by A. W. SCHAWLOW and C. H. TOWNES: Phys. Rev. **112**, 1940 (1958) (compare p. 4 of this article). See also V. M. FAIN and YA. J. KHANIN: Soviet Phys. JETP **14**, 1069 (1962).

[3] C. H. TOWNES, in: Advances in Quantum Electronics (ed. J. R. SINGER), p. 3. New York and London 1961.

for the multiple-mode case is difficult and has not yet been performed, at least to our knowledge*. Therefore we discuss only the most important features of this case: as explained in VI.9 (see also VII.2), the light-modes cause both a time-dependent and a time-independent response of the atomic system. Provided the spacing between the frequencies of the different modes is big enough**:

$$|\Omega_{\lambda_1} - \Omega_{\lambda_2}| > 1/T; \quad \lambda_1 \neq \lambda_2 \tag{VII.3.9}$$

the time-dependent response can be neglected. We neglect further the possibility of frequency mixing, an effect which is very small, anyhow. With these simplifications the real and imaginary parts of Eqs. (VII.2.6) lead to two sets of equations, which determine the photon density of the modes and the frequency shift.

α) *Equations for the photon densities of M modes* (from the real part)

$$\left.\begin{aligned}
\frac{3}{2}\bar{n}_1 \cdot \frac{\varkappa_1}{d_{thr}(\Delta_1)} + \bar{n}_2 \frac{\varkappa_2}{d_{thr}(\Delta_2)} + \cdots + \bar{n}_M \frac{\varkappa_M}{d_{thr}(\Delta_M)} &= \frac{\varrho}{4T}\left(1 - \frac{d_{thr}(\Delta_1)}{d_0}\right), \\
\bar{n}_1 \frac{\varkappa_1}{d_{thr}(\Delta_1)} + \frac{3}{2}\bar{n}_2 \frac{\varkappa_2}{d_{thr}(\Delta_2)} + \cdots + \bar{n}_M \frac{\varkappa_M}{d_{thr}(\Delta_M)} &= \frac{\varrho}{4T}\left(1 - \frac{d_{thr}(\Delta_2)}{d_0}\right) \\
&\vdots
\end{aligned}\right\} \tag{VII.3.10}$$

(M equations).

$d_{thr}(\Delta_\lambda)$ is defined in (VII.3.6), where ω is supplemented with the index λ.

These equations are identical with the rate-equations (see VIII.1), and will be discussed in VIII.2. There it will be shown that the Eqs. (VII.3.10) determine the *coexistence of modes in a homogeneously broadened line due to a space-dependent inversion.*

β) *Equations for the frequency shift* [from the imaginary part of Eqs. (VII.2.6), using (VII.3.10)]

$$\Omega_\lambda = \frac{\gamma\omega_\lambda + \varkappa_\lambda \nu}{\gamma + \varkappa_\lambda} \tag{VII.3.11}$$

i.e. all modes show the power-independent pulling, known from single-mode action.

VII.4. The solid-state laser with an inhomogeneously broadened Gaussian line. Single- and two-mode operation[1]. The essential difference between the preceding paragraph and the present one is that we must not only sum up over the atomic positions but also over the transition energy (or transition frequency).

a) One mode. As an example we choose a Gaussian distribution for ν. It should be noted that other distributions can be treated similarly, or in the case of a Lorentzian line shape, even more simply. Compare VII.4c.

The sums over the atomic indices μ then lead to the following expressions

$$\left.\begin{aligned}
\varkappa + i(\Omega - \omega) = \frac{\varrho A d_0}{2} \int \left\{ \frac{\gamma - i(\Omega - \nu)}{\gamma^2 + (\Omega - \nu)^2} - 3\gamma T A \bar{n} \frac{\gamma - i(\Omega - \nu)}{[\gamma^2 + (\Omega - \nu)^2]^2} \right\} \times \\
\times \frac{e^{-\left(\frac{\nu - \nu_0}{\alpha}\right)^2} d\nu}{\alpha\sqrt{\pi}}
\end{aligned}\right\} \tag{VII.4.1}$$

* A special case of particular interest, which has been treated, is that of mode locking in a multimode laser (compare VII.13).

** This condition was derived in VI.9 for two modes. It holds in general also for M modes, as can be deduced from Eqs. (VII.2.6).

[1] H. HAKEN and H. SAUERMANN: l.c.[1] on p. 182.

where

$$A = \frac{4\pi \nu_0^2 |\vartheta_{12}|^2}{\hbar\omega}$$

ν_0: frequency at the line center.

In Eq. (VII.4.1) we have already performed the integrals over $\boldsymbol{x}$. Further it was assumed that the vector of polarization $\boldsymbol{e}$ is parallel to the atomic dipole moment $\boldsymbol{\vartheta}_{12}$. The integral over ν is evaluated under the assumption that the inhomogeneous linewidth is large compared to the homogeneous Lorentzian linewidth, which is described by the imaginary part γ of ν.

Using the result of the evaluation of the integral and splitting Eq. (VII.4.1) into real and imaginary parts, we find the following equations:

α) *Equation for the frequency shift*:

$$\Omega - \omega = -\frac{\varrho A\, d_0}{\alpha} e^{-\delta^2} \Phi(\delta) + \bar{n}\, C \tag{VII.4.2}$$

where

$$\Phi(\delta) = \int_0^\delta e^{u^2}\, du \tag{VII.4.3}$$

and

$$C = \frac{3}{2} \frac{A^2 \varrho\, d_0\, T \sqrt{\pi}\, \delta\, e^{-\delta^2}}{2\alpha^2} \tag{VII.4.4}$$

and

$$\delta = \frac{\Omega - \nu_0}{\alpha}. \tag{VII.4.5}$$

A is defined in (VII.4.1), N is the total number of atoms, d_0 is given by (VII.1.34a), ϱ is the density of laser atoms, T is defined in (VII.1.34b), α is the inhomogeneous width [compare Eq. (VII.4.1)].

The first term on the right-hand side of (VII.4.2) is peculiar to the inhomogeneous Gaussian broadening[2]. Note that this term is pump dependent on account of its factor d_0. It gives rise to a *pulling* of the *mode frequency* ω in the unloaded cavity to the center of the emission line ν_0.

The next term depends on the *photon density* (or, in other words, on the *field intensity* in the cavity), and describes a frequency pushing*. In order to determine its size, we insert the value of $\bar{n}$, which is anticipated from Eq. (VII.4.9) below:

$$\Omega - \omega = -\varkappa \frac{2\, d_0}{d_{thr}(\delta)\sqrt{\pi}} \left\{ \Phi(\delta) - \delta \frac{\gamma}{\alpha} \sqrt{\pi} \left(1 - \frac{d_{thr}(\delta)}{d_0} \right) \right\} \tag{VII.4.6}$$

where $d_{thr}(\delta)$ denotes the threshold inversion, which is given by

$$d_{thr}(\delta) = \frac{\hbar\, \alpha\, \varkappa\, \omega\, e^{\delta^2}}{2\sqrt{\pi^3}\, \varrho |\nu_0|^2 |\vartheta_{12}|^2}. \tag{VII.4.7}$$

From (VII.4.6) we see that the repulsive term is small compared to the first on account of its factor γ/α.

---

* This term has been predicted by Bennett, l.c. [2] by a qualitative argument "holeburning of one line on itself", which, however, left the sign of this term open.

Holeburning was first described in 1947 by Bloembergen in the field of magnetic resonance (see the references in N. Bloembergen: Nuclear Magnetic Relaxation. New York: W. A. Benjamin Inc. 1961).

[2] W. R. Bennett: Phys. Rev. **126**, 580 (1962); — J. Appl. Optics, Suppl. on Optical Masers (Dec. 1962).

The

β) *Equation for the photon density* follows as the real part from (VII.4.1), and reads:

$$\varkappa = \frac{\varrho A d_0}{2} \left\{ \frac{\sqrt{\pi}}{\alpha} e^{-\delta^2} - 3 A T \bar{n} \frac{\sqrt{\pi}\, e^{-\delta^2}}{2\gamma \cdot \alpha} \right\}. \tag{VII.4.8}$$

Solving this equation for $\bar{n}$ we arrive at

$$\bar{n} = \frac{2\gamma}{3AT} \left(1 - \frac{d_{thr}(\delta)}{d_0}\right) = \frac{\gamma}{\alpha} \frac{\sqrt{\pi}\, \varrho\, d_{thr}(0)}{3\varkappa T} \left(1 - \frac{d_{thr}(\delta)}{d_0}\right). \tag{VII.4.9}$$

This equation allows us to determine the photon density (or light intensity) as a function of the pump rate, which occurs in d_0 and T [compare Eqs. (VII.1.34a) and (VII.1.34b)]. Eq. (VII.4.8) can also be derived in the rate-equation approximation.

b) Two modes. We again consider the case of fixed atoms at random lattice sites. Replacing the sums over the atomic indices by integrals, which are again evaluated up to order γ/α and splitting the equations for the two amplitudes B_1^* and B_2^* into real and imaginary parts, we obtain the following equations:

α) *Equations for the photon densities* $\bar{n}_\lambda$. They are linear equations in $\bar{n}_1$, $\bar{n}_2$:

$$\bar{n}_1 P_{11} + \bar{n}_2 P_{12} = 1 - \frac{d_{thr}(1)}{d_0}, \tag{VII.4.10}$$

$$\bar{n}_1 P_{21} + \bar{n}_2 P_{22} = 1 - \frac{d_{thr}(2)}{d_0}, \tag{VII.4.11}$$

where

$$P_{11} = P_{22} = \frac{3AT}{2\gamma}. \tag{VII.4.12}$$

The expressions for P_{12} and P_{21} are given by

$$P_{12} = A T f_{12}; \quad P_{21} = A T f_{21}, \tag{VII.4.13}$$

where the f's are complicated functions of γ, $\Omega_1 - \Omega_2$, T, α and δ_1, δ_2. For their explicit form the reader is referred to the original paper[1]. The detailed discussion shows that the photon numbers of the two modes influence one another notably only if the frequency separation $\Delta\Omega = \Omega_1 - \Omega_2$ is of the order of the natural linewidth γ. For $\left(\frac{\gamma}{\Delta\Omega}\right)^2 < 1$ we can therefore use the expression (VII.4.9) for $\bar{n}_\lambda$. The coupling between the two photon numbers decreases roughly as $\left(\frac{\gamma}{\Delta\Omega}\right)^2$.

As we will show immediately the mutual influence of the frequencies is larger and decreases only as $\frac{\gamma}{\Delta\Omega}$ if $|\Delta\Omega| > \gamma$ in the same order of approximation.

β) *Equations for the frequency shifts*

$$\Omega_1 - \omega_1 = -\frac{\varrho A d_0}{\alpha} e^{-\delta_1^2} \Phi(\delta_1) + \bar{n}_1 C_1 + \bar{n}_2 M_{12}. \tag{VII.4.14}$$

The first two sum terms are identical with those of the expression (VII.4.2) for the frequency shift in single mode operation [we have merely to replace in (VII.4.2) and in (VII.4.5) δ by δ_1 and to put $C_1 = C$]. The last sum term shows that the frequency shift of the first mode depends also on the intensity of the second mode.

The coefficient M_{12} has a rather complicated structure, so that the reader is referred to the original paper. We quote here the result for a large frequency separation $(\Delta\Omega)^2 \gg \gamma^2$. In order to put the relative size of the various contributions in evidence, we rewrite (VII.4.14) in the form [with $(\Delta\Omega)^2 \gg \gamma^2$!]:

$$\Omega_1 - \omega_1 = \varkappa \frac{d_0}{d_{thr}(\delta_1)} \{g_1 + \bar{n}_1 g_1' + \bar{n}_2(g_2 + g_3)\} \tag{VII.4.15}$$

where

$$g_1 = -\frac{2}{\sqrt{\pi}} \Phi(\delta_1), \quad g_1' = \frac{3AT\delta_1}{\alpha}, \tag{VII.4.16}$$

$$g_2 = \frac{2AT}{\Delta\Omega} e^{\delta_1^2 - \delta_2^2}, \tag{VII.4.17}$$

$$\left.\begin{aligned} g_3 = \frac{AT}{1+T^2\Delta\Omega^2} \Bigg\{ & \frac{1}{\Delta\Omega}(1 + e^{\delta_1^2-\delta_2^2}) + \frac{2\delta_1}{\alpha} + \\ & + \frac{2T}{\sqrt{\pi}}\left(e^{\delta_1^2-\delta_2^2}\Phi(\delta_2) - \Phi(\delta_1)\right) + \frac{2T\Delta\Omega}{\sqrt{\pi}\,\alpha}\left(1 - 2\int_0^{\delta_1}\Phi(u)\,du\right)\Bigg\}. \end{aligned}\right\} \tag{VII.4.18}$$

The first two terms in (VII.4.15) which are proportional to the expressions (VII.4.16) are exactly the same as those of one mode which, as we have discussed above, give rise to a pulling of the line toward the center of the emission line. The following term $\sim$(VII.4.17) is brought about by a time independent depletion of the excited atomic states due to their interaction with the other mode. It is thus a holeburning effect as has been described in a macroscopic language by BENNETT[3].

The group $\sim$(VII.4.18) consists of terms which stem from the time dependent response of the atomic system. Whereas for a homogeneously broadened line the latter group can be important for a frequency repulsion, in the case of an inhomogeneously broadened line the first group will be more important in general. The size of the terms $\sim$(VII.4.18) depends in a crucial manner on the effective relaxation time T under which the inversion of the atomic system is achieved. The other group of terms $\sim$(VII.4.16) and $\sim$(VII.4.17) does not depend on this relaxation time, since the n's are inversely proportional to T. One might therefore hope to find systems in which the measurement of line shifts leads to a direct determination of the relaxation time.

c) Lorentzian line shape. If we choose a Lorentzian distribution

$$w(\nu) = \frac{\alpha}{\pi} \frac{1}{\alpha^2 + (\nu - \nu_0)^2} \tag{VII.4.19}$$

for the center frequencies of the active atoms, all integrals over the transition energy can be evaluated exactly with the help of the calculus of residues. In the limit $\alpha \to 0$, $w(\nu)$ tends to $\delta(\nu - \nu_0)$, so that we regain the results of the homogeneously broadened line in this case. On the other hand, if $\alpha \gg \gamma$, the following formulas correspond to the situation which was treated in *a)* and *b)* for a Gaussian. As the calculations are straightforward, we merely quote the results.

For a single mode we get

$$\bar{n} = \frac{\dfrac{\alpha+\gamma}{3AT}\left(1 - \dfrac{d_{thr}(\delta)}{d_0}\right)}{\dfrac{\alpha}{2\gamma} + \dfrac{1}{1+\delta_1^2}}, \tag{VII.4.20}$$

$$\Omega = \omega - \varkappa\delta \left\{ 1 + \frac{\left(\dfrac{d_0}{d_{thr}} - 1\right)}{1 + \dfrac{2\gamma}{\alpha(1+\delta^2)}} \right\} \tag{VII.4.21}$$

---

[3] W. R. BENNETT: l.c. [2] on p. 188.

where we have used as a new definition $\delta = \frac{\omega - \nu_0}{\alpha + \gamma}$ and $d_{thr}(\delta) = \frac{2\varkappa(\alpha+\gamma)(1+\delta^2)}{\varrho A}$, which is the threshold inversion. A is defined in (VII.4.1).

In two-mode operation we find for the frequency shifts:

$$\left.\begin{aligned} \Omega_1 - \omega_1 = {} & -\varkappa\delta_1 + \varkappa \frac{d_0}{d_{thr}(0)} \left\{ -\frac{\tilde{n}_1 \delta_1 \alpha}{2\gamma(1+\delta_1^2)} + \right. \\ & + \frac{\alpha\tilde{n}_2 \left[\Delta\left(1 + \frac{2}{1+\alpha/\gamma}\right) - \delta_1 \right]}{3\gamma(1+\delta_2^2)(1+\Delta^2)} + \frac{\tilde{n}_2}{3(1+T^2\Delta\Omega^2)} \times \\ & \times \left. \left[\frac{T\Delta\Omega - \delta_2}{1+\delta_2^2} + \frac{T\Delta\Omega + \delta_1}{1+\delta_1^2} + \frac{\alpha}{\gamma} \frac{T\Delta\Omega(1+\Delta\delta_2) + \Delta - \delta_2}{(1+\delta_2^2)(1+\Delta^2)} \right] \right\} \end{aligned}\right\} \quad \text{(VII.4.22)}$$

where

$$\tilde{n} = \frac{3\bar{n}AT}{\gamma+\alpha} \quad \text{and} \quad \Delta = \frac{\Omega_1 - \Omega_2}{2\gamma} = \frac{\Delta\Omega}{2\gamma}.$$

Inspection of (VII.4.22) shows that in the range $\alpha \approx \gamma$, which was not covered by the results of *a* and *b*, no new physical effects occur.

VII.5. The solid-state laser with an inhomogeneously broadened line: multimode action.

a) Normal multimode action. For a discussion of this case we start with the general Eqs. (VII.2.6). With a few exceptions (frequency locking etc.) discussed below, these equations can be simplified in the following way:

From the discussion of the exact laser equations we know that the phases of the exact B's fluctuate due to the noise inherent both in the resonator and in the atomic system. In the *truly steady state* we may keep only those terms in the Eqs. (VII.2.6), which do not vanish after a phase-average (over a time long compared to the inverse line-width). We *assume* now that the phases of the B's are uncorrelated. This assumption has been discussed in detail in VI.9 for the example of two modes, from which we extrapolate that it is fulfilled if the frequency difference is big enough

$$|\Omega_\lambda - \Omega_{\lambda'}| > \frac{1}{T}, \quad (\lambda \neq \lambda'). \quad \text{(VII.5.1)}$$

We multiply Eq. (VII.2.6) with B_λ and perform the phase-average. Only those terms $B_\lambda B_{\lambda_1} B^*_{\lambda_2} B^*_{\lambda_3}$ survive for which*

$$\lambda_2 = \lambda_1 \quad \text{and} \quad \lambda = \lambda_3 \quad \text{(VII.5.2)}$$

or

$$\lambda_3 = \lambda_1 \quad \text{and} \quad \lambda = \lambda_2 .$$

With these simplifications the real and imaginary parts of (VII.2.6) read:

$$2\varkappa_\lambda = C^0_\lambda - C_{\lambda 1}\bar{n}_1 - C_{\lambda 2}\bar{n}_2 \ldots - C_{\lambda M}\bar{n}_M, \quad \text{(VII.5.3)}$$

$$\Omega_\lambda = \omega_\lambda + F_\lambda + I_{\lambda 1}\bar{n}_1 + I_{\lambda 2}\bar{n}_2 + \ldots I_{\lambda M}\bar{n}_M, \quad \text{(VII.5.4)}$$

where

$$C^0_\lambda = 2\sum_\mu d_{0\mu}|g_{\mu\lambda}|^2 \frac{\gamma}{(\Omega_\lambda - \nu_\mu)^2 + \gamma^2} \quad \text{(VII.5.5)}$$

* We assume that the mode frequencies are non-degenerate.

is the (unsaturated) gain of mode λ in the inverted material and the $C_{\lambda\lambda'}$'s are given by

$$\left.\begin{aligned} C_{\lambda\lambda'} = 8 \sum_{\mu} d_{0\mu} |g_{\mu\lambda}|^2 |g_{\mu\lambda'}|^2 \frac{T\gamma^2}{((\Omega_\lambda - \nu_\mu)^2 + \gamma^2)((\Omega_{\lambda'} - \nu_\mu)^2 + \gamma^2)} + \\ + 4(1 - \delta_{\lambda\lambda'}) \sum_{\mu} d_{0\mu} |g_{\mu\lambda}|^2 |g_{\mu\lambda'}|^2 \frac{1}{\left(\frac{1}{T}\right)^2 + (\Omega_\lambda - \Omega_{\lambda'})^2} \operatorname{Re}\{\cdots\} \end{aligned}\right\} \quad \text{(VII.5.6)}$$

with

$$\operatorname{Re}\{\cdots\} = \operatorname{Re}\left\{\frac{\left(\frac{1}{T} - i(\Omega_\lambda - \Omega_{\lambda'})\right)(\Omega_{\lambda'} - \Omega_\lambda + 2i\gamma)}{(\Omega_\lambda - \nu_\mu - i\gamma)^2(\Omega_{\lambda'} - \nu_\mu + i\gamma)}\right\} \quad \text{(VII.5.7)}$$

Re means real part.

The first sum in (VII.5.6) stems from the time-independent response of the atomic system, whereas the second one is due to the time-dependent response. A closer inspection of the sums shows that the C's decrease rapidly with increasing distance $|\Omega_\lambda - \Omega_{\lambda'}|$, so that for $|\Omega_\lambda - \Omega_{\lambda'}| > \gamma, \frac{1}{T}$ we find $C_{\lambda\lambda'} \approx O$ for $\lambda \neq \lambda'$. Under these circumstances the photon numbers n_λ do not influence each other and "burn individual holes" into the atomic lines.

Thus, it is straightforward to calculate first the $\bar{n}$'s from (VII.5.3). The result (for a Gaussian line) is given in (VII.4.9). On the other hand, the I's decrease more slowly, as has been pointed out in the two-mode case (see p. 185), so that the Ω's are not only shifted from the frequencies ω in the unloaded cavity due to the F's but also by the holeburning effect of the other modes.

The F's and D's read:

$$F = \sum_{\mu} d_{0,\mu} |g_{\mu\lambda}|^2 \frac{(\nu_\mu - \Omega_\lambda)}{(\Omega_\lambda - \nu_\mu)^2 + \gamma^2} \quad \text{(VII.5.8)}$$

and

$$\left.\begin{aligned} I_{\lambda\lambda'} = 4 \sum_{\mu} d_{0,\mu} |g_{\mu\lambda}|^2 |g_{\mu\lambda'}|^2 \frac{T\gamma(\Omega_\lambda - \nu_\mu)}{[(\Omega_\lambda - \nu_\mu)^2 + \gamma^2][(\Omega_{\lambda'} - \nu_\mu)^2 + \gamma^2]} + \\ + 2(1 - \delta_{\lambda\lambda'}) \sum_{\mu} d_{0\mu} |g_{\mu\lambda}|^2 |g_{\mu\lambda'}|^2 \frac{1}{\left(\frac{1}{T}\right)^2 + (\Omega_\lambda - \Omega_{\lambda'})^2} \operatorname{Im}\{\cdots\} \end{aligned}\right\} \quad \text{(VII.5.9)}$$

with

$$\operatorname{Im}\{\cdots\} = \operatorname{Im}\left\{\frac{\left(\frac{1}{T} - i(\Omega_\lambda - \Omega_{\lambda'})\right)(\Omega_{\lambda'} - \Omega_\lambda + 2i\gamma)}{(\Omega_\lambda - \nu_\mu - i\gamma)^2(\Omega_{\lambda'} - \nu_\mu + i\gamma)}\right\} \quad \text{(VII.5.10)}$$

Im means imaginary part.

F is identical with the first term on the right-hand side of Eq. (VII.4.2) (the explicit form applies if $\gamma \ll \alpha$). $I_{\lambda\lambda'}$ is the same expression which occurs as the coefficient of n_2 in the Eq. (VII.4.15) of the two-mode case. We have merely to replace 1 by λ and 2 by λ'.

b) Combination tones[1,2]. We now consider the case where the cooperation of laser modes leads to additional mode-frequencies, which is not covered by the above treatment (*a*), but can also be deduced from Eqs. (VII.2.6). These equations contain on their right hand sides all possible frequencies of the form $\Omega_{\lambda_2} + \Omega_{\lambda_3} - \Omega_\lambda$. Even if "originally" only two modes oscillate, new frequencies $2\Omega_2 - \Omega_1$ and

[1] For the gas laser, combination tones were treated in detail by W. E. LAMB, JR.: Phys. Rev. **134**, A 1429 (1964). For the solid-state laser, the present treatment is new.

[2] Combination tones in gas-lasers were observed by W. J. WITTEMAN and J. HAISMA: Phys. Rev. Letters **12**, 287 (1964). — W. J. WITTEMAN: Phys. Rev. **143**, 316 (1966) and others.

$2\Omega_1 - \Omega_2$ appear, which are again close to resonance. Consequently, these two new frequencies exist in the cavity and can appear in the output. Well below threshold for normal three-frequency oscillation, the amplitude b_3^* of the new mode will be much smaller than b_1^* and b_2^*, so that powers of b_3^* can be neglected.

Let us take as an example $\Omega_3 = 2\Omega_2 - \Omega_1$: We choose in Eq. (VII.2.6) $\lambda = 3$, $\lambda_1 = 1$ and $\lambda_2 = \lambda_3 = 2$. We then obtain after integration:

$$\left.\begin{aligned} b_3^* = \Big\{ & i\Omega_3 - i\omega_3 + \varkappa + i \sum_\mu d_{0\mu} |g_{\mu 3}|^2 \frac{1}{(\Omega_3 - \nu_\mu - i\gamma)} - \\ & - A_1 n_1 - A_2 n_2 \Big\}^{-1} (b_2^*)^2 b_1 A_3, \qquad (n_i = V \bar{n}_i) \end{aligned}\right\} \tag{VII.5.11}$$

where

$$\left.\begin{aligned} A_1 &= \tilde{A}(3,1,1,3) + \tilde{A}(3,1,3,1), \\ A_2 &= \tilde{A}(3,2,2,3) + \tilde{A}(3,2,3,2), \\ A_3 &= \tilde{A}(3,1,2,2) \end{aligned}\right\} \tag{VII.5.12}$$

with

$$\left.\begin{aligned} \tilde{A}(\lambda,\lambda_1,\lambda_2,\lambda_3) = & \sum_\mu 2 d_{0\mu} g^*_{\mu\lambda} g^*_{\mu\lambda_1} g_{\mu\lambda_2} g_{\mu\lambda_3} \times \\ & \times \frac{1}{\left[\frac{1}{T} + i(\Omega_{\lambda_2} - \Omega_{\lambda_1})\right]} \frac{1}{[\Omega_{\lambda_2} + \Omega_{\lambda_3} - \Omega_{\lambda_1} - \nu_\mu - i\gamma]} \times \\ & \times \left(\frac{1}{\Omega_{\lambda_2} - \nu_\mu - i\gamma} - \frac{1}{\Omega_{\lambda_1} - \nu_\mu + i\gamma} \right). \end{aligned}\right\} \tag{VII.5.13}$$

The negative real part of the denominator (i.e. the curly bracket) in (VII.5.11) represents the gain of mode 3. The expression (VII.5.11) holds as long as this gain is negative, i.e. mode 3 is still below the actual laser threshold. Otherwise normal three-mode operation is reached, which may be treated as outlined above, (*a*). In order to obtain the combination tone $2\Omega_2 - \Omega_1$ experimentally one should tune the cavity, so that ω_2 is slightly above the atomic transition frequency ν_0, thereby making Ω_1 a little nearer resonance than ω_3. The inversion should be high enough for two-mode operation, but still below threshold for genuine three-mode operation. This phenomenon occurs of course only, if $\tilde{A}(3,1,2,2) \neq 0$, which depends on the spatial behavior of the $g_{\mu\lambda}$'s.

c) Frequency locking[3]. When the cavity tuning is gradually changed in normal three-frequency operation so that $(\Omega_2 - \Omega_1) - (\Omega_3 - \Omega_2)$ becomes small (typically 1 kc/sec for *gas lasers*!), a frequency jump occurs. For an explanation, we proceed as follows:

We write

$$b_\lambda^* = r_\lambda e^{i\Omega_\lambda t + i\varphi_\lambda} \tag{VII.5.14}$$

where we admit that r_λ and φ_λ are slowly varying functions. Inserting the ansatz (VII.5.14) into (VII.2.6) leads to coupled differential equations for r_λ and φ_λ, which can be rearranged to yield a differential equation for the new variable Ψ:

$$\Psi \equiv (2\Omega_2 - \Omega_1 - \Omega_3)\, t + (2\varphi_2 - \varphi_1 - \varphi_3) \tag{VII.5.15}$$

[3] For the gas-laser, frequency locking has been treated by W. E. LAMB: l.c. [1] on p. 192; — Frequency locking phenomena are well known in radio physics, see e.g. the books quoted in ref. 3 on p. 171. The first who seems to have observed frequency locking phenomena with clocks was HUYGENS as quoted in: MINORSKY, Nonlinear Oscillations, p. 438. In lasers, frequency locking has been observed by JAVAN (quoted in LAMB's paper) and by others.

namely[4]

$$\dot{\Psi} = \xi + \alpha \sin \Psi + \beta \cos \Psi \tag{VII.5.16}$$

where $\xi = 2\xi_2 - \xi_1 - \xi_3$. ξ_λ is responsible for the single-mode frequency pulling and is represented by the first sum term on the right-hand side of Eq. (VII.4.2).

α and β are slowly varying quantities, which depend on the r's. If we neglect this latter time dependence, Eq. (VII.5.16) can be solved by separation of the variables in an implicit form:

$$t = \int_{\Psi_0}^{\Psi} \frac{dx}{\xi + \alpha \sin x + \beta \cos x} \tag{VII.5.17}$$

where Ψ_0 is the value of Ψ at $t = 0$. The character of $\Psi(t)$ depends critically on whether

$$(\alpha^2 + \beta^2)^{\frac{1}{2}} < |\xi| \quad \text{or} \quad (\alpha^2 + \beta^2)^{\frac{1}{2}} > |\xi|. \tag{VII.5.18}$$

In the first case the integrand of (VII.5.17) has no singularity. By expanding the integrand into a series of powers of $\sin x$ and $\cos x$ we see that the integral behaves like $C\Psi$ + pulsations. Apart from these pulsations and minor corrections we then have $\Psi = \xi t + \text{const.}$ where

$$\xi = 2\xi_2 - \xi_1 - \xi_3 \equiv 2\Omega_2 - \Omega_1 - \Omega_3 \tag{VII.5.19}$$

which coincides with the behavior of normal three mode operation. If, on the other hand, $(\alpha^2 + \beta^2) > \xi^2$ the integral diverges, i.e. $t \to \infty$ when Ψ reaches the value

$$\Psi = \arcsin \left(\xi / \sqrt{\alpha^2 + \beta^2}\right). \tag{VII.5.20}$$

Since Ψ has no more a linear dependence on t, according to (VII.5.20) we find

$$\Omega_2 - \Omega_1 = \Omega_3 - \Omega_2. \tag{VII.5.21}$$

The sudden jump of beat notes $\Omega_2 - \Omega_1$ and $\Omega_3 - \Omega_2$ with cavity tuning can now be understood as follows: Let the laser be first in normal three-frequency operation with two distinct beat notes $\Omega_3 - \Omega_2$ and $\Omega_2 - \Omega_1$, so that $\xi = -\Omega_3 - \Omega_1 + 2\Omega_2$ is somewhat greater than $(\alpha^2 + \beta^2)^{\frac{1}{2}}$. As the middle cavity frequency is tuned closer to the atomic resonance frequency ν_0, the separation of beat note frequencies $|\xi|$ decreases. When $|\xi|^2$ reaches $(\alpha^2 + \beta^2)$, a quick transition to the locked state occurs and only one beat note remains.

VII.6. Equations of motion for the gas laser[1]. The essential difference compared to the solid state laser is given here by the motion of the gas atoms. Consequently, the coordinate of a single atom is now given by $\boldsymbol{x}_\mu + \boldsymbol{v}_\mu t$. We allow further for arbitrary angles $\Theta_{\mu\lambda}$ between the vector of polarization of the light mode λ and the dipole moment of the μ-th atom. The interaction constant $g_{\mu\lambda}$ between the mode λ and the atom μ thus takes the form

$$g_{\mu\lambda} = -\frac{ig}{2} \left(e^{i\boldsymbol{k}_\lambda \boldsymbol{x}_\mu + i\boldsymbol{k}_\lambda \boldsymbol{v}_\mu t} - \text{c.c.}\right) \cos \Theta_{\mu\lambda}. \tag{VII.6.1}$$

---

[4] For an explicit, exact solution of this equation, see the formulas following (VII.9.36).

[1] The gas-laser has been treated in an adequate manner, which takes the motion of atoms into account by H. HAKEN and H. SAUERMANN: Z. Physik **176**, 47 (1963), and W. E. LAMB JR.: Phys. Rev. **134**, 1429 (1964). Their treatments are different in form, but equivalent. In our article we follow the version of HAKEN and SAUERMANN.

The summation over μ runs in the corresponding equations over the positions $\boldsymbol{x}_\mu$, the velocities $\boldsymbol{v}_\mu$ (for which a Boltzmann-distribution is assumed) and over all angles Θ_μ.

The equations of motion can now be taken directly from those referring to fixed atoms, (VII.1.32)—(VII.1.34), if the replacement (VII.6.1) is made. ν_μ is simply to be identified with the center frequency ν_0. (The Doppler broadening is automatically taken care of by the explicit consideration of the atomic motion).

We represent still another way of introducing the atomic motion due to LAMB[1], which is different in form from, but completely equivalent to the procedure VII.6.1) of HAKEN and SAUERMANN outlined above.

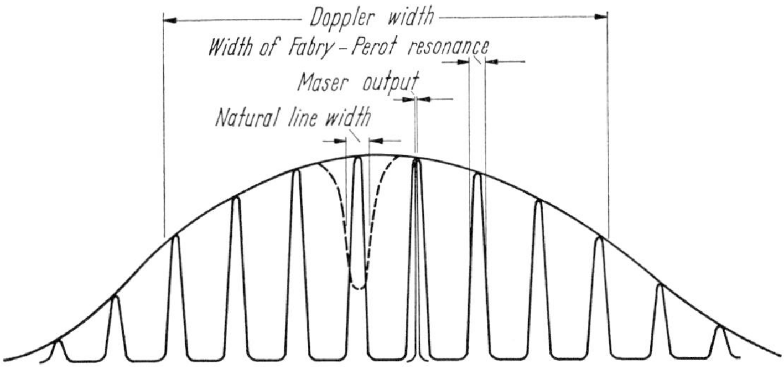

Fig. 41. **Spectral-linewidth factors in gas lasers.** [From O. H. HERRIOTT: J. Opt. Soc. Am. **52**, 31 (1962).]

Consider an atom excited by some process at time t_0 to its upper level. Let the atom be at the position $\boldsymbol{x}_0$ at t_0 and have the velocity $\boldsymbol{v}$ so that at time $t > t_0$ the atom will be at

$$\boldsymbol{x} = \boldsymbol{x}_0 + \boldsymbol{v}(t - t_0). \tag{VII.6.2}$$

The corresponding density matrix $\varrho(\boldsymbol{x}_0, t_0, \boldsymbol{v}; t)$ obeys the equation

$$\dot{\varrho} = \frac{-i}{\hbar}[H, \varrho] - \frac{1}{2}(\Gamma\varrho + \varrho\Gamma) \tag{VII.6.3}$$

where

$$\Gamma = \begin{pmatrix} \gamma_a & 0 \\ 0 & \gamma_b \end{pmatrix}$$

γ_a, γ_b are the decay constants of the corresponding levels. The field in H has the argument as defined in (VII.6.2): $H = H(\boldsymbol{x}_0 + \boldsymbol{v}(t - t_0), t)$. Furthermore, $\varrho(\boldsymbol{x}_0, t_0, \boldsymbol{v}; t)$ has to obey the initial condition

$$\varrho(\boldsymbol{x}_0, t_0, \boldsymbol{v}; t_0) = \begin{pmatrix} 1 & 0 \\ 0 & 0 \end{pmatrix}. \tag{VII.6.4}$$

Because the light mode interacts with a macroscopic polarization, $\boldsymbol{P}(\boldsymbol{x}, t)$, we combine the contributions of all atoms which arrive at $\boldsymbol{x}$ at time t, no matter when or where they are excited. We note that

$$\boldsymbol{P}(\boldsymbol{x}, t) = \left(\boldsymbol{\vartheta}_{12}\bar{\varrho}_{12}(\boldsymbol{x}, t) + \boldsymbol{\vartheta}_{21}\bar{\varrho}_{21}(\boldsymbol{x}, t)\right) \tag{VII.6.5}$$

where $\boldsymbol{\vartheta}_{ij}$ is the dipole moment matrix element of a single atom and $\bar{\varrho}_{ij}$ the density matrix element for the corresponding transition. Instead of deriving an expression for $\boldsymbol{P}(\boldsymbol{x}, t)$, we therefore derive $\bar{\varrho}(\boldsymbol{x}, t)$.

We assume that the excitation processes obey a Poisson distribution. Then $\bar{\varrho}(\boldsymbol{x}, t)$ is obtained by integrating over the time $(t-t_0)$ since the last excitation collision, where the probability distribution $\frac{1}{T_p} e^{-(t-t_0)/T_p}$ must be used.

We then find

$$\left.\begin{aligned}\bar{\varrho}(\boldsymbol{x}, t) = \frac{1}{T_p} \int_{-\infty}^{t} e^{-(t-t_0)/T_p}\, d t_0 \int d^3 x_0 \int d^3 v\, \tilde{\lambda}(\boldsymbol{x}_0, t_0, \boldsymbol{v}) \times \\ \times \varrho(\boldsymbol{x}_0, t_0, \boldsymbol{v}, t)\, \delta\big(\boldsymbol{x}-\boldsymbol{x}_0-\boldsymbol{v}(t-t_0)\big)\end{aligned}\right\} \tag{VII.6.6}$$

$\tilde{\lambda}(\boldsymbol{x}_0, t_0, \boldsymbol{v})$ is a mean excitation number, whose significance becomes especially clear in Eq. (VII.6.11). Since $\varrho(\boldsymbol{x}_0, t_0, \boldsymbol{v}, t)$ decays exponentially on account of the Γ in Eq. (VII.6.3) the exponential function $e^{-(t-t_0)/T_p}$ can be neglected, if $\frac{1}{T_p} < \gamma_a, \gamma_b$.

We now demonstrate the *equivalence between the formulation of* HAKEN-SAUERMANN *and* LAMB. In order to demonstrate the equivalence of the Eqs. (VII.6.3)—(VII.6.6) to (VII.1.32)—(VII.1.34), *where the matrix element* (VII.6.1) is used, we write $\bar{\varrho}(\boldsymbol{x}, t)$ in the form

$$\bar{\varrho}(\boldsymbol{x}, t) = \int d^3 v\, \bar{\varrho}(\boldsymbol{x}, \boldsymbol{v}, t) \tag{VII.6.7}$$

with

$$\left.\begin{aligned}\bar{\varrho}(\boldsymbol{x}, \boldsymbol{v}, t) = \frac{1}{T_p} \int_{-\infty}^{t} e^{-(t-t_0)/T_p}\, d t_0 \int d^3 x_0\, \tilde{\lambda}(\boldsymbol{x}_0, t_0, \boldsymbol{v}) \times \\ \times \varrho(\boldsymbol{x}_0, t_0, \boldsymbol{v}; t)\, \delta\big(\boldsymbol{x}-\boldsymbol{x}_0-\boldsymbol{v}(t-t_0)\big)\end{aligned}\right\} \tag{VII.6.8}$$

and investigate the time variation of $\bar{\varrho}(\boldsymbol{x}, \boldsymbol{v}, t)$:

$$\left.\begin{aligned}\frac{d\bar{\varrho}(\boldsymbol{x}, \boldsymbol{v}, t)}{dt} = {} & \frac{\Lambda(\boldsymbol{x}, t, \boldsymbol{v})}{T_p} - \frac{1}{T_p}\bar{\varrho} - \boldsymbol{v}\, \mathrm{grad}_{\boldsymbol{x}}\, \bar{\varrho}(\boldsymbol{x}, \boldsymbol{v}, t) + \\ & + \int_{-\infty}^{t} d t_0\, e^{-(t-t_0)/T_p} \frac{\tilde{\lambda}}{T_p}\big(\boldsymbol{x}-\boldsymbol{v}(t-t_0), t_0, \boldsymbol{v}\big) \times \\ & \times \Big\{-\frac{i}{\hbar}\big[H(\boldsymbol{x}, t), \varrho\big(\boldsymbol{x}-\boldsymbol{v}(t-t_0), t_0, \boldsymbol{v}, t\big)\big] - \\ & - \frac{1}{2}(\Gamma\varrho + \varrho\Gamma)\Big\}.\end{aligned}\right\} \tag{VII.6.9}$$

$\Lambda(\boldsymbol{x}, t, \boldsymbol{v})$ is given by $\tilde{\lambda}(\boldsymbol{x}, t, \boldsymbol{v})\begin{pmatrix}1 & 0\\ 0 & 0\end{pmatrix}$.

Evidently, H no longer contains t_0 so that H can be put in front of the integral: Thus we find finally

$$\left.\begin{aligned}\frac{d\bar{\varrho}(\boldsymbol{x}, \boldsymbol{v}, t)}{dt} + \boldsymbol{v}\, \mathrm{grad}_{\boldsymbol{x}}\, \bar{\varrho}(\boldsymbol{x}, \boldsymbol{v}, t) = \frac{\Lambda(\boldsymbol{x}, t, \boldsymbol{v}) - \bar{\varrho}(\boldsymbol{x}, \boldsymbol{v}, t)}{T_p} - \\ - \frac{i}{\hbar}\big[H(\boldsymbol{x}, t), \bar{\varrho}(\boldsymbol{x}, \boldsymbol{v}, t)\big] - \frac{1}{2}\big(\Gamma\bar{\varrho}(\boldsymbol{x}, \boldsymbol{v}, t) + \bar{\varrho}\Gamma\big).\end{aligned}\right\} \tag{VII.6.10}$$

In the theory of fluids, (VII.6.10) would correspond to the local description. By making the substitution $\boldsymbol{x} = \boldsymbol{x}' + \boldsymbol{v}t'$ (and afterwards dropping the prime) we end up with the substantial description:

$$\left.\begin{aligned}\frac{d\bar{\varrho}(\boldsymbol{x}+\boldsymbol{v}t,\boldsymbol{v},t)}{dt}&=\frac{\Lambda(\boldsymbol{x}+\boldsymbol{v}t,t,\boldsymbol{v})-\bar{\varrho}(\boldsymbol{x}+\boldsymbol{v}t,\boldsymbol{v},t)}{T_p}-\\&-\frac{i}{\hbar}[H(\boldsymbol{x}+\boldsymbol{v}t,t),\bar{\varrho}(\boldsymbol{x}+\boldsymbol{v}t,\boldsymbol{v},t)]-\frac{1}{2}(\Gamma\varrho(\boldsymbol{x}+\boldsymbol{v}t,\boldsymbol{v},t)+\varrho\Gamma).\end{aligned}\right\}\quad\text{(VII.6.11)}$$

When we split the matrix $\bar{\varrho}$ into its elements, we obtain from (VII.6.11) the Eqs. (VII.1.33) for $\varrho_{12,\mu}$ and the conjugate complex for $\varrho_{21,\mu}$, where $\bar{\varrho}_{ik}(\boldsymbol{x}+\boldsymbol{v}t,\boldsymbol{v},t)$ is replaced by $\varrho_{ik,\mu}(t)$ and, most important, $g_{\mu,\lambda}$ has the form (VII.6.1). The decay constants following from (VII.6.11) can be identified with the one of (VII.1.33): $\frac{1}{T_p}+\frac{1}{2}(\gamma_a+\gamma_b)=\gamma_{21}$. When we take the difference of the equations for the diagonal elements of ϱ, Eq. (VII.1.34) results for $d_\mu=\varrho_{22,\mu}-\varrho_{11,\mu}$ again with $g_{\mu,\lambda}$ of the form (VII.6.1). When the pump and decay constants are identified with each other it turns out that LAMB's formulation with aid of (VII.6.3) is in so far more restricted than (VII.1.34) as the longitudinal and transverse decay times are in a fixed ratio to each other, whereas (VII.1.33) and (VII.1.34) still allow additional phase destroying processes, so that γ_{21} may be larger than $1/T_p$ (see also the discussion on p. 58). On the other hand, Eq. (VII.1.34) with (VII.1.34a) and (VII.1.34b) describes a two-level system with $\varrho_{11}+\varrho_{22}=1$. For a truly three-level system the Eq. (IV.6.6) must be used instead (see also HAKEN and SAUERMANN, l.c. [1] on p. 182).

Except for these two restraints the equations of the atomic motion (VII.6.11), (VII.1.33) and (VII.1.34) are completely equivalent. Because the equivalence of the corresponding field equations can be demonstrated even more simply, we drop that proof.

VII.7. Single- and two-mode operation in gas lasers[1,2].

a) Single-mode operation. The method of solution consists again in the iteration procedure described in VII.2. We have to repeat it, however, because the coupling coefficients depend also on time. Making the hypothesis

$$b^*(t)=B^*\,e^{i\Omega t}\qquad\text{(VII.7.1)}$$

we obtain after the above iteration steps in the same approximation:

$$[\varkappa+i(\Omega-\omega)]\,B^*\,e^{i\Omega t}=i\sum_\mu g_\mu^*(t)\,\varrho_{12,\mu}(t)\qquad\text{(VII.7.2)}$$

where the $\varrho_{12,\mu}(t)$ has in the second approximation the following structure:

$$\left.\begin{aligned}\varrho_{12,\mu}(t)&=B^*\,e^{i\Omega t}(c_+\,e^{ikx_\mu+ivt}+c_-\,e^{-i(kx_\mu+vt)})+\\&+\bar{n}B^*\,e^{i\Omega t}[d_+\,e^{+i(kx_\mu+vt)}+d_-\,e^{-i(kx_\mu+vt)}+f_+\,e^{3ikx_\mu+3ivt}+\\&+f_-\,e^{-3ikx_\mu-3ivt}+j_+\,e^{ikx_\mu+ivt}+j_-\,e^{-ikx_\mu-ivt}]\end{aligned}\right\}\quad\text{(VII.7.3)}$$

---

[1] We follow the treatment of H. HAKEN and H. SAUERMANN: l.c. [1] on p. 182; — For an equivalent approach see W. E. LAMB, l.c. [1] on p. 192. — F. ARONOWITZ: Phys. Rev. **139**, A 635 (1965), treated single-mode action in a ring laser and included isotope effects; — A more accurate treatment (the polarization is calculated by perturbation theory up to the fifth order) of single-mode operation has been given by K. UEHARA and K. SHIMODA: Japan. J. Appl. Phys. **4**, 921 (1965). A higher order calculation was also made by W. CULSHAW: Phys. Rev. **164**, 329 (1967); — Finally, H. GREENSTEIN: Phys. Rev. **175**, 438 (1968) has calculated the nonlinear polarization without utilizing perturbation techniques, so that his theory retains validity for laser fields of arbitrary intensity. This paper treats the gain profile, power-tuning characteristics and the frequency of oscillation; — The theory of nonlinear effects in a gas laser amplifier of weak and strong signals has been developed by A. DIENES: Phys. Rev. **174**, 400, 414 (1968).

[2] For a general survey on gas lasers (experiment and theory) see the book of C. G. B. GARRETT: Gas Lasers. New York: McGraw-Hill Book Co. 1967.

$c_\pm, d_\pm, f_\pm, j_\pm$ are complex constants, independent of space and time, which are given by

$$c_\pm = \pm \frac{i\, d_0\, g}{2} \cos\Theta \frac{1}{\Omega - \nu_0 - i\gamma \pm \nu}; \tag{VII.7.4}$$

$$\left.\begin{aligned} d_\pm &= \pm\gamma T d_0 \frac{g}{2} |g|^2 \cos^3\Theta \times \\ &\times \left\{\frac{1}{(\Omega-\nu_0+\nu)^2+\gamma^2} + \frac{1}{(\Omega-\nu_0-\nu)^2+\gamma^2}\right\} \frac{1}{i(\Omega-\nu_0-i\gamma\pm\nu)} \end{aligned}\right\} \tag{VII.7.5}$$

$$f_\pm = \mp \frac{d_0|g|^2 g \cos^3\Theta\, (\gamma \pm i\nu)}{2i\left[\frac{1}{T} \pm 2i\nu\right]} \cdot \frac{1}{[(\Omega-\nu_0)^2 - (\nu \mp i\gamma)^2]\,[\Omega-\nu_0-i\gamma\pm 3\nu]}, \tag{VII.7.6}$$

$$j_\pm = \pm \frac{d_0(\gamma \pm i\nu) \cos^3\Theta\, |g|^2 g}{2i\left[\frac{1}{T} \pm 2i\nu\right]} \cdot \frac{1}{[(\Omega-\nu_0)^2 - (\nu \mp i\gamma)^2]\,[\Omega-\nu_0-i\gamma\pm\nu]} \tag{VII.7.7}$$

$\nu = kv$ where v is the velocity of atom μ.

When we multiply (VII.7.3) by g_μ^* [see (VII.7.2)] and sum over the coordinates, the time dependence with $e^{i\nu t}$ and $e^{i3\nu t}$ drops out on account of the orthogonality properties so that the right hand side has exactly the same time dependence as required for b^* by the ansatz (VII.7.1).

Performing the integrals over $\boldsymbol{x}$ and the average over Θ explicitly*, and assuming a symmetric velocity distribution $w(\nu)$ as well as a homogeneous spatial distribution of the active atoms, we obtain:

$$\left.\begin{aligned} \varkappa + i(\Omega-\omega) &= \frac{\varrho A\, d_0}{6} \int \frac{\gamma + i(\nu+\nu_0-\Omega)}{(\Omega-\nu-\nu_0)^2+\gamma^2} \times \\ &\times \left\{1 - \frac{3\gamma T \bar{n} A}{5[(\Omega-\nu_0-\nu)^2+\gamma^2]} - \frac{3\gamma T \bar{n} A}{5[(\Omega-\nu_0+\nu)^2+\gamma^2]}\right\} w(\nu)\, d\nu + \\ &+ \frac{\varrho A^2 d_0 \bar{n}}{10} \int \frac{\nu + i\gamma}{\left(\frac{1}{T} - 2i\nu\right)(\Omega-\nu_0-\nu-i\gamma)^2(\Omega-\nu_0+\nu+i\gamma)} w(\nu)\, d\nu. \end{aligned}\right\} \tag{VII.7.8}$$

ϱ is the density of atoms and A is defined in Eq. (VII.4.1). The terms containing A under the first integral arise from a static depletion of excited atomic states, while the second integral arises from the time-dependent response of the atomic system. This can be seen most easily by following the single steps of our iteration procedure.

Assuming a Gaussian velocity distribution, it can be shown, using constants typical for a *He-Ne*-Laser, that the last integral is one order of magnitude smaller than the first one and therefore may be neglected. Up to order γ/α we are then left with the following two equations, if we split them into real and imaginary parts

α) *Equation for the photon density*

$$\varkappa = \frac{\sqrt{\pi}\, \varrho A\, d_0\, e^{-\delta^2}}{6\alpha} \left\{1 - \frac{3}{5} A \bar{n} T \left[\frac{1}{2\gamma} + \frac{\gamma}{2(\gamma^2 + (\Omega-\nu_0)^2)}\right]\right\} \tag{VII.7.9}$$

* We treat the gas as an ensemble of atoms whose dipole moments are random. We thus neglect the possibility, that each single atom is capable of all three dipole orientations or, mathematically speaking, we neglect density matrix elements of the form $\varrho_{\mu\nu}$ where μ and ν both refer to the degenerate p-states.

from which we determine the photon density:

$$\bar{n} = \frac{\frac{10\gamma}{3AT}\left(1 - \frac{3d_{thr}(\delta)}{d_0}\right)}{1 + \frac{\gamma^2}{\gamma^2 + (\Omega - \nu_0)^2}}. \tag{VII.7.10}$$

The photon density shows a dip[3] if plotted as a function of frequency. It is interesting to note that this dip is brought about by the fact that the atoms move in both directions, so resulting in two holes being burned into the inhomogeneously broadened line at two symmetric points of the line shape. If we have fixed atoms instead, no such dip occurs.

β) Equation for the frequency shift

$$\Omega - \omega = -\frac{\varrho A d_0}{3\alpha} e^{-\delta^2} \Phi(\delta) + \bar{n} C' \tag{VII.7.11}$$

where A, δ, and $\Phi(\delta)$ are defined in Eq. (VII.4.1), (VII.4.5) and (VII.4.3), respectively.

C' is given by:

$$C' = \frac{A^2 \varrho d_0 T \sqrt{\pi}}{2\alpha} e^{-\delta^2} \left\{ \frac{1}{10} \cdot \frac{(\Omega - \nu_0)}{(\gamma^2 + (\Omega - \nu_0)^2)} + \frac{1}{5} \frac{\delta}{\alpha} \right\}. \tag{VII.7.12}$$

We compare this result with the expression (VII.4.2) for *fixed* atoms and a single direction of polarization. The first sum term on the right-hand side in (VII.7.11) agrees with the corresponding one in (VII.4.2) except for a factor $\frac{1}{3}$ which stems from the integration over the polarization (only $\frac{1}{3}$ of the atoms participates in the laser process on the average). This term represents power-dependent mode pulling and stems from the Doppler shape of the line.

With respect to the term proportional to $\bar{n}$, which comes from the time independent atomic inversion, we observe the following:

The part of C' which stems from the second sum term in the curly bracket agrees with the total expression C [Eq. (VII.4.4)] except for a numerical factor which stems from the integration over the polarization angles again. This second term in C' is in general, i.e. for not too strong detuning ($\delta \ll 1$), much smaller than the first sum term, which describes the frequency pushing being due to the existence of two holes burned into the inversion (note, that a *standing* wave interacts with atoms whose frequencies are shifted both by $+kv$ and $-kv$!). The frequency pushing becomes dominant if the mode is tuned to the center of the line to within about a natural linewidth.

Inserting $\bar{n}$ according to Eq. (VII.7.10) into (VII.7.11) yields as a final result for the frequency shift:

$$\left.\begin{aligned} \Omega = \omega - \varkappa \frac{2}{\sqrt{\pi}} \int_0^\delta e^{u^2} du + \varkappa \left(\frac{d_0}{3 d_{thr}(\delta)} - 1 \right) \times \\ \times \left(- \frac{2}{\sqrt{\pi}} \int_0^\delta e^{u^2} du + \frac{\gamma (\Omega - \nu_0)}{2\gamma^2 + (\Omega - \nu_0)^2} \right). \end{aligned}\right\} \tag{VII.7.13}$$

b) Two-mode operation. The procedure described in the foregoing leads to two equations for the two laser modes. The equation for mode 1 reads

$$(\varkappa + i(\Omega_1 - \omega_1)) B_1^* e^{i\Omega_1 t} = i \sum_\mu g_{\mu 1}^*(t) \varrho_{12,\mu}(t) \tag{VII.7.14}$$

[3] This detuning dip (also called Lamb dip) was observed experimentally by R. A. McFarlane, W. R. Bennett Jr., and W. E. Lamb Jr.: Appl. Phys. Letters **2**, 189 (1963). — A. Szöke and A. Javan: Phys. Rev. Letters **10**, 521 (1963); — Phys. Rev. **145**, 137 (1966)

where

$$\left.\begin{aligned}\varrho_{12,\mu}(t) = \{&B_1^* e^{i\Omega_1 t}(c_+^{(1)} e^{ik_1 x + ik_1 v t} + c_-^{(1)} e^{-ik_1 x - ik_1 v t}) + \\ &+ B_2^* e^{i\Omega_2 t}(c_+^{(2)} e^{ik_2 x + ik_2 v t} + c_-^{(2)} e^{-ik_2 x - ik_2 v t}) \times \\ &\times \{1 - n_1(l_+^{(1)} + l_-^{(1)}) - n_2(l_+^{(2)} + l_-^{(2)})\}.\end{aligned}\right\} \quad \text{(VII.7.15)}$$

x and v on the right hand side of Eq. (VII.7.15) still depend on the index μ. $c_\pm^{(\lambda)}$ is defined in (VII.7.4), where Ω, Θ and ν are still to be supplemented with the index λ, whereas $l_\pm^{(\lambda)}$ is given by:

$$l_\pm^{(\lambda)} = \frac{\gamma T |g|^2 \cos^2 \Theta_\lambda}{(\nu_0 - \Omega_\lambda \pm k_\lambda v)^2 + \gamma^2}. \quad \text{(VII.7.16)}$$

Eq. (VII.7.15) contains only the time-independent part of the inversion. In evaluating the sum over μ, i.e. integration over $\boldsymbol{x}$, as well as averaging over the orientations of the atomic dipoles, all terms cancel which have a time dependence other than $e^{i\Omega_1 t}$. Noting that

$$\overline{\cos^2 \Theta} = \tfrac{1}{3}, \quad \overline{\cos^4 \Theta} = \tfrac{1}{5}, \quad \overline{\cos^2 \Theta_1 \cos^2 \Theta_2} = \tfrac{1}{15}(1 + 2\cos^2 \alpha) \quad \text{(VII.7.17)}$$

(α is here the angle between the polarization vectors of the two modes) we are then led to

$$\left.\begin{aligned}\varkappa + i(\Omega_1 - \omega_1) = &\frac{\varrho A d_0}{6} \int \frac{i}{\nu_0 + k_1 v - \Omega_1 + i\gamma} \times \\ &\times \left\{1 - \frac{3\gamma T \bar{n}_1 A}{5[(\nu_0 - \Omega_1 + k_1 v)^2 + \gamma^2]} - \frac{3\gamma T \bar{n}_1 A}{5[(\nu_0 - \Omega_1 - k_1 v)^2 + \gamma^2]} - \right. \\ &\left. - \frac{(1 + 2\cos^2\alpha)\gamma T \bar{n}_2 A}{5[(\nu_0 - \Omega_2 - k_2 v)^2 + \gamma^2]} - \frac{(1 + 2\cos^2\alpha)\gamma T \bar{n}_2 A}{5[(\nu_0 - \Omega_2 + k_2 v)^2 + \gamma^2]}\right\} w(v)\, dv\end{aligned}\right\} \quad \text{(VII.7.18)}$$

and a corresponding equation with reversed indices. The difference between k_1 and k_2 in expression (VII.7.18) is entirely negligible, as may be shown by decomposition into partial fractions. These integrals have been evaluated up to order γ/α, which leads to the following set of equations when split into real and imaginary parts:

α) *Equations for the photon densities*

$$\left.\begin{aligned}\frac{3 d_{thr}(1)}{d_0} = 1 &- \frac{3 A T \bar{n}_1}{10\gamma}\left(1 + \frac{\gamma^2}{\gamma^2 + (\Omega_1 - \nu_0)^2}\right) - \\ &- \frac{A T \bar{n}_2}{10\gamma}\left(\frac{4\gamma^2}{4\gamma^2 + \Delta\Omega^2} + \frac{4\gamma^2}{4\gamma^2 + (\Omega_1 + \Omega_2 - 2\nu_0)^2}\right) \cdot (1 + 2\cos^2\alpha)\end{aligned}\right\} \quad \text{(VII.7.19)}$$

and a second equation with indices 1 and 2 reversed.

Treating Ω_1 and Ω_2 as given parameters, the solutions of these equations read:

$$n_1 = \frac{10\gamma}{3AT} \times \frac{\left(1 - \frac{3 d_{thr}(1)}{d_0}\right)\left(1 + \frac{\gamma^2}{\gamma^2 + (\Omega_2 - \nu_0)^2}\right) - \left(\frac{1 + 2\cos^2\alpha}{3}\right) \times \left(1 - \frac{3 d_{thr}(2)}{d_0}\right)\left(\frac{4\gamma^2}{4\gamma^2 + \Delta\Omega^2} + \frac{4\gamma^2}{4\gamma^2 + (\Omega_1 + \Omega_2 - 2\nu_0)^2}\right)}{\left(1 + \frac{\gamma^2}{\gamma^2 + (\Omega_1 - \nu_0)^2}\right)\left(1 + \frac{\gamma^2}{\gamma^2 + (\Omega_2 - \nu_0)^2}\right) - \frac{1}{9} \times (1 + 2\cos^2\alpha)^2 \left(\frac{4\gamma^2}{4\gamma^2 + \Delta\Omega^2} + \frac{4\gamma^2}{4\gamma^2 + (\Omega_1 + \Omega_2 - 2\nu_0)^2}\right)^2}. \quad \text{(VII.7.20)}$$

$\bar{n}_2$ can be obtained from (VII.7.20) simply by exchanging the indices.

Let us discuss the following two cases:

If $\alpha=0$, i.e. the two modes are polarized in parallel, formula (VII.7.20) holds as long as the two modes are placed unsymmetrically with respect to the center of the line. If they are in a symmetrical position within the accuracy of a natural linewidth, then both modes are fed by identical atoms, since each mode exists on both sides of the Doppler line. As long as one neglects the different spatial distributions, as is done in our present approximation, neglecting terms of order $(\gamma/\alpha)^2$, one obtains an unstable situation in which either one or the other mode can exist. This is reflected mathematically in (VII.7.20) by the fact that the photon numbers become infinite. If $\alpha=\pi/2$, i.e. the polarization vectors are orthogonal, (VII.7.20) gives a stable configuration in each case, even in the symmetric one. The latter result agrees with experimental observations reported by TANG and STATZ[4]. It is also compatible with W. R. Bennett's[5] experiments on frequency shifts, because there a polaroid was always used, when observing beat notes between different modes.

β) Equations for the frequency shifts

$$\Omega_1-\omega_1=\Delta\Omega\ (\text{single mode})+\bar{n}_2 L \tag{VII.7.21}$$

where $\Delta\Omega$ (single mode) is given by the right-hand side of Eq. (VII.7.11) (where δ and Ω are to be supplemented with the index 1).

L is given by:

$$L=(1+2\cos^2\alpha)\,\frac{d_0\sqrt{\pi}\,\varrho A^2 T e^{-\delta_2^2}}{60\alpha}\left(\frac{2\Delta\Omega}{4\gamma^2+\Delta\Omega^2}+\frac{2(\Omega_1+\Omega_2-2\nu_0)}{4\gamma^2+(\Omega_1+\Omega_2-2\nu_0)^2}\right). \tag{VII.7.22}$$

The expressions for mode 2 can be obtained again simply by exchanging the indices.

The size of the frequency-pushing effect depends on the size of the angle between the polarization vectors of the two modes.

VII.8. Some exactly solvable problems. The semiclassical Eqs. (VII.1.32) to (VII.1.34) allow an exact treatment in the case of single-mode operation (or if the frequencies Ω_λ of several modes coincide so that there is no time-dependent atomic response). We treat the

a) Single-mode operation in solid state lasers for example in detail.

We put

$$\dot{d}_\mu=0 \tag{VII.8.1}$$

and

$$b^*=B^*\,e^{i\Omega t} \tag{VII.8.2}$$

where B^* is time-independent. We insert (VII.8.2) into the Eq. (VII.1.33) for the atomic dipole-moment and integrate, which yields

$$\varrho_{12,\mu}=\frac{-i g_\mu b^* d_\mu}{i(\Omega-\nu_\mu)+\gamma} \tag{VII.8.3}$$

where d_μ is still an undetermined parameter.

---

[4] C. L. TANG and H. STATZ: Phys. Rev. **128**, 1013 (1962). In this paper also a theoretical account is given of some questions concerning polarization effects.

[5] W. R. BENNETT: l.c. [2] on p. 188.

We now introduce (VII.8.3) into the Eq. (VII.1.34) for the atomic inversion, d_μ. Using (VII.8.1) we then find

$$d_\mu = \frac{d_0}{1 + \dfrac{4\,T\gamma |g_\mu|^2\, n}{(\Omega - \nu_\mu)^2 + \gamma^2}} \tag{VII.8.4}$$

so that the "parameter" d_μ in (VII.8.3) is now fixed. The "photon number" n is given by $B^* B$. Inserting (VII.8.3) [with d_μ given by (VII.8.4)] into the field Eq. (VII.1.32) we obtain finally

$$b^* \left(i(\Omega - \omega) + \varkappa\right) = \sum_\mu \frac{|g_\mu|^2\, d_\mu\, b^*}{i(\Omega - \nu_\mu) + \gamma}\,. \tag{VII.8.5}$$

After division of both sides of (VII.8.5) by b^*, and after splitting the remaining equation into real and imaginary parts, we find two coupled equations for the frequency shift and the photon number:

$$\Omega - \omega = -\sum_\mu \frac{(\Omega - \nu_\mu)\,|g_\mu|^2 d_0}{(\Omega - \nu_\mu)^2 + \gamma^2 + 4\,T\gamma |g_\mu|^2\, n}\,, \tag{VII.8.6}$$

$$\varkappa = \gamma \sum_\mu \frac{|g_\mu|^2\, d_0}{(\Omega - \nu_\mu)^2 + \gamma^2 + 4\,T\gamma |g_\mu|^2 n}\,. \tag{VII.8.7}$$

The solution of the coupled (nonlinear) differential equations is thus reduced to the solution of the Eqs. (VII.8.6) and (VII.8.7), for which we now give examples:

α) *Homogeneously broadened line* ($\nu_\mu = \nu$). The factor $(\Omega - \nu_\mu)$ in the sum of Eq. (VII.8.6) can be taken in front of the sum, which agrees now with that on the right hand side of Eq. (VII.8.7). By the elimination of this sum from Eqs. (VII.8.6) and (VII.8.7) we find quickly

$$\Omega = \frac{\gamma\omega + \varkappa\nu}{\gamma + \varkappa}\,. \tag{VII.8.8}$$

This well-known frequency pulling is thus an exact result. For the solution of the remaining Eq. (VII.8.7) we distinguish between running and standing waves.

1. *Running waves* ($|g_\mu|^2 = |g|^2$). Because the expression under the sum in Eq. (VII.8.7) does not depend on μ, the sum can be replaced by the factor N (total number of atoms):

$$\varkappa = \gamma N |g|^2 d_0 \frac{1}{(\Omega - \nu)^2 + \gamma^2 + 4\,T\gamma |g|^2 n} \tag{VII.8.9}$$

or, after solving for n

$$n = \frac{d_0 N}{4\varkappa T} - \frac{(\Omega - \nu)^2 + \gamma^2}{4\,T\gamma |g|^2}\,. \tag{VII.8.10}$$

2. *Standing waves in axial direction* ($g_\mu = g\sqrt{2}\sin k x_\mu$). The equation for the photon number now reads

$$\varkappa = \gamma \frac{|g|^2 d_0}{L}\, 2N \int_0^L \frac{\sin^2 k x}{(\Omega - \nu)^2 + \gamma^2 + 4\,T\gamma\, n\, 2|g|^2 \sin^2 k x}\, dx \tag{VII.8.11}$$

where the sum has been replaced by an integration over the axial coordinate x, which runs from O to L. L is the length of the cavity.

The integration can be exactly performed, so that Eq. (VII.8.11) reads finally

$$\varkappa = \frac{d_0 N}{4 T n} \left\{ 1 - \frac{1}{\sqrt{1 + \frac{4 T \gamma 2 |g|^2 n}{(\Omega - \nu)^2 + \gamma^2}}} \right\}. \tag{VII.8.12}$$

Its solution can best be found by a graphical plot.

β) *Inhomogeneously broadened line, running waves.* The Eqs. (VII.8.6) and (VII.8.7) take for a Gaussian distribution of ν the form

$$\varkappa = \gamma |g|^2 d_0 \int \frac{e^{-(\nu-\nu_0)^2/\alpha^2}}{(\Omega - \nu)^2 + \gamma^2 + 4 T \gamma |g|^2 n} \frac{1}{\alpha \sqrt{\pi}} d\nu \tag{VII.8.13}$$

and

$$(\Omega - \omega) = - |g|^2 d_0 \int \frac{(\Omega - \nu)\, e^{-(\nu-\nu_0)^2/\alpha^2}}{(\Omega - \nu)^2 + \gamma^2 + 4 T \gamma |g|^2 n} \frac{1}{\alpha \sqrt{\pi}} d\nu. \tag{VII.8.14}$$

We discuss the special case of exact resonance ($\omega = \nu_0$) in more detail:

The right-hand side of Eq. (VII.8.14) vanishes on account of the odd symmetry of the integrand, while the integration in Eq. (VII.8.13) can be done in closed form:

$$\varkappa = \frac{\gamma |g|^2 d_0 \sqrt{\pi}}{\alpha\, a} e^{a^2/\alpha^2} \left(1 - \Phi\left(\frac{a}{\alpha}\right)\right) \tag{VII.8.15}$$

where

$$a = \sqrt{\gamma^2 + 4 T \gamma |g|^2 n} \tag{VII.8.16}$$

and

$$\Phi(u) = \frac{2}{\sqrt{\pi}} \int_0^u e^{-v^2} dv$$

is the error integral.

Eq. (VII.8.15) allows us to determine the photon number n as a function of the inversion d_0 (which would be achieved without laser action). The evaluation can again best be made by a graphical plot.

Our procedure can be extended to standing waves or to other line shapes (e.g. a Lorentzian, in which case also the off-resonance case can be treated explicitly).

We conclude this chapter with a remark about the exact treatment of a

b) Single-mode in the gas laser. If the single-mode is represented by a *running wave*, this problem can be reduced to that of an inhomogeneously broadened Gaussian line with fixed atoms, which was treated above. The coupling coefficient g_μ reads:

$$g_\mu = \text{const.}\ e^{ikx_\mu + ikv_\mu t}.$$

It occurs in the field-Eq. (VII.1.32) in connection with $\varrho_{12,\mu}$. We therefore introduce $e^{ikv_\mu t} \cdot \varrho_{12,\mu} = \tilde{\varrho}_{12,\mu}$ as a new variable. It can then be most easily checked, that $\tilde{\varrho}_{12,\mu}$ and $d_\mu \equiv \tilde{d}_\mu$ obey the same equations as the atomic operators of the inhomogeneously broadened line with fixed atoms, with a transition frequency $\nu_\mu = \nu_0 + k v_\mu$.

VII.9. External fields. External fields act in the first place on the matter of which the laser consists, but do not directly affect the electromagnetic field*.

* Such a direct interaction would require the violation of the superposition law of electromagnetic fields in vacuo. (This superposition law is a consequence of the linearity of Maxwell's equations.) As is shown in quantum electrodynamics, nonlinearities indeed exist in the vacuum, e.g. the scattering of light by light, but these effects are extremely small, so that they can be completely neglected in the present context. It should be noted, however, that the observation of those effects would be of great importance for quantum electrodynamics.

The latter is, however, indirectly influenced, because the properties of the matter determine those of the laser light. Let us consider the example of a solid state laser:

The "matter" consists of

a) the laser-active impurity atoms,

b) the host lattice (or, more generally speaking, of the solid state-matrix),

c) the "cavity walls" like mirrors, Kerr cells, etc.

We list some of the main effects:

a) The laser-active atoms are directly affected by external electric and magnetic fields. Their energy levels may be split, and their wavefunctions may be altered, as is known from the usual Stark and Zeeman effect. It must be noted that in

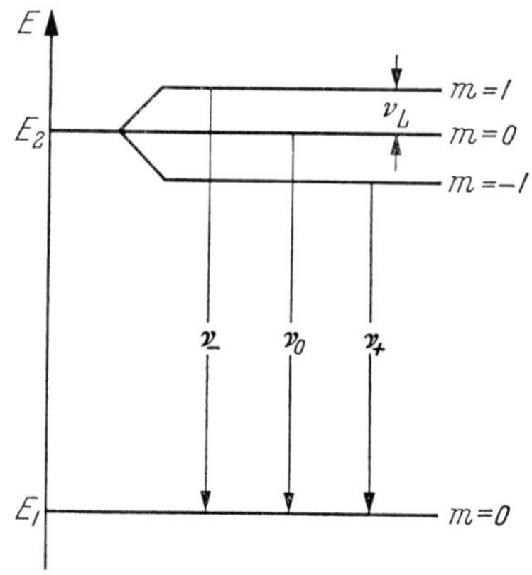

Fig. 42. Atomic transitions (From H. Pelikan, ref. [1] on p. 205).

the corresponding calculations often the crystalline field must be taken into account. In gas lasers, on the other hand, the inclusion of collision effects may be necessary.

All these effects can in principle be included in the density matrix Eq. (VII.1.63), where H now depends on the external fields. The method of treating the laser Eqs. (VII.1.32)—(VII.1.34) remains basically the same: all one has to do is to use the density matrix based on the new wavefunctions and energy levels in the presence of the external fields.

b) The solid-state matrix is, in general, only weakly influenced, although one might think of changes in the index of refraction, conductivity, etc. by external means. Such effects would best be taken care of by the field Eq. (VII.1.5), by using an effective dielectric constant and an effective conductivity.

a—b) There are "mixed" effects, in that the external field acts indirectly on the impurity atom via the host lattice. (Example: The quasi-Stark effect).

c) the "cavity walls".

The reflectivity of the endfaces may be altered by ultrasound (diffraction grating), by Kerr cell switches etc. In this way *loss modulation* is achieved. These effects can be taken into account by using an effective conductivity, σ, in the field equations.

We illustrate our survey (which is by no means complete) by two typical examples:

1. The effect of a longitudinal magnetic field on the single spatial-mode output of a laser,

2. the effect of loss-modulation (generation of ultrashort pulses). (See VII.10, VII.11.)

a) The effect of a longitudinal magnetic field[1, 2] *on the single spatial-mode output.* The degenerate atomic levels of the active laser atoms split up if a magnetic field is applied. For the sake of simplicity we consider a $I=1 \rightarrow I=0$ atomic transition. There are three transitions with different frequencies and polarization. The transition $\Delta m=0$ is linearly polarized parallel to the magnetic field and has the frequency ν_0, while the transitions $\Delta m = \pm 1$ are right and left handed circularly polarized in a plane perpendicular to the magnetic field and have the frequencies ν_- and ν_+. If the magnetic field is applied parallel to the laser axis ($=z$-axis), the polarization of the linearly polarized wave is parallel to the z-axis, while the circularly polarized modes are polarized in a plane parallel to the x-y-plane. Since the direction of propagation is perpendicular to the plane of polarization, only the circularly polarized modes propagate in the z-direction and may thus be amplified by the laser process.

Now let us consider a laser operating without magnetic field in a single mode at a frequency ν_0 and with arbitrary polarization. When the magnetic field is applied, we expect a left and a right handed circularly polarized mode at frequen-

---

[1] The effect of magnetic fields on the laser has been studied both experimentally and theoretically by a number of authors:

A) Theory: W. Culshaw and J. Kannelaud: Phys. Rev. **136**, A 1209 (1964); **141**, 237 (1966) (longitudinal field, observation of frequency locking, theoretical treatment as $I=\frac{1}{2} \rightarrow I=\frac{1}{2}$ transition), Phys. Rev. **145**, 258 (1966); **156**, 308 (1967) (longitudinal and transverse field, $I=0 \leftrightarrow I=1$ transition). — N. N. Rozanov and A. V. Tulub: Dokl. Akad. Nauk SSSR **165**, 1280 (1965) (give amplitude and frequency-determining equations for axial magnetic fields, $I=0 \leftrightarrow I=1$ transition). — R. L. Fork and M. Sargent III: Phys. Rev. **139**, A 617 (1965); and in Proc. of the Internat. Conference on the Physics of Quantum Electronics, ed. by P. L. Kelley, B. Lax, and P. E. Tannenwald, p. 611—619. New York: McGraw-Hill Book Co 1966 (discuss amplitude and frequency-determining equations for axial magnetic fields, $I=0 \leftrightarrow I=1$ transition). — M. I. D'Yakonov: Zh. Eksperim. i Teoret. Fiz. 49, 1169 (1965) [theory, weak longitudinal field, arbitrary angular momenta of the atoms; cavity can have a loss-anisotropy (see also text above)]. — C. V. Heer and R. D. Graft: Phys. Rev. **140**, A 1088 (1965) (theory: field at an arbitrary angle to the laser axis, atoms with arbitrary angular momenta, hfs, and isotopic abundance. Isotropic cavity losses.) — M. I. D'Yakonov and V. E. Perel: Opt. i Spektroskopiya **20**, 472 (1966), and review article by M. I. D'Yakonov and S. A. Fridrikhov: Usp. Fiz. Nauk **90**, 565 (1966) (arbitrary field orientation and angular momenta, inclusion of spontaneous emission). — M. I. D'Yakonov and V. I. Perel: Zh. Eksperim. i Teoret. Fiz. **50**, 448 (1966) (qualitative description of longitudinal and transverse field, use of hole-burning model). — G. Durand: Ann. Inst. Poincaré A **4**, 263 (1966); — IEEE J. Quant. Electronics **QE-2**, 448 (1966) (transverse field, $0 \leftrightarrow 1$ transition). — H. Pelikan: Phys. Letters **21**, 652 (1966); — Z. Physik **201**, 523 (1967) (longitudinal field, $I=0 \leftrightarrow I=1$ transition, cavity has a loss-anisotropy, locking effects). — M. Sargent, III, W. E. Lamb Jr., W. J. Tomlinson, and R. L. Fork: Bull. Am. Soc. **12**, 90 (1967) (longitudinal field, cavity has a loss-anisotropy, locking effects). — M. Sargent, III, W. E. Lamb Jr., and R. L. Fork: Phys. Rev. **164**, 436, 450 (1967) (field at arbitrary angle to the maser axis, anisotropic cavity loss and resonance, atoms with arbitrary angular momenta, hfs, and isotopic abundance, computer programs available).

Related topics: D. Polder and W. van Haeringen: Phys. Letters **19**, 380 (1965). — W. van Haeringen: Phys. Letters **24** A, 65 (1967); — Phys. Rev. **158**, 256 (1967) (arbitrary angular momenta, zero field). — W. M. Doyle and M. B. White: Phys. Rev. **147**, 359 (1966); — Phys. Rev. Letters **17**, 467 (1966) (zero field, cavity anisotropy).

Experimental: J. Kannelaud and W. Culshaw: Appl. Phys. Letters **9**, 120 (1966) ($I=1 \leftrightarrow I=0$ transition in Xe). — H. de Lang and G. Bouwhuis: Phys. Letters **20**, 383 (1966) ($I=2 \rightarrow I=2$ transition in Ne). — W. Culshaw and J. Kannelaud: Phys. Rev. **145**, 257 (1966). — W. J. Tomlinson and R. L. Fork: Phys. Rev. **164**, 466 (1967) ($I=1 \rightarrow I=0$, and $I=1 \rightarrow I=2$ transition in He-Ne, in particular frequency locking).

[2] In our representation we essentially follow the work of Pelikan, l. c. [1] above.

cies ν_+, ν_-, which are separated by the atomic Zeeman splitting. Experimentally, however, something different is found. At weak fields the circularly polarized modes have *equal* frequencies, i.e. frequency locking occurs and they combine to a linearly polarized mode. The plane of polarization has a definite angle to the x-axis and may rotate from zero up to $\pm\pi/4$ with increasing magnetic field. Furthermore the rotation depends on the laser intensity, the asymmetry in the cavity and the cavity detuning. If the magnetic field exceeds a critical value, the linearly polarized mode changes into circularly polarized modes with different frequencies. In higher magnetic fields there may exist a further locking region, where the modes have equal frequencies. The polarization is again linear and rotates with the magnetic field from $-\pi/4$ up to $+\pi/4$.

In order to treat these effects properly, one has to allow for an anisotropy of the cavity losses which we will do below.

b) The field equations. We proceed in close analogy to VII.1, VII.2, but apply the following modification. Since only the circularly polarized waves are amplified by the resonator, we do not expand the vector potential in linearly but in circularly polarized modes. Assuming only one left and one right-handed circularly polarized mode above threshold, we write the vector potential in complex form

$$\left.\begin{aligned}\mathfrak{A} &\equiv A_x + i A_y\\ &= \sqrt{\frac{8\pi\hbar c^2}{V}}\left\{\frac{r_+}{\sqrt{\Omega_+}}\, e^{i(\Omega_+ t+\varphi_+)} A_+(\boldsymbol{x}) + \frac{r_-}{\sqrt{\Omega_-}}\, e^{-i(\Omega_- t+\varphi_-)} A_-(\boldsymbol{x})\right\}\end{aligned}\right\} \tag{VII.9.1}$$

or

$$\mathfrak{A} = \sqrt{\frac{8\pi\hbar c^2}{V}}\left\{\frac{b_+(t)}{\sqrt{\Omega_+}} A_+(\boldsymbol{x}) + \frac{b_-(t)}{\sqrt{\Omega_-}} A_-(\boldsymbol{x})\right\} \tag{VII.9.2}$$

where

$$b_\pm(t) = r_\pm\, e^{\pm i(\Omega_\pm t+\varphi_\pm)}, \tag{VII.9.3}$$

$$A_\pm(\boldsymbol{x}) = \sin k_\pm z; \quad k_\pm = \frac{\omega_\pm}{c} \tag{VII.9.4}$$

$\omega_\pm$ are the cavity resonances.

A_x and A_y are the components of the vector potential in x- and y-direction. $r_+\, e^{i(\Omega_+ t+\varphi_+)}$ describes the left handed circularly polarized laser mode with time-independent frequency Ω_+, and $r_-\, e^{-i(\Omega_- t+\varphi_-)}$ the right-handed circularly polarized mode with frequency Ω_- · $r_\pm$ are real amplitudes. Since we have split off the term $e^{\pm i\Omega_\pm t}$ which varies rapidly with time, we may assume as usual that the amplitudes $r_\pm$ and phases $\varphi_\pm$ are slowly varying functions of time. The number of right and left-handed circularly polarized photons is given by $n_- = r_-^2$, $n_+ = r_+^2$.

In order to describe the experiments properly, we must allow for different cavity losses for waves polarized in x- and y-direction i.e. $\varkappa_x \neq \varkappa_y$. Therefore we generalize Eq. (VII.1.4) into*

$$-\Delta A_i + \frac{1}{c^2}\ddot{A}_i + \frac{2\varkappa_i}{c^2}\dot{A}_i = \frac{4\pi}{c}\dot{P}_i \tag{VII.9.5}$$

$i = x, y$.

Introducing the average damping constant $2\varkappa = \varkappa_x + \varkappa_y$ and the difference of the damping constants $2\Delta\varkappa = \varkappa_y - \varkappa_x$ we find an equation for $\mathfrak{A} = A_x + i A_y$

$$-\Delta\mathfrak{A} + \frac{1}{c^2}\ddot{\mathfrak{A}} + \frac{2\varkappa}{c^2}\dot{\mathfrak{A}} - \frac{2\Delta\varkappa}{c^2}\dot{\mathfrak{A}}^* = \frac{4\pi}{c}\dot{P}_+ \tag{VII.9.6}$$

* If $\varkappa$ is isotropic, (VII.9.5) is equivalent to Ep. (VII.1.5) because $\boldsymbol{E} = -\frac{1}{c}\dot{\boldsymbol{A}}$.

where

$$\mathfrak{A}^* = A_x - iA_y; \quad P_+ = P_x + iP_y. \tag{VII.9.7}$$

The average value P_+ may be calculated from the electrical dipole operator

$$\vartheta^{\pm} = e(x \pm iy) \tag{VII.9.8}$$

by means of

$$P_+ = \frac{N}{V}\, Tr(\varrho\,\vartheta^+). \tag{VII.9.9}$$

ϱ is the statistical operator with the matrix elements* $\varrho_{1,2m}$. The indices 1 and $2, m$ correspond to the indices in Fig. 42. N/V is density of the active atoms which we assume to be constant. If the selection rules for ϑ^+ are taken into account ($\vartheta^+_{1,2m} = \vartheta^+_{12}\,\delta_{m,1}$; $\vartheta^+_{2m,1} = \vartheta^+_{21}\,\delta_{m,-1}$), the polarization is

$$P_+ = \frac{N}{V}\,(\varrho_{1,2-1}\,\vartheta^+_{21} + \varrho_{21,1}\,\vartheta^+_{12}). \tag{VII.9.10}$$

Inserting Eqs. (VII.9.2) and (VII.9.4) in Eq. (VII.9.6), using the orthogonality relations for A_+ and A_-

$$\int A_- A_+\,dV = \frac{V}{2}\,\delta_{k+,k-}, \tag{VII.9.11}$$

and the rotating wave approximation (see III.4) (i.e. taking into account the slow variation of $r_\pm$ and $\varphi_\pm$, we neglect terms like $\ddot{r}_\pm$, $\ddot{\varphi}_\pm$) we find equations for the amplitudes r_+ and r_-

$$\left.\begin{aligned} &\dot{r}_+ - i(\omega_+ - \Omega_+ - \dot{\varphi}_+)\,r_+ + \varkappa\,r_+ - \Delta\varkappa r_-\;e^{i[(\Omega_- - \Omega_+)t + \varphi_- - \varphi_+]}\,\delta_{k_+,k_-} \\ &\quad = \nu_+\,\frac{N}{V}\,e^{-i(\Omega_+ t + \varphi_+)} \int g^+_{21k_+}\,\varrho_{1,2-1}\,dV, \end{aligned}\right\} \tag{VII.9.12}$$

$$\left.\begin{aligned} &\dot{r}_- + i(\omega_- - \Omega_- - \dot{\varphi}_-)\,r_- + \varkappa\,r_- - \Delta\varkappa r_+\;e^{i[(\Omega_- - \Omega_+)t + \varphi_- - \varphi_+]}\,\delta_{k_+,k_-} \\ &\quad = \nu_-\,\frac{N}{V}\,e^{i(\Omega_- t + \varphi_-)} \int g^+_{12k_-}\,\varrho_{21,1}\,dV \end{aligned}\right\} \tag{VII.9.13}$$

where

$$g^+_{12k\pm} = \sqrt{\frac{2\pi}{\hbar V \Omega_\pm}}\,\vartheta^+_{12}\sin k_\pm z; \quad g^+_{21k\pm} = \sqrt{\frac{2\pi}{\hbar V \Omega_\pm}}\,\vartheta^+_{21}\sin k_\pm z. \tag{VII.9.14}$$

Eqs. (VII.9.12) and (VII.9.13) show that the terms

$$\Delta\varkappa r_\pm\;e^{i[(\Omega_- - \Omega_+)t + \varphi_- - \varphi_+]} \tag{VII.9.15}$$

appear only if $k_+ = k_-$, i.e. if the left and right-handed circularly polarized laser modes arise from the same spatial mode. We assume that the corresponding cavity resonances are degenerate, i.e. the resonances of the left- and right-handed circularly polarized oscillations of the unloaded resonator are equal. In weak magnetic field the splitting of the atomic levels is small compared to the frequency difference of two adjacent cavity resonances (compare Fig. 43, case 1). Fig. 43 shows the special case where ω_2 is in resonance with the degenerate atomic frequency ν_0. Thus both laser modes arise from the same cavity resonance $\omega_2 = \omega_+ = \omega_-$, and we have to treat Eqs. (VII.9.12) and (VII.9.13) with the terms

$$\Delta\varkappa r_\pm\;e^{i[(\Omega_- - \Omega_+)t + \varphi_- - \varphi_+]}. \tag{VII.9.16}$$

---

* More exactly, $\varrho_{1,2m}$ should read $\varrho_{1m',2m}$. Because the lower state 1 has only the magnetic quantum number $m' = 0$, we have dropped that index everywhere.

In higher magnetic fields (case 2) the left-handed circularly polarized laser mode may arise from the resonances ω_1 and ω_2 and the right-handed circularly polarized mode from ω_2 and ω_3. If the pumping is high enough, there are more than two modes above threshold. This case is more complicated and will not be treated here. In case 3 the atomic frequencies agree approximately with two different

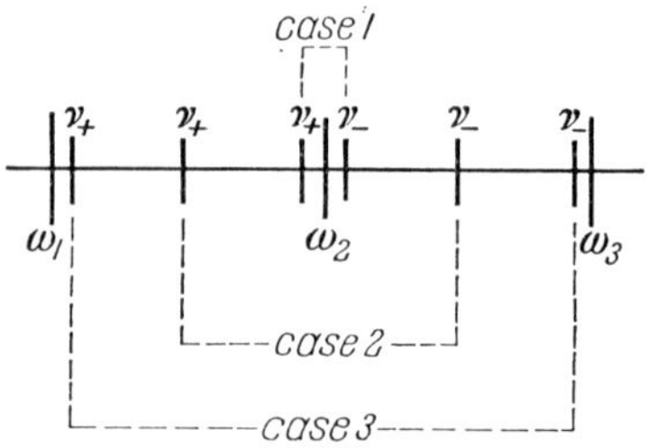

Fig. 43. Comparison of the atomic frequencies and resonator frequencies for different magnetic fields. (After H. PELIKAN, ref. [1].)

cavity resonances ω_1 and ω_3. The laser modes then should arise from $\omega_+ = \omega_1$ and $\omega_- = \omega_3$, respectively. In this case $\delta_{k+,k-} = 0$ and the terms

$$\Delta \varkappa r_\pm \, e^{i[(\Omega_- - \Omega_+)t + \varphi_- - \varphi_+]} \tag{VII.9.17}$$

vanish. Thus we have in higher magnetic fields no coupling of the modes

$$r_+ \, e^{i(\Omega_+ t + \varphi_+)} \quad \text{and} \quad r_- \, e^{-i(\Omega_- t + \varphi_-)}$$

produced by the asymmetrical damping.

c) The matter equations. The laser-active atoms are described by the density matrix equation (compare also IV.7).

$$\frac{d\varrho}{dt} = \left(\frac{\partial \varrho}{\partial t}\right)_{coh} + \left(\frac{\partial \varrho}{\partial t}\right)_{incoh}. \tag{VII.9.18}$$

$\left(\frac{\partial \varrho}{\partial t}\right)_{coh}$ is given by

$$-\frac{i}{\hbar}[H, \varrho],$$

where

$$H = \begin{pmatrix} E_1 & H^{int}_{1,2-1} & H^{int}_{1,20} & H^{int}_{1,21} \\ H^{int}_{2-1,1} & E_2 - \hbar\nu_L & & 0 \\ H^{int}_{20,1} & & E_2 & \\ H^{int}_{21,1} & 0 & & E_2 + \hbar\nu_L \end{pmatrix} \tag{VII.9.19}$$

E_1 and E_2 are the energies of the unsplit levels.

Because of the selection rules for dipole transitions, the matrix elements $H^{int}_{2m,2m'}$ vanish. The interaction of an electron with the electromagnetic field is determined by the operator

$$H^{int} = -\frac{e}{mc}\,\boldsymbol{p}\,\boldsymbol{A} = -\frac{e}{2mc}\,(p^- \mathfrak{A} + p^+ \mathfrak{A}^*) \tag{VII.9.20}$$

where

$$p^\pm = p_x \pm i p_y \tag{VII.9.21}$$

$\boldsymbol{p}$ is the usual momentum operator.

The matrix elements of the interaction operator are

$$H^{int}_{a,b} = -\frac{e}{2mc}\left(p^-_{ab}\mathfrak{A} + p^+_{ab}\mathfrak{A}^*\right) = -\frac{i\nu_{ab}}{2c}\left(\vartheta^-_{ab}\mathfrak{A} + \vartheta^+_{ab}\mathfrak{A}^*\right) \qquad \text{(VII.9.22)}$$

with

$$\nu_{ab} = \frac{E_a - E_b}{\hbar} \quad \text{and} \quad p^{\pm}_{ab} = \frac{m}{e}\, i\nu_{ab}\,\vartheta^{\pm}_{ab}. \qquad \text{(VII.9.22a)}$$

$\left(\frac{\partial \varrho}{\partial t}\right)_{incoh}$ describes all incoherent processes, like the pump etc., where we treat the following pump and decay scheme: level 1 and the three sublevels of 2 are pumped from a ground state 0, which acts as a reservoir (and does not appear *explicitly* in the equations). We denote the relaxation times T_j after which the levels $j = 1, 2$ come to equilibrium by $1/\gamma_j$. The equilibrium or "unsaturated" values of $\varrho_{jm,jm}$ are denoted by $\varrho^0_{jm,jm}$. The damping constants of the nondiagonal elements $\varrho_{1,2m}$ and $\varrho_{2m,2l}$ are denoted by γ_{12} and γ_2 respectively. Thus the components of $\left(\frac{\partial \varrho}{\partial t}\right)_{incoh}$ read

$$\left.\begin{aligned}
\left(\frac{\partial}{\partial t}\varrho_{11}\right)_{incoh} &= \gamma_1\left(\varrho^0_{11} - \varrho_{11}\right)\\
\left(\frac{\partial}{\partial t}\varrho_{2l,2m}\right)_{incoh} &= \gamma_2\left(\varrho^0_{2m,2m}\,\delta_{lm} - \varrho_{2l,2m}\right)\\
\left(\frac{\partial}{\partial t}\varrho_{1,2m}\right)_{incoh} &= -\gamma_{12}\,\varrho_{1,2m}
\end{aligned}\right\} \qquad \text{(VII.9.23)}$$

where

$$\gamma_{12} = \tfrac{1}{2}(\gamma_1 + \gamma_2).$$

The Eqs. (VII.9.18) represent a system of coupled differential equations for the components of the density matrix. It is solved along similar lines to the equations in Chap. VII.2 expanding ϱ in terms up to third order in the electric field ($\sim r$). Inserting the corresponding expressions for $\varrho_{1,2-1}$ and $\varrho_{21,1}$ into (VII.9.12), (VII.9.13), and splitting the resulting equations into real and imaginary parts, we find equations for the amplitudes and frequencies

$$\dot r_{\pm} = \Delta\varkappa r_{\mp}\cos\Psi\,\delta_{k+,k-} + \alpha_{\pm} r_{\pm} - \beta_{\pm} r^3_{\pm} - \Theta_{\pm} r_{\pm} r^2_{\mp}, \qquad \text{(VII.9.24)}$$

$$\Omega_{\pm} + \dot\varphi_{\pm} = \omega_{\pm} \pm \Delta\varkappa\frac{r_{\mp}}{r_{\pm}}\sin\Psi\,\delta_{k+,k-} + \xi_{\pm} - \eta_{\pm} r^2_{\pm} - \tau_{\pm} r^2_{\mp} \qquad \text{(VII.9.25)}$$

where

$$\Psi = (\Omega_- - \Omega_+)\,t + \varphi_- - \varphi_+. \qquad \text{(VII.9.26)}$$

The constants $\alpha_{\pm}, \beta_{\pm}, \Theta_{\pm}, \xi_{\pm}, \eta_{\pm}, \tau_{\pm}$ are given by:

$$\alpha_{\pm} = -\varkappa + \frac{1}{2}\frac{\nu^2_{\pm}}{\Omega_{\pm}} N f d^0_{2\mp1}\frac{\gamma_{12}}{\gamma^2_{12} + (\nu_{\pm} - \Omega_{\pm})^2}, \qquad \text{(VII.9.27)}$$

$$\left.\begin{aligned}
\beta_{\pm} = \frac{3}{4}\frac{\nu^4_{\pm}}{\Omega^2_{\pm}} N f^2 d^0_{2\mp1}\frac{\gamma_{12}}{[\gamma^2_{12} + (\nu_{\pm} - \Omega_{\pm})^2]^2}\times\\
\times\left(\frac{\gamma_{12}\gamma_2 + (\nu_{\pm} - \Omega_{\pm})^2}{\gamma^2_2 + (\nu_{\pm} - \Omega_{\pm})^2} + \frac{\gamma_{12}\gamma_1 + (\nu_{\pm} - \Omega_{\pm})^2}{\gamma^2_1 + (\nu_{\pm} - \Omega_{\pm})^2}\right),
\end{aligned}\right\} \qquad \text{(VII.9.28)}$$

$$\left.\begin{aligned}
\Theta_{\pm} = \frac{3}{8}\frac{\nu^2_+\nu^2_-}{\Omega_+\Omega_-} N f^2 \frac{1}{\gamma^2_{12} + (\nu_{\mp} - \Omega_{\mp})^2}\times\\
\times\left\{2\frac{\gamma_{12}[\gamma_{12}\gamma_1 + (\nu_{\mp} - \Omega_{\mp})^2]\,d^0_{2\pm1}}{[\gamma^2_{12} + (\nu_{\pm} - \Omega_{\pm})^2]\,[\gamma^2_1 + (\nu_{\mp} - \Omega_{\mp})^2]} + \frac{\gamma_2\,d^0_{2\pm1}}{\gamma^2_2 + (\nu_{\pm} - \Omega_{\pm})^2} +\right.\\
\left. + \frac{\gamma^2_{12}\gamma_2 - (\nu_+ - \Omega_+)(\nu_- - \Omega_-)\gamma_2 + \gamma_{12}(\nu_{\mp} - \Omega_{\mp})(\nu_+ - \Omega_+ + \nu_- - \Omega_-)}{[\gamma^2_{12} + (\nu_{\pm} - \Omega_{\pm})^2]\cdot[\gamma^2_2 + (\nu_{\mp} - \Omega_{\mp})^2]}\,d^0_{2\mp1}\right\},
\end{aligned}\right\} \qquad \text{(VII.9.29)}$$

$$\xi_\pm = \frac{1}{2}\frac{\nu_\pm^2}{\Omega_\pm} N f d^0_{2\mp1} \frac{\nu_\pm - \Omega_\pm}{\gamma_{12}^2 + (\nu_\pm - \Omega_\pm)^2}, \tag{VII.9.30}$$

$$\left.\begin{aligned}\eta_\pm = \frac{3}{4}\frac{\nu_\pm^4}{\Omega_\pm^2} N f^2 d^0_{2\mp1} \frac{\nu_\pm - \Omega_\pm}{[\gamma_{12}^2 + (\nu_\pm - \Omega_\pm)^2]^2} \times \\ \times\left(\frac{\gamma_{12}\gamma_2 + (\nu_\pm - \Omega_\pm)^2}{\gamma_2^2 + (\nu_\pm - \Omega_\pm)^2} + \frac{\gamma_{12}\gamma_1 + (\nu_\pm - \Omega_\pm)^2}{\gamma_1^2 + (\nu_\pm - \Omega_\pm)^2}\right),\end{aligned}\right\} \tag{VII.9.31}$$

$$\left.\begin{aligned}\tau_\pm = &\frac{3}{8}\frac{\nu_+^2\nu_-^2}{\Omega_+\Omega_-}\frac{N f^2}{\gamma_{12}^2 + (\nu_\mp - \Omega_\mp)^2} \times \\ &\times\left\{\frac{2(\nu_\pm - \Omega_\pm)[\gamma_{12}\gamma_1 + (\nu_\mp - \Omega_\mp)^2]\, d^0_{2\pm1}}{[\gamma_{12}^2 + (\nu_\pm - \Omega_\pm)^2][\gamma_1^2 + (\nu_\mp - \Omega_\mp)^2]} + \frac{(\nu_\pm - \Omega_\pm)\, d^0_{2\pm1}}{\gamma_2^2 + (\nu_\pm - \Omega_\pm)^2} + \right. \\ &\left. + \frac{\gamma_{12}\gamma_2(\nu_+ - \Omega_+ + \nu_- - \Omega_-) - (\nu_\mp - \Omega_\mp)[\gamma_{12}^2 - (\nu_+ - \Omega_+)(\nu_- - \Omega_-)]}{[\gamma_{12}^2 + (\nu_\pm - \Omega_\pm)^2][\gamma_2^2 + (\nu_\mp - \Omega_\mp)^2]}\, d^0_{2\mp1}\right\}\end{aligned}\right\} \tag{VII.9.32}$$

where

$$f = \frac{2\pi}{\hbar V}|\vartheta|^2. \tag{VII.9.33}$$

(For a $J=1\to J=0$ transition is $|\vartheta_{12}^+|^2 = |\vartheta_{12}^-|^2 = |\vartheta|^2$).

The constants $\Theta_\pm$ and $\tau_\pm$ are valid for the case where both laser modes arise from the same cavity resonance. If they arise from different cavity resonances, we have to replace the factor $\frac{3}{8}$ by $\frac{1}{4}$. $d^0_{2\pm1}$ is the unsaturated population difference between the excited sublevel 2, +1 (or 2, −1) and the ground level 1: $d^0_{2\pm1} = \overset{0}{\varrho}_{2\pm1,2\pm1} - \overset{0}{\varrho}_{1,1}$.

d) Solution of the amplitude and frequency-determining Eqs. (VII.9.24), (VII.9.25). The Eqs. (VII.9.24) for the temporal change of the amplitudes are coupled with those (VII.9.25) of the frequencies, via the term $\Delta\varkappa r_\pm \cos\Psi$.

In a good approximation we may neglect this coupling, so that the amplitudes $r_\pm$ may be assumed time-independent [note, that all coefficients ($\alpha, \ldots$) are constants].

The amplitude equations then simplify to

$$\beta_+ r_+^2 + \Theta_+ r_-^2 = \alpha_+, \tag{VII.9.34}$$

$$\beta_- r_-^2 + \Theta_- r_+^2 = \alpha_- \tag{VII.9.35}$$

which are two coupled equations for the photon numbers,

$$r_\pm^2 = n_\pm.$$

The solution of these equations is straightforward. The physically interesting phenomena are described by the frequency equations (VII.9.25). The time-independent terms on the right-hand side of Eq. (VII.9.25) represent shifts from the frequency of the unloaded cavity $\omega_\pm$ by pulling ($\xi_\pm$) and pushing $-(\eta_\pm r_\pm^2 + \tau_\pm r_\mp^2)$.

The term $\Delta\varkappa\frac{r_\mp}{r_\pm}\sin\Psi$ introduces a coupling between the frequencies $(\Omega_+ + \dot\varphi_+)$ and $(\Omega_- + \dot\varphi_-)$. In order to discuss the role of this coupling we substract the Eq. (VII.9.25) with (+) from that with (−) and obtain

$$\dot\Psi = \Delta\Omega - a\,\delta_{k-,k+}\sin\Psi \tag{VII.9.36}$$

where

$$a = \Delta\varkappa\left(\frac{r_+}{r_-} + \frac{r_-}{r_+}\right) \tag{VII.9.37*}$$

* We have corrected signs according to a correction given by H. PELIKAN: Z. Physik **211**, 418 (1968).

and

$$\Delta\Omega = \omega_- - \omega_+ + \xi_- - \xi_+ - (\eta_- r_-^2 - \eta_+ r_+^2) - (\tau_- r_+^2 - \tau_+ r_-^2). \qquad \text{(VII.9.38)}$$

The physical meaning of Ψ is the following: By the superposition of a left- and right-handed circularly polarized mode we find an elliptically polarized mode, and the polarization ellipsoid rotates with half the beat frequency Ψ of the circularly polarized modes.

Eq. (VII.9.36) is characteristic for the occurrence of frequency locking [see also (VII.5c)], which depends on whether:

$$\text{(I)} \quad |a| < |\Delta\Omega| \quad \text{or} \quad \text{(II)} \quad |a| > |\Delta\Omega|.$$

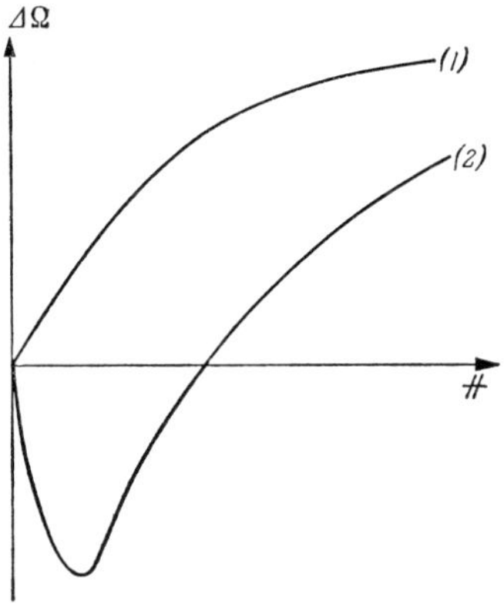

Fig. 44.

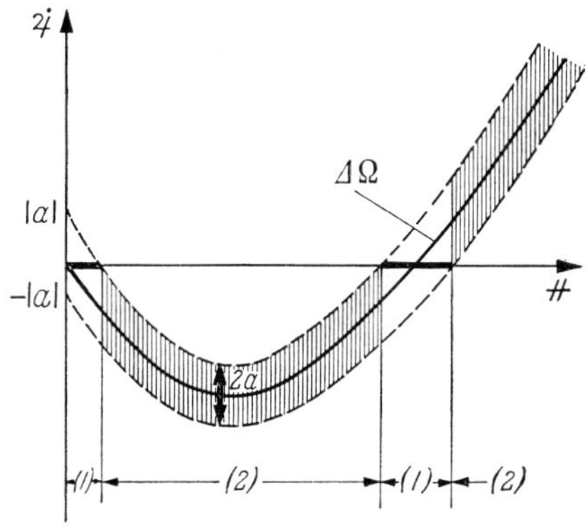

Fig. 45.

Fig. 44. $\Omega_- - \Omega_+$ as a function of the magnetic field. (After H. Pelikan.)

Fig. 45. The beat frequency as a function of the magnetic field. (*1*) ... locking regions, (*2*) ... regions with frequency modulated modes. (After H. Pelikan.)

Since the coefficients $\xi_\pm$, $\eta_\pm$, $\tau_\pm$ in Eq. (VII.9.38) depend on the magnetic field strength via the frequencies, experimentalists may pass from region (I) to region (II) and vice versa by changing the magnetic field (see Fig. 44).

Region (I):

$$|a| < |\Delta\Omega|.$$

If $|a| < |\Delta\Omega|$ the solution of (VII.9.36) is

$$\Psi = 2 \arctan\left[\frac{1}{\Delta\Omega}\left\{a + \sqrt{\Delta\Omega^2 - a^2} \tan \sqrt{\Delta\Omega^2 - a^2}\, \frac{t}{2}\right\}\right]. \qquad \text{(VII.9.39)}$$

Differentiating Ψ with respect to time we find the beat frequency

$$\dot{\Psi} = \frac{(\Delta\Omega)^2 - a^2}{\Delta\Omega + a \sin\left(\sqrt{(\Delta\Omega)^2 - a^2}\, t + \alpha\right)} \qquad \text{(VII.9.40)}$$

where

$$\sin \alpha = \frac{a}{\Delta\Omega}. \qquad \text{(VII.9.41)}$$

Thus the beat frequency is a function of time and varies between the curves $\Delta\Omega \pm a$ (Fig. 45).

Region II:
The solution of Eq. (VII.9.36) behaves quite differently if $|a| > \Delta\Omega$. The integration gives

$$t = \frac{1}{\sqrt{a^2 - (\Delta\Omega)^2}} \ln \frac{\Delta\Omega \tan \frac{\Psi}{2} - a - \sqrt{a^2 - (\Delta\Omega)^2}}{\Delta\Omega \tan \frac{\Psi}{2} - a + \sqrt{a^2 - (\Delta\Omega)^2}} \tag{VII.9.42}$$

t has a singularity, i.e. $t \to \infty$ if Ψ reaches the value Ψ_0 determined by

$$\tan \frac{\Psi_0}{2} = \frac{a - \sqrt{a^2 - (\Delta\Omega)^2}}{\Delta\Omega}. \tag{VII.9.43}$$

This means Ψ reaches Ψ_0 asymptotically for $t \to \infty$. Thus the time derivative of Ψ is zero. The left and right-handed circularly polarized modes have the same frequency, which means that frequency locking occurs. The relative phase of the modes $\Psi = (\Omega_- t + \varphi_-) - (\Omega_+ t + \varphi_+)$ has the constant value Ψ_0. We see that there should be no locking if $a = 0$, i.e. in the symmetric case $\Delta\varkappa = 0$. As $\Omega_- - \Omega_+$ is a function of the magnetic field we may pass from $|a| > |\Delta\Omega|$ to $|a| < |\Delta\Omega|$ by varying the magnetic field. In small magnetic fields there is $|a| > |\Delta\Omega|$ and we find the first locking region. In higher magnetic fields, if $|a| < |\Delta\Omega|$, the beat frequency is a function of time and varies around $\Delta\Omega$ with the amplitude a. If $\Delta\Omega$ passes through zero a second time we will find a second locking region. The width of the locking regions depends on the asymmetry of the damping constants $\Delta\varkappa$ and increases with $\Delta\varkappa$.

A left- and a right-handed circularly polarized mode with the same frequency may be combined to a generally elliptically polarized mode. The mode is linearly polarized if the circularly polarized modes have the same amplitudes. The angle of the major axis of the ellipsoid at the x-axis is half the relative phase $\Psi_0/2$ of the circularly polarized modes. Thus we will find elliptically polarized modes in the locking regions. The major axis of the ellipsoid is at the angle $\Psi_0/2$ to the x-axis. $\Psi_0/2$ is determined by Eq. (VII.9.43). Using the identity

$$\tan \frac{x}{2} = \frac{1 - \sqrt{1 - \sin^2 x}}{\sin x} \tag{VII.9.44}$$

we find

$$\sin \Psi_0 = \frac{\Delta\Omega}{a} \tag{VII.9.45}$$

As $\Delta\Omega$ depends on the magnetic field, the angle of the major axis of the ellipsoid at the x-axis varies with the magnetic field. In zero magnetic field $\Delta\Omega = 0$ and thus $\Psi_0/2 = 0$. The major axis is oriented in the direction where the mode has most gain. With increasing magnetic field, $\Delta\Omega$ and thus $|\Psi_0/2|$ increases, i.e. the magnetic field causes a rotation out of the energetically preferred direction $\Psi_0/2 = 0$. Since $|\sin \Psi_0|$ may not exceed 1 for real Ψ_0 the major axis should rotate only up to $\pm\pi/4$ with increasing magnetic field. If $\Delta\Omega$ exceeds $|a|$ we find $|\sin \Psi_0| > 1$, and Ψ_0 is complex. As we have seen earlier, there is no locking in this case, and the polarization ellipsoid rotates with half the beat frequency of the circularly polarized modes. $\Delta\Omega$ depends on the laser intensity r_+^2 and r_-^2 and on the cavity tuning (the constants $\xi_\pm, \eta_\pm, \tau_\pm$ are functions of $\omega_\pm$). Furthermore a is proportional to the asymmetry of the cavity $\Delta\varkappa$. That is why we expect a dependence of the angle of rotation on the laser intensity, the cavity tuning and the asymmetry of the cavity.

VII.10. Ultrashort optical pulses: the principle of mode locking. In the usual production of giant pulses often several modes participate, which have usually no phase relation, however. The total electric field* $E(t)$ may be written as the superposition of the electric fields $E_\lambda(t)$ of the single modes

$$E(t) = \sum_\lambda E_\lambda(t). \tag{VII.10.1}$$

We first assume that the phases of the fields

$$E_\lambda \sim e^{i\varphi_\lambda} \tag{VII.10.2}$$

are uncorrelated, so that

$$\overline{E_\lambda E_{\lambda'}} = 0 \quad (\lambda \neq \lambda'). \tag{VII.10.3}$$

The total intensity then becomes

$$I = \overline{|E(t)|^2} = \sum_\lambda I_\lambda(t) \tag{VII.10.4}$$

where

$$I_\lambda = |E_\lambda|^2 \tag{VII.10.5}$$

is the intensity of the single mode. When we have N modes with about equal intensity

$$|E_\lambda|^2 \approx |E_0|^2 \tag{VII.10.6}$$

the total intensity is given by

$$I = N|E_0|^2. \tag{VII.10.7}$$

We now consider the case that the modes *have fixed phases and frequencies* with a constant separation ω_0 (compare Fig. 46).

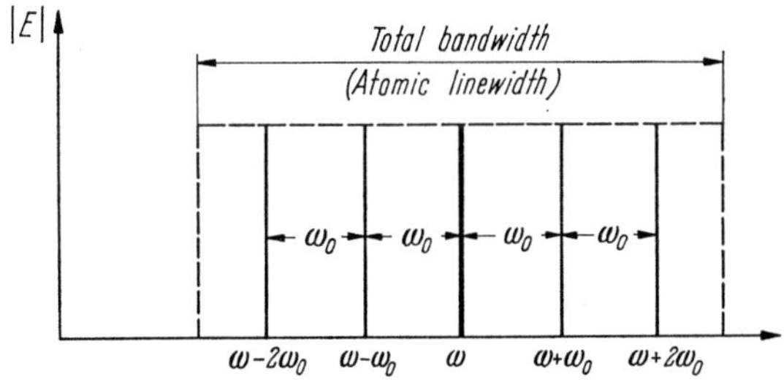

Fig. 46. A model for frequency locking: modes with equidistant frequencies and equal amplitudes within a total bandwidth.

We further assume for simplicity that the modes have equal amplitudes. The total electric field of the phase-locked modes may then be written as the superposition

$$E_{PL}(t) = \sum_\lambda E_0\, e^{i(\omega + \lambda\omega_0)t} \tag{VII.10.8}$$

where the summation runs over

$$\lambda = -\frac{N-1}{2}, \ldots, \lambda = \frac{N-1}{2}. \tag{VII.10.9}$$

* We assume E always polarized in one direction, so that we need not write it as a vector.

The summation over λ is easily performed so that E_{PL} takes the form

$$E_{PL}(t) = E_0\, e^{i\omega t} \frac{\sin \frac{N}{2}\omega_0 t}{\sin \frac{\omega_0}{2} t}. \tag{VII.10.10}$$

From (VII.10.10) we obtain for the total intensity

$$I_{PL}(t) = |E_0|^2 \frac{\sin^2 \frac{N}{2}\omega_0 t}{\sin^2 \frac{\omega_0}{2} t}. \tag{VII.10.11}$$

The properties of $I_{PL}(t)$ are:
The *maxima are reached* at

$$t=0, \quad t=\frac{2\pi}{\omega_0} \tag{VII.10.12}$$

and generally at

$$\boxed{t = m\cdot\frac{2\pi}{\omega_0}; \quad m: \text{integer}}. \tag{VII.10.13}$$

The *peak intensity is given* by

$$\boxed{I_{PL} = |E_0|^2 \cdot N^2 = N \cdot I_{unlocked}}. \tag{VII.10.14}$$

The intensity is thus increased by the factor of the participating modes N compared to the case in which only unlocked modes are used.

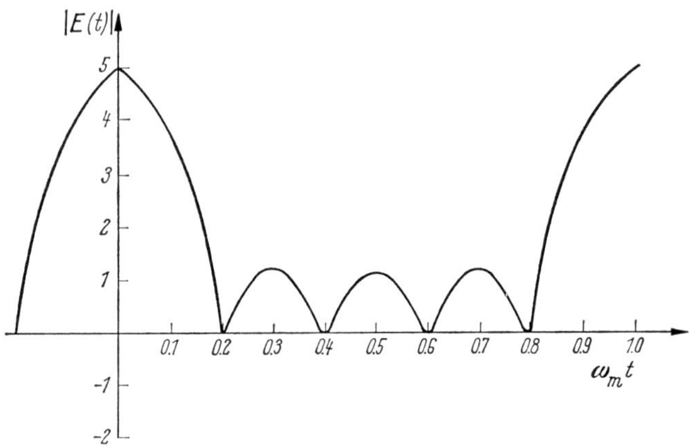

Fig. 47. This fig. shows a plot of the absolute value of the field strength as a function of time for five phase locked modes (after YARIV, l.c. [6] on p. 2).

The *pulse width* (that is the time until the intensity reaches its first zero) is given by

$$\tau = \frac{2\pi}{\omega_0 N}. \tag{VII.10.15}$$

The pulse width is therefore proportional to the inverse of the number of participating modes and of the frequency separation between the modes or, in other words, is given by the inverse total band width. An example of the intensity as a function of time for the case of five modes is represented in Fig. 47.

The important question to be discussed is, how to obtain *mode locking*.

Mode locking is well known in the field of radio engineering[1] where an oscillator is locked in its frequency ω_{os} to the frequency ω_s of an external signal. Frequency locking occurs if the frequency separation $\delta\omega$ is small enough (compare Fig. 48).

An experimental verification in lasers where one gas laser was locked to another was given by STOVER and STEINER[2]. A theoretical treatment with respect to lasers has been given by UCHIDA[3] and TANG and STATZ[4].

This case does not apply directly, however, to the case under consideration because a finite frequency separation ω_0 is wanted (compare Fig. 49).

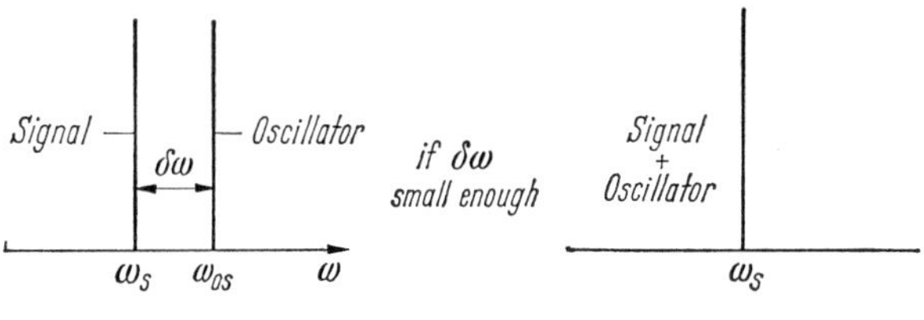

Fig. 48.

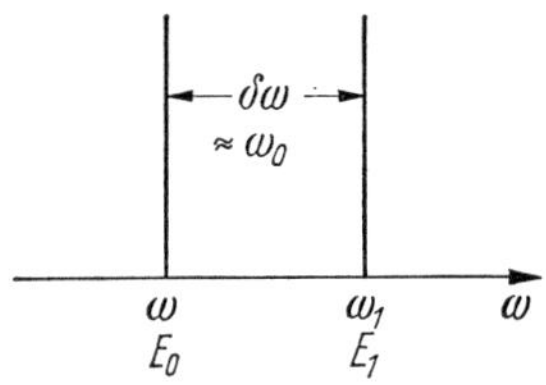

Fig. 49.

Fig. 48. An example for frequency locking. If the frequency distance $\delta\omega$ between the signal frequency ω_S and the oscillator frequency ω_{OS} is small enough the oscillator acquires the frequency ω_S.

Fig. 49. When the modes E_0 and E_1 must be locked together the frequency separation $\delta\omega$ must be bridged by the generation of sidebands of ω.

One has therefore to bridge the gap between the mode at ω_1 and at ω by the generation of sidebands of ω:

$$\omega \pm \omega_0, \quad \omega \pm 2\omega_0, \ldots \tag{VII.10.16}$$

so that

$$\omega_1 \approx \omega + \omega_0 \tag{VII.10.17}$$

holds in particular. This may be achieved in different ways:

a) Loss modulation by an externally driven modulator[5]. The principle of loss modulation may be seen from Fig. 49.

The modulator element is externally driven at a frequency which is about equal to the axial mode spacing. It consists e.g. of a Kerr cell or a diffraction grating driven by ultrasound. The modulator causes a time- and space-dependent

[1] For a detailed representation see the books: GARDNER, F. M.: Phaselock techniques. New York-London-Sydney: John Wiley & Sons 1966, and A. J. VITERBI: Principles of coherent communication. New York: McGraw-Hill Book Co. 1966.

[2] H. L. STOVER and H. W. STEINER: Appl. Phys. Letters **8**, 91 (1966).

[3] T. UCHIDA: IEEE J. Quant. Electronics **3**, 7 (1967).

[4] C. L. TANG and H. STATZ: J. Appl. Phys. **38**, 323 (1967); — For a treatment of classical oscillators (using vacuum tubes) see: ADLER, R.: Proc. IRE **34**, 351 (1946).

[5] For an analogy to microwave circuits see p. 217. Laser mode locking by this mechanism was treated theoretically by DIDOMENICO JR., M.: J. Appl. Phys. **35**, 2870 (1964). — A. YARIV: J. Appl. Phys. **36**, 388 (1965) and others (see footnote 1 on p. 217); — It was first observed by HARGROVE, L. E., R. L. FORK, and M. A. POLLACK: Appl. Phys. Letters **5**, 4 (1964); — For a proposal for a related device („Auskoppelmodulation") see K. GÜRS and R. MÜLLER: Phys. Letters **5**, 179 (1963); — A detailed theoretical treatment of different kinds of interval modulation in multimode laser oscillators is given by A. YARIV: J. Appl. Phys. **36**, 388 (1965). Both he and DIDOMENICO use a linear theory.

modulation of the cavity loss $\varkappa$ by a contribution

$$q \sin \omega_0 t \tag{VII.10.18}$$

(see also below).

Under such circumstances the mode with the electric field E_1 is subject to a driving force proportional to

$$\begin{array}{ccc} E_0(t) & \cdot & q \sin \omega_0 t \\ \downarrow & & \downarrow \\ \omega & + & \omega_0 \end{array} \tag{VII.10.19}$$

where $E_0(t)$ is the field of the mode at ω. This driving force, which is a substitute for the external signal considered above, oscillates at a frequency

$$\omega_1' = \omega + \omega_0 . \tag{VII.10.20}$$

b) Loss modulation by a saturable absorber[6]. A saturable absorber may also act as modulator element. It acts as follows: At low light intensities only a small fraction of the laser light is transmitted through the absorber. At high intensities the absorber becomes transparent. Now let us consider a wave packet which bounces back and forth between the two mirrors at a repetition time* $t_1 = \frac{2L}{c'}$. At these time intervals the absorber opens in a self-adjusted manner, thus modulating the cavity loss at a circular frequency $\omega = \frac{2\pi}{t_1}$.

c) Gain modulation[7]. In each laser its gain is modulated by the nonlinear polarization (i.e. by the population pulsations: see VI.9; VII.5c). The nonlinear part of the polarization has in its lowest order the form

$$P \sim E_i E_j E_k . \tag{VII.10.21}$$

Now consider three modes according to Fig. 50.

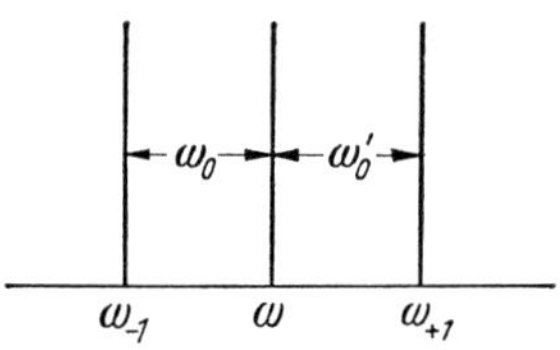

Fig. 50. Three modes E_{-1}, E_0 and E_{+1}, whose frequencies are nearly equally spaced produce sidebands of nearly the right frequency separation via the nonlinear polarization.

---

* c' is an effective velocity of light $\frac{L}{c'} = \frac{L_1}{c_1} + \frac{L_2}{c_2}$; $L_1 + L_2 = L$ where L is the total distance between the mirrors, L_1, L_2 the single light path with and without the active material, respectly, c_1, c_2 are the corresponding velocities of light.

[6] For experimental observations see e.g.: PENNEY, A. W., JR., and H. A. HEYNAU: Appl. Phys. Letters **9**, 257 (1966). — DEMARIA, A. J., D. A. STETZER, and H. A. HEYNAU: Appl. Phys. Letters **8**, 174 (1966). — STETZER, D. A. and A. J. DEMARIA: Appl. Phys. Letters **9**, 118 (1966); — For a theoretical treatment see e.g.: E. GARMIRE and A. YARIV: IEEE J. Quant. Electronics **QE-3**, 222 (1967). — S. E. SCHWARZ: IEEE J. Quant. Electronics **QE-4**, 509 (1968). — N. G. BASOV, P. G. KRIUKOV, V. S. LETOKHOV, and YU. V. SENATZKII: IEEE J. Quant. Electronics **QE-4**, 606 (1968). — S. A. AKHMANOV, A. S. CHIRKIN, K. N. DRABOVICH, A. I. KOVRIGIN, R. V. KHOKHLOV, and A. P. SUKHORUKOV: IEEE J. Quant. Electronics **QE-4**, 598 (1968). — V. S. LETOKHOV: Zh. Eksperim. i Teoret. Fiz. **55**, 1077 (1968). — V. I. BESPALOV and E. YA. DAUME: Zh. Eksperim. i Teoret. Fiz. **55**, 1321 (1968), and others.

[7] See also Chap. VI.9, VII.5c and, in particular, VII.13 (self-pulsing), where further references are given. — For a discussion in particular of the influence and magnitude of this mechanism see: H. STATZ: J. Appl. Phys. **38**, 4648 (1967).

The mode with the electric field E_{+1} is subjected to a driving force at the frequency $\omega+\omega_0$, where the driving force due to the nonlinear polarization is given by

$$\left.\begin{array}{c} E_0(t) \times \underbrace{E_{-1}^*(t)\, E_0(t)} \\ \downarrow \qquad\qquad \downarrow \\ \omega_1 \approx \omega \;+\; \underbrace{(\omega-\omega_{-1})}_{\omega_0} \end{array}\right\} \tag{VII.10.22}$$

Under these circumstances it may be shown (see VII.5c) that the frequencies of all three modes are locked together, so that

$$\Psi = 2\varphi_2 - \varphi_1 - \varphi_3 \tag{VII.10.23}$$

becomes a constant. φ_j are the single-mode phases (including the frequency dependent part).

d) Frequency modulation[8]. Here a periodic perturbation with frequency ω_0 acts in a similar way as the modulated loss.

e) Analogy to microwave circuits[9]. According to CUTLER the feedback loop of the generated pulse is arranged to Fig. 51.

The decisive new element is the expander, which has the following properties:

1. emphazising the highest amplitude of the recirculating pulse,
2. reducing the lower amplitudes,
3. discriminating against noise and reflections,
4. acting to shorten the pulse until the pulse width is limited by the frequency response of the circuit.

A laser can be constructed having analogue elements according to Fig. 52.

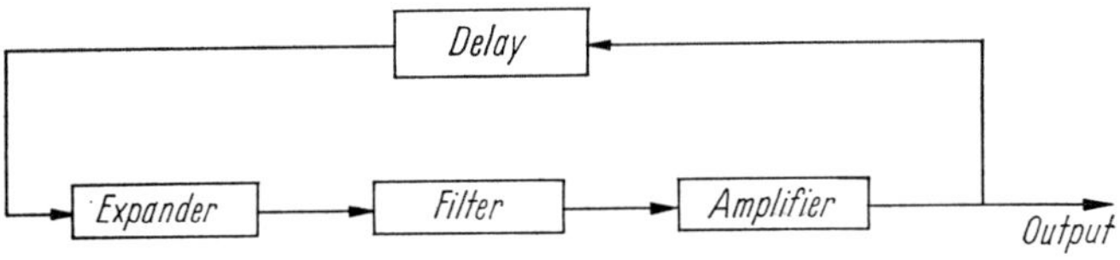

Fig. 51. Feedback loop of the microwave circuit according to CUTLER.

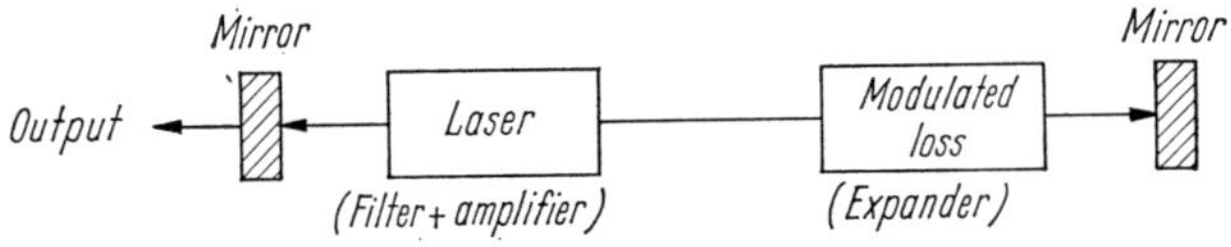

Fig. 52. Experimental set-up of the loss modulated laser.

VII.11. Ultrashort optical pulses: detailed treatment of loss modulation[1]. We start from the wave Eq. (VII.1.8), for the electric field, E. The effective con-

[8] S. E. HARRIS and O. P. McDUFF: IEEE J. Quant. Electronics **QE-1**, 245 (1965).

[9] C. C. CUTLER: Proc. IRE **43**, 140 (1955).

[1] This problem was treated theoretically by M. DI DOMENICO JR.: J. Appl. Phys. **35**, 2870 (1964). — A. YARIV: J. Appl. Phys. **36**, 388 (1965). — M. H. CROWELL: IEEE J. Quant. Electronics **QE-1**, 12 (1965), who use a linear theory; — A nonlinear theory, taking into account gain saturation, was developed by H. HAKEN and M. PAUTHIER: IEEE J. Quant. Electronics **QE-4**, 454 (1968). (Homogeneously broadened line). O. P. McDUFF and S. E. HARRIS: IEEE J. Quant. Electronics **QE-3**, 101 (1967) (inhomogeneously broadened line).

ductivity, σ, which describes all cavity losses, depends on the space coordinate $\boldsymbol{x}$ (such a dependence is e.g. caused by the finite extension of the modulator).

Thus $\sigma_{\lambda\lambda'} \neq 0$ for different modes λ, λ'. [See (VII.1.9).]

The modulator is assumed to modulate σ, at a frequency ω_m, so that

$$\sigma(\boldsymbol{x}, t) = \sigma^0 + \sigma^1(\boldsymbol{x})\,(1 - \cos\omega_m t). \tag{VII.11.1}$$

In the following, ω_m is chosen close to the axial mode spacing, $\Delta\Omega$.

We put

$$\left.\begin{aligned} 2\pi\,\sigma^0_{\lambda\lambda} &= \varkappa, \\ \sigma^0_{\lambda\lambda'} &= 0 \quad \text{for} \quad \lambda \neq \lambda', \\ 2\pi\,\sigma^1_{\lambda\lambda} &= \varkappa_a, \\ 2\pi\,\sigma^1_{\lambda,\lambda\pm1} &= 2\varkappa_c. \end{aligned}\right\} \tag{VII.11.2}$$

All other $\sigma_{\lambda,\lambda'}$'s are assumed zero (this corresponds to the assumption that only synchronous terms are kept in the equation of motion, see below). The $\varkappa$'s are treated as constants, which do not depend on λ.

In VII.1 the rotating wave approximation (III.4), and the use of dimensionless amplitudes b^*, b [see (VII.1.25)], (VII.1.26) had led to the Eq. (VII.1.32).

With the assumptions (VII.11.1), (VII.11.2), the Eq. (VII.1.32) with the additional term (VII.1.32a) may be specialized:

$$\dot{b}^*_\lambda(t) = (i\omega_\lambda - \varkappa - \varkappa_a)\,b^*_\lambda + 2\varkappa_c\cos\omega_m t\,(b^*_{\lambda+1} + b^*_{\lambda-1}) + (G_\lambda - i\psi_\lambda)\,b^*_\lambda \tag{VII.11.3}$$

$$\lambda = 0, \pm1, \pm2, \ldots$$

where

$$(G_\lambda - i\psi_\lambda)\,b^*_\lambda \equiv i\sum_\mu g^*_{\mu\lambda}\,\varrho_{12,\mu}(t). \tag{VII.11.3a}$$

The real constants G_λ and ψ_λ represent the saturated gain and the phase shift, respectively.

In the following we treat axial modes only, so that

$$\omega_\lambda = \omega_0 + \lambda\Delta\omega$$

and we let the mode frequency ω_0 coincide with the atomic line center:

$$\omega_0 = \nu_0. \tag{VII.11.4}$$

We put

$$b^*_\lambda(t) = B^*_\lambda(t)\,e^{i(\omega_0+\lambda\omega_m)t} \tag{VII.11.5}$$

and apply the rotating wave approximation with respect to $e^{i\omega_m t}$. This leaves us with

$$\dot{B}^*_\lambda = (i\lambda\Delta\omega_m - \varkappa - \varkappa_a)\,B^*_\lambda + \varkappa_c(B^*_{\lambda+1} + B^*_{\lambda-1}) + (G_\lambda - i\psi_\lambda)\,B^*_\lambda \tag{VII.11.6}$$

where

$$\Delta\omega_m = \Delta\omega - \omega_m \tag{VII.11.7}$$

represents the difference between the mode-spacing in the unloaded cavity and the frequency of the loss-modulation. The Eqs. (VII.11.6) are, in general, nonlinear in the field amplitudes, b^*, because the gain G_λ and frequency shift ψ_λ depends on the B's.

We consider the steady state.

Two solutions are known:

a) homogeneously broadened line: explicit solution if many modes are coupled together (HAKEN and PAUTHIER, l.c. [1] on p. 217),

b) inhomogeneously broadened line: numerical solution by means of a computer (McDUFF and HARRIS, l.c. [1] on p. 217).

We indicate the solution a), following HAKEN and PAUTHIER. The explicit expression for $(G_\lambda - i\psi_\lambda)$ may be found from Eqs. (VII.11.3a) and (VII.1.33), which yields, according to (VII.14.7)

$$i\sum_\mu g^*_{\mu\lambda}\varrho_{12,\mu} = \sum_{\mu\lambda'} \frac{g^*_{\mu\lambda} g_{\mu\lambda'}(N_{2\mu} - N_{1\mu})}{i(\Omega_{\lambda'} - \nu_\mu) + \gamma} b^*_{\lambda'}. \tag{VII.11.8}$$

We use running waves, so that the saturated inversion $d_\mu = (N_{2,\mu} - N_{1,\mu}) = d$ does not depend on the atomic position (x_μ) (see below).

Due to the orthogonality of the g's, the sum over λ' reduces to a single term:

$$\frac{d}{i(\Omega_\lambda - \nu_0) + \gamma} \sum_\mu |g_{\mu\lambda}|^2 b^*_\lambda = (G_\lambda - i\psi_\lambda)\, b^*_\lambda \tag{VII.11.9}$$

so that

$$G_\lambda = \frac{\gamma}{(\Omega_\lambda - \nu_0)^2 + \gamma^2} \overline{|g_{\mu\lambda}|^2}\, D, \tag{VII.11.10}$$

$$i\psi_\lambda = \frac{i(\Omega_\lambda - \nu_0)}{(\Omega_\lambda - \nu_0)^2 + \gamma^2} \overline{|g_{\mu\lambda}|^2}\, D \tag{VII.11.11}$$

where

$$\overline{|g_{\mu\lambda}|^2} = \frac{1}{N} \sum_\mu |g_{\mu\lambda}|^2, \tag{VII.11.12}$$

is a constant independent of λ, and $D = Nd$.

The saturated inversion $(N_2 - N_1)_\mu$ has the general form (VIII.1.12) for an N-level atom. We choose as a specific example a three-level atom with its upper transition being laser-active. $(N_2 - N_1)_\mu$ is then explicitly given by (VIII.1.30). (The indices 1, 2, 3 must be substituted by 0, 1, 2, respectively.)

$$d = (N_2 - N_1) = \frac{w_{02}}{w_{02} + w_2} \frac{1}{1 + R} \tag{VII.11.13}$$

where

$$R = \frac{1}{w_{02} + w_2} \sum_{\lambda'} W^0_{21,\lambda'} n_{\lambda'}; \qquad n_\lambda = |B_\lambda|^2. \tag{VII.11.14}$$

Note that $(N_2 - N_1)$ is independent of λ.

G_λ thus takes the form

$$G_\lambda = \frac{\gamma}{(\Omega_\lambda - \nu_0)^2 + \gamma^2} \frac{N w_{02}}{w_{02} + w_2} \frac{1}{1 + R}. \tag{VII.11.15}$$

Thus the only dependence of G_λ and ψ_λ on λ stems from the factors

$$\frac{1}{(\Omega_\lambda - \nu_0)^2 + \gamma^2} \quad \text{and} \quad \frac{(\Omega_\lambda - \nu_0)}{(\Omega_\lambda - \nu_0)^2 + \gamma^2} \tag{VII.11.16}$$

respectively.

We now proceed to the solution of the Eq. (VII.11.6). Assuming a steady state we put $\dot{B}^*_\lambda = 0$*.

---

* If $\omega_0 \neq \nu_0$ we would have to put $B^*_\lambda = e^{i\varphi t} \tilde{B}^*_\lambda$ with $\dot{\varphi} = \dot{\tilde{B}}^*_\lambda = 0$, but $\varphi \neq 0$.

Now let us anticipate that the loss modulation couples many modes together. In this case we may proceed as follows: We introduce a new function $B'(\lambda')$ which is continuous on λ' and which coincides with B_λ for the discrete set $\lambda' = \lambda$. In the following we will put λ as an argument whenever λ and B are treated as continuous variables, so that by dropping the primes in $B'(\lambda')$ no confusion will arise.

When many modes are coupled together, $B(\lambda)$ changes very little when λ is replaced by $\lambda \pm 1$, so that we may approximate as usual:

$$B^*(\lambda+1) + B^*(\lambda-1) \approx 2B^*(\lambda) + \frac{d^2 B^*(\lambda)}{d\lambda^2}. \tag{VII.11.17}$$

Note that the replacement (VII.11.17) holds to an excellent degree. If, for example, the total number of modes is 10, the error made by omitting the higher derivatives is of the order of one or two percent, as may be checked by using the explicit result (VII.11.29).

Our second approximation consists in expanding the saturated gain and the frequency pulling into a power series of λ

$$G_\lambda \approx G_0 + \lambda^2 G^{(2)} \quad \text{where} \quad G^{(2)} = -\left(\frac{\omega_m}{\gamma}\right)^2 G_0 \tag{VII.11.18}$$

according to (VII.11.10) and

$$\psi_\lambda \approx \lambda \psi^{(1)} \tag{VII.11.19}$$

according to (VII.11.11).

Note that both G_λ and Ψ_λ are functions of all field amplitudes, so that (VII.11.6) (VII.11.10) and (VII.11.11) imply a self-consistency requirement. It will be checked in detail on p. 223 and found justified for a series of experimental conditions. Under these assumptions the Eq. (VII.11.6) takes the form

$$\left[-\frac{1}{2}\frac{d^2}{d\lambda^2} + \frac{1}{2}\frac{\lambda^2}{\lambda_m^4} + i\Phi\lambda\right] B^*(\lambda) = W B^*(\lambda) \tag{VII.11.20}$$

where we have used the abbreviations*

$$\frac{1}{\lambda_m^4} = -\frac{G^{(2)}}{\varkappa_c} = \left(\frac{\omega_m}{\gamma}\right)^2 \frac{G_0}{\varkappa_c}, \tag{VII.11.21}$$

$$\Phi = \frac{1}{2\varkappa_c}\left[-\Delta\omega_m + \psi^{(1)}\right] \tag{VII.11.22}$$

and

$$W = \frac{G_0 - \varkappa - \varkappa_a}{2\varkappa_c} + 1. \tag{VII.11.23}$$

Eq. (VII.11.20) is that of a displaced quantum mechanical harmonic oscillator with the following correspondence:

$$\left.\begin{array}{lcl}
\text{harmonic oscillator:} & & \text{present case:} \\
\hbar = m = 1 & & \\
\text{frequency } \omega & \leftrightarrow & \dfrac{1}{\lambda_m^2} \\
\text{real displacement } x_0 & \leftrightarrow & \text{complex quantity } i\Phi\lambda_m^4 \\
\text{energy} & & \\
W = \omega\left(M + \dfrac{1}{2}\right) - \dfrac{x_0^2}{2}\omega^2 & \leftrightarrow & W = \dfrac{1}{\lambda_m^2}\left(M + \dfrac{1}{2}\right) + \dfrac{\Phi^2\lambda_m^4}{2}
\end{array}\right\} \tag{VII.11.24}$$

* As we will see below, λ_m^4 is a measure of the number of participating modes.

where M is as usual a positive integer starting with 0. Because the eigenvalues must have the form (VII.11.23), we find using (VII.11.24) the relation

$$\frac{G_0 - \varkappa - \varkappa_a}{\varkappa_c} + 2 = \frac{2}{\lambda_m^2}\left(M + \frac{1}{2}\right) + \Phi^2 \lambda_m^4. \qquad \text{(VII.11.25)}$$

When we express λ_m in Eq. (VII.11.25) by G_0 [compare Eq. (VII.11.21)] we obtain an equation for the saturated gain alone:

$$\frac{G_0 - \varkappa - \varkappa_a}{2\varkappa_c} + 1 = \left(M + \frac{1}{2}\right)\frac{\omega_m}{\gamma}\sqrt{\frac{G_0}{\varkappa_c}} + \frac{\Phi^2}{2}\left(\frac{\gamma}{\omega_m}\right)^2 \frac{\varkappa_c}{G_0}. \qquad \text{(VII.11.26)}$$

Note that this equation is also an equation for the number of participating modes*.

In order to demonstrate the dependence of G_0 or λ_m on the parameters of the system, we solve (VII.11.26) by an iteration technique, where the right hand side of Eq. (VII.11.26) is considered as a small quantity:

We find in the lowest order

$$G_0 = \varkappa + \varkappa_a - 2\varkappa_c \qquad \text{(VII.11.27)}$$

and in first order

$$\left.\begin{aligned} G_0 = \varkappa + \varkappa_a - 2\varkappa_c + 2\frac{\omega_m}{\gamma}\varkappa_c\left(\frac{\varkappa + \varkappa_a - 2\varkappa_c}{\varkappa_c}\right)^{\frac{1}{2}}\left(M + \frac{1}{2}\right) + \\ + \Phi^2 \varkappa_c \left(\frac{\gamma}{\omega_m}\right)^2 \frac{\varkappa_c}{\varkappa + \varkappa_a - 2\varkappa_c} \end{aligned}\right\} \qquad \text{(VII.11.27a)}$$

λ_m is determined by inserting the explicit expression of G_0 into (VII.11.21). Using (VII.11.27), we obtain in lowest order

$$\lambda_m = \left(\frac{\gamma}{\omega_m}\right)^{\frac{1}{2}}\left(\frac{\varkappa_c}{\varkappa + \varkappa_a - 2\varkappa_c}\right)^{\frac{1}{4}}. \qquad \text{(VII.11.28)}$$

From Eq. (VII.11.27a) it follows that, with increasing modulator driving strength $\varkappa_c$, a smaller gain per mode, G_0, results, but that the number of modes $\sim \lambda_m \sim (G_0)^{-\frac{1}{4}}$ increases, so that more modes are coupled together.

We now turn to the pulse shape. Because the field amplitudes are solutions of the equation of a displaced quantum mechanical harmonic oscillator, the solutions of (VII.11.20) are

$$B^*(\lambda') = C\, e^{\xi^2/2} H_M(\xi) \qquad \text{(VII.11.29)}$$

where

$$H_M(\xi) = \frac{(-1)^M}{2^{\frac{M}{2}} \pi^{\frac{1}{4}}} e^{\xi^2} \frac{d^M e^{-\xi^2}}{d\,\xi^M} \qquad \text{(VII.11.30)}$$

are Hermitian polynomials with

$$\xi = \frac{\lambda'}{\lambda_m}; \qquad \lambda' = \lambda + i\Phi\lambda_m^4. \qquad \text{(VII.11.31)}$$

There are a series of different mode configurations possible, according to $M = 0, 1, 2, 3, \ldots$.

The solution with the smallest pulse width is that for $M = 0$:

$$B^*(\lambda) = C\, e^{\frac{1}{2\lambda_m^2}(\lambda + i\Phi\lambda_m^4)^2} \qquad \text{(VII.11.32)}$$

C is a normalization constant which is determined by the saturated gain G_0**.

---

* See footnote* on p. 220.

** G_0 is fixed by the parameters of the system [see Eq. (VII.11.27a)]. On the other hand G_0 depends on C via the sum R [see (VII.11.15) and (VII.11.14)], in which $n_\lambda = b_\lambda^* b_\lambda \sim C^2$ occurs.

The somewhat lengthy calculation yields

$$|C|^2 = \frac{w_{02}+w_2}{W^0_{21,0}\,\gamma^2}\left(\frac{1}{2G_0}\,\frac{N\,w_{02}}{w_{02}+w_2}\,W^0_{21,0}-1\right)(\Sigma)^{-1} \tag{VII.11.33}$$

where we quote Σ for two special cases:

$$\Sigma = \frac{\lambda_m\,M!}{\gamma^2+\omega_m^2\,\lambda_m^2\,(M+\frac{1}{2})} \qquad \Phi=0, \tag{VII.11.34}$$

$$\Sigma = e^{\Phi^2\lambda_m^6}\,\frac{\lambda_m}{\gamma^2+\frac{1}{2}\,\omega_m^2\,\lambda_m^2} \qquad M=0. \tag{VII.11.35}$$

a) Pulse shape and pulse width. The electric field strength of the total pulse is proportional to

$$\sim e^{i\omega_0 t}\sum_\lambda B_\lambda^*\,e^{i\omega_m\lambda t} \tag{VII.11.36}$$

provided it is measured at a point where all $\boldsymbol{u}$'s [see (VII.1.6)] are equal.

From (VII.11.36) it is evident that the pulse repetition time is given by $t_0 = 2\pi/\omega_m$. When (VII.11.36) is converted into an integral, it can be easily evaluated to yield

$$E_{tot}\sim C\cdot e^{i\omega_0 t}\,e^{\Phi\lambda_m^4\omega_m t}\,i^M\,\lambda_m(2\pi)^{\frac{1}{2}}\,H_M(\omega_m t\lambda_m)\,e^{-\frac{1}{2}(\omega_m\lambda_m t)^2}. \tag{VII.11.37}$$

The quantities C, Φ, λ_m, H_m are given by the Eqs. (VII.11.33), (VII.11.22), (VII.11.21), (VII.11.30), respectively.

For $M=0$ we find a Gaussian distribution.

The result (VII.11.37) shows that other mode configurations ($M\geqq 1$) are also possible, which have an internal structure and a bigger pulse width. Which configuration is realized depends on the initial "preparation" of the pulse by spontaneous emission (which gives rise to random phases) and on their stability. We define the *pulse width* by

$$\overline{\Delta t} = \overline{\left((t-\bar{t})^2\right)^{\frac{1}{2}}} \tag{VII.11.38}$$

where the average is defined by

$$\overline{t^l} = \frac{A_l}{A_0} \tag{VII.11.39}$$

where

$$A_l = \int_{\text{single spike}} t^l\left|\sum_\lambda B_\lambda^*\,e^{i\omega_m t}\right|^2 dt, \qquad l=0,1,2,\ldots. \tag{VII.11.40}$$

The final result reads:

$$\overline{\Delta t} = \frac{1}{\sqrt{2}\,\lambda_m\,\omega_m}\left\{1+\frac{d^2}{d\eta^2}\ln\left(\sum_{\mu=0}^{M}\binom{M}{\mu}^2(M-\mu)!\,\eta^{2\mu}\right)\right\}^{\frac{1}{2}} \tag{VII.11.41}$$

where

$$\eta = \sqrt{2}\,\Phi\,\lambda_m^3. \tag{VII.11.42}$$

We quote two special cases of (VII.11.41):

1. $M=0$

$$\overline{\Delta t} = \frac{1}{\sqrt{2}\,\lambda_m\,\omega_m}. \tag{VII.11.43}^*$$

which is independent of the detuning (in the framework of our approximation)

* Note added in proof: Recent experiments by A. SIEGMANN and coworkers (private communication) substantiate (VII.11.43) with (VII.11.28).

2. $\Phi \to 0$: (no detuning)

$$\overline{\Delta t} = \frac{(M+\frac{1}{2})^{\frac{1}{2}}}{\lambda_m \omega_m}. \qquad \text{(VII.11.44)}$$

b) Discussion of the results and of the range of validity. The quantity of principal interest is, of course, the pulse width (VII.11.41). Because the general formula is rather complicated, we confine our discussion to the case of no detuning, $\Phi = 0$, (i.e. $\eta = 0$).

We have $\bar{t} = 0$, and $\overline{\Delta t}$ is given by (VII.11.44). Since λ_m is a measure of the number of phase-locked modes, and thus $\lambda_m \omega_m$ measures the total frequency width covered by these modes, we obtain the well-known result that the spread of the wave packet in time is inversely proportional to its width in frequency. As is seen from formula (VII.11.28), λ_m increases with increasing modulator driving strength (which agrees qualitatively with the results valid for an inhomogeneous line), but it decreases with detuning and higher mode configurations, $M \geqq 1$. $\overline{\Delta t}$ depends more strongly on M via $(M+\frac{1}{2})^{\frac{1}{2}}$ than via λ_m.

It should be noted that M cannot be bigger than the number of participating modes, because in a completely exact theory, where the modes are treated discontinuously via Eq. (VII.11.3), the number of new mode configurations is equal to that of the original modes which participate in the process. (For analogy see e.g. Debey's theory of the specific heat, where the lattice vibrations are treated by a continuum theory, but a cut-off is introduced so that the number of lattice modes equals the number of atoms).

The above theory can be expected to hold nearly exactly for low M, but it has only a qualitative meaning for M's which are about equal to the number of modes.

We now turn to the question of validity for low M's. Two approximations were made above.

1. The discontinuous index $\lambda = 0, \pm 1, \ldots$ was replaced by a continuous variable. This requires that $B(\lambda)$ changes only a little when λ is replaced by $\lambda + 1$. As is evident from Eqs. (VII.11.29), (VII.11.31) B depends on λ/λ_m and is a smooth function of its argument, at least for small M's. Therefore our approach is well justified for

$$\lambda_m > 1. \qquad \text{(VII.11.45)}$$

2. The expressions for the gain and the frequency pulling or pushing were expanded into a power series with respect to the frequency distance $\Omega_\lambda - \nu_0 \equiv \lambda \Delta\Omega$, or more exactly, to

$$\lambda \frac{\Delta\Omega}{\gamma}$$

so that

$$\lambda_m \frac{\Delta\Omega}{\gamma} < 1. \qquad \text{(VII.11.46)}$$

We must insert the expression (VII.11.28) for λ_m into (VII.11.45) and (VII.11.46). Combining these two inequalities and rearranging them we obtain

$$\left(\frac{\Delta\Omega}{\gamma}\right)^2 < \frac{\varkappa + \varkappa_a - 2\varkappa_c}{\varkappa_c} < \left(\frac{\gamma}{\Delta\Omega}\right)^2. \qquad \text{(VII.11.47)}$$

Since we must have at least two participating modes, $\Delta\Omega/\gamma < 1$. Because $\varkappa_c < \varkappa_a, \varkappa$, the left-hand side of (VII.11.47) is fulfilled. The right-hand side is equivalent to the postulate

$$\varkappa_c > \frac{\varkappa + \varkappa_a}{2 + \left(\frac{\gamma}{\Delta\Omega}\right)^2}. \qquad \text{(VII.11.48)}$$

E.g. for $(\gamma/\Delta\Omega)=10$ this inequality requires that the additional loss $\varkappa_c$, introduced by the loss modulation, is only bigger than a few per cent of the total loss.

The most striking result of the analysis is the occurrence of a variety of "supermodes" with higher pulse width than the fundamental new mode configurations $M=0$. The important question arises whether the higher "supermodes" are peculiar to the homogeneously broadened line, or whether they can also be expected for the inhomogeneously broadened line. Because the procedure is rather general it is believed that the phenomenon of "supermodes" should also occur in inhomogeneously broadened lasers.

c) Numerical application. The best known example of a homogeneously broadened line is ruby (above 77° K). At room temperature its linewidth is about 4 Å, from which we obtain for the linewidth γ an order of magnitude of $10^{12}\,\text{sec}^{-1}$. For an external cavity of 100 cm length the frequency separation of adjacent axial modes is $\Delta\Omega\approx 10^9\,\text{sec}^{-1}$.

The critical ratio $\frac{\Delta\Omega}{\gamma}$ is then of the order of 10^{-3}. From the above consideration it follows that the additional loss $\varkappa_c$, introduced by the loss modulation need only be bigger than 10^{-6} times the total loss in order for our theory to be applicable, which is always achieved in practical cases.

VII.12. Super-radiance[1]. Spin and photo echo.

a) Definition of super-radiant states. The word super-radiance is used nowadays with two different meanings:

1. If the gain of a laser is so high that no resonator with reflecting ends is needed, one obtains a "super-radiant" emission.

In such an experiment the atoms are (as usual in lasers) incoherently excited by the pump. If we visualize the atoms as dipoles, the pump creates dipoles in random directions.

2. A coherent field can cause the atomic dipoles to rotate in phase at the beginning. The dipoles can be (magnetic) spins as well. The corresponding phenomena were actually first observed in spin resonance experiments[2]. These in-phase dipoles act as a macroscopic dipole, which causes a coherent emission of radiation. The term "super-radiance" has been coined by DICKE[1] for the latter situation (2).

Because case 2) contains some aspects which are of basic importance (mainly for the spin and photo echo), we present this problem in more detail, although case 1) is also important, especially for applications.

We base our analysis on a *system of two-level atoms* or a *system of $\frac{1}{2}$ spins* in a constant magnetic field $\boldsymbol{H}_0$ and use the fact that the corresponding Hamiltonians have exactly the same mathematical properties (compare III.6).

The Hamiltonian of the spin-system is composed of a sum of single-spin Hamiltonians (III.6.5) and of that of the free electromagnetic field. We take into account only a single mode which represents a running wave (or a standing

---

[1] R. H. DICKE: Phys. Rev. **93**, 99 (1954), whom we will follow especially in the first part, has given a quantum mechanical treatment. This phenomenon is also known as radiation damping effect and has been treated classically: N. BLOEMBERGEN and R. V. POUND: Phys. Rev. **95**, 8 (1954); — Various aspects of radiation damping have been treated by S. BLOOM: J. Appl. Phys. **28**, 800 (1957). — C. GREIFINGER and G. BIRNBAUM: IRE Trans. Electron. Devices **6**, 288 (1959). — R. M. BEVENSEE: Proc. IEEE (Corres.) **51**, 215 (1963). — J. R. SINGER, and W. WANG: Phys. Rev. Letters **6**, 351 (1961). — J. C. KEMP: Phys. Rev. Letters **7**, 21 (1961). — A. YARIV: J. Appl. Phys. **31**, 740 (1960). — J. C. KEMP: J. Appl. Phys. **30**, 1451 (1959).

[2] For details see A. ABRAGAM: Principles of Nuclear Magnetism. London: Oxford Univ. Press 1961.

wave with infinite wavelength). As has been shown in VI.4, the spatial dependence of the coupling coefficients between the field and spins can be eliminated by the transformation (VI.4.6). We start therefore directly with

$$H = \hbar\,\omega\, b^+ b + \hbar \nu_0 \sum_\mu \sigma_{z,\mu} + \hbar g\, b^+ \sum_\mu \sigma_\mu^- + \hbar g\, b \sum_\mu \sigma_\mu^+ . \qquad \text{(VII.12.1)}$$

We denote further

$$\hbar\, \nu_0 \sum_\mu \sigma_{z,\mu} \quad \text{by} \quad H^0 . \qquad \text{(VII.12.2)}$$

The eigenfunctions of H^0 can be represented as a product of eigenfunctions of $\hbar \nu_0 \sigma_{z,\mu}$ (with all μ's different). These eigenfunctions are those in which the spin μ is pointing up (in z-direction) (eigenvalue $\frac{1}{2}\hbar\nu_0$) or down (eigenvalue $-\frac{1}{2}\hbar\nu_0$). The total eigenfunction with spin up at $\mu=1$, spin down at $\mu=2$ etc. is denoted by

$$\Psi = (\uparrow\downarrow\cdots) . \qquad \text{(VII.12.3)}$$

This function is highly degenerate: When there are N spins, of which N_+ are up and N_- down, the degree of degeneracy is

$$\frac{N!}{N_+!\,N_-!} . \qquad \text{(VII.12.4)}$$

This degeneracy can be eliminated by a weak interaction or an initial perturbation. We explain the effect of this lifting on the radiation first by considering only two spins. There are four states:

$$\underset{(1)}{\uparrow\uparrow}, \quad \underset{(2)}{\uparrow\downarrow}, \quad \underset{(3)}{\downarrow\uparrow}, \quad \underset{(4)}{\downarrow\downarrow} . \qquad \text{(VII.12.5)}$$

The states (1), (2) and (3) can radiate. Particularly (2) and (3) have the same transition rate to the ground state $= C|M|^2$, where $M = g\langle\downarrow\sigma^-\uparrow\rangle$ is the optical transition matrix element. When the degeneracy between (2) and (3) is removed, a symmetric and an antisymmetric wave function emerge, which we denote symbolically by:

$$\Psi_s = \frac{1}{\sqrt{2}}(\uparrow\downarrow + \downarrow\uparrow) \quad \text{and} \quad \Psi_a = \frac{1}{\sqrt{2}}(\uparrow\downarrow - \downarrow\uparrow) . \qquad \text{(VII.12.6)}$$

The optical transition matrix elements with the wave functions (VII.12.6) read

$$(S)\; g\langle(\downarrow\downarrow)(\sigma_1^- + \sigma_2^-)\frac{1}{\sqrt{2}}(\uparrow\downarrow + \downarrow\uparrow)\rangle \quad \textit{for the symmetric initial state} \qquad \text{(VII.12.7)}$$

and

$$(A)\; g\langle(\downarrow\downarrow)(\sigma_1^- + \sigma_2^-)\frac{1}{\sqrt{2}}(\uparrow\downarrow - \downarrow\uparrow)\rangle \quad \textit{for the antisymmetric initial state} . \qquad \text{(VII.12.8)}$$

Because σ^- flips a spin from its up directions to its down direction and gives zero if applied to a spin in the down direction we find for the transition probability

$$(S) \qquad W = C\left|\frac{2M}{\sqrt{2}}\right|^2 = 2|M|^2 C \qquad \text{(VII.12.9)}$$

(with the same constant C as above)

$$(A) \qquad W = 0 . \qquad \text{(VII.12.10)}$$

Thus there is either no radiation at all ("light trapping"), or the emission rate is doubled in a single step.

In the following we wish to construct especially those states of the N-spin system, which give a particularly high emission rate.

To this end we construct those eigenfunctions of H^0 which are simultaneously eigenfunctions of the "total spin" R^2, where

$$R^2 = R_x^2 + R_y^2 + R_z^2 \quad \text{with} \quad \sum_\mu \sigma_\mu^\pm = R^\pm$$

$$R_z = \sum_\mu \sigma_{z,\mu}, \quad R_x = \frac{1}{2} \sum_\mu (\sigma_\mu^+ + \sigma_\mu^-), \quad R_y = \frac{1}{2i} \sum_\mu (\sigma^+ - \sigma^-). \quad \text{(VII.12.11)}$$

The procedure is well known from spin theory (composition of spins), so that we remind the reader only briefly of the method: R^2 possesses the eigenvalues $r(r+1)$ where r is a half integer, which runs from $\frac{1}{2}$ to $\frac{N}{2}$ (if N is odd), or an integer, $0 \leqq r < \frac{N}{2}$, if N is even. R_z has the eigenvalues m, where m is also a half integer (or integer) with $|m| \leqq r$.

The construction runs as follows: we start with the state $(\uparrow\uparrow \cdots \uparrow)$ (all spins up) which is an eigenfunction of R^2 and R_z with $r = \frac{N}{2}$, $m = \frac{N}{2}$. By application of R^- on $(\uparrow \cdots \uparrow)$ we obtain a new eigenfunction with $r = \frac{N}{2}$, $m = \frac{N}{2} - 1$, which is given by a symmetric superposition of functions in which only one spin is flipped.

$$\Psi_{\frac{N}{2}, \frac{N}{2}-1} = (\downarrow\uparrow \cdots \uparrow) + (\uparrow\downarrow\uparrow \cdots \uparrow) + \cdots (\uparrow \cdots \uparrow\downarrow\uparrow \cdots \uparrow) + \cdots (\downarrow \cdots \downarrow\uparrow). \quad \text{(VII.12.12)}$$

There are, however, N such functions with only one spin flipped so that we can form $(N-1)$ further linear combinations, which are chosen to be orthogonal to (VII.12.12). It can be shown that these $(N-1)$ functions are again eigenfunctions of R^2 and R_z, but with $r = \frac{N}{2} - 1$, $m = \frac{N}{2} - 1$. These will be distinguished by a further index k. This procedure can now be continued by applying R^- on these new functions and so on. A list of these possibilities is given in Fig. 53.

Most important for our present analysis is the following selection rule:

$$R^\pm \Psi_{r,m} = \sqrt{(r \mp m)(r \pm m + 1)}\, \Psi_{r,m\pm 1}. \quad \text{(VII.12.13)}$$

(Note that only functions with the same degeneracy index k are connected. Consequently we drop k.)

The optical matrix element is determined by:

$$M = \hbar g \langle \Psi_{r',m'}^*\, R^\pm\, \Psi_{r,m} \rangle = \hbar g \sqrt{(r \mp m)(r \pm m + 1)}\, \delta_{m',m\pm 1}\, \delta_{rr'}. \quad \text{(VII.12.14)}$$

We now determine the spontaneous emission rate in analogy to the single atom (compare V.3).

$$\hbar g \sum_\mu \sigma_\mu^- \cdot b^+ = \hbar g\, R^- b^+ \quad \text{(VII.12.15)}$$

is the interaction Hamiltonian responsible for the emission of a single photon. Using (VII.12.14) we find for the spontaneous emission rate

$$I = I_0 (r + m)(r - m + 1) \quad \text{(VII.12.16)}$$

where I_0 equals the spontaneous emission rate of a single atom $(r = m = \frac{1}{2})$ into a single mode (we do not treat here a continuum of modes).

For $r = m = N/2$ we find

$$I = N I_0 \qquad \text{(VII.12.17)}$$

i.e. all atoms radiate with no interference effect. If, however, $r \approx N/2$ but $m \ll N/2$:

$$I \sim I_0 N^2. \qquad \text{(VII.12.18)}$$

The states which lead to a radiation rate of the form (VII.12.18) are called the superradiant states. Because their emission rate goes with N^2, it must be due to

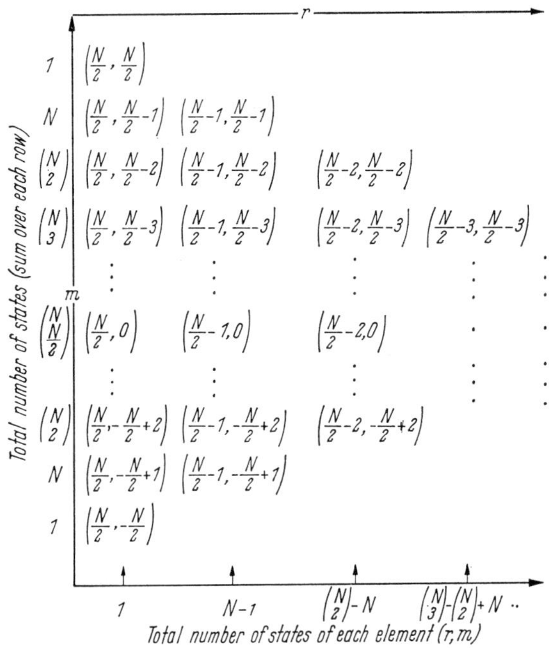

Fig. 53. A list of all possible states (r, m). The last row gives the multiplicity of each element e.g. $\left(\frac{N}{2}-2, \frac{N}{2}-2\right)$ has the multiplicity $\binom{N}{2}-N$, i.e. there are $\binom{N}{2}-N$ different states with $r=\frac{N}{2}-2$, $m=\frac{N}{2}-2$.

a cooperative ("in phase") emission. These states represent indeed a macroscopic spin, which rotates in the x, y-plane. This high degree of internal coherence of the spin system is reflected by the high symmetry of the corresponding wave-functions. For instance, the $(r = N/2, m = 0)$-wave function reads

$$(R^-)^{N/2}\, \Psi(\uparrow \cdots \uparrow) \sim \Sigma \Psi(\uparrow\uparrow \cdots \uparrow\downarrow\downarrow \cdots) \qquad \text{(VII.12.19)}$$

where the sum runs over *all permutations with $N/2$ up- and $N/2$-down spins*. The above emission process (VII.12.18) leads after each emission of a light quantum to a state of essentially the same structure so that a whole cascade of light quanta is emitted. As we will see below, the process can also be described classically which is an even more adequate description than that of a single light quantum emission.

It is important to note that the above phenomenon has nothing to do with stimulated emission. Indeed, the net gain in the stimulated emission process is determined by the genuine stimulated emission rate minus the absorption rate which are both proportional to the number n of photons present. From (VII.12.13) we obtain immediately

$$\left.\begin{aligned} &|\langle \Psi^*_{r,m-1} R^- \Psi_{r,m}\rangle|^2 - |\langle \Psi^*_{r,m+1} R^+ \Psi_{r,m}\rangle|^2 \\ &\qquad \sim (r+m)(r-m+1) - (r-m)(r+m+1) = N_+ - N_- \end{aligned}\right\} \tag{VII.12.20}$$

i.e. the usual expression for the net gain in stimulated emission. Or in other words: the new states do not change the rate.

b) Generation of super-radiant states. We first seek how we can mathematically construct states of the structure (VII.12.19) by means of functions which represent suitably oriented *single* spins. To this end we choose a wavefunction which describes a spin (μ) oriented in the x—y-plane, which has as a wave function

$$\frac{1}{\sqrt{2}}\,((\uparrow)+(\downarrow))_\mu. \tag{VII.12.21}$$

We assume that all spins have the same direction, so that their total wave function reads

$$\prod_\mu \left\{\frac{1}{\sqrt{2}}\,((\uparrow)+(\downarrow))_\mu\right\}. \tag{VII.12.22}$$

When we evaluate the product we find a linear combination of functions $[\uparrow\uparrow\cdots\downarrow\downarrow\cdots\uparrow\downarrow\cdots\uparrow]$ all having the same relative phase.

Consider for instance $N=3$: Then

$$\prod_\mu \{\ldots\} \sim (\uparrow\uparrow\uparrow) + \sqrt{3}\left(\frac{1}{\sqrt{3}}\,(\downarrow\uparrow\uparrow + \uparrow\downarrow\uparrow + \uparrow\uparrow\downarrow)\right) + \cdots \tag{VII.12.23}$$

where $1/\sqrt{3}$ takes care of the normalization.

When there are N_+ spins up and N_- spins down, this normalization coefficient is $\left(\sqrt{\frac{N!}{N_+!\,N_-!}}\right)^{-1}$, so that the probability of finding a state with N_+ spins up is given by $\frac{N!}{N_+!\,N_-!}$ which has a pronounced peak for $N_+ = N_- = \frac{N}{2}$, so that we in fact obtain the super radiant state (VII.12.19). The whole analysis remains valid, if we use the correct time-dependent spin states:

$$\frac{1}{\sqrt{2}}\left\{(\uparrow)\,e^{-i\frac{\nu_0}{2}t+i\delta} + (\downarrow)\,e^{+i\frac{\nu_0}{2}t}\right\} \tag{VII.12.24}$$

provided that phase δ is the same for all spins. We then find an ensemble of spins rotating in phase.

The *generation of states* (VII.12.24) can be achieved *experimentally* as shown by the following *theoretical* considerations:

α) Classical treatment of the spin motion[3]. We first have a static magnetic field $\boldsymbol{H}_0$ in the z-direction, so that the spin precesses around the z-axis with an angular frequency

$$\nu_0 = \frac{e}{2mc}\,H_0. \tag{VII.12.25}$$

[3] See footnote [2] on p. 224.

We now add a rotating field $\boldsymbol{H}=(H\cos\nu t, H\sin\nu t, 0)$ where we choose $\nu=\nu_0$. In a rotating coordinate system moving with a frequency ν_0, $\boldsymbol{H}$ appears constant. The equation of motion for the angular momentum $\boldsymbol{J}=\boldsymbol{J}_{\mathrm{spin}}$ then reads:

$$\frac{d\boldsymbol{J}}{dt}=(\boldsymbol{M}\times\boldsymbol{H})_{\mathrm{rot}}. \tag{VII.12.26}$$

$\boldsymbol{M}$: magnetic moment of the spin.

According to (VII.12.26) the spin is being erected. If the magnetic field is switched off at the right moment (see below), the spin is exactly in its horizontal position ("a 90° pulse has been given to the spin"). As we will show in the next section, Eq. (VII.12.26) is not only a classical one but is also valid for the *spin operator*.

β) Quantum theoretical treatment[4]. We now treat a single spin and assume that it is coupled to a strong driving field so that we can neglect its interaction with the mode formerly considered. The SCHRÖDINGER equation reads:

$$\{\hbar\nu_0\sigma_z+\hbar g\sigma^+F_0e^{-i\omega t}+\hbar g^*\sigma^-F_0^*e^{i\omega t}\}\Psi=i\hbar\dot{\Psi} \tag{VII.12.27}$$

where $F_0e^{-i\omega t}$ represents the external, time dependent magnetic field. We choose $\omega=\nu_0$. Inserting

$$\Psi=e^{+i\frac{\nu_0}{2}t}c_1(t)\cdot(\downarrow)+e^{-i\frac{\nu_0}{2}t}c_2(t)(\uparrow) \tag{VII.12.28}$$

into (VII.12..27) yields two differential equations for c_1, c_2:

$$\left.\begin{aligned} g c_1 F_0 &= i\dot{c}_2,\\ g^* c_2 F_0^* &= i\dot{c}_1. \end{aligned}\right\} \tag{VII.12.29}$$

We assume that at the beginning ($t=0$) the spin points downwards, so that

$$c_1(0)=1,\qquad c_2(0)=0. \tag{VII.12.30}$$

The solution of (VII.12.29) is then given by

$$\left.\begin{aligned} c_2 &= e^{i\varphi_0}\sin\nu t,\\ c_1 &= \cos\nu t \end{aligned}\right\} \tag{VII.12.31}$$

where $\nu=|gF_0|$ and $e^{i\varphi_0}=-iF_0g/\nu$.

The normalized solution of (VII.12.27) reads therefore

$$\Psi(t)=\left\{e^{-i\frac{\nu_0}{2}t}e^{i\varphi_0}\sin(\nu t)(\uparrow)+e^{+i\frac{\nu_0}{2}t}\cos(\nu t)(\downarrow)\right\}. \tag{VII.12.32}$$

When the field F_0 is switched off at $\nu t_0=\pi/4$ we obtain precisely the solution (VII.12.24) required for the super-radiant state.

Another way of treating the spin motion can be done in the Heisenberg picture. We first proceed from the Schrödinger representation of Eq. (VII.12.27) to the *interaction* representation which removes the term with σ_z and replaces $F_0e^{-i\omega t}$ by $F_0e^{-i\omega t}e^{+i\nu_0 t}$. The equations of motion for the spin flip operators

[4] See e.g. D. J. BLOCHINZEW: Foundations of quantum mechanics. Moscow-Leningrad 1949.

σ^+, σ^- and σ_z are then given by

$$\left.\begin{aligned} \dot{\sigma}^- &= i g F(t)\, 2\sigma_z, \\ \dot{\sigma}^+ &= -i g^* F^*(t)\, 2\sigma_z, \\ \dot{\sigma}_z &= i\left(\sigma^- g^* F^*(t) - \sigma^+ g F(t)\right) \end{aligned}\right\} \tag{VII.12.33}$$

where

$$F(t) = F_0\, e^{-i\omega t + i\nu_0 t}. \tag{VII.12.34}$$

We first consider $F(t)$ as a completely general function of the time and then specialize it in an appropriate way. Instead of the flip operators we may introduce the x and y components of the spin operators by the relations

$$\left.\begin{aligned} \sigma_x &= \frac{1}{2}(\sigma^+ + \sigma^-), \\ \sigma_y &= \frac{1}{2i}(\sigma^+ - \sigma^-). \end{aligned}\right\} \tag{VII.12.35}$$

Using these relations we may rewrite Eqs. (VII.12.33) in the form

$$\left.\begin{aligned} \dot{\sigma}_x &= i(gF - g^*F^*)\sigma_z, \\ \dot{\sigma}_y &= -(gF + g^*F^*)\sigma_z, \\ \dot{\sigma}_z &= -i(gF - g^*F^*)\sigma_x + \sigma_y(gF + g^*F^*). \end{aligned}\right\} \tag{VII.12.36}$$

We now assume g *real* and put

$$\boldsymbol{J} = \{\sigma_x, \sigma_y, \sigma_z\}\hbar \quad \text{and} \quad \boldsymbol{M} = g\boldsymbol{J}. \tag{VII.12.37}$$

We find exactly the form (VII.12.26) *if we make the following identifications* [see also (III.6.5)]

$$\left.\begin{aligned} H_x(t) &= -\left(F^*(t) + F(t)\right), \\ H_y(t) &= i\left(F^*(t) - F(t)\right) \end{aligned}\right\} \tag{VII.12.38}$$

where H_x and H_y refer to the rotating frame. This becomes particularly clear if we decompose F_0 of Eq. (VII.12.34) into a real amplitude and a phase factor:

$$F_0 = f_0(t)\, e^{-i\varphi_0(t)} \tag{VII.12.39}$$

so that

$$\left.\begin{aligned} H_x &= -2f_0(t)\cos[(\omega - \nu_0)t + \varphi_0(t)], \\ H_y &= -2f_0(t)\sin[(\omega - \nu_0)t + \varphi_0(t)]. \end{aligned}\right\} \tag{VII.12.40}$$

In contrast to the treatment in Eqs. (VII.12.27)—(VII.12.32), $f_0(t)$ may now depend on the time.

Let us now treat the special case where the magnetic field rotates at the same speed as the rotating frame, which means that the vector (H_x, H_y) has a constant direction. This requires $\nu_0 = \omega$ and $\varphi_0(t) = \text{const}$. We choose $\varphi_0 = 0$ so that $H_y = 0$.

In order to discuss the Eqs. (VII.12.36), we convert them into classical ones [see (VII.12.26)] by taking expectation values. We put

$$\langle\sigma_z\rangle = \tfrac{1}{2}\cos\Theta \tag{VII.12.41}$$

where Θ is the angle between the direction of the expectation value of the spin vector and the z-axis. A simple analysis shows that one is allowed to put

$$\langle\sigma_x\rangle = 0 \tag{VII.12.42}$$

and

$$\langle\sigma_y\rangle = \tfrac{1}{2}\sin\Theta \tag{VII.12.43}$$

provided (VII.12.42) is fulfilled at an initial time. Inserting (VII.12.41), (VII.12.42) and (VII.12.43) into the averaged Eqs. (VII.12.36), we directly obtain an equation for Θ alone in the form

$$\dot{\Theta} = -2g f_0 = -g(F^* + F) = g H_x. \tag{VII.12.44}$$

From (VII.12.44) we can deduce the angle by which the spin is rotated under the action of the external field during the time from t_0 to t_1:

$$\left.\begin{aligned}\Theta(t_1) - \Theta(t_0) &= -g\int_{t_0}^{t_1} 2f_0\,dt = -g\int_{t_0}^{t_1}(F^* + F)\,dt\\ &= g\int_{t_0}^{t_1} H_x\,dt.\end{aligned}\right\} \tag{VII.12.45}$$

We finally discuss the relation between the angle Θ and the coefficients c_j of the wave function determined in the Schrödinger representation above. To this end we evaluate the expectation value of σ_z with respect to the wave function (VII.12.28) which yields immediately

$$\langle\sigma_z\rangle = \tfrac{1}{2}(|c_2|^2 - |c_1|^2) \tag{VII.12.46}$$

which according to (VII.12.41)

$$= \tfrac{1}{2}\cos\Theta. \tag{VII.12.47}$$

It should be noted that all the relations hold also if averages over statistical ensembles are taken, because these equations are all linear.

c) Classical description of super-radiant emission[5]. We treat R and R_z as classical variables. The energy represented by H^0 can then be written in the form

$$E = E_0 R_z = E_0 R\cos\Theta \tag{VII.12.48}$$

where Θ is the angle (R, R_z).

The decrease of energy per second, $E_0 R\sin\Theta\cdot\dot{\Theta}$, must be equal to the radiated energy which, according to (VII.12.16), is

$$I = I_0(R + R_z)(R - R_z + 1) \approx I_0 R^2\sin^2\Theta \tag{VII.12.49}$$

so that

$$E_0\dot{\Theta} = I_0 R\sin\Theta$$

with the solution

$$\sin\Theta = \operatorname{sech}(\alpha t); \qquad \alpha = I_0 R/E_0 \tag{VII.12.50}$$

which fulfills the initial condition $\Theta = 90°$ for $t = 0$ (i.e. a super radiant initial state).

d) The spin-echo experiment[6]. We apply first a 90°-impulse to the spin system so that all spins start a precession in the x–y-plane, forming a super radiant

[5] Compare R. H. Dicke: Footnote [1] on p. 224.
[6] R. L. Hahn: Phys. Rev. **80**, 580 (1950).

state. The time of their coherent radiation is limited by the radiation itself, but also by the different speed of precession of the single spins, i.e. by different ν_0's. So far we have neglected any homogeneous or inhomogeneous line broadening which cause a spread of ν_0's. Let us assume that the line is essentially inhomogeneously broadened so that each spin (μ) possesses its individually fixed frequency $\nu_{0,\mu}$ out of an interval $\Delta\nu_0$. Then after a time $t_0 \sim \frac{1}{\Delta\nu_0}$ the spins are completely out of phase and the emission stops.

If a 180° pulse is applied at t_1 the direction of each spin is inverted, but all still precess in the same direction. As can be seen by studying Fig. 54, after the

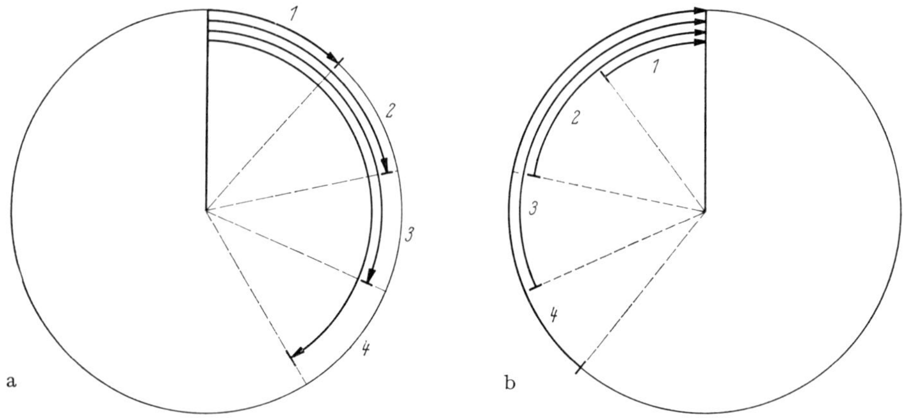

Fig. 54a and b. After application of a 90° pulse the spins show all in the same direction (in the present picture in the vertical one). Due to their different speeds of precession, caused by different magnetic fields, they reach after a time t the indicated positions. The 180° pulse inverts the spins which are still precessing in the same direction. Thus after the same time t they come again into a completely inphase position (Fig. b). A more elementary explanation which would correspond to inverting the static magnetic field is the following: After a time t the spins are caused to precess in the opposite direction, thus meeting again after a time t completely in phase in the vertical direction.

same time t_1 they come again to a completely in phase position and once again start to radiate strongly. The time t' in which such a return to the same initial state can be achieved is, however, limited by the homogeneous broadening $\Delta\nu_h$ which causes an irreversible mixture of phases $\left(\text{so that } t' < \frac{1}{\Delta\nu_h}\right)$.

e) The photo-echo experiment[7]. In order to explain why the spin-echo experiment possesses an analogue in the optical range, we have to exploit the formal analogy between a spin system and a two-level system (compare III.6) more closely. In order to make things as transparent as possible we assume that the two-level atom possesses a lower state with *s*- and an upper state with *p*-symmetry.

The ground state Φ_0 thus describes the *s*-state and the state $a_2^+ a_1 \Phi_0$ the *p*-state.

A superposition $\Phi = c_1 \Phi_0 + c_2 a_2^+ a_1 \Phi_0$ describes a state as indicated in Fig. 55c.

[7] N. A. Kurnit, I. D. Abella, and R. S. Hartmann: Phys. Rev. Letters **13**, 567 (1964);— Phys. Rev. **141**, 391 (1966). — We follow here a presentation of our lecture about coherence at Stuttgart, Summer 1964. According to C. L. Tang and H. Statz: Appl. Phys. Letters **10**, 145 (1967), another well-known nuclear magnetic resonance effect — the transient nutation effect — possesses an optical analog.

We can now establish the following correspondence:

Spin direction	Spin function	Electron wave function (in the notation of second quantization)	Electron wave function in Schrödinger's notation	Spatial distribution
$\downarrow$	$(\downarrow)$	$\boldsymbol{\Phi}_0$	$\varphi_1(\boldsymbol{x})$	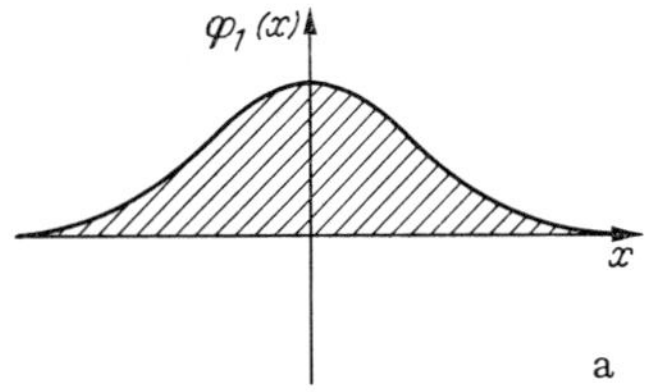 a
$\uparrow$	$(\uparrow)$	$a_2^+ a_1 \boldsymbol{\Phi}_0$	$\varphi_2(\boldsymbol{x})$	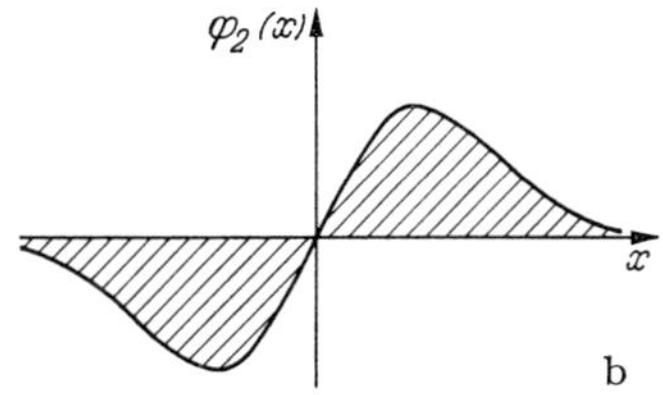 b
$\rightarrow$	$c_1(\downarrow)+c_2(\uparrow)$ $\lvert c_1\rvert^2=\lvert c_2\rvert^2$	$c_1\boldsymbol{\Phi}_0+c_2 a_2^+ a_1 \boldsymbol{\Phi}_0$	$c_1\varphi_1(\boldsymbol{x})+c_2\varphi_2(\boldsymbol{x})$	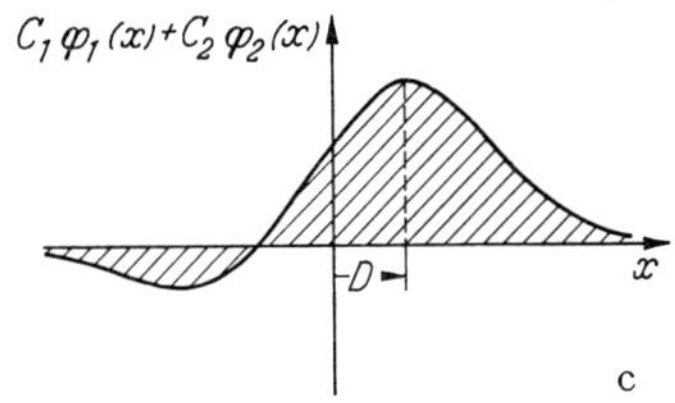 c

Fig. 55a—c.

The last state is connected with an electric dipole moment, as is the spin state with a magnetic dipole moment in the x-direction. When the relative phase between c_1 and c_2 changes, the spin rotates in the x—y-plane. Similarly the electric dipole moment changes its size and finally reverses its sign. Because the phase difference goes with $\nu_0 t$, the dipole moment oscillates in x-direction with the same frequency. While ν_0 is prescribed in the spin case by the constant magnetic field $\boldsymbol{H}_0$ in z-direction, it is given in the atomic case by the atomic excitation frequency $\nu_0=\frac{E_2-E_1}{\hbar}$. In order to generate a super radiant state, one has to excite the atoms first with a strong, coherent light-field which is in resonance with the atomic transition. In order to achieve a 90° pulse it must be switched off after a time $t_0=\frac{\pi}{4|g F_0|}$. With the aid of (III.3.3), (III.3.5), (III.6.4) we determine $F_0 \hbar g e^{-i\omega t}$ to be

$$F_0 \hbar g e^{-i\omega t}=-\frac{e}{m c}\int \varphi_2^* A_x^{(+)}(t)\, \boldsymbol{p}_x\, \varphi_1\, dV \qquad \text{(VII.12.51)}$$

with $\boldsymbol{A}^{(+)}(t)=\boldsymbol{A}^{(+)}(0)\, e^{-i\omega t}$.

The vector potential $\boldsymbol{A}(t)=\boldsymbol{A}^+(0)\, e^{-i\omega t}+\text{c.c.}$ is treated as a classical field. When we evaluate (VII.12.51) in the dipole approximation, using the relation in footnote* on p. 28, and $\boldsymbol{E}^{(+)} = -\frac{1}{c}\dot{\boldsymbol{A}}^{(+)} = \frac{i\omega}{c}\boldsymbol{A}^{(+)}$, we obtain

$$F_0 \hbar g e^{-i\omega t} = -\boldsymbol{\vartheta}_{21} \boldsymbol{E}^{(+)}, \tag{VII.12.52}$$

where $\boldsymbol{\vartheta}_{21}$ is the atomic electric dipole matrix element and $\boldsymbol{E}^{(+)}$ the positive frequency part of the electric field.

After this 90° pulse the electric dipoles start to oscillate in phase, thus strongly emitting light. Due to the rather broad linewidth, they quickly come out of phase and must be put back in phase by a 180° pulse.

In the optical range the homogeneous linewidths are unfortunately large so that the time intervals between the 180° pulse must be very small.

f) A further analogy between a spin $\frac{1}{2}$ system and a two-level system: the fictitious spin. The formal relation between these two systems, which has been exhibited in Chap. III.6 and in the foregoing section, can be used to transcribe the results of Sect. *b) β)* to a two-level atom. Starting from the formal correspondence (III.6.1):

$$\sigma^+ \leftrightarrow a_2^+ a_1, \qquad \sigma^- \leftrightarrow a_1^+ a_2, \qquad \sigma_z \leftrightarrow \tfrac{1}{2}(a_2^+ a_2 - a_1^+ a_1) \tag{VII.12.53}$$

we introduce in analogy to (VII.12.35), but in a completely formal way the operators

$$\left.\begin{aligned} I_x &= \frac{1}{2}(a_2^+ a_1 + a_1^+ a_2), \\ I_y &= \frac{1}{2i}(a_2^+ a_1 - a_1^+ a_2), \\ I_z &= \frac{1}{2}(a_2^+ a_2 - a_1^+ a_1) \end{aligned}\right\} \tag{VII.12.54}$$

which we might call fictitious spin operators. We now observe that the equations of motion for the σ's and the $(a_i^+ a_k)$'s have exactly the same form (see III.6). Thus we can immediately express the Eqs. (VII.12.36) in terms of the fictitious spin operators (VII.12.54). Using the correspondence (VII.12.52) and the relation (VII.12.34) we obtain instead of (VII.12.36)

$$\left.\begin{aligned} \dot{I}_x &= -\frac{i}{\hbar}\left(\boldsymbol{\vartheta}_{21}\boldsymbol{E}^{(+)}(t)\, e^{i\nu_0 t} - \boldsymbol{\vartheta}_{12}\boldsymbol{E}^{(-)}(t)\, e^{-i\nu_0 t}\right) I_z, \\ \dot{I}_y &= \frac{1}{\hbar}\left(\boldsymbol{\vartheta}_{21}\boldsymbol{E}^{(+)}(t)\, e^{i\nu_0 t} + \boldsymbol{\vartheta}_{12}\boldsymbol{E}^{(-)}(t)\, e^{-i\nu_0 t}\right) I_z, \\ \dot{I}_z &= \frac{i}{\hbar}\left(\boldsymbol{\vartheta}_{21}\boldsymbol{E}^{(+)}(t)\, e^{i\nu_0 t} - \boldsymbol{\vartheta}_{12}\boldsymbol{E}^{(-)}(t)\, e^{-i\nu_0 t}\right) I_x \\ &\quad - \frac{1}{\hbar}\left(\boldsymbol{\vartheta}_{21}\boldsymbol{E}^{(+)}(t)\, e^{i\nu_0 t} + \boldsymbol{\vartheta}_{12}\boldsymbol{E}^{(-)}(t)\, e^{-i\nu_0 t}\right) I_z. \end{aligned}\right\} \tag{VII.12.55}$$

In (VII.12.55) the factors $e^{i\nu_0 t}$ and $e^{-i\nu_0 t}$ occur because we use the rotating frame or, in the language of usual quantum theory, because we use the interaction representation. In analogy to (VII.12.38) we introduce "effective" electric fields by the definitions

$$\left.\begin{aligned} \widetilde{E}_{\mathit{eff},x}(t) &= \frac{1}{|\boldsymbol{\vartheta}_{12}|}\left(\boldsymbol{\vartheta}_{21}\boldsymbol{E}^{(+)}(t)\, e^{i\nu_0 t} + \boldsymbol{\vartheta}_{12}\boldsymbol{E}^{(-)}(t)\, e^{-i\nu_0 t}\right), \\ \widetilde{E}_{\mathit{eff},y}(t) &= \frac{i}{|\boldsymbol{\vartheta}_{12}|}\left(\boldsymbol{\vartheta}_{21}\boldsymbol{E}^{(+)}(t)\, e^{i\nu_0 t} - \boldsymbol{\vartheta}_{12}\boldsymbol{E}^{(-)}(t)\, e^{-i\nu_0 t}\right), \\ \widetilde{E}_{\mathit{eff},z}(t) &= 0. \end{aligned}\right\} \tag{VII.12.56}$$

Note that the direction of the vector $(\widetilde{E}_{eff,x}, \widetilde{E}_{eff,y})$ depends on ϑ_{21} and need not coincide with the direction of the vector $(E_x^{(+)}, E_y^{(-)})$, so that one should perhaps call (VII.12.56) "fictitious" fields. The wiggle $\sim$ should remind the reader that $\widetilde{\boldsymbol{E}}_{eff}$ is taken in the interaction representation. With (VII.12.56), Eqs. (VII.12.55) can be cast in the form

$$\dot{\boldsymbol{I}} = \frac{|\vartheta_{21}|}{\hbar} (\boldsymbol{I} \times \widetilde{\boldsymbol{E}}_{eff}) \tag{VII.12.57}$$

which has a complete formal resemblance* to the case of a spin in a magnetic field [compare Eq. (VII.12.26)].

Eq. (VII.12.57) describes the rotation of the fictitious spin under the action of a fictitious force $\widetilde{\boldsymbol{E}}_{eff}$. To complete the analogy we define a fictitious angle Θ, by which the fictitious spin is rotated. We now interpret the right-hand side of Eq. (VII.12.46) as population difference $d = N_2 - N_1$ (N_j: occupation number of level j), so that Θ is defined by

$$\left.\begin{aligned} \cos\Theta &= N_2 - N_1, \\ \tfrac{1}{2}\cos\Theta &= \langle I_z \rangle \end{aligned}\right\} \tag{VII.12.58}$$

and [compare (VII.12.43)]

$$\tfrac{1}{2}\sin\Theta = \langle I_y \rangle. \tag{VII.12.59}$$

We assume the same situation as above in Sect. *b)* β) p. 231], namely $\langle I_x \rangle = 0$ and $\widetilde{E}_{eff,y} = 0$, i.e. the phase of $\widetilde{\boldsymbol{E}}^{(+)}$ in $\boldsymbol{E}^{(+)} = \widetilde{\boldsymbol{E}}^{(+)} e^{-i\omega t}$ is chosen in such a way that $\vartheta_{21} \widetilde{\boldsymbol{E}}^{(+)}$ is real. Then $F_0 g$ [see (VII.12.52)] is real and all reality conditions about gF_0 which were used in Sect. *b)* β) are fulfilled. This allows us to transcribe the result (VII.12.45) to an electron in a two-level atom:

$$\Theta(t_1) - \Theta(t_0) = \frac{1}{\hbar} \int_{t_0}^{t_1} (\vartheta_{21} \widetilde{\boldsymbol{E}}^{(+)} + \vartheta_{12} \widetilde{\boldsymbol{E}}^{(-)})\, dt. \tag{VII.12.60}$$

Note that according to (VII.12.39) and (VII.12.52), $\widetilde{\boldsymbol{E}}^{(+)}$ may still have a *time-dependent real* amplitude. Eq. (VII.12.60) plays an important role in the discussion of pulse propagation (see Chap. VII.13).

We finally indicate how (VII.12.57) changes if we pass from the interaction picture to the Heisenberg picture. We further introduce damping terms in Eqs. (VII.12.33).

Performing all the steps again starting from Eqs. (VII.12.33), which now contain the additional terms:

$$\left.\begin{aligned} &(-i\nu_0 - \gamma)\,\sigma^-, \\ &(i\nu_0 - \gamma)\,\sigma^+, \\ &\frac{d_0 - \sigma_z}{T} \end{aligned}\right\} \tag{VII.12.61}$$

in the corresponding equations, we find instead of Eq. (VII.12.57)

$$\langle \dot{\boldsymbol{I}} \rangle = \frac{|\vartheta_{21}|}{\hbar} (\langle \boldsymbol{I} \rangle \times \boldsymbol{E}_{eff}) + \begin{Bmatrix} -\gamma \langle I_x \rangle, \\ -\gamma \langle I_y \rangle, \\ \dfrac{d_0 - \langle I_z \rangle}{T} \end{Bmatrix} + \begin{Bmatrix} -\nu_0 \langle I_y \rangle, \\ \nu_0 \langle I_x \rangle, \\ 0 \end{Bmatrix} \tag{VII.12.62}$$

where $\boldsymbol{E}_{eff}$ is now defined by omitting the factors $e^{\pm i\nu_0 t}$ in Eq. (VII.12.56).

---

* This resemblance is well known [see R. P. FEYNMAN, F. L. VERNON, JR., and R. W. HELLWARTH: J. Appl. Phys. **28**, 49 (1957)] and has been used in the framework of discussing pulse-propagation (see Chap. VII.13).

Obviously, the last column can be incorporated into $\boldsymbol{E}_{eff}$, if we define

$$\boldsymbol{E}_{eff,\,total} = \begin{Bmatrix} \boldsymbol{E}_{eff,x}\,, \\ \boldsymbol{E}_{eff,y}\,, \\ -\hbar\nu_0/|\boldsymbol{\vartheta}_{21}|\,. \end{Bmatrix} \tag{VII.12.63}$$

Because the introduction of the damping terms (VII.12.61) violates the quantum mechanical consistency (compare Chap. IV.4b), the Eqs. (VII.12.62) no longer refer to operators but rather to their quantum statistical averages, which are defined by $\langle \boldsymbol{I} \rangle = trace\ (\boldsymbol{I}\varrho)$ where ϱ is the density matrix (see Chap. IV.7).

In spin resonance theory, Eqs. (VII.12.62) are identical with Bloch's equations* with the longitudinal relaxation time $T_1 = T$ and the transverse time $1/T_2 = \gamma$.

VII.13. Pulse propagation in laser-active media. In most parts of the present article we consider a laser with mirrors at its ends so that standing waves arise. In this chapter we investigate the behaviour of a wave running in one direction, either in an infinitely extended laser-active material or in a ring laser, causing a periodicity condition for field, polarization, and inversion. In accordance with the semiclassical approach, we treat these quantities as *classical* variables. Depending on the initial conditions and physical parameters like the pump rate, several physically quite different situations may occur. The following typical solutions are known so far:

1. One-dimensional case. a) At a low pump rate, only *damped field oscillations* occur.

b) Above a certain pump threshold a *stationary solution* is possible. A field and a polarization wave propagate jointly through the material. Their amplitudes and the atomic inversion are time and space independent constants if the field is in resonance with the atomic transition.

c) If the pump exceeds a certain second threshold, the stationary solution becomes unstable[1] and breaks into *pulses*[2].

d) The material is not pumped steadily, but only before the pulse starts. Thus the laser-active atoms are first in the upper state and then, after a light pulse has passed through the material, are in their lower state. This pulse is called a "π-pulse"[3]. The word "π-pulse" becomes clear if we think of the analogy between

---

* F. Bloch: Phys. Rev. **70**, 460 (1946).

[1] R. Graham and H. Haken: Z. Physik **213**, 420 (1968). H. Risken and K. Nummedal: Phys. Letters **26** A, 275 (1968).

[2] Self-pulsing in lasers was investigated by H. Risken and K. Nummedal: J. Appl. Phys. **39**, 4662 (1968). — P.W. Smith: IEEE J. Quantum Electronics **QE-3**, 627 (1967). The latter author uses rate equations with a time-dependent transition probability. For possibly an experimental observation see e.g.: S.L. Shapiro, M.A. Duguay and L.B. Kreuzer: Appl. Phys. Letters **12**, 36 (1968).

[3] π-pulse propagation has been investigated for the case where the pulse width is long compared to the lifetime or to the decay time of the atoms in the framework of rate equations by L.M. Franz and J.S. Nodvik [J. Appl. Phys. **34**, 2346 (1963)] and by Basov et al. [JETP **23**, 16 (1966)], and by theories in which the pulse may be narrowed compared to the atomic decay times assuming homogeneous broadening: J.P. Wittke and P.J. Warter: J. Appl. Phys. **35**, 1668 (1964). — F.T. Arecchi and R. Bonifacio: IEEE J. Quantum Electronics **QE-1**, 169 (1965). — C.L. Tang and B.D. Silverman [in "Physics of Quantum Electronics" ed. by P.E. Kelley, B. Lax, and P.E. Tannenwald (New York: McGraw Hill Book Comp. 1966)] treated the inhomogeneously broadened amplifier assuming a thin medium. — J.A. Armstrong and E. Courtens: IEEE J. Quantum Electronics **QE-4**, 411 (1968). — F.T. Arecchi, V. De Giorgo and S.G. Someda: Phys. Letters **27** A, 588 (1968).

a 2-level system and a spin $\frac{1}{2}$ system and the fact that a field may rotate the "spin" from its upper to its lower position (compare Chap. VII.12.b.β and VII.12.f)

e) The atoms are first in their lower state. Then a light pulse brings them continuously into their upper state and consecutively into their lower state so that the fictitious spin was rotated by 2π ("2π-pulse")[4]. In such a case no absorption occurs, although the field is in resonance with the atomic transition. This effect is called selfinduced transparency.

2. *Three-dimensional case.* In the three-dimensional case the additional effect of self-focussing may happen. Further effects are self-modulation, self-steepening and spectral development of light in small scale trapped filaments[5]. In our article we treat in more or less detail the cases a)—e).

a—c) Steady state and self-pulsing[6].

α) *The basic equations.* We treat a system of two-level atoms with a homogeneously broadened line and consider the polarization as a macroscopic quantity. Confining our analysis to a ring laser with waves running in *one* direction, we use the *wave picture* and the *slowly varying amplitude approximation.* We decompose the field E and the polarization P (which are assumed to be polarized parallel to each other and perpendicular to the laser axis z) as follows:

$$\left.\begin{aligned} E(z,t) &= \hat{E}^{(-)}(z,t)\, e^{i\omega_0 t - ikz} + \text{conj. compl.}, \\ P(z,t) &= \hat{P}^{(-)}(z,t)\, e^{i\omega_0 t - ikz} + \text{conj. compl.} \end{aligned}\right\} \qquad \text{(VII.13.1)}$$

ω_0 is assumed to be in resonance with the atomic transition frequency. According to Chap. VII.1.c.γ $\hat{E}^{(-)}$, $\hat{P}^{(-)}$ and the inversion density D obey the Eqs. (VII.1.51), (VII.1.52), (VII.1.53). Field, polarization and inversion are subject to a periodic boundary condition

$$\left.\begin{aligned} \hat{E}^{(-)}(z+L,t) &= \hat{E}^{(-)}(z,t), \\ \hat{P}^{(-)}(z+L,t) &= \hat{P}^{(-)}(z,t), \\ D(z+L,t) &= D(z,t) \end{aligned}\right\} \qquad \text{(VII.13.2)}$$

where L is the laser cavity length. The losses of the field are assumed to be homogeneously distributed over the laser and time-independent, so that the con-

[4] The possibility of 2π-pulses was first found and discussed by S.L. McCall and E.L. Hahn: Phys. Rev. Letters **18**, 908 (1967), who treated an inhomogeneously broadened material initially in its groundstate and having an infinitely sharp homogeneous linewidth ($T_2=\infty$). — The homogeneously broadened attenuator or self-induced transparency were also treated by: G.L. Lamb Jr.: Phys. Letters **25** A, 181 (1967). — E. Courtens: Phys. Rev. Letters **21**, 3 (1968). — C.K. Rhodes, A. Szöke, and A. Javan: Phys. Rev. Letters **21**, 1151 (1968). — E. Courtens and A. Szöke: Phys. Letters **28** A, 296 (1968).

[5] N. Bloembergen and P. Lallemand: Phys. Rev. Letters **16**, 81 (1966). — P. Lallemand: Appl. Phys. Letters **8**, 276 (1966). — K. Shimoda: J. Appl. Phys. (Japan) **5**, 615 (1966). — H.P.H. Grieneisen and C.A. Sacchi: Bull. Am. Phys. Soc. **12**, 686 (1967). — R.G. Brewer: Phys. Rev. Letters **19**, 8 (1967). — F. Shimizu: Phys. Rev. Letters **19**, 1097 (1967). — A.C. Cheung, D.M. Rank, R.Y. Chiao, and C.H. Townes: Phys. Rev. Letters **20**, 786 (1968). — For a comprehensive discussion of small scale selftrapped filaments see: R.G. Brewer, J.R. Lifshitz, E. Garmire, R.Y. Chiao, and C.H. Townes: Phys. Rev. **166**, 326 (1968). — F. De Martini, C.H. Townes, T.K. Gustafson, and P.L. Kelley: Phys. Rev. **164**, 312 (1967). — There exist a great number of experimental and theoretical papers on self-focusing, e.g. G.A. Askar'yan: Soviet Phys. JETP **15**, 1088 (1962). — R.Y. Chiao, E. Garmire, and C.H. Townes: Phys. Rev. Letters **13**, 479 (1964). — K. Grob and M. Wagner: Phys. Rev. Letters **17**, 819 (1966). — C.S. Wang: Phys. Rev. **173**, 908 (1968). — Many further references may be found in the review article S.A. Akhmanov, A.P. Sukkorukov, and R.V. Khokhlov: Soviet Phys.-Usp. **10**, 609 (1968).

[6] We follow essentially the paper of H. Risken and K. Nummedal: l.c.[2] on p. 236.

ductivity σ occurring in Eq. (VII.1.51) is a constant. We abbreviate as usual:

$$2\pi\sigma = \varkappa \quad \text{and put} \quad \frac{1}{T} = \gamma_{\parallel}.$$

β) Stationary solution. The stationary ("*cw*") solution of Eqs. (VII.1.51), (VII.1.52), (VII.1.53) is found by putting

$$\frac{\partial \hat{P}^{(-)}}{\partial t} = \frac{\partial \hat{E}^{(-)}}{\partial t} = \frac{\partial \hat{E}^{(-)}}{\partial z} = \frac{\partial D}{\partial t} = 0. \tag{VII.13.3}$$

The resulting algebraic equations are readily solved:

$$\left.\begin{aligned} D_{cw} &= \frac{\gamma\varkappa\hbar}{|\vartheta_{12}|^2\, 2\pi\,\omega_0}, \\ \hat{E}^{(-)}_{cw} &= \frac{1}{2}\left[(\gamma_{\parallel}\hbar\omega_0\, 2\pi/\varkappa)\,(D_0 - D_{cw})\right]^{\frac{1}{2}} e^{i\Theta}, \\ \hat{P}^{(-)}_{cw} &= i\left(\varkappa/(2\pi\,\omega_0)\right)\hat{E}^{(-)}_{cw} \end{aligned}\right\} \tag{VII.13.4}$$

Θ is an arbitrary phase.

The saturated inversion density D_{cw} is identical with the threshold inversion density D_c. In (VII.13.4) it is assumed that the unsaturated inversion density D_0 is bigger than D_{cw}. For $D_0 < D_c$ only damped solutions exist. Bearing in mind that the total field amplitude $E(z, t)$ is given by (VII.13.1), we see that the stationary solution is represented by a *single running mode* in accordance with the considerations of Chap. VIII.1.c. For further considerations it is convenient to use

γ) Normalized amplitudes defined by

$$\left.\begin{aligned} &\overline{E} = \frac{\hat{E}^{(-)}}{\hat{E}^{(-)}_{cw}}; \quad \overline{P} = \frac{\hat{P}^{(-)}}{\hat{P}^{(-)}_{cw}}; \quad \overline{D} = \frac{D}{D_{cw}}; \\ &l = \frac{D_0 - D_{cw}}{D_{cw}}; \quad l_c = \frac{D_{\text{self-pulsing}} - D_{cw}}{D_{cw}}. \end{aligned}\right\} \tag{VII.13.5}$$

Because it turns out that the relative phase between $\hat{E}$ and $\hat{P}$ is stable (as long as fluctuations are neglected, also the total phase is stable), it suffices to solve the amplitude equations, which follow from (VII.1.51)—(VII.1.53) by use of (VII.13.5):

$$\left.\begin{aligned} |\dot{\overline{P}}| + \gamma|\overline{P}| &= \gamma|\overline{E}|\overline{D}, \\ \dot{\overline{D}} + \gamma_{\parallel}\overline{D} &= \gamma_{\parallel}(l + 1 - l|\overline{E}|\,|\overline{P}|), \\ c\frac{\partial|\overline{E}|}{\partial z} + |\dot{\overline{E}}| + \varkappa|\overline{E}| &= \varkappa|\overline{P}|. \end{aligned}\right\} \tag{VII.13.6}$$

δ) Stability of the stationary solution $|\overline{E}| = |\overline{P}| = |\overline{D}| = 1$ *of the amplitude equations.* To this end we apply the conventional stability criterion: We decompose the time and space dependent quantities $|\overline{E}|$, $|\overline{P}|$ and $\overline{D}$ into the stationary solution and small additional quantities.

$$\left.\begin{aligned} |\overline{E}| &= 1 + \delta|\overline{E}|, \\ |\overline{P}| &= 1 + \delta|\overline{P}|, \\ \overline{D} &= 1 + \delta\overline{D}. \end{aligned}\right\} \tag{VII.13.7}$$

Inserting (VII.13.7) into the Eqs. (VII.13.6) and linearizing them with respect to the small quantities, we obtain for them three linear homogeneous differential equations with constant coefficients. They can be solved by the ansatz

$$\begin{pmatrix} \delta|\overline{E}| \\ \delta|\overline{P}| \\ \delta\overline{D} \end{pmatrix} = \begin{pmatrix} \delta|\overline{E}|(0,0) \\ \delta|\overline{P}|(0,0) \\ \delta\overline{D}(0,0) \end{pmatrix} \cdot \exp(i\alpha z/c + \beta t) + \text{c.c.} \qquad \text{(VII.13.8)}$$

where α and β are connected by an algebraic equation. Due to the boundary conditions (VII.13.2), α must be real and takes on only the discrete values

$$\alpha_n = 2\pi(c/L)\,n; \quad n = 0, \pm 1, \pm 2, \ldots. \qquad \text{(VII.13.2a)}$$

The stationary solution is unstable if any of the roots β of the algebraic equation has a positive real part. A discussion of that equation shows that at least one of the β's has a positive real part if the unsaturated inversion density D_0 is bigger than a certain critical value:

$$D_0 > D_{self\text{-}pulsing} = D_c\{5 + 3\gamma_\parallel/\gamma + 2\sqrt{4 + 6(\gamma_\parallel/\gamma) + 2(\gamma_\parallel/\gamma)^2}\} \qquad \text{(VII.13.9a)}$$

and if at least one of the discrete α_n values obeys the inequality

$$\left.\begin{aligned} &\alpha_{min} < |\alpha_n| < \alpha_{max}, \\ &\alpha_{\substack{max\\min}} = \sqrt{\frac{\gamma_\parallel}{2}(3\gamma l - \gamma_\parallel \pm R)}\left\{1 - \frac{2\varkappa}{\gamma(l-2) - \gamma_\parallel \pm R}\right\}, \\ &R = \sqrt{l^2\gamma^2 - 2(4\gamma + 3\gamma_\parallel)l\gamma + \gamma_\parallel^2}. \end{aligned}\right\} \qquad \text{(VII.13.9b)}$$

Here $D_c \equiv D_{cw}$ is the critical inversion density (VII.13.4) for laser action to occur. If the laser is not a ring laser, but infinitely extended without boundary conditions, the additional requirement (VII.13.9b) must be dropped.

The instability leads to a build-up of the pulse. Besides a special case which will be treated below, the solution of Eqs. (VII.13.6) can be obtained only by means of a computer.

ε) *Transient build-up of the pulse.* The above Eqs. (VII.13.6) were solved with a computer by transforming them into difference equations which are correct to second order differences. As initial values a (small) instable solution of the linearized equation was chosen. Fig. 56 shows how the amplitudes of the waves become saturated and how the waveform of the electric field is changed from a pure sinusoidal form to a more peaked waveform. The maximum and minimum values of the electric field are compared with the values that follow from the linearized equations in Fig. 57. In the same figure, the transient velocity (defined as the velocity of the peak of the electric field amplitude) is also plotted as a function of the number of round trips N. Notice that the pulse velocity decreases with increasing pulse height, but always remains larger than the velocity c in the host material. The final pulse has the form $f(t - z/v)$ and thus describes a pulse travelling with the velocity v through the cavity.

Usually the laser oscillation does not start at $|\overline{E}| \approx |\overline{P}| \approx \overline{D} \approx 1$. If instead we assume a fast Q switching of the laser resonator, the initial conditions are given by $\overline{D} = l + 1$, $|\overline{E}| = |\overline{P}| = 0$. Because noise is neglected in the present treatment, the solutions would remain $|\overline{E}| = |\overline{P}| = 0$ for all times. In order to get the laser started, one must therefore assume a small initial disturbance of the electric field, e.g. a Gaussian. The transient build-up of the pulse from such a small

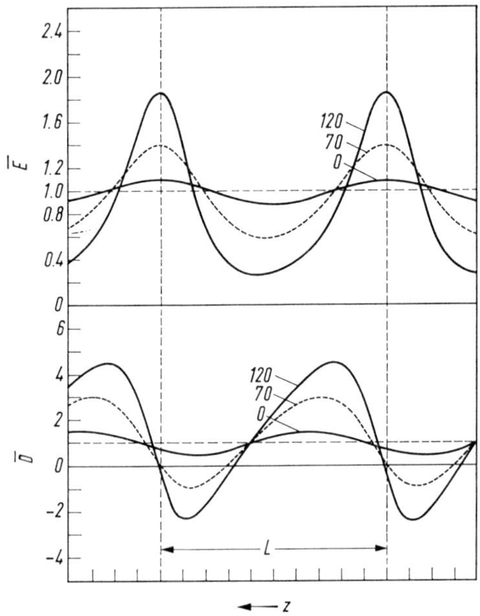

Fig. 56. The transient waveform of the electric field $\bar{E}$ and inversion $\bar{D}$ as a function of z for different times $t = NL/c$. (N is the number of round trips that light, travelling with the velocity c through the cavity, makes in the time t). The direction of z is to the left so that the pulse travels to the left, consistent with Fig. 56 and Fig. 57. The parameters are $\gamma_{\|} = \frac{1}{2}\gamma$, $\varkappa = \frac{1}{10}\gamma$, $l = 15$ and $L = 2\pi c/(3.2\,\gamma)$. Thus, α is in the instable region. As initial conditions we have chosen an instable solution (VII. 13.8) with $\delta|\bar{E}|\,(0, 0) = 0.1$. In the figure, the pulses are shifted in z in such a way that the maximum values always coincide. [After H. Risken and K. Nummedal: J. Appl. Phys. **39**, 10, 4662 (1968).]

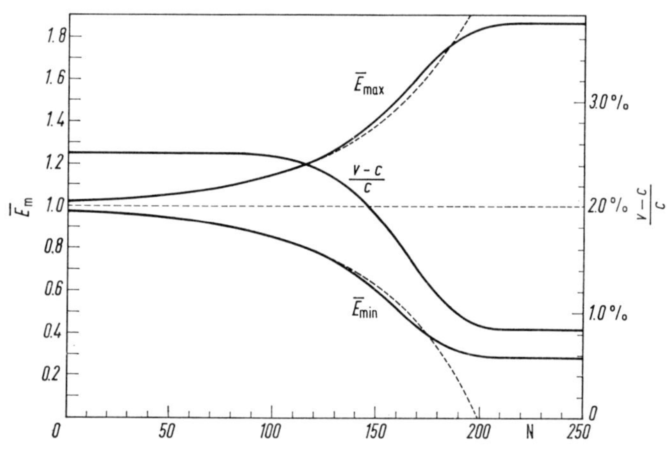

Fig. 57. The maximum and minimum amplitudes of the electric field together with the relative excess velocity $(v-c)/c$ as a function of the time $t = NL/c$. v is the velocity of the maximum value of the electric field. The parameters are $\gamma_{\|} = \frac{1}{2}\gamma$, $\varkappa = \frac{1}{10}\gamma$, $l = 15$, and $L = 2\pi c/(3.2\,\gamma)$. Thus, α is in the instable region. As initial conditions we have chosen an instable solution with $\delta|\bar{E}|\,(0, 0) = 0.02$. The time variation of $\bar{E}_{max}$ and $\bar{E}_{min}$ that follows from the linear theory is dotted in for comparison. [After H. Risken and K. Nummedal: J. Appl. Phys. **39**, 10, 4662 (1968).]

disturbance is shown in Fig. 58. After a few round trips, the variables reach the approximate *cw* solutions $\overline{E} \approx \overline{P} \approx \overline{D} \approx 1$, but for the parameters of Fig. 58, these values are not stable. Hence, a pulse builds up with the same final shape as in Fig. 56. In the first few round trips the field distribution is rather complicated. After these few round trips, the field variables have the form of a travelling wave pulse $f(t-z/v, t)$, which slowly changes its amplitude, its velocity, and its pulse shape (see Figs. 57 and 56). For similar initial conditions as in Fig. 58 but with a pump parameter l below the critical value l_c, the transient solution shows in the beginning a similar behavior as in Fig. 58. But then after a few round trips the stationary *cw* values $|\overline{E}| = |\overline{P}| = \overline{D} = 1$ are reached. Risken and Nummedal also calculated the transient phases Φ of the field (i.e. $\hat{\overline{E}}^{(-)} = |\hat{\overline{E}}^{(-)}| e^{-\Phi}$) and Ψ of the polarization as solutions of the more general Eqs. (VII.1.51), (VII.1.53). As

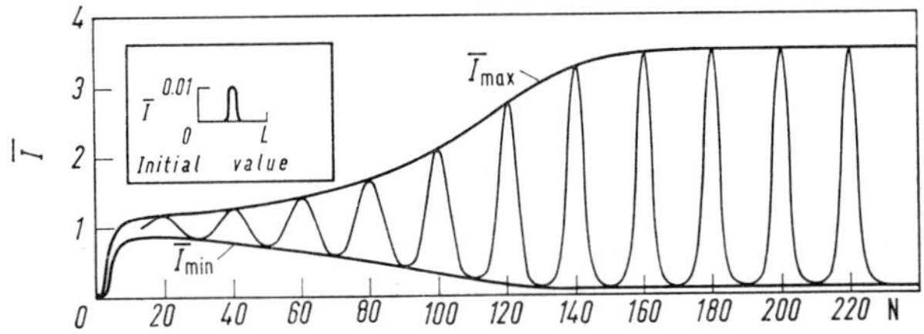

Fig. 58. The transient build-up of the intensity from a small Gaussian disturbance as a function of the time $t = NL/c$ in the instable region $l = 15$, $L = 2\pi c/(3.2\,\gamma)$, $\gamma_{\|} = \frac{1}{2}\gamma$ and $\varkappa = \frac{1}{10}\gamma$. The initial values are given by $\overline{E}(z, 0) = 0.1 \exp[-100(z/L - \frac{1}{2})^2]$, $\overline{P}(z, 0) = 0$, $\overline{D}(z, 0) = l + 1$. Only every 20th pulse is shown with a width 20 times the actual width. ($\overline{I} = |\overline{E}|^2$) [After H. Risken and K. Nummedal: J. Appl. Phys. **39**, 10, 4662 (1968).]

can be seen in their paper, a constant phase $\Phi = \Psi$ is reached after about 10—15 round trips, thus showing a high stability of the phases in the nonlinear region. One is therefore justified in omitting the phases in Eq. (VII.13.6) for the cases shown in Figs. 56, 57, 58 as long as the initial phase is zero.

ζ) *Steady state pulse.* The transient solutions of Sect. VII.13. ε show that there exists a steady state pulse of the form

$$|\overline{E}| = E(t - z/v); \quad |\overline{P}| = \overline{P}(t - z/v); \quad \overline{D} = \overline{D}(t - z/v) \qquad \text{(VII.13.10)}$$

of the Eq. (VII.13.6) where v is the velocity of the pulse. Putting (VII.13.10) into the partial differential Eqs. (VII.13.6) leads to a set of ordinary differential equations:

$$\overline{P}' + \gamma \overline{P} = \gamma \overline{E}\,\overline{D}, \qquad \text{(VII.13.11)}$$

$$\overline{D}' + \gamma_{\|} \overline{D} = \gamma_{\|}(l + 1 - l\overline{E}\,\overline{P}), \qquad \text{(VII.13.12)}$$

$$\varepsilon \overline{E}' + \overline{E} = \overline{P} \qquad \text{(VII.13.13)}$$

where the prime means differentiation with respect to $t - z/v$ and where ε is connected with the velocity v through the relation:

$$\varepsilon = (1/\varkappa)\,[1 - (c/v)],\ v/c = 1/(1 - \varepsilon\varkappa). \qquad \text{(VII.13.14)}$$

The periodic boundary conditions (VII.13.2) require the variables $\overline{P}, \overline{D}$ and $\overline{E}$ to be periodic in L/v. Note that ε is fixed by the boundary conditions. The solutions of Eqs. (VII.13.11)—(VII.13.13) therefore depend on the parameters L/v, l and

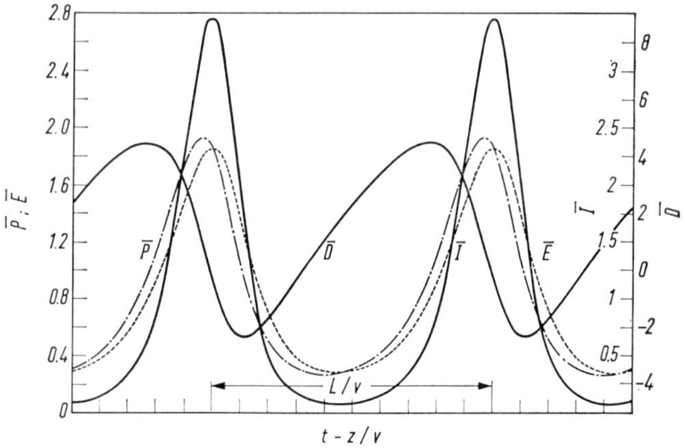

Fig. 59. Steady state fields $\bar{E}$, $\bar{P}$ (dotted) and $\bar{D}$, $\bar{I}=\bar{E}^2$ (solid) *vs* $t-z/v$ as obtained from Eqs. (VII.13.11)–(VII.13.13) from the Runge-Kutta integration. Parameters $\gamma_{\|}=\frac{1}{2}\gamma$, $l=15$, $L/v=2\pi/(3.47\,\gamma)$. Direction of propagation is from right to left along the z axis. The one period average intensity is $\langle\bar{I}\rangle=0.973$, the average inversion $\langle\bar{D}\rangle=1.40$. The field intensity is normalized with respect to the *cw* value. [After H. RISKEN and K. NUMMEDAL: J. Appl. Phys. **39**, 10, 4662 (1968).]

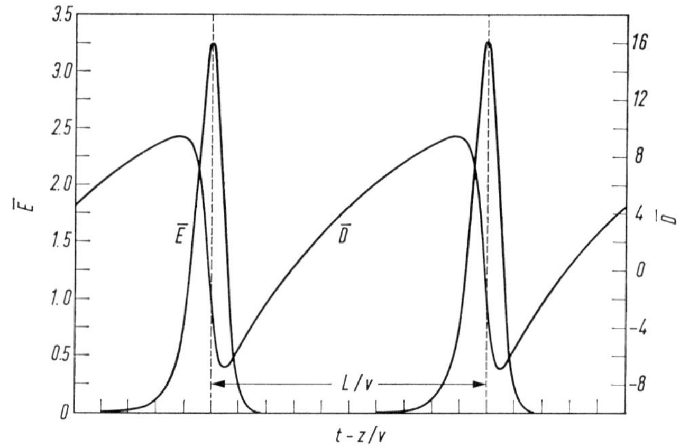

Fig. 60. The steady state pulses of the electric field $\bar{E}$ and inversion $\bar{D}$ are plotted *vs* $(t-z/v)$ for $\gamma_{\|}=\frac{1}{2}\gamma$, $l=15$, but for $L/v=2\pi c/(2.02\,\gamma)$ (small α). The average inversion $\langle\bar{D}\rangle=3.01$, the average intensity $\langle\bar{I}\rangle=0.866$ and the 2nd harmonic efficiency factor $(\langle\bar{I}^2\rangle)^{\frac{1}{2}}=5.96$. [After H. RISKEN and K. NUMMEDAL: J. Appl. Phys. **39**, 10, 4662 (1968).]

$\gamma_{\|}/\gamma$ only and not on $\varkappa$, explicitly. Fig. 59 shows one set of solutions of Eqs. (VII.13.11)–(VII.13.13).

An example of a pulse shape with high amplitude is shown in Fig. 60. Fig. 61 shows the spectrum of this pulse. The numerical analysis shows that the pulse velocity v is always larger than the light velocity in the host material.

Eq. (VII.13.14) gives the connection between ε and the pulse velocity. The steady state pulse has a velocity larger than the velocity of light in the host material (as long as $\varepsilon\varkappa<1$). If the resonator losses approach zero ($\varkappa\rightarrow 0$) the velocity v becomes equal to the velocity c. Assuming that the active atoms are located in vacuum, the relation $v<c$ seems at a first glance to contradict the

principle of relativity that no energy can travel faster than the speed of light in vacuum. It must be borne in mind, however, that a completely periodic process is described (see also below). Thus no real energy transport occurs. Another explanation is offered by RISKEN and NUMMEDAL. It can be seen from Eq. (VII.13.13) that the photons (light) are produced in the front edge of the pulse ($\overline{P} > \overline{E}$ see Fig. 59); they travel with the velocity $(v-c)$ through the pulse and are finally absorbed at the trailing edge of the pulse ($\overline{P} < \overline{E}$), thus never exceeding the velocity of light. (According to McCALL and HAHN[4] the reverse is true for a pulse travelling in an inactive medium.)

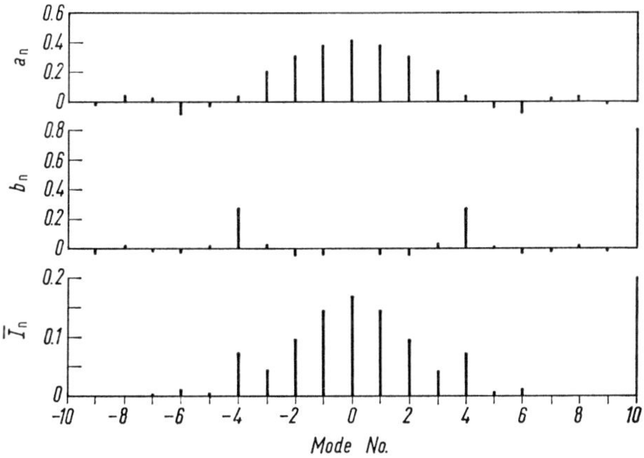

Fig. 61. The spectrum of the E field as shown in Fig. 60; a_n is the coefficient of the cosine term belonging to the n-th mode; b_n is the corresponding coefficient of the sine term. The Fourier integration interval is between the two points of maximum $\overline{E}$ field; b_n is therefore related to the asymmetry of the pulse. The mode intensity $I_n = a_n^2 + b_n^2$ is also plotted. The mode frequencies are $n \cdot 2\pi v/L = n \cdot 2\gamma v/c$, $n = 0, \pm 1, \pm 2, \ldots \pm 10$. Thus, many modes outside the Lorentzian line are excited. [After H. RISKEN and K. NUMMEDAL: J. Appl. Phys. **39**, 10, 4662 (1968).]

So far we have only considered the case of one pulse in the cavity. Obviously, the case of several pulses in the cavity is a simple extension of the previous theory if one allows the cavity length L to be 2, 3, ... times the previous pulse length. The analysis shows that we have a complicated multistable operation where stable regions for the *cw* solution still exist. For very large cavity lengths $L > 2\pi c/(\alpha_{max} - \alpha_{min})$, these stable regions of the *cw* solutions disappear and therefore these solutions are no longer possible*.

η) *A simplified model.* The most surprising result of the foregoing analysis is $v > c$. In order to elucidate the reason for this results we treat the limiting case where v is close to c, i.e. where the parameter ε in Eq. (VII.13.13) is small, i.e.

$$\varepsilon \ll 1/\gamma, \quad \varepsilon \ll 1/[\overline{E}_{max} \sqrt{\gamma \gamma_{\parallel}} l]. \tag{VII.13.15}$$

* In the treatment presented above the influence of noise (e.g. spontaneous emission) on the pulse is neglected. The neglect of spontaneous emission is allowed if the intensity of the electric field is large compared to the intensity of the electric field at laser threshold. The minimum energy of the pulse becomes very small for sharp pulses (one finds $\overline{I}_{min} = \overline{I}_{max} \exp(-\overline{I}_{max})$. Therefore, the spontaneous emission (or other noise forces) may influence the pulse during its minimum if $\overline{I}_{max}$ is very large. The effect will be that the repetition frequency is not sharp but will have a certain linewidth. For larger noise forces, the pulse breaks into two (or more) smaller pulses where the minimum field is much higher. (See also Chap. VI.13.b for the transient build-up of a laser amplitude under the influence of noise.)

Under this assumption Eq. (VII.13.13) can be approximated by

$$E \approx \bar{P} - \varepsilon \bar{P}'. \tag{VII.13.16}$$

We put Eq. (VII.13.16) into (VII.13.11) and (VII.13.12) and take into account only linear terms in ε. It turns out to be convenient to treat instead of the quantity $\bar{P}$ its logarithm $V = \ln \bar{P}$. Because the steady state solution implies $\bar{P} = 1$, V is a small quantity near the steady state. The equation for V reads

$$\left.\begin{aligned} V'' + \gamma_\parallel [1 + \varepsilon \gamma l (2 - 3e^{2V}) - \varepsilon V'] V' \\ + \gamma \gamma_\parallel l (e^{2V} - 1)(1 - \varepsilon \gamma) = 0 \end{aligned}\right\} \tag{VII.13.17}$$

and the corresponding one for the inversion $\bar{D}$

$$\bar{D} - 1 = V'(1 + \varepsilon \gamma + \varepsilon V')/\gamma. \tag{VII.13.18}$$

Consider a small amplitude V so that Eq. (VII.13.17) can be linearized with respect to V. It is then evident that we always find a damped solution, unless

$$\varepsilon = 1/(\gamma \cdot l). \tag{VII.13.19}$$

Because the periodicity condition (VII.13.2) implies that V is not damped, the result (VII.13.19) is a direct consequence of the periodicity condition and thus the requirement $v > c$ (see VII.13.14).

The condition (VII.13.19) is in good agreement with the exactly calculated ε parameter.

ϑ) *The special case* $v = c$. Waiving the periodicity condition we may look for solutions in which $v = c$. To this class of solutions belong those which were given in the literature to treat the π pulse. We further neglect the damping term $\gamma_\parallel V$ but keep the pump term $\gamma_\parallel l$. The Eq. (VII.13.17) then reduces to

$$V'' + \nu_0^2 (e^{2V} - 1)/2 = 0 \tag{VII.13.20}$$

where the small amplitude frequency ν_0 is given by

$$\nu_0 = (2\gamma\gamma_\parallel l)^{\frac{1}{2}}. \tag{VII.13.21}$$

The "energy balance" of (VII.13.20) leads to the first integral

$$\frac{\sqrt{2}}{\nu_0} V' = \sqrt{\frac{\gamma}{\gamma_\parallel l}} (\bar{D} - 1) = \sqrt{k - (e^{2V} - 2V)} \tag{VII.13.22}$$

where the integration constant is related to the minimum and maximum amplitudes by

$$k = \exp(2V_{max}) - 2V_{max} = \exp(2V_{min}) - 2V_{min}. \tag{VII.13.23}$$

The relation (VII.13.22) is plotted in Fig. 62 as a function of $\bar{P} = \exp(V)$. This figure shows that for large amplitudes ($\bar{P}_{max} \gg 1$), $\bar{P}$ and $\sqrt{\gamma/(\gamma_\parallel l)}\,(\bar{D} - 1)$ lie on a semicircle which is the case for π-pulses (see next section).

Furthermore it turns out that the expression [compare (VII.12.60)]

$$\Theta = \frac{1}{\hbar} \int_{period} (\boldsymbol{\vartheta}_{21} \hat{\boldsymbol{E}}^{(+)} + \boldsymbol{\vartheta}_{12} \hat{\boldsymbol{E}}^{(-)})\, dt \tag{VII.13.24)*}$$

is equal to π for large amplitudes (π-pulses) whereas for small amplitudes Θ is equal to $\pi\sqrt{2}$.

* We use $\hat{E}_{eff}$ instead of $\tilde{E}_{eff}$ because we have split off the space dependence according to (VII.13.1).

Taking into account only the exponential term under the square root in (VII.13.22) and putting $k = \exp(2V_{max})$, we find the following pulse shape of $\bar{E} = e^V$

$$\bar{E}(t - z/v) = \bar{E}_{max}/\cosh[\bar{E}_{max} \sqrt{\gamma\gamma_{\|} l}\,(t - z/v)]. \qquad \text{(VII.13.25)}$$

This reciprocal cosh form was found by ARECCHI and BONIFACIO[3] (see next section) for a pulse propagating through an inverted medium and by McCALL and HAHN[4] for a pulse travelling through an inactive (not inverted) medium (see Sect. VII.13.e).

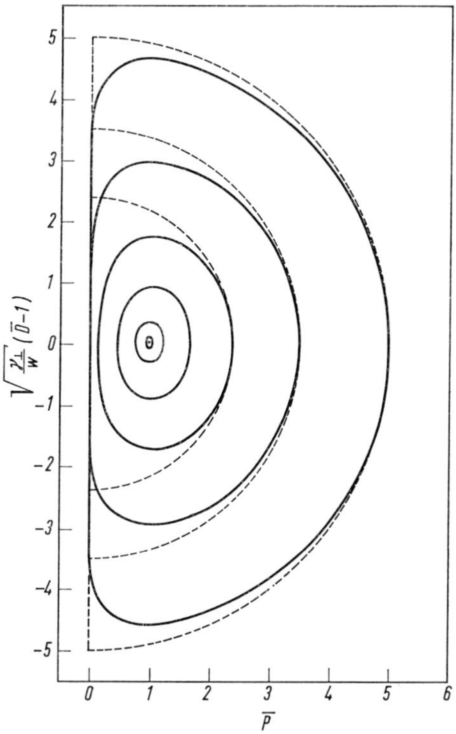

Fig. 62. A plot of the inversion versus polarization $\bar{P}$ as given by Eq. (VII.13.22). In the case of the π pulse (no pumping) the system starts at $\bar{P} = 0$ and the maximum inversion then falls down along the dotted half cycles and ends up again at $\bar{P} = 0$ and $\bar{D}_{\min}$. If pumping is included the cycle is repeated, as is shown by the solid lines. ($W = \gamma_{\|} l$) (After RISKEN, unpublished results.)

d) The π-pulse. Several authors[3] have investigated the case where a light pulse propagates through an infinitely extended medium thereby bringing the atoms to their ground states. The main difference between their treatment and the previous one is that they disregarded the continuous pumping and took other boundary conditions. ARECCHI and BONIFACIO[3] have found that the steady state light pulse propagating through an inverted medium is of the form

$$\left.\begin{aligned} \hat{E}^{(-)}(z, t) &= E_{max}\,\mathrm{sech}\,[\alpha(t - z/c)], \\ D(z, t) &= D_{max}\,\mathrm{tgh}\,[\alpha(t - z/c)]. \end{aligned}\right\} \qquad \text{(VII.13.26)}$$

To explain the notation "π-pulse", we invoke the formal analogy between a two-level system and a spin $\frac{1}{2}$. In the case of spins we have shown in Chap. VII.12

that the angle Θ between the spin and the z-axis is rotated by an applied (magnetic) field in resonance with the spin. Transcribing the spin system into a 2-level system according to Chaps. III.6 and VII.12, Θ is connected with the occupation numbers of the upper and lower level by the equation [compare (VII.12.58)].

$$\cos\Theta = N_2 - N_1 = d. \qquad \text{(VII.13.27)}$$

The occupation numbers and thus the inversion d are functions of a space point where the atom is situated, so that Θ itself is a function of the space point. The angle Θ_f after the field has acted on the atom or, in other words, after the pulse has passed by, is given by the equation [see (VII.12.60)]

$$\Theta_f - \Theta_i = \frac{1}{\hbar}\int_{-\infty}^{+\infty} (\boldsymbol{\vartheta}_{21}\hat{E}^{(+)} + \boldsymbol{\vartheta}_{12}\hat{\boldsymbol{E}}^{(-)})\,dt. \qquad \text{(VII.13.28)}$$

$\boldsymbol{\vartheta}_{21}$ is the atomic dipole matrix element [in the derivation of (VII.13.28) (see Chap. VII.12) we have chosen the phase of $\hat{E}^{(+)}$ in such a way that $\boldsymbol{\vartheta}_{21}\hat{\boldsymbol{E}}^{(+)}$ becomes real]. By comparison of (VII.13.28) with (VII.13.27) we see immediately, that the integral in (VII.13.28) must acquire the value π to make the inversion d change from $+1$ to -1.

e) The 2π-pulse. (Self-induced transparency). Here the medium is first in its ground state. By the light pulse it is then continuously brought to its upper atomic state and finally back to its lower atomic state (McCall and Hahn [4]). The 2π-pulse is thus realized when the integral in Eq. (VII.13.28) assumes the value 2π. Of course, the electric field is in turn altered by the action of the atomic dipole moments so that the right-hand side is again a function of Θ. We now assume an inhomogeneously broadened line, so that the wave equation for the electric field in the slow amplitude approximation is given by (VII.1.51).

If we further assume that the transverse relaxation time $1/\gamma$ is neglible compared with the duration time of the pulse, it is possible to eliminate the electric field from the Eqs. (VII.1.51) and (VII.12.62) and to end up with a simple equation for Θ:

$$\frac{d\Theta}{dz} = -\frac{\alpha}{2}\sin\Theta \qquad \text{(VII.13.29)}$$

where

$$\alpha = 4\pi^2|\boldsymbol{\vartheta}_{12}|^2\omega g(0)\,N/(\eta\hbar c V).$$

$g(\Delta\omega)$ is the density of states (per unit frequency) of the inhomogeneously broadened atomic line as a function of detuning $\Delta\omega = \omega - \nu_0$. η is the index of refraction of the host material. The other quantities are defined as usual in this book.

In deriving (VII.13.29) the losses of the field were neglected. Otherwise the term $-\varkappa\Theta$ must be added to the right-hand side of Eq. (VII.13.29).

(VII.13.29) has the solution

$$\tan\frac{1}{2}\Theta(z) = (\tan\frac{1}{2}\Theta_0)\exp(-\frac{1}{2}\alpha z). \qquad \text{(VII.13.30)}$$

Steady state solutions are given by

$$\Theta = n_0 2\pi \qquad (n_0\text{: integer number}).$$

The detailed treatment of McCall and Hahn [4] shows: for initial pulse areas $\Theta_0 < \pi$ below the critical area $\Theta(z) \sim \int \hat{E}\,dt$ at $\Theta = \pi$, the pulse area diminishes toward $\Theta(t) = 0$ for increasing z.

Above the critical area π, the pulse increases in area toward the limit 2π.

A given arbitrary input with $\Theta_0 = n_0 2\pi$ divides into n separate 2π pulses, after traversing some distance into the medium. This 2π pulse has the form (VII.13.25).

The pulse velocity v is decreased compared to empty space as expressed by the formula

$$\frac{1}{v} = \frac{\eta}{c} + \frac{\alpha \tau_0^2}{2\pi g(0)} \int_{-\infty}^{+\infty} \frac{g(\Delta\omega)\, d(\Delta\omega)}{1 + (\Delta\omega\tau_0)^2} \tag{VII.13.31}$$

τ_0 is the final pulse width, which may be expressed by

$$\frac{1}{\tau_0} = \frac{|\vartheta_{12}|}{\hbar\sqrt{2}} (\hat{E}^{(+)} + \hat{E}^{(-)}). \tag{VII.13.32}$$

In summary, the most striking features of the 2π-pulse are:

1. a short pulse of coherent light above a critical input energy for a given pulse width, τ, can pass through an optically resonant medium as though it were transparent, but below the critical energy the pulse energy is absorbed.

2. The pulse velocity is decreased, for instance, in ruby a pulse velocity of 10^8 cm/sec was found.

VII.14. Derivation of rate equations[1]. We base our derivation on the semiclassical Eqs. (VII.1.32)—(VII.1.34) which we repeat here in a somewhat more general form, applicable to four-level atoms. The laser action is assumed to occur

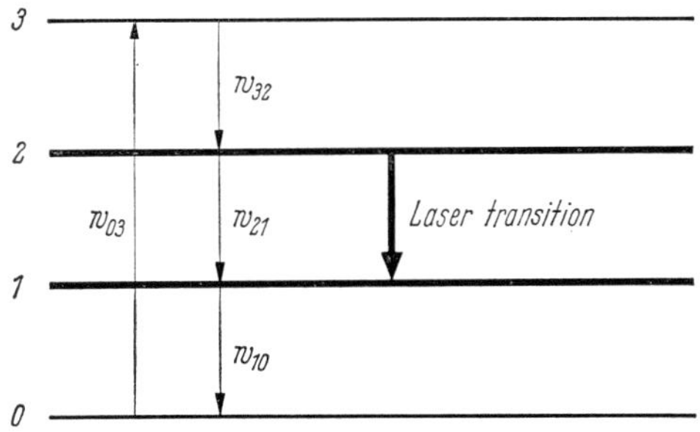

Fig. 63. Pump and decay scheme of the laser treated by Eqs. (VII.14.1)—(VII.14.6).

between levels 2 and 1. The levels 0 and 3 serve as auxiliary levels for the pumping process. The pumping process and the decay scheme is that of Fig. 63, although our treatment is also applicable to the general scheme as well. The initial equations read:

$$\dot{b}_\lambda^+ = (i\omega_\lambda - \varkappa_\lambda)\, b_\lambda^+ + i \sum_\mu g_{\mu\lambda}^* \varrho_{12,\mu} \tag{VII.14.1}$$

$$\dot{\varrho}_{12,\mu} = (i\nu_\mu - \gamma)\, \varrho_{12,\mu} - i \sum_\lambda g_{\mu\lambda} (N_{2,\mu} - N_{1,\mu})\, b_\lambda^+, \tag{VII.14.2}$$

[1] Rate equations for a single mode have been derived by C. L. Tang: J. Appl. Phys. **34**, 2935 (1963), who also discusses the range of their applicability; — Multi-mode rate equations were derived in a fully quantum mechanical fashion by V. Arzt, H. Haken, H. Risken, H. Sauermann, C. Schmid, and W. Weidlich: Z. Physik **197**, 207 (1966); — A detailed discussion of the noise sources occurring in rate equations in a quantum mechanical treatment has been given by D. E. McCumber: Phys. Rev. **141**, 306 (1966), and their derivation from quantum mechanical Langevin equations by W. Brunner: Ann. d. Physik **20**, 53 (1967) — M. Lax: IEEE J. Quantum Electronics **3**, 37 (1967).

$$\dot{N}_{0,\mu} = N_{1,\mu} w_{10} - N_{0,\mu} w_{03}, \tag{VII.14.3}$$

$$\dot{N}_{1,\mu} = N_{2,\mu} w_{21} - N_{1,\mu} w_{10} - \left\{ i \varrho_{21,\mu} \sum_\lambda g_{\mu\lambda} b_\lambda^+ - i \varrho_{12,\mu} \sum_\lambda g_{\mu\lambda}^* b_\lambda \right\} \tag{VII.14.4}$$

$$\dot{N}_{2,\mu} = N_{3,\mu} w_{32} - N_{2,\mu} w_{21} + \{\cdots\}, \tag{VII.14.5}$$

$$\dot{N}_{3,\mu} = N_{0,\mu} w_{03} - N_{3,\mu} w_{32}. \tag{VII.14.6}$$

The quantities occurring in Eqs. (VII.14.1) and (VII.14.2) are explained in VII.1. The $N_{i,\mu}$'s are the occupation numbers of the i'th level of the μ'th atom. w_{ik} is the transition rate from level i to level k, caused by incoherent sources (pump, spontaneous decay, radiationless transitions).

We put $b_\lambda^+ = B_\lambda^+ e^{i\Omega_\lambda t}$ and assume that B_λ^+ changes only slowly in time (compared to γ), and also that $N_{2,\mu} - N_{1,\mu}$ changes slowly. The integration of (VII.14.2) then yields:

$$\varrho_{12,\mu} = \sum_\lambda \frac{-i g_{\mu\lambda} (N_{2,\mu} - N_{1,\mu}) b_\lambda^+}{i(\Omega_\lambda - \nu_\mu) + \gamma}. \tag{VII.14.7}$$

We now insert (VII.14.7) into Eq. (VII.14.1), multiply this equation by b_λ and add the complex conjugate expression to the result. This leaves us with

$$\dot{n}_\lambda = -2\varkappa_\lambda n_\lambda + \sum_\mu \sum_{\lambda'} \left\{ \frac{g_{\mu\lambda}^* g_{\mu\lambda'} (N_{2,\mu} - N_{1,\mu}) b_{\lambda'}^+ b_\lambda}{i(\Omega_{\lambda'} - \nu_\mu) + \gamma} + \text{conj. complex} \right\} \tag{VII.14.8}$$

(where $n_\lambda = b_\lambda^+ b_\lambda$).

We now intend to neglect all terms with $b_{\lambda'}^+ b_\lambda \, (\lambda \neq \lambda')$. This can be justified as follows: As we have shown in VI.7 the phases of the b^+'s fluctuate. As long as no phase locking occurs (see e.g. VI.14γ, VII.5c), the phase fluctuations of the different b's are uncorrelated. Thus the mixed terms $(\lambda \neq \lambda')$ on the right-hand side of Eq. (VII.14.8) vanish, if an average is taken over the phases. The *rate equation for the photon numbers* thus reads:

$$\dot{n}_\lambda = -2\varkappa_\lambda n_\lambda + \sum_\mu \frac{2\gamma |g_{\mu\lambda}|^2}{(\Omega_\lambda - \nu_\mu)^2 + \gamma^2} (N_{2,\mu} - N_{1,\mu}) n_\lambda \tag{VII.14.9}$$

(VII.14.3) and (VII.14.6) already have the required form of equations between occupation numbers.

When we insert (VII.14.7) into (VII.14.4), (VII.14.5) and apply the same reasoning, we find:

$$\dot{N}_1 = N_{2,\mu} w_{21} - N_{1,\mu} w_{10} + \sum_\lambda \frac{2\gamma |g_{\mu\lambda}|^2}{(\Omega_\lambda - \nu_\mu)^2 + \gamma^2} (N_{2,\mu} - N_{1,\mu}) n_\lambda, \tag{VII.14.10}$$

$$\dot{N}_2 = N_{3,\mu} w_{32} - N_{2,\mu} w_{21} - \sum_\lambda \frac{2\gamma |g_{\mu\lambda}|^2}{(\Omega_\lambda - \nu_\mu)^2 + \gamma^2} (N_{2,\mu} - N_{1,\mu}) n_\lambda. \tag{VII.14.11}$$

The total set of rate equations is represented by Eqs. (VII.14.9), (VII.14.3), (VII.14.6), (VII.14.10) *and* (VII.14.11).

The rate equations allow us to determine photon numbers and atomic occupation numbers. They do not contain any information on frequency shifts. For this reason one usually uses the frequencies in the unloaded cavity, ω_λ, rather than the Ω_λ's in the active material.

VIII. Rate equations and their applications.

VIII.1. Formulation of rate equations and solution for the steady state (especially: threshold condition, pump power requirement, single versus multimode laser action).

a) The rate equations[1]. We consider M cavity modes which are distinguished by an index λ, where λ refers not only to the wavelength, but also to all other spatial properties of the modes. These cavity modes interact with N atoms at random lattice sites $\boldsymbol{x}_\mu$. Each of the atoms is supposed to possess L levels. The optical transition will occur between the levels j and k. The other levels participate in the pumping process.

The rate equations can be divided into two groups:

α) The field equations,

β) The matter equations.

α) *The field equations* describe the rate of change of the mean photon number n_λ of the mode λ under the influence of losses and gain.

1. The losses can be caused by transmission through the end faces, refraction, scattering by impurity centers, crystal inhomogeneities etc. These effects are taken care of by attributing to the mode λ a decay time $t_\lambda = 1/(2\varkappa_\lambda)$. The loss rat is thus given by

$$-2\varkappa_\lambda n_\lambda. \tag{VIII.1.1}$$

2. The gain is due to spontaneous and induced emission. The spontaneous decay rate from the level j at the atom μ into the mode λ is given by

$$N_{j,\mu} \cdot W^0_{jk,\lambda,\mu} \tag{VIII.1.2}$$

where $N_{j,\mu}$ is the corresponding occupation number, and $W^0_{jk,\lambda,\mu}$ is given in the dipole approximation by

$$W^0_{jk,\lambda,\mu} = \frac{2\gamma}{\gamma^2 + (\nu_\mu - \omega_\lambda)^2} |g_{\mu\lambda}|^2 \tag{VIII.1.3}$$

where

$$g_{\mu\lambda} = i\sqrt{\frac{2\pi\,\omega_\lambda}{\hbar}}\, \boldsymbol{e}_\lambda \boldsymbol{\vartheta}_{kj}\, U_\lambda(\boldsymbol{x}_\mu) \tag{VIII.1.4}$$

(or $|g_{\mu\lambda}|^2 = g^2 V |U_\lambda(\boldsymbol{x}_\mu)|^2$, where g does not depend on λ, if we restrict our analysis to a single polarization).

The quantities occurring in (VIII.1.3), (VIII.1.4) have the following meaning:

γ: homogeneous part of the linewidth, } of the transition

ν_μ: transition frequency of the atom μ, } $j \to k$

ω_λ: mode frequency,

$\boldsymbol{e}_\lambda$: polarization vector of the light mode,

$\boldsymbol{\vartheta}_{kj}$: dipole matrix element,

U_λ: mode amplitude at position $\boldsymbol{x}_\mu$,

V: volume of the „cavity".

---

[1] The first to use rate equations for photons was, of course, EINSTEIN; — In connection with masers, they were first applied by H. STATZ and G. A. DE MARS, in: Quantum Electronics, ed. by C. H. TOWNES, p. 530—537. New York: Columbia University Press 1960; — Rate equations have been used by a great number of authors, see e.g. the monographs on lasers quoted on p. 1—2, in particular the book of BIRNBAUM. The general treatment of VIII.1 b seems to be new, however; — Rate equations in quantized form were introduced by K. SHIMODA, H. TAKAHASI, and C. H. TOWNES: Proc. Phys. Soc. Japan **12**, 686 (1957). See also the footnote[1] on p. 247.

The induced emission rate can be obtained from (VII.1.2) by multiplication by the photon number n_λ.

Because the photons are absorbed at a rate $N_{k,\mu} W^0_{jk,\lambda,\mu} n_\lambda$, the net gain is given by

$$n_\lambda W^0_{jk,\lambda,\mu}(N_j - N_k)_\mu . \tag{VIII.1.5}$$

We thus obtain the field equations

$$\dot{n}_\lambda = -2\varkappa_\lambda n_\lambda + n_\lambda \sum_\mu W^0_{jk,\lambda,\mu}(N_j - N_k)_\mu + \sum_\mu W^0_{jk,\lambda,\mu} N_{j,\mu} . \tag{VIII.1.6}$$

β) *The matter equations* can be derived quite similarly.

They read:

1. For the levels participating in the laser process

$$\dot{N}_{j,\mu} = \sum_{l \neq j} w_{lj,\mu} N_{l,\mu} - N_{j,\mu} \sum_{l \neq j} w_{jl,\mu} - \sum_\lambda n_\lambda W^0_{jk,\lambda,\mu}(N_j - N_k)_\mu , \tag{VIII.1.7}$$

$$\dot{N}_{k,\mu} = \sum_{l \neq k} w_{lk,\mu} N_{l,\mu} - N_{k,\mu} \sum_{l \neq k} w_{kl,\mu} + \sum_\lambda n_\lambda W^0_{jk,\lambda,\mu}(N_j - N_k)_\mu . \tag{VIII.1.8}$$

2. For the other levels which are used by the pumping process:

$$\dot{N}_{m,\mu} = \sum_{l \neq m} w_{lm,\mu} N_{l,\mu} - N_{m,\mu} \sum_{l \neq m} w_{ml,\mu} . \tag{VIII.1.9}$$

The Eqs. (VIII.1.7)—(VIII.1.9) are not independent of each other. If we sum up these equations over all L levels of the atom μ, we obtain

$$\frac{d}{dt}\left(\sum_{l=1}^{L} N_{l,\mu}\right) = 0 ,$$

so that

$$\sum_{l=1}^{L} N_{l,\mu} = \text{constant}.$$

The fact that the sum over the occupation numbers $N_{l,\mu}$ of a single atom is a time-independent constant is a consequence of the conservation of the electron number in a single atom. Hence we have for the single-electron problem under consideration

$$\sum_{l=1}^{L} N_{l,\mu} = 1 \tag{VIII.1.10}$$

$w_{lm,\mu}$ is the transition rate from level l to level m of atom μ caused by

1. external pump light
2. collisions of the second kind
3. excitation collisions with electrons etc. as well as by all nonradiative and radiative spontaneous transitions.

In the following we drop the term $\sum_\mu W^0_{jk,\lambda,\mu} N_{j,\mu}$ in Eq. (VIII.1.6) since this expression describes the generation of "incoherent photons" (For a deeper motivation and a detailed discussion see VI.7).

b) Treatment of the steady state. The main task consists in the determination of the photon number as a function of the pumping, and especially in the derivation of a threshold condition. We first discuss the general procedure:

In the steady state we put

$$\dot{n}_\lambda = \dot{N}_{l,\mu} = 0. \qquad \text{(VIII.1.11)}$$

From (VIII.1.7), (VIII.1.8), (VIII.1.9) we obtain for the inversion $\Delta N_{jk,\mu} = (N_j - N_k)_\mu$ an expression of the form

$$\Delta N_{jk,\mu} = \frac{A_\mu}{B_\mu + (n_1 W^0_{jk,1,\mu} + n_2 W^0_{jk,2,\mu} + \cdots + n_M W^0_{jk,M,\mu})\, C_\mu} \qquad \text{(VIII.1.12)}$$

A_μ, B_μ, C_μ do not depend on the n_λ's (but still contain the transition rates, w_{lm}).

Eq. (VIII.1.12) can be derived as follows: On account of (VIII.1.10) we need only consider $L-1$ Eqs. (VIII.1.7)—(VIII.1.9). We drop e.g. the equation with $m=k$ (VIII.1.8) and introduce the new variables: $\Delta N_{jk} = N_j - N_k$ and $N_k + N_j$ which replace the variables N_j and N_k. The n_λ's then appear only in a single equation, where they act as factors of ΔN_{jk} in the form

$$\sum_\lambda n_\lambda W^0_{jk,\lambda,\mu}. \qquad \text{(VIII.1.13)}$$

On account of the general theorem about the solution of linear equations by means of determinants we can write

$$\Delta N_{jk} = \frac{D_1}{D} \qquad \text{(VIII.1.14)}$$

where the total determinant D depends on (VIII.1.13) at most in a linear fashion, because (VIII.1.13) occurs only in a single element of the entire D. Furthermore, D_1 is obtained from D in such a way that the column containing (VIII.1.13) is cancelled and replaced by the inhomogeneity introduced by Eq. (VIII.1.10). Consequently, D_1 does not depend on the n_λ's.

For an explicit evaluation of the A's, B's, and C's it is advisable to treat the 3 and 4-level systems explicitly. The *photon numbers* (and thus the *emitted coherent power*) can be determined as follows:

We assume that M modes show laser actions so that $n_\lambda > 0$ for $\lambda = 1 \ldots M$.

After dividing the Eqs. (VIII.1.6) by n_λ, we find*

$$2\varkappa_\lambda = \sum_\mu W^0_{jk,\lambda,\mu}(N_j - N_k)_\mu \qquad \text{(VIII.1.15)}$$

or on account of (VIII.1.12)

$$2\varkappa_\lambda = \sum_\mu \frac{W^0_{jk,\lambda,\mu} A_\mu}{B_\mu + C_\mu \sum_{\lambda'} n_{\lambda'} W^0_{jk,\mu,\lambda'}}. \qquad \text{(VIII.1.16)}$$

These are M equations for the M unknowns n_λ, for which the condition $n_\lambda > 0$ holds. If it turns out that some $n_{\lambda'} \leq 0$, these $n_{\lambda'}$'s *and* the corresponding Eq. (VIII.1.16) must be dropped.

In order to obtain an explicit solution of Eq. (VIII.1.16) it is in general necessary to treat special cases.

We start with

c) The completely homogeneous case.

α) *General formulation.* We assume that the atomic line is *homogeneously* broadened so that the transition frequencies ν_μ are equal $(=\nu_0)$. The modes will possess a space-independent intensity, i.e. we consider running waves. Then the optical transition rate does not depend on μ. Finally, it is assumed that the

* Note that we have dropped the last term in (VIII.1.6).

pumping process (and the decay rates) are spatially homogeneous. Under these assumptions, A, B, C do not depend on the atomic index μ so that (VIII.1.16) simplifies:

$$\sum_{\lambda'} n_{\lambda'} W^0_{jk,\lambda'} = \frac{N W^0_{jk,\lambda} A}{2\varkappa_\lambda C} - \frac{B}{C} \qquad \lambda = 1 \dots M. \tag{VIII.1.17}$$

Because the left-hand side does not depend on the mode-index λ, we find that

$$\frac{W^0_{jk,\lambda}}{\varkappa_\lambda} \quad \textit{must be independent of } \lambda. \tag{VIII.1.18}$$

Consequently, the Eqs. (VIII.1.17) can be solved in a consistent way only for those modes for which the condition (VIII.1.18) is fulfilled. From (VIII.1.17) it can be deduced that the mode for which $\frac{W^0_{jk,\lambda}}{\varkappa_\lambda}$ has a maximum starts the laser action first. If this maximum is assumed for only one mode, then this mode remains the only laser mode for every pump strength.

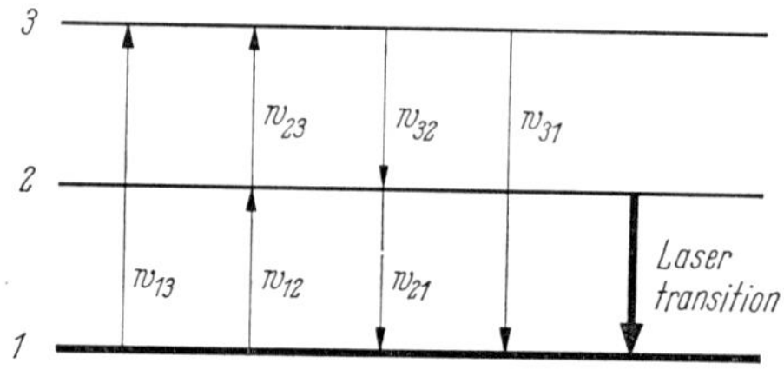

Fig. 64. Pump and decay scheme of the 3-level system treated in VIII.1.c.β.

In the completely homogeneous case only a single mode [or degenerate modes in the sense of (VIII.1.18) *with maximum* $W/\varkappa$] *can oscillate*[2].

The *laser condition* is obtained from (VIII.1.15) by keeping only the equation for the mode λ with the highest ratio (VIII.1.18). Dropping the index λ for that mode and letting $n \to 0$, we obtain with (VIII.1.3)

$$2\varkappa = (N_j - N_k)_{threshold} \cdot N \frac{2\gamma |g|^2}{\gamma^2 + (\omega - \nu_0)^2}. \tag{VIII.1.19}$$

Note that we have evidently

$$(N_j - N_k) = (N_j - N_k)_{threshold},$$

i.e. in the completely homogeneous case the saturated inversion $(N_j - N_k)$ at arbitrary photon number equals the threshold inversion.

β) *3-Level system, the lower transition is laser-active.* We specialize $j = 2$, $k = 1$, $m = 3$.

[2] H. Haken and H. Sauermann: Z. Physik **173**, 261 (1963). — C. L. Tang, H. Statz, and G. A. de Mars: J. Appl. Phys. **34**, 2289 (1963); — This statement is true up to a pump rate which causes an unsaturated inversion less than about 9 times the threshold inversion. Above this value a certain type of phase locking occurs, the rate equations cease to be valid, and undamped, ultrashort pulses are formed in the laser; — For the derivation of the condition for this new state see: R. Graham and H. Haken: Z. Physik **213**, 420 (1968). — H. Risken and K. Nummedal: Phys. Letters A **26**, 275 (1968); — For a detailed treatment of this new state, see: H. Risken and K. Nummedal: J. Appl. Phys. **39**, 4662 (1968), and Chap. VII.13 of this article.

We allow transitions with rates w_{lm} between all levels and find readily the form (VIII.1.12) for $(N_2-N_1)_\mu$ where

$$\left.\begin{aligned}A_\mu&=\{w_{13}\,w_{32}-w_{31}\,w_{23}-(w_{31}+w_{32})\,(w_{21}-w_{12})\}_\mu,\\B_\mu&=\{w_{31}(w_{23}+w_{21}+w_{12})+w_{13}(w_{32}+w_{23}+w_{21})+w_{32}(w_{21}+w_{12})+\\&\quad+w_{23}\,w_{12}\}_\mu,\\C_\mu&=\{2w_{31}+2w_{32}+w_{13}+w_{23}\}_\mu.\end{aligned}\right\}\quad\text{(VIII.1.20)}$$

We specialize these expressions for cases of practical interest: Let the temperature be so low that $\hbar\nu_{32}\gg kT$, so that $w_{23}\approx 0$ and $\hbar\nu_{21}\gg kT$, so that $w_{12}\approx 0$.

Finally we assume that the recombination is so quick that w_{32} is bigger than all other transition rates. This yields:

$$\varDelta N_\mu\equiv(N_2-N_1)_\mu=\left(\frac{w_{13}-w_{21}}{w_{13}+w_{21}+\sum\limits_\lambda 2n_\lambda\,W^0_{21,\lambda}}\right)_\mu. \qquad \text{(VIII.1.21)}$$

Because at threshold all $n_\lambda=0$, we obtain for the inversion at threshold:

$$(N_2-N_1)_{\mu,thr}=\left(\frac{w_{13}-w_{21}}{w_{13}+w_{21}}\right)_\mu. \qquad \text{(VIII.1.22)}$$

If $(N_2-N_1)_\mu$ is independent of μ it follows from (VIII.1.15)

$$(N_2-N_1)_{thr}\cdot N\cdot W^0_{21,1}=2\varkappa_1. \qquad \text{(VIII.1.23)}$$

In this case (N_2-N_1) does not change due to (VIII.1.15), when the pump power is increased so that all additional pump power serves to increase the emitted coherent power $P_{coh}=2\varkappa_1 n_1\hbar\nu_{21}$. P_{coh} is obtained by resolving (VIII.1.21) with respect to $n_1(\neq 0)$ and substituting $(N_2-N_1)_{thr}$ for (N_2-N_1) according to Eq. (VIII.1.23)

$$P_{coh}=\tfrac{1}{2}\{w_{13}-w_{21}-(w_{13}+w_{21})\,\varDelta N_{thr}\}\cdot N\cdot\hbar\nu_{21}. \qquad \text{(VIII.1.24)}$$

On the other hand, we can determine the pump power necessary to obtain the threshold by putting $P_{coh}=0$:

γ) *Pump power at threshold*:

$$P_{pump,\,thr}\equiv N\,\hbar\,\nu_{31}\,w_{13,thr}=\hbar\,\nu_{31}\,N\,w_{21}\,\frac{1+\varDelta N_{thr}}{1-\varDelta N_{thr}}. \qquad \text{(VIII.1.25)}$$

The considerations following (VIII.1.23) are only valid, if $\varDelta N$ does not depend on μ. Some of the deviations in the general case are discussed in VIII.2.

δ) *3-Level-system, the upper transition is laser-active.* We are again interested in (N_3-N_2) which can be obtained from (N_2-N_1) [Eq. (VIII.1.20)] by cyclic permutation $(1\to 2,\ 2\to 3,\ 3\to 1)$. We then obtain

$$(N_3-N_2)_\mu=\left(\frac{A}{B+C\,\Sigma n_\lambda\,W^0_{32,\lambda}}\right)_\mu \qquad \text{(VIII.1.26)}$$

where

$$\left.\begin{aligned}A_\mu&=\{w_{21}\,w_{13}-w_{12}\,w_{31}-(w_{12}+w_{13})\,(w_{32}-w_{23})\}_\mu,\\B_\mu&=\{w_{12}(w_{31}+w_{32}+w_{23})+w_{21}(w_{13}+w_{31}+w_{32})+\\&\quad+w_{13}(w_{32}+w_{23})+w_{31}\,w_{23}\}_\mu,\\C_\mu&=\{2w_{12}+2w_{13}+w_{21}+w_{31}\}_\mu.\end{aligned}\right\}\quad\text{(VIII.1.27)}$$

The expression (VIII.1.27) can be considerably simplified in important special cases.

Assuming $w_{12} \gg w_{23}$; $w_{32} \ll w_{21}$ (which is necessary for the laser process);

$$\left.\begin{array}{ll} w_{12} \gg w_{13} & \\ w_{23} \ll w_{32} & \text{(satisfied for } \hbar\nu_{32} \gg kT) \\ w_{12} = w_{21}\, e^{-\hbar\nu_{21}/kT} & \end{array}\right\} \tag{VIII.1.28}$$

we obtain

$$\left.\begin{aligned} \Delta N_{32,\mu} &\equiv (N_3 - N_2)_\mu \\ &= \left\{\frac{w_{13} - w_3\, e^{-\hbar\nu_{21}/kT}}{w_{13} + w_3 + \sum_\lambda W^0_{32,\lambda}\, n_\lambda + e^{-\hbar\nu_{21}/kT}\,(w_3 + 2\sum_\lambda W^0_{32,\lambda}\, n_\lambda)}\right\}_\mu \end{aligned}\right\} \tag{VIII.1.29}$$

where $w_3 = w_{31} + w_{32}$.

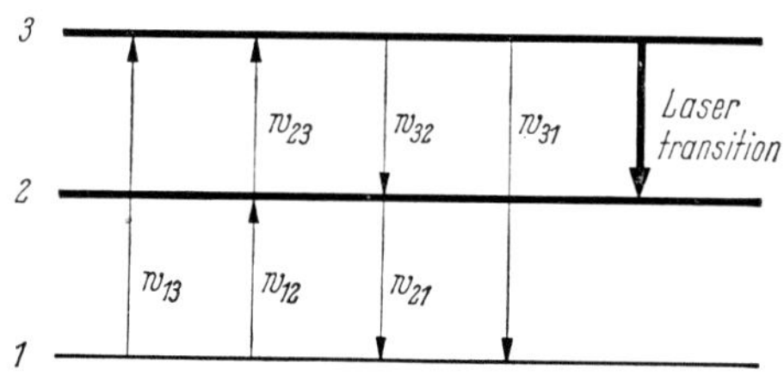

Fig. 65. Pump and decay scheme of the 3-level system treated in VIII.1.c.δ.

For $\hbar\nu_{21}/kT \gg 1$ it reduces to

$$(N_3 - N_2)_\mu = \left\{\frac{w_{13}}{w_{13} + w_3 + \sum_\lambda W^0_{32,\lambda}\, n_\lambda}\right\}_\mu . \tag{VIII.1.30}$$

Confining ourselves to *complete homogeneity* (μ-independence), and to single-mode operation, it follows from (VIII.1.29)

$$W^0_{32,1}\, n_1 = \frac{w_{13} - w_3\, e^{-\hbar\nu_{21}/kT} - \Delta N_{32}(w_{13} + w_3 + e^{-\hbar\nu_{21}/kT}\, w_3)}{\Delta N_{32}(1 + 2\, e^{-\hbar\nu_{21}/kT})} . \tag{VIII.1.31}$$

In it we have to put

$$\Delta N_{32} = (\Delta N_{32})_{thr} = \frac{2\varkappa_1}{N W^0_{32,1}} . \tag{VIII.1.32}$$

The power output is obtained from (VIII.1.31) by multiplying both sides by $(\Delta N_{32})_{thr} N \cdot \hbar\nu_{32}$. According to (VIII.1.32) the left-hand side is just

$$= P_{coh} = 2\varkappa_1\, \hbar\nu_{32}\, n_1$$

so that

$$P_{coh} = \frac{w_{13} - w_3\, e^{-\hbar\nu_{21}/kT} - (\Delta N_{32})_{thr}(w_{13} + w_3 + e^{-\hbar\nu_{21}/kT}\, w_3)}{(\Delta N_{32})_{thr}(1 + 2\, e^{-\hbar\nu_{21}/kT})}\, N\, \hbar\, \nu_{32} . \tag{VIII.1.33}$$

The pump power at threshold can be obtained by putting $P_{coh} = 0$:

$$P_{pump,\,thr} = N\, \hbar\nu_{31}\, w_{13,\,thr} = \hbar\nu_{31}\, N\, w_3\, \frac{(\Delta N_{32})_{thr}(1 + e^{-\hbar\nu_{21}/kT}) + e^{-\hbar\nu_{21}/kT}}{1 - \Delta N_{32}} . \tag{VIII.1.34}$$

For $\hbar\nu_{21}/kT \gg 1$ it reduces to

$$P_{pump,\,thr} = \hbar\nu_{31}\, N\, \frac{(\Delta N_{32})_{thr}\, w_3}{1 - (\Delta N_{32})_{thr}} . \tag{VIII.1.35}$$

A comparison between (VIII.1.35) and (VIII.1.25) shows that the pump rate is essentially smaller for the upper laser transition than for the lower transition. If on the other hand we put $\hbar\nu_{21} \ll kT$ we obtain

$$P_{pump,\,thr} \equiv N\,\hbar\,\nu_{31}\,w_{13,\,thr} = \hbar\nu_{31}\,N\,w_3\,\frac{1+2(\Delta N_{32})_{thr}}{1-(\Delta N_{32})_{thr}}\,. \tag{VIII.1.36}$$

ε) *4-Level system; laser action between the two middle levels.* The most general case leads to very lengthy formulas for ΔN_{32}, so that it is advisable to treat special cases directly. It often happens that w_{43} is bigger than w_{14} and the other rates. In this case level 4 is kept almost empty so that the system reduces to a 3-level system where the upper transition shows laser action.

VIII.2. The coexistence of modes on account of spatial inhomogeneities or an inhomogeneously broadened line. We now treat the case where different atoms have a different inversion, i.e. ΔN_{jk} depends explicitly on either the atomic position or the atomic transition energy (or both). Specifically, we consider the following cases:

a) the modes depend on the space points, for instance like a standing wave (but the line is homogeneously broadened),

b) the pumping is spatially inhomogeneous (homogeneous line),

c) the line is inhomogeneously broadened.

We base our treatment of these problems on the following *approximation for the determination of the photon-numbers.*

We have seen above that photon numbers n_λ can be obtained "in principle" by solving the equations (VIII.1.6) and (VIII.1.12). This requires, however, that the sums over μ be performed, which can be done explicitly only in very special cases (compare e.g. VII.8aα2). We therefore expand the right-hand side of Eq. (VIII.1.12) into a power series with respect to $\sum_{\lambda'} n_{\lambda'} W^0_{jk,\lambda',\mu}$ where we usually confine ourselves to terms linear in the photon number. As long as the laser does not operate too far above threshold, the photon numbers are still small, so that this approximation is well justified (and can be substantiated by numerical examples). We thus obtain from (VIII.1.6), (VIII.1.12) the following set of linear, inhomogeneous equations for n_λ (with the supplementary condition $n_\lambda > 0$ and $\dot{n}_\lambda = 0$):

$$\sum_{\mu,\lambda'} \frac{A_\mu C_\mu}{B^2_\mu}\,W^0_{jk,\lambda,\mu}\,W^0_{jk,\lambda'\mu}\,n_{\lambda'} = -2\varkappa_\lambda + \sum_\mu W^0_{jk,\lambda,\mu}\,\frac{A_\mu}{B_\mu}\,. \tag{VIII.2.1}$$

From (VIII.2.1) again the threshold condition follows by putting all n_λ's equal zero.

In the following we treat the threshold condition and the determination of the photon numbers separately according to a—c.

a) Homogeneous line, but space-dependent modes (represented by standing waves)[1]. We treat, as an explicit example, modes whose space dependence is given by $\sin k_\lambda x$ so that according to (VIII.1.3)

$$W^0_{jk,\lambda,\mu} \sim \sin^2 k_\lambda\,x_\mu\,. \tag{VIII.2.2}$$

Let us consider single-mode operation first. The expressions for (ΔN_{jk}) show that the inversion is unaltered where the mode has its node. Thus a second mode with a different wavelength can still "eat" from those portions of the crystal in which the inversion is not appreciably diminished by the first mode.

[1] See also H. Haken and H. Sauermann: l. c. [2] on p. 252. — C. L. Tang, H. Statz, and G. A. de Mars: l. c. [2] on p. 252.

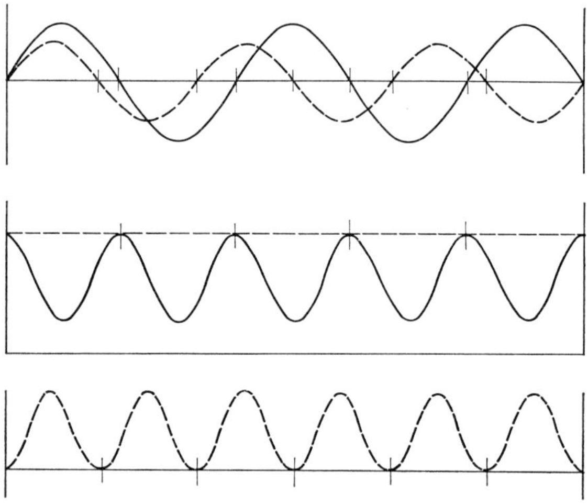

Fig. 66. This figure shows how to visualize the coexistence of two modes. The upper part exhibits the field amplitude of mode 1 and mode 2 having different wave lengths. The middle part shows by its dashed line the unsaturated inversion. The solid line represents the saturated inversion caused by mode 1. The effect of the saturation is largely exaggerated in this figure. The lower part shows the intensity distribution of the second mode. As can be seen, there are appreciable parts where the mode intensity is high but where the saturated inversion is not appreciably lowered compared to the unsaturated inversion, so that the second mode can still be appreciably supported.

We first confine ourselves to the one-dimensional case, as indicated by (VIII.2.2), or to modes with k vectors which differ only in one component.

We then obtain using (VIII.1.3) and (VIII.2.2)

$$\left.\begin{aligned} &\sum_\mu W^0_{jk,\lambda,\mu} = W_\lambda, \\ &N\sum_\mu W^0_{jk,\lambda,\mu} W^0_{jk,\lambda',\mu} = \tfrac{3}{2} W_\lambda^2 \qquad \text{for} \quad \lambda=\lambda', \\ &\qquad\qquad = W_\lambda W_{\lambda'} \qquad \text{for} \quad \lambda\neq\lambda' \\ &\text{where} \quad W_\lambda = \frac{g^2\, 2\gamma}{(\nu_0-\omega_\lambda)^2+\gamma^2} N. \end{aligned}\right\} \tag{VIII.2.3}$$

The system of Eqs. (VIII.2.1) takes thus the form

$$3 W_\lambda n_\lambda + 2\sum_{\lambda'\neq\lambda} W_{\lambda'} n_{\lambda'} = \underbrace{2\left(\frac{B}{C} - \frac{2\varkappa_\lambda}{d_0 W_\lambda}\,\frac{B}{C}\right)N}_{=l_\lambda} \tag{VIII.2.4}$$

where we have put $\frac{A}{B}=d_0$.

Introducing $W_\lambda n_\lambda$ as new variables, we observe that (VIII.2.4) has a completely symmetric form, which immediately allows the explicit solution

$$\left.\begin{aligned} W_\lambda n_\lambda &= l_\lambda - \frac{2(l_1+\cdots+l_M)}{2M+1} = \frac{2BN}{C} \times \\ &\times\left(\frac{1}{2M+1} - \frac{2\varkappa_\lambda}{d_0 W_\lambda} + \frac{2\left(\frac{2\varkappa_1}{W_1}+\cdots+\frac{2\varkappa_M}{W_M}\right)}{(2M+1)\,d_0}\right) \end{aligned}\right\} \tag{VIII.2.5}$$

where M is the number of modes.

For the further evaluation of Eq. (VIII.2.5) we treat two examples:

α) *Axial modes with a different frequency distance from the line center.* Let the axial modes be discriminated from the non-axial modes by a great difference between their losses so that we can confine ourselves to the axial modes alone. These modes shall have the same losses, $\varkappa$, but they are still discriminated by their different frequencies, ω_λ. We put $\omega_\lambda = \nu_0 + m\delta$, m: integer, and obtain as the condition for $n_{\lambda_0} > 0$ (where $\omega_{\lambda_0} = \nu_0 + m_0\,\delta$ and where m_0 is the index for the mode which is farest away from resonance)

$$\frac{1}{2M+1} - \frac{\varkappa}{\gamma\, g^2\, d_0}\,(\gamma^2 + m_0^2\,\delta^2) + \frac{2\dfrac{\varkappa}{d_0\, g^2}\,\gamma\, M + \dfrac{2\varkappa\,\delta^2}{d_0\,\gamma\, g^2}\,\Sigma m^2}{2M+1} > 0 \tag{VIII.2.6}$$

or

$$1 - \frac{\varkappa\,\gamma^2}{\gamma\, g^2\, d_0} - \frac{\varkappa}{\gamma\, g^2\, d_0}\,\delta^2\{m_0^2(2M+1) - 2\Sigma m^2\} > 0. \tag{VIII.2.7}$$

Because we have assumed a symmetric position of the mode frequencies with respect to the atomic line center,

$$m_0 = \frac{M-1}{2} \tag{VIII.2.8}$$

Thus we can write Eq. (VIII.2.7) finally in the form

$$1 - \frac{\varkappa\,\gamma^2}{\gamma\, g^2\, d_0} - \frac{\varkappa\,\delta^2}{\gamma\, g^2\, d_0}\left\{m_0^2(4m_0+3) - \frac{2}{3}\,m_0(m_0+1)(2m_0+1)\right\} > 0. \tag{VIII.2.9}$$

The higher the inversion d_0 and the smaller the frequency difference (compared to the linewidth γ), the bigger m_0 is.

β) *Different losses.* We assume that the mode with the lowest $\varkappa$ lies exactly at resonance, and that the other modes which are lying closest to the resonance are non-axial modes. For the sake of simplicity we assume further that all these modes are situated in the same plane in k-space. Because the spatial modulation varies now in a single direction (perpendicular to the laser axis), a spatially inhomogeneous inversion result, so that again a factor $\frac{3}{2}$ occurs in Eq. (VIII.2.3). (The modes are assumed to have radial symmetry).

We can neglect the change in frequency, but we have to take into account that the losses $\varkappa_\lambda$ increase with an increasing angle between the propagation direction of the modes and the laser axis.

We put therefore

$$\varkappa_m = \varkappa(1+q)^{m-1} \qquad (Mq \ll 1) \tag{VIII.2.10}$$

Inserting (VIII.2.10) into Eq. (VIII.2.5) yields approximately for the number of modes, M:

$$M^2 \lesssim \frac{d_0 W}{2\varkappa q}\left(1 - \frac{2\varkappa}{d_0 W}\right). \tag{VIII.2.11}$$

When we express the inversion d_0 by the threshold inversions d_{thr} in the form $d_0 = d_{thr}(1+\eta)$, Eq. (VIII.2.11) can be further simplified:

$$M^2 \lesssim \frac{\eta}{q}. \tag{VIII.2.12}$$

If for instance $q=10^{-2}$, i.e. if the losses differ by only one percent, the coexistence of modes becomes possible by increasing the inversion by a few percent.

The coexistence of modes can be weakened by cross relaxation between atoms at different space points or by diffusion processes so that the spatial dependence of the inversion is flattened.

b) Spatially inhomogeneous pumping, homogeneously broadened line[2]. We start with formula (VIII.2.1) and distinguish between:

α) Running waves, i.e. $W^0_{jk,\lambda,\mu}$ does not depend on λ and μ,

β) Standing waves, i.e. $W^0_{jk,\lambda,\mu}$ *does* depend on λ and μ.

α) *Running waves.* Because the pumping cannot distinguish between the modes (which enter into the equations only through $W^0_{jk,\lambda,\mu}$), we find the same situation as discussed for the homogeneously broadened line (compare p. 252), i.e. only one mode can show laser action*.

β) *Standing waves.* For an illustration of the essential mechanism, which may either enhance or suppress multimode laser action, we treat the three-level system (laser transition between the two upper levels), with the simplification of Eq. (VIII.1.30). From the comparison between (VIII.1.30) and (VIII.1.12) we obtain

$$\left.\begin{aligned} A_\mu &= w_{13,\mu}, \\ B_\mu &= w_{13,\mu}+w_3, \\ C_\mu &= 1. \end{aligned}\right\} \tag{VIII.2.13}$$

We assume that only $w_{13,\mu}$, which is determined by the pumplight depends explicitly, but weakly on μ so that we may put $w_{13,\mu}=w_{13,0}(1+\delta_\mu)$ with $|\delta_\mu|\ll 1$. (For $w_{13,0}\ll w_3$ this condition is not necessary, however).

Thus Eq. (VIII.2.1) reduces to

$$\left.\begin{aligned} &\sum_{\lambda'} n_{\lambda'} \sum_\mu W^0_{32,\lambda,\mu} W^0_{32,\lambda',\mu} \frac{w_{13,0}}{(w_{13,0}+w_3)^2}\left\{1+\delta_\mu\left(1-\frac{2w_{13,0}}{(w_{13,0}+w_3)}\right)\right\} \\ &\qquad = -2\varkappa_\lambda + \sum_\mu W^0_{32,\lambda,\mu}\frac{w_{13,0}}{w_{13,0}+w_3}\left\{1+\delta_\mu\left(1-\frac{w_{13,0}}{w_{13,0}+w_3}\right)\right\}. \end{aligned}\right\} \tag{VIII.2.14}$$

In order to show most explicitly the way in which modes may be suppressed (or enhanced) we consider two modes, namely standing waves with wave-numbers k_1, k_2 and $\delta_\mu=\delta\cos[2(k_1-k_2)x_\mu]$, where x_μ is the position of atom μ. After an evaluation of the sums over μ we find [with aid of the formulas (VIII.2.3)]

$$\left.\begin{aligned} &3W_1 n_1 + W_2 n_2(2+F) = l_1 \quad \text{where} \quad F\sim\delta, \\ &W_1 n_1(2+F) + 3W_2 n_2 = l_2 \end{aligned}\right\} \tag{VIII.2.15}$$

with the solution

$$\left.\begin{aligned} &W_1 n_1 = \frac{1}{\tilde D}\{3l_1-(2+F)\,l_2\}; \quad W_2 n_2 = \frac{1}{\tilde D}\{3l_2-(2+F)\,l_1\}, \\ &\tilde D = 5-4F-F^2 \end{aligned}\right\} \tag{VIII.2.16}$$

l_λ was defined in Eq. (VIII.2.4).

---

* At a very high inversion a new situation occurs, however. The single mode becomes unstable and mode locking produces ultrashort pulses (see GRAHAM and HAKEN: Z. Physik **213**, 420 (1968) and in particular H. RISKEN and K. NUMMEDAL: J. Appl. Phys. **39**, 4662 (1968).

[2] We represent here unpublished results of the present author. Our analysis has been carried further in the meantime by K. KAUFMANN and W. WEIDLICH: Z. Physik, **217**, 113 (1968).

Provided $l_1 > l_2$ and $F > O\,(\delta > O)$ one establishes readily that mode 2 remains suppressed up to a higher pump power than without the inhomogeneous pumping term. This can also be seen directly by considering the limiting case $F \to 1$ where the equations are identical with those of running waves. In this case all modes but one are suppressed. On the other hand, with reversed sign of δ the coexistence of modes is supported.

The extension of these considerations to other pump light distributions and other mode distributions is obvious.

c) *Inhomogeneously broadened line.* The coexistence of modes has been treated in detail in Chap. VII.5a [see also the discussion following Eq. (VII.2.11)]. There we have shown that modes, whose frequency separation $|\Omega_{\lambda'} - \Omega_\lambda|$ is bigger than the homogeneous part γ of the atomic width, are supported by energetically different sets of atoms and are thus (nearly) independent of each other. Although the modes are independent of each other, in general not all of them show laser action. This stems from the fact that different modes may have a different threshold, because their threshold depends on the cavity loss and the mode frequency distance from the center of the inhomogeneously broadened line [compare Eq. (VII.4.7)].

VIII.3. Laser cascades[1,2]. So far we have treated laser action connected with a single atomic transition. As has been found experimentally in *He—Ne* and pure *Ne*-lasers, there may also occur two and three-step laser cascades of the type shown in Fig. 67. The rate equations allow this problem to be treated in a good

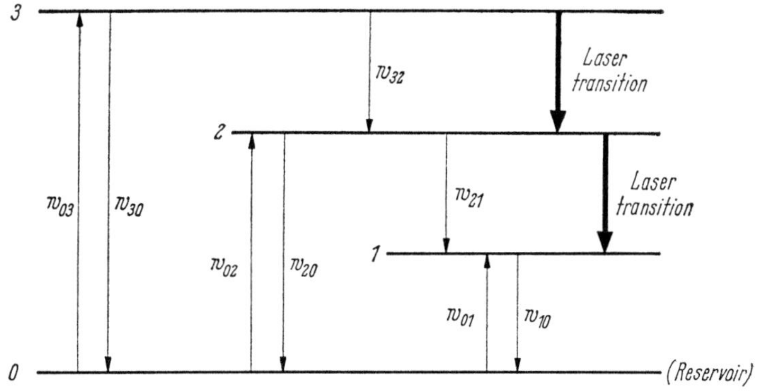

Fig. 67. Pump and decay scheme of the laser cascade treated in this chapter.

[1] Laser cascades have been observed by R. A. McFarlane, W. L. Lamb, C. K. N. Patel, and C. G. B. Garrett: Third Internat. Conference on Quantum Electronics, ed. by N. Bloembergen and P. Grivet. Paris: 1963 Dunod Cie. — R. Cagnard, R. der Agobian, R. Echard et J. L. Otto: Compt. Rend. **257**, 1044 (1963). — R. der Agobian, J. L. Otto, R. Echard et R. Cagnard: Compt. Rend. **257**, 3844 (1963). — H. J. Gerritsen and P. V. Goerdertier: Appl. Phys. Letters **4**, 20 (1964). — R. Grudzinski, M. Paillette et J. Becrelle: Compt. Rend. **258**, 1452 (1964). — J. L. Otto, R. Cagnard, R. Echard et R. der Agobian: Compt. Rend. **258**, 2779 (1964). — W. L. Faust, R. A. McFarlane, C. K. N. Patel, and C. G. B. Garrett: Phys. Rev. **133**, A 1476 (1964). — R. der Agobian, R. Cagnard, R. Echard et J. L. Otto: Compt. Rend. **258**, 3661 (1964). — W. R. Bennett Jr.: Bull. Am. Phys. Soc. **9**, 500 (1964). — V. N. Smiley: Appl. Phys. Letters **4**, 123 (1964). — R. der Agobian, J. L. Otto, R. Cagnard et R. Echard: Compt. Rend. **259**, 85 (1964). — R. der Agobian, J. L. Otto, R. Cagnard et R. Echard: Compt. Rend. **259**, 323 (1964). — R. der Agobian, J. L. Otto, R. Cagnard et R. Echard: Compt. Rend. **260**, 473 (1965), and others.

[2] A theoretical treatment equivalent to the use of rate equations has been given by H. Haken, R. der Agobian and M. Pauthier: Phys. Rev. **140**, A 437 (1965).

approximation, where we give the example of a two-step cascade. Because the treatment of fixed atoms is considerably simpler than that of moving atoms and because there are only minor differences in the results of a gas laser and a solid laser with an inhomogeneously broadened line, we treat fixed atoms. We consider four levels (compare Fig. 67) and allow either homogeneous or inhomogeneous broadening.

Level 0 serves as a pump reservoir, whereas the cascade starts at levels 3, and using level 2 as an intermediate step, ends up at level 1. There are now two laser modes, 1 and 2, connected with the transitions $(2\rightarrow 1)$ and $(3\rightarrow 2)$. The corresponding *field equations* are in the generalization of (VIII.1.6):

$$\frac{dn_1}{dt} = -2\varkappa_1 n_1 + \sum_\mu W^0_{21,1,\mu}(N_2 - N_1)_\mu n_1, \tag{VIII.3.1}$$

$$\frac{dn_2}{dt} = -2\varkappa_2 n_2 + \sum_\mu W^0_{32,2,\mu}(N_3 - N_2)_\mu n_2. \tag{VIII.3.2}$$

The pump (which is achieved by inelastic collisions with atoms or electrons in the present case) is assumed to connect level 0 with 1, 2 and 3 (see Fig. 67). We further allow a direct recombination (especially by spontaneous emission) from level 3 to 2 and 2 to 1, with transition rates $w_{32} = 1/T_2'$ and $w_{21} = 1/T_1'$ respectively.

a) The *matter equations* read:

$$\frac{dN_3}{dt} = N_0 w_{03} - N_3(w_{30} + w_{32}) - (N_3 - N_2)\, n_2 W^0_{32,2}, \tag{VIII.3.3}*$$

$$\left.\begin{aligned}\frac{dN_2}{dt} &= N_0 w_{02} - N_2(w_{20} + w_{21}) + N_3 w_{32} + \\ &\quad + (N_3 - N_2)\, n_2 W^0_{32,2} - (N_2 - N_1)\, n_1 W^0_{21,1},\end{aligned}\right\} \tag{VIII.3.4}$$

$$\frac{dN_1}{dt} = N_0 w_{01} - N_1 w_{10} + N_2 w_{21} + (N_2 - N_1)\, n_1 W^0_{21,1}. \tag{VIII.3.5}$$

The equation for dN_0/dt follows from (VIII.3.3)—(VIII.3.5) using

$$N_0 + N_1 + N_2 + N_3 = 1 \tag{VIII.3.6}$$

and need not be written down.

A completely exact treatment of the present problem would lead to a most lengthy calculation, which we may circumvent, however, in the case of practical interest where the occupation numbers N_3, N_2, N_1 of the cascading levels are considerably smaller than that of the pump reservoir. We can then replace N_0 everywhere by 1 in (VIII.3.3)—(VIII.3.5). In the following we use the abbreviations

$$(w_{30} + w_{32}) = \frac{1}{T_3}; \qquad (w_{20} + w_{21}) = \frac{1}{T_2}; \qquad w_{10} = \frac{1}{T_1}$$

and

$$\left.\frac{w_{03}}{w_{30} + w_{32}} = d_3; \qquad \frac{w_{02}}{w_{20} + w_{21}} = d_2; \qquad \frac{w_{01}}{w_{10}} = d_1.\right\} \tag{VIII.3.7}$$

In the steady state all time derivatives are equal to zero, so that we are left with algebraic equations. One calculates first $(N_3 - N_2)$ and $(N_2 - N_1)$ from (VIII.3.3) to (VIII.3.5).

* The index μ is dropped everywhere.

In the next step of our analysis we have to insert these expressions into (VIII.3.1) and (VIII.3.2) and to perform the sum over μ. As we noticed in VIII.2, this summation or integration presents essential difficulties if the quantities $n_\lambda W^0_{jk,\lambda,\mu}$ are in the denominator.

These expressions are small enough for moderate pumping, however, so that we can expand $(N_3 - N_2)$ and $(N_2 - N_1)$ in powers of n_λ, where we confine ourselves to expressions linear in n_λ. We thus obtain for the

b) Homogeneously broadened line and standing waves (modes in axial direction) the equation for mode 1:

$$\varkappa_1 = \frac{\gamma_{21}}{(\Delta\hat{\Omega}_1)^2 + \gamma_{21}^2} \left\{ g_1^2 N A + \frac{3}{2} n_1 N \widetilde{W}_{21} g_1^4 B + n_2 g_1^2 g_2^2 N \widetilde{W}_{32} C \right\} \qquad \text{(VIII.3.8)}$$

where

$$\left.\begin{aligned} A &= \left[-d_1 + d_2\left(1 - \frac{T_1}{T_1'}\right) + d_3\left(\frac{T_2}{T_2'} - \frac{T_1 T_2}{T_1' T_2'}\right)\right], \\ B &= \left(T_1 + T_2 - T_2 \frac{T_1}{T_1'}\right)\left[d_1 - d_2\left(1 - \frac{T_1}{T_1'}\right) - d_3 \frac{T_2}{T_2'}\left(1 - \frac{T_1}{T_1'}\right)\right], \\ C &= T_2\left(1 - \frac{T_1}{T_1'}\right)\left(1 - \frac{T_3}{T_2'}\right)\left[-d_2 + d_3\left(1 - \frac{T_2}{T_2'}\right)\right]. \end{aligned}\right\} \qquad \text{(VIII.3.9)}$$

Equation for mode 2:

$$\varkappa_2 = \frac{\gamma_{32}}{(\Delta\hat{\Omega}_2)^2 + \gamma_{32}^2} \left\{ g_2^2 N A' + n_1 N \widetilde{W}_{21} g_1^2 g_2^2 B' + \frac{3}{2} n_2 N \widetilde{W}_{32} C' g_2^4 \right\} \qquad \text{(VIII.3.10)}$$

where

$$\left.\begin{aligned} A' &= -d_2 + d_3\left(1 - \frac{T_2}{T_2'}\right), \\ B' &= -d_1 T_2 + d_2 T_2\left(1 - \frac{T_1}{T_1'}\right) + d_3 T_2^2 \frac{1}{T_2'}\left(1 - \frac{T_1}{T_1'}\right), \\ C' &= d_2\left(T_2 + T_3 - T_3 \frac{T_2}{T_2'}\right) - d_3\left(1 - \frac{T_2}{T_2'}\right)\left(T_2 + T_3 - T_3 \frac{T_2}{T_2'}\right). \end{aligned}\right\} \qquad \text{(VIII.3.11)}$$

We have used the following abbreviations:

γ_{jk}: linewidth of the optical transition $j \to k$

$$\Delta\hat{\Omega}_1 = (\nu_{21} - \Omega_1); \qquad \Delta\hat{\Omega}_2 = (\nu_{32} - \Omega_2) \qquad \text{(VIII.3.12)}$$

ν_{jk} is the (circular) transition frequency between the levels j and k. N: number of atoms. g_1 and g_2 are proportional to the optical matrix elements and are given by

$$g_k = -\frac{e}{m} \sqrt{2\pi/\hbar\omega_\lambda} \left| \int \varphi_{k+1}^* \boldsymbol{p} \varphi_k \, dV \boldsymbol{e}_\lambda \right| \sqrt{1/V} \qquad \text{[compare (III.5.8)]}$$

$$\widetilde{W}_{jk} = \frac{W_{jk,\lambda,\mu}}{|g_{j,k,\lambda,\mu}|^2}. \qquad \text{[compare (VIII.1.3)]}$$

The factors $\frac{3}{2}$ stem from the spatially inhomogeneous inversion (see VIII.2) and are replaced by 1 for running waves.

c) Inhomogeneously broadened line and standing waves. We assume now that there is a spread of transition frequencies ν_{21} and ν_{32}. This requires summing up in Eqs. (VIII.3.8) and (VIII.3.10) not only over the atomic position but also taking a weighted average over the spread of frequencies. To this end we assume independent Gaussian distributions for ν_{21} and ν_{32}.

In the *gas laser* this Gaussian distribution is, of course, brought about by the Maxwellian velocity distribution in one dimension. Consider an atom with velocity v in its upper state, whose transition is in resonance with mode 2. After the transition the atom can face two different situations:

1. Its velocity v is such that there is no resonance between the transition $2 \to 1$ and the mode 1 (or the set of modes 1). Then the atom must wait for a wide-angle scattering, which randomizes the velocity distribution. Then we have to apply two independent Gaussian distributions for ν_{32} and ν_{21} (having the same width, of course).

2. Its velocity v is such that the lower transition can follow immediately. Then the spread of ν_{32} and ν_{21} is simultaneously determined by a single Gaussian distribution, so that only *one* integral occurs in (VIII.3.8), (VIII.3.10).

We assume here that the situation 1) dominates. Instead of Eq. (VIII.3.8) for mode 1 we now obtain:

$$\left.\begin{aligned}\varkappa_1 = \iint \frac{d\nu_{21}\, d\nu_{32}}{\pi \alpha_1 \alpha_2} \exp\left[-\left(\frac{\nu_{21}-\nu_{21}^0}{\alpha_1}\right)^2\right] \exp\left[-\left(\frac{\nu_{32}-\nu_{32}^0}{\alpha_2}\right)^2\right] \times \\ \times \frac{\gamma_{21}}{(\Omega_1-\nu_{21})^2+\gamma_{21}^2} \{\cdots\}\end{aligned}\right\} \qquad \text{(VIII.3.13)}$$

where the bracket $\{\cdots\}$ is the same as in Eq. (VIII.3.8). The integrals were evaluated in VII.4.

One thus obtains readily:

$$\left.\begin{aligned}\varkappa_1 = g_1^2\, N\, A \frac{\pi^{\frac{1}{2}}}{\alpha_1} \exp(-\delta_1^2) + n_1\, N \left(\frac{3}{2}\right) g_1^4 \cdot B\, \pi^{\frac{1}{2}} \exp(-\delta_1^2)/\gamma_{21}\, \alpha_1 + \\ + n_2\, g_1^2\, g_2^2 \cdot N \cdot C \frac{2\pi}{\alpha_1 \alpha_2} \exp(-\delta_1^2) \exp(-\delta_2^2).\end{aligned}\right\} \qquad \text{(VIII.3.14)}$$

The quantities A, B, C, are the same as defined in (VIII.3.9) and

$$\delta_1 = \frac{\Omega_1 - \nu_{21}^0}{\alpha_1}; \qquad \delta_2 = \frac{\Omega_2 - \nu_{32}^0}{\alpha_2}. \qquad \text{(VIII.3.15)}$$

If we consider optical transitions close to resonance, we may replace the exponential factors by 1. As one may note, the former factor $1/\gamma_{21}$ for resonance in front of the bracket in (VIII.3.8) is now replaced by $\pi^{\frac{1}{2}}/\alpha_1$. Different factors occur, however, for the n's.

In a completely analogous manner we obtain for mode 2:

$$\left.\begin{aligned}\varkappa_2 = g_2^2 \cdot N \cdot A' \frac{\pi^{\frac{1}{2}}}{\alpha_2} \exp(-\delta_2^2) + n_1\, N\, g_1^2\, g_2^2\, B' \cdot \frac{2\pi}{\alpha_1 \alpha_2} \times \\ \times \exp[-(\delta_1^2+\delta_2^2)] + n_2\, N\, C'\, g_2^4 \left(\frac{3}{2}\right) \frac{\pi^{\frac{1}{2}}}{\gamma_{32}\, \alpha_2} \exp(-\delta_2^2)\end{aligned}\right\} \qquad \text{(VIII.3.16)}$$

where A', B', and C' are defined in (VIII.3.11).

d) Discussion of an example: Because the coefficients of the n's occurring in Eqs. (VIII.3.14) and (VIII.3.16) are rather complicated and since there are cascades in which only the upper level is pumped from the reservoir, let us consider this case in more detail.

The occurrence of a cascade depends critically on the relative size of the w_{ik}'s. As an example we use w_{ik}'s which seem adequate for the cascade in pure neon:

$w_{21} < w_{10}$; w_{20} negligible compared to w_{21}.

We then have:

$$\left.\begin{aligned}
&A=\frac{w_{32}}{w_{21}}\left(1-\frac{w_{21}}{w_{10}}\right)d_3>0; \qquad B=-d_3\frac{w_{32}}{w_{21}^2}\left(1-\frac{w_{21}}{w_{10}}\right)<0,\\
&C=\frac{1}{w_{21}}\left(1-\frac{w_{21}}{w_{10}}\right)\frac{w_{30}}{w_{30}+w_{32}}\left(1-\frac{w_{32}}{w_{21}}\right)d_3\gtrless 0,\\
&A'=\left(1-\frac{w_{32}}{w_{21}}\right)d_3\gtrless 0, \qquad B'=\frac{w_{32}}{w_{21}^2}\left(1-\frac{w_{21}}{w_{10}}\right)d_3>0,\\
&C'=-d_3\left(1-\frac{w_{32}}{w_{21}}\right)\frac{w_{30}+w_{21}}{w_{30}+w_{32}}\,\frac{1}{w_{21}}\gtrless 0.
\end{aligned}\right\} \tag{VIII.3.17}$$

The ambiguity in the sign of C, A' and C' stems from the ambiguity in the relative size of w_{32} and w_{21}.

1. $w_{32}>w_{21}$: Then $C<0$, $A'<0$, $C'>0$. (VIII.3.16) can be put into the form

$$\frac{\varkappa_2}{d_3}+\alpha'=\beta' n_1+\gamma' n_2 \tag{VIII.3.18}$$

where $\alpha', \beta', \gamma'>0$ and do not depend on the pumping occupation d_3. This formula shows that in this case, $w_{32}>w_{21}$, no cascade can occur. If the inversion $\sim d_3$ increases, the total output of either n_1 or n_2 or both must decrease.

2. $w_{32}<w_{21}$: Then $C>0$, $A'>0$, $C'<0$.

Eq. (VIII.3.14) for mode 1 takes the form

$$\left.\begin{aligned}
&\alpha-\frac{\varkappa_1}{d_3}=n_1\beta-\gamma n_2\\
&\text{with}\qquad \alpha,\beta,\gamma>0.
\end{aligned}\right\} \tag{VIII.3.19}$$

and Eq. (VIII.3.16) for mode 2:

$$\left.\begin{aligned}
&\alpha'-\frac{\varkappa_2}{d_3}=\gamma' n_2-\beta' n_1\\
&\alpha',\beta',\gamma'>0
\end{aligned}\right\} \tag{VIII.3.20}$$

(VIII.3.19) and (VIII.3.20) are to be solved with the condition $n_1, n_2>0$, otherwise one of the eqs. must be dropped.

We discuss Eqs. (VIII.3.19) and (VIII.3.20) by a graphical plot (Fig. 68). Fig. 68 demonstrates three typical situations, according to three different pump rates. The dashed lines represent Eq. (VIII.3.19) the solid lines Eq. (VIII.3.20). For low pumping we obtain the lines I. Their crossing point lies in the third quadrant, i.e. both n's are negative and no laser cascade occurs. Therefore, we check whether single-mode action may happen, by dropping one of the Eqs. (VIII.3.19), (VIII.3.20). In the representation of Fig. 68 this means that we have to see whether the dashed line allows for $n_1>0$, $n_2=0$ or the solid line for $n_2>0$, $n_1=0$, which in the present example is not the case. Therefore no laser action occurs.

Consider now the lines II, where a somewhat increased pumping rate is assumed. The crossing point lies in the second quadrant and no cascade can show up. However, the solid line cuts the n_2-axis at a positive value, which indicates, that we are now (slightly) above threshold for the upper laser transition. At a still higher pumping rate, we obtain finally the lines III, having a crossing point for $n_1>0$, $n_2>0$. Here the full cascade occurs. Fig. 68 allows us to deduce a condition for the occurrence of a cascade. The crossing point shifts with increasing d_3 (pumping) from the 3rd to the first quadrant only, if the slope of the dashed lines is steeper than that of the solid lines or, if

$$\frac{\beta}{\beta'}>\frac{\gamma}{\gamma'}.$$

It is a remarkable feature of Eqs. (VIII.3.19) and (VIII.3.20) that n_1 and n_2 play a completely symmetric role. Thus, not only can laser action of the upper transition cause laser action of the lower one, but the inverse can also happen. Therefore the question arises, which of the two transitions happens first and may,

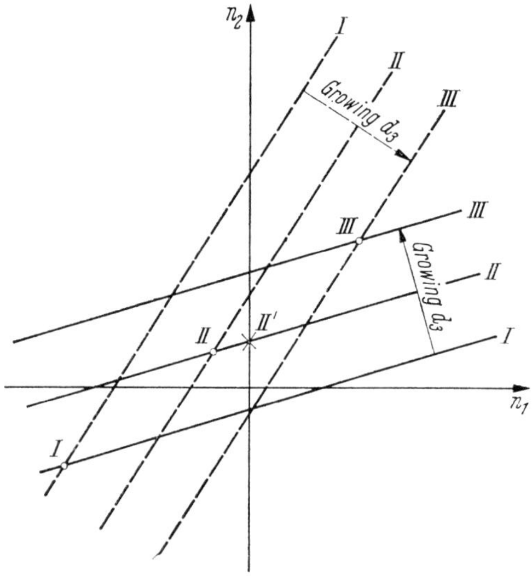

Fig. 68. Schematic plot of n_2 versus n_1. The dashed lines represent equation (VIII.3.19) the solid lines Eq. (VIII.3.20) for different values of the inversion parameter d_3. The points I, II, III represent situations where no mode shows laser action, mode 2 shows laser action, and both modes show laser action in a cascade, respectively.

at higher pumping, trigger the other transition. Inspection of Eqs. (VIII.3.19), (VIII.3.20) shows that the line which lasers first is the one for which whose corresponding Eq. (VIII.3.19) or (VIII.3.20) has an inhomogeneous term bigger than that of the remaining equation.

VIII.4. Solution of the time-dependent rate equations. Relaxation oscillations[1,2].

a) The 3-level system with laser action between the two lower levels. The basic equations are given by (VIII.1.6), (VIII.1.7)–(VIII.1.9). In the time-dependent

[1] Relaxation oscillations of lasers were observed by R. J. COLLINS, D. F. NELSON, A. L. SCHAWLOW, W. BOND, C. G. B. GARRETT, and W. KAISER: Phys. Rev. Letters **5**, 303 (1960) immediately after the detection of laser action. Since then, a great number of papers on this subject has appeared; — For references see e.g. the book of Röss, l. c. 6 on p. 2.

[2] Time-dependent rate equations were introduced into maser theory by H. STATZ and G. A. DE MARS, in: Quantum Electronics, ed. by C. H. TOWNES, p. 530—537. New York: Columbia University Press 1960; — They formulated these equations to account for the undamped spiking in the output of the microwave ruby maser oscillator. — D. M. SINNET: J. Appl. Phys. **33**, 1578 (1962) and G. MAKKOV: J. Appl. Phys. **33**, 202 (1962), have shown, however, that these equations have no isolated periodic solution (i.e. no limit cycle); — For a detailed and rigorous proof see: E. HOFELICH-ABATE and F. HOFELICH: Z. Physik **211**, 142 (1968). — MAKHOV proposed to add further terms to account for the pulsed mode of the ruby laser: see MAKHOV, l.c. [2], and G. MAKHOV and O. RISGIN, in: Quantum Electronics, ed. by P. GRIVET and N. BOEMBERGEN, part 2, p. 1121—1129. Paris: Dunod; New York: Columbia University Press 1964; — Rate equations extended by an intensity-dependent loss term for photons to include the effect of a saturable absorber, have a stable limit cycle under certain conditions: see: E. HOFELICH-ABATE and F. HOFELICH: Phys. Letters **26** A, 426 (1968); — J. Appl. Phys. **39**, 4823 (1968).

case their solution becomes very difficult due to the nonlinear terms $n_\lambda(N_2-N_1)$, so that one has to solve the equations either by approximations or by computer. In the following we confine ourselves to a single mode, spatially homogeneous pumping and inversion and a homogeneously broadened line. Therefore we drop everywhere the mode index λ and the atomic index μ. We put $w_{23}\approx 0$, $w_{12}\approx 0$.

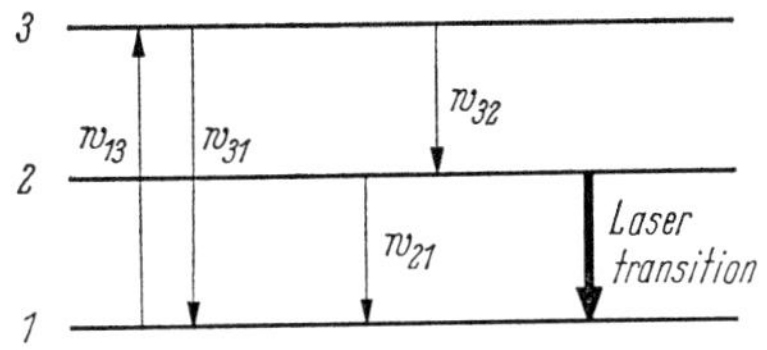

Fig. 69. Pump and decay scheme for the 3-level system treated in Sect. VIII.4.a.

The *field equation* reads

$$\dot{n} = -2\varkappa n + n N W_{21}^0 \Delta N + W_{21}^0 N_2 N \tag{VIII.4.1}$$

whereas we can reduce the matter equations due to $N_1+N_2+N_3=1$ to two equations

$$\dot{N}_1 = w_{31} - w_{31}(N_1+N_2) - w_{13} N_1 + n W_{21}^0 \Delta N + W_{21,tot}^0 N_2 + w_{21} N_2, \tag{VIII.4.2}$$

$$\dot{N}_2 = w_{32} - w_{32}(N_1+N_2) - n W_{21}^0 \Delta N - W_{21,tot}^0 N_2 - w_{21} N_2. \tag{VIII.4.3}$$

$W_{21,tot}^0$ is the spontaneous emission rate into all field modes. In Eqs. (VIII.1.7) to (VIII.1.9) it is implicitly contained in w_{21}.

Using $\Delta N = N_2 - N_1$ and $N_0 = N_2 + N_1$ as new variables, these equations can be given the form

$$\left.\begin{aligned}(\Delta N)^\cdot = {} & w_{32} - w_{31} - (w_{32}-w_{31}) N_0 + \tfrac{1}{2} w_{13}(N_0 - \Delta N) - \\ & - 2n W_{21}^0 \Delta N - (W_{21,tot}^0 + w_{21})(N_0 + \Delta N),\end{aligned}\right\} \tag{VIII.4.4}$$

$$\dot{N}_0 = w_{31} + w_{32} - (w_{31}+w_{32}) N_0 - w_{13} N_1. \tag{VIII.4.5}$$

N_0 adjusts itself almost adiabatically provided w_{32} is big:

$$N_0 = 1 - w_{13} N_1/(w_{31}+w_{32}). \tag{VIII.4.6}$$

Thus we obtain approximately

$$\left.\begin{aligned}(\Delta N)^\cdot = {} & \big(w_{13} - (W_{21,tot}^0 + w_{21})\big) N_0 - 2n W_{21}^0 \Delta N - \\ & - (W_{21,tot}^0 + w_{13} + w_{21}) \Delta N.\end{aligned}\right\} \tag{VIII.4.7}$$

In it N_0 can be considered as a constant in a good approximation. Note that N_2 which occurs in (VIII.4.1) is now, of course, to be replaced by

$$\tfrac{1}{2}(N_0 + \Delta N) \approx \tfrac{1}{2}(1+\Delta N).$$

b) 3-Level system, laser action between the two upper levels. The *field equation* reads

$$\dot{n} = -2\varkappa n + n N W_{32}^0 \Delta N + W_{32}^0 N_3 \cdot N \tag{VIII.4.8}$$

whereas the matter equations are reduced to

$$\dot{N}_2 = -w_{21} N_2 + n(N_3 - N_2) W^0_{32} + (W^0_{32,tot} + w_{32}) N_3, \quad \text{(VIII.4.9)}$$

$$\dot{N}_3 = w_{13}(1 - N_2 - N_3) - w_{31} N_3 - n(N_3 - N_2) W^0_{32} - (W^0_{32,tot} + w_{32}) N_3. \quad \text{(VIII.4.10)}$$

We introduce $\Delta N = N_3 - N_2$ and $N_0 = N_2 + N_3$ as new variables thus obtaining

$$\left.\begin{aligned}(\Delta N)^{\cdot} &= w_{13}(1 - N_0) + w_{21}\tfrac{1}{2}(N_0 - \Delta N) - w_{31}\tfrac{1}{2}(N_0 + \Delta N) - \\ &\quad - 2n\,\Delta N\, W^0_{32} - (W^0_{32,tot} + w_{32})(N_0 + \Delta N),\end{aligned}\right\} \quad \text{(VIII.4.11)}$$

$$\dot{N}_0 = w_{13} - N_0(w_{13} + \tfrac{1}{2}w_{21} + \tfrac{1}{2}w_{31}) + \tfrac{1}{2}(w_{21} - w_{31})\Delta N \quad \text{(VIII.4.12)}$$

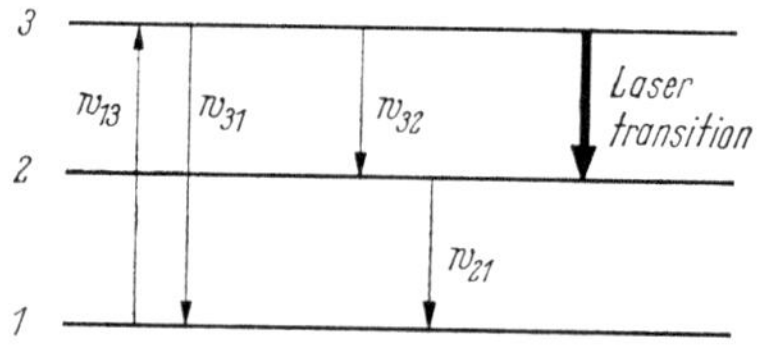

Fig. 70. Pump and decay scheme for the 3-level system treated in Sect. VIII.4.b.

w_{21} is assumed to be very big so that N_0 follows adiabatically:

$$N_0 = \frac{w_{13} + \frac{1}{2}(w_{21} - w_{31})\Delta N}{w_{13} + \frac{1}{2}w_{21} + \frac{1}{2}w_{31}}. \quad \text{(VIII.4.13)}$$

(Note that in this case $N_2 \approx 0$ but $w_{21} N_2 \neq 0$).

On account of $N_2 \approx 0$ it is sufficient to consider Eq. (VIII.4.10) for N_3 alone and to put $N_3 \approx \Delta N$:

$$(\Delta N)^{\cdot} = w_{13} - \Delta N(W^0_{32,tot} + w_{32} + w_{13} + w_{31}) - n\,\Delta N\, W^0_{32}. \quad \text{(VIII.4.14)}$$

For $w_{13} \approx w_{31} \ll W^0_{32,tot} + w_{32}$ this equation reduces finally to

$$(\Delta N)^{\cdot} = w_{13} - \Delta N(W^0_{32,tot} + w_{32}) - n\,\Delta N\, W^0_{32}. \quad \text{(VIII.4.15)}$$

c) 4-Level system. Provided that only the following transitions occur: $1 \to 4 \to 3 \to 2 \to 1$ and that further w_{43} and w_{21} are very big compared to all other transition rates, the corresponding system of equations can be reduced quite similarly to the 3-level system (with the upper transition lasing) to an equation of the type (VIII.4.14).

The terms $W^0_{21} N_2 N$ and $W^0_{32} N_3 N$ in the field Eqs. (VIII.4.1) and (VIII.4.8) respectively, which describe the spontaneous emission, can be neglected during the laser process. They serve merely as the trigger for laser action.

d) Approximate solution for small oscillations[3]. The Eqs. (VIII.4.1), (VIII.4.7), (VIII.4.8), and (VIII.4.15) have exactly the same structure. We consider therefore as examples (VIII.4.8) and (VIII.4.15) (upper transition).

[3] This sort of approximation has been performed by a number of authors: R. DUNSMUIR: J. Electron. Control **10**, 453 (1961). — W. KAISER, C. G. B. GARRETT, and D. L. WOOD: Phys. Rev. **123**, 766 (1962). — P. P. SOROKIN, M. J. STEVENSON, J. R. LANKARD, and G. D. PETTIT: Phys. Rev. **127**, 503 (1962). — H. STATZ, C. LUCK, C. SCHAFER, and M. CIFTAN, in: Advances in Quantum Electronics (J. SINGER, ed.), p. 342. New York: Columbia Univ. Press 1961. — K. GÜRS: Z. Naturforsch. **17** a, 990 (1962); **18** a, 510, 1363 (1963). For a recent detailed treatment of the spiking regime see e.g. A.M. SAMSON and L.A. KOTOMZEVA: Optika i Spektroskopiya **24**, 756 (1968).

We put

$$\Delta N \approx N_3 = N_3^0 + \delta N_3, \qquad \text{(VIII.4.16)}$$

$$n = n_0 + \delta n \qquad \text{(VIII.4.17)}$$

where the stationary solutions n_0 and N_3^0 are fixed by

$$-2\varkappa + N\, W_{32}^0\, N_3^0 = 0, \qquad \text{(VIII.4.18)}$$

$$w_{13} - N_3^0 (W_{32,\,tot}^0 + w_{32}) - n_0\, N_3^0\, W_{32}^0 = 0. \qquad \text{(VIII.4.19)}$$

Inserting (VIII.4.16) and (VIII.4.17) into (VIII.4.8) and (VIII.4.15) using (VIII.4.18) and (VIII.4.19) and linearizing we obtain after a slight rearrangement the following equations

$$\delta \dot n = \delta N_3\, n_0 \frac{2\varkappa}{N_3^0}, \qquad \text{(VIII.4.20)}$$

$$\delta \dot N_3 = -\frac{\delta N_3}{N_3^0}\, w_{13} - \delta n\, N_3^0\, W_{32}^0. \qquad \text{(VIII.4.21)}$$

In order to solve this linear homogeneous system of differential equations with constant coefficients we put

$$\delta n = A\, e^{\alpha t}, \qquad \delta N_3 = B\, e^{\alpha t} \qquad \text{(VIII.4.22)}$$

α has then to obey the equation

$$\alpha^2 + \alpha \frac{w_{13}\, N\, W_{32}^0}{2\varkappa} + \left(W_{32}^0\, N\, w_{13} - (W_{32,\,tot}^0 + w_{32})\, 2\varkappa\right) = 0. \qquad \text{(VIII.4.23)}$$

According to the two roots of Eq. (VIII.4.23), δn can be written in the form

$$A_1\, e^{-(\Gamma - i\omega_r)t} + A_2\, e^{-(\Gamma + i\omega_r)t} \qquad \text{(VIII.4.24)}$$

where

$$\frac{1}{\Gamma} = \frac{2 w_{thr}}{w_{13} (W_{32,\,tot}^0 + w_{32})} \qquad \text{(VIII.4.25)}$$

and

$$\omega_r = \sqrt{-\frac{w_{13}^2 (W_{32,\,tot}^0 + w_{32})^2}{4 w_{thr}^2} + \left(\frac{w_{13}}{w_{thr}} - 1\right) \cdot 2\varkappa (W_{32,\,tot}^0 + w_{32})} \qquad \text{(VIII.4.26)}$$

w_{thr} is given by

$$w_{thr} = \frac{(W_{32,\,tot}^0 + w_{32})\, 2\varkappa}{N\, W_{32}^0}.$$

As is evident from (VIII.4.24), the photon number returns after a deviation δn from its equilibrium back to its equilibrium value n_0 whereby the relaxation oscillations with the frequency ω_r occur.

VIII.5. The giant pulse laser. Let us consider a resonator whose quality factor* Q is kept very low at the beginning so that no laser action can take place. Then the pumping drives the population inversion to a higher value than that which can be maintained during the laser process. If now the Q is suddenly increased, the feedback causes laser action at a considerably higher output power level than can be achieved in steady-state operation. This method, called Q-spoiling[1],

* For the definition of Q see footnote[1] on p. 10.

[1] This method, known in connection with the problem of inverting spins in the microwave region [R. F. Chester and D. I. Bolef: Proc. I.R.E. (Corres.) **45**, 1287 (1957). — G. Feher, J. P. Gordon, E. Buehler, E. A. Gere, and C. D. Thurmond: Phys. Rev. **109**, 221 (1958)], was proposed by R. W. Hellwarth, in: Advances in Quantum Electronics (J. Singer, ed.), p. 334, New York: Columbia Univ. Press 1961 for the laser.

can be experimentally verified by rotating mirrors (or prisms or shutter discs)[2] so that the reflectivity of one mirror changes periodically with time. The switching can also be achieved by a Kerr cell[3], which is particularly suitable where the light of the laser-active atoms is already polarized (especially of ruby-crystals, for example).

Instead of changing the loss, we can also change the gain. For example in ruby, we can cause a quasi-STARK effect, which splits the atomic line, thus giving rise to an increased effective linewidth and a lowered gain. When the electric field is released, the laser pulse builds up[4].

a) Semiquantitative treatment. The whole time evolution of this process can be treated by means of the Eqs. (VIII.4.1) and (VIII.4.7). In the beginning ($t<0$) the Q-value is kept low (i.e. $2\varkappa_0$ is kept high).

Under a steady pumping, $\Delta\tilde{N}\equiv N\Delta N$ and n assume steady state values

$$\Delta\tilde{N}_i,\ n_i.$$

Note, that $\Delta\tilde{N}_i$ is essentially determined by *spontaneous emission.* At $t=0$ we switch the resonator suddenly to a high Q (low $2\varkappa_1$); because the losses are low, we expect the laser pulse to build up so quickly by induced emission that the spontaneous emission rate (as well as nonradiative decay rates) can be neglected. Furthermore, we neglect the further pumping. The laser equations then take the form

$$\frac{dn}{dt}=\underbrace{(W_{21}^0\,\Delta\tilde{N}-2\varkappa_1)}_{\alpha}\,n, \tag{VIII.5.1}$$

$$\frac{d(\Delta\tilde{N})}{dt}=-2W_{21}^0\,n\,\Delta\tilde{N}. \tag{VIII.5.2}$$

According to (VIII.5.2) the inversion $\Delta\tilde{N}$ decreases rather slowly for small n's, so that we may replace $\Delta\tilde{N}$ on the right hand sides of (VIII.5.1) (VIII.5.2) by its initial value $\Delta\tilde{N}_i$.

The solutions of (VIII.5.1) and (VIII.5.2) then read:

$$n=n_i\,e^{\alpha t}, \tag{VIII.5.3}$$

$$\Delta\tilde{N}(t)=\Delta\tilde{N}_i\left(1+2W_{21}^0\,n_i\frac{1}{\alpha}\,(1-e^{\alpha t})\right). \tag{VIII.5.4}$$

In the first phase of the pulse, n increases exponentially with the gain constant α.

On account of (VIII.5.2) or (VIII.5.4) the inversion decreases. The increase stops certainly at a time t_1, when $\Delta\tilde{N}(t_1)\approx 0$ (it stops due to the loss term $2\varkappa_1$ somewhat, but not appreciably, earlier). Neglecting 1 compared to $e^{\alpha t}$, it follows from (VIII.5.4)

$$n_i\,e^{\alpha t_1}=\frac{\alpha}{2W_{21}^0}=n_{\max} \tag{VIII.5.5}$$

or

$$t_1=\frac{1}{\alpha}\ln\frac{\alpha}{2W_{21}^0\,n_i}. \tag{VIII.5.6}$$

---

[2] R. J. COLLINS and P. KISLIUK: J. Appl. Phys. **33**, 2009 (1962).

[3] F. J. McCLUNG and R. W. HELLWARTH: J. Appl. Phys. **33**, 828 (1962).

[4] This method has been developed by W. KAISER and H. LESSING: Z. Physik **176**, 525 (1963); Appl. Phys. Letters **2**, 206 (1963); — A number of other methods have also been applied [e.g. strong magnetic fields: H. C. NEDDERMANN, Y. C. KIANG, and F. C. UNDERLEITNER: Proc. I.R.E. **50**, 1687 (1962)].

(VIII.5.5) incidentally represents the number of photons present at this time, t_1. After the time t_1, we may assume $\Delta\tilde{N}\approx 0$. The photon number then decays more or less exponentially

$$n(t)=n_{\max}\cdot e^{-2\varkappa_1(t-t_1)}. \tag{VIII.5.7}$$

These arguments give at least rough estimates of the quantities t_1, $n_{\max}$ and $n(t)$. For a more detailed analysis the Eqs. (VIII.5.1) and (VIII.5.2) must be solved in the vicinity of $t=t_1$ by a computer (whereas the slopes are well represented by (VIII.5.3) and (VIII.5.7). This has been performed by WAGNER and LENGYEL[5], who have also obtained several exact relations, and whom we will follow now:

b) Quantitative treatment[5]*. We assume again that the switching takes place so quickly that the population inversion is not appreciably changed. It is further assumed that during the pulse we can neglect the effects of pumping and spontaneous emission. We investigate the coupled equations for the temporal development of the photon number of a single mode and of the atomic inversion, starting from Eqs. (VIII.5.1) and (VIII.5.2).

Multiplying (VIII.5.1) by 2 and adding it to (VIII.5.2) yields:

$$\frac{d}{dt}(\Delta\tilde{N}+2n)=-4\varkappa n. \tag{VIII.5.8}$$

Integrating and neglecting $n_{initial}$ and n_{final} yields:

$$\Delta\tilde{N}_i-\Delta\tilde{N}_f=2\int_0^\infty 2\varkappa n\,dt. \tag{VIII.5.9}$$

This equation states that the total photon number radiated equals half the difference of the inversion of the initial and final state. (VIII.5.9) is equivalent to the conservation of energy.

Dividing (VIII.5.1) by (VIII.5.2) yields a differential equation for $\frac{dn}{d\Delta\tilde{N}}$ which has the solution

$$n-n_i=-\frac{1}{2}(\Delta\tilde{N}-\Delta\tilde{N}_i)+\frac{\varkappa}{W_{21}^0}\ln\left(\frac{\Delta\tilde{N}}{\Delta\tilde{N}_i}\right). \tag{VIII.5.10}$$

From (VIII.5.1) it is seen that $\frac{2\varkappa}{W_{21}^0}$ equals the threshold inversion $\Delta\tilde{N}_p$. Since $n_f=n_i=0$, the final population inversion is, after a rearrangement of (VIII.5.10), given by

$$\frac{\Delta\tilde{N}_f}{\Delta\tilde{N}_i}=\exp\left\{\frac{\Delta\tilde{N}_i}{\Delta\tilde{N}_p}\left[\frac{\Delta\tilde{N}_f}{\Delta\tilde{N}_i}-1\right]\right\}. \tag{VIII.5.11}$$

While $\frac{\Delta\tilde{N}_i}{\Delta\tilde{N}_p}$ is given, $\frac{\Delta N_f}{\Delta\tilde{N}_i}$ is to be determined, which can be done numerically (compare Fig. 71).

The total energy output of the pulse is

$$E=\tfrac{1}{2}(\Delta\tilde{N}_i-\Delta\tilde{N}_f)\,\hbar\omega \qquad [\text{compare (VIII.5.9)}] \tag{VIII.5.12}$$

---

* Recently, this problem has been solved exactly by E. HOFELICH-ABATE and F. HOFELICH: Z. Physik **209**, 13 (1968) where the effect of spontaneous emission is included.

[5] W. G. WAGNER and B. A. LENGYEL: J. Appl. Phys. **34**, 2040 (1963).

For high inversion (for instance $\Delta\tilde{N}_i/\Delta\tilde{N}_p > 4$) it is seen from Fig. 71, that $\Delta\tilde{N}_f \ll \Delta\tilde{N}_i$, so that

$$E \approx \tfrac{1}{2}\Delta\tilde{N}_i\hbar\omega. \tag{VIII.5.13}$$

If on the other hand the inversion is slight ($\Delta\tilde{N}_i \approx \Delta\tilde{N}_f$), one finds essentially

$$E \approx (\Delta\tilde{N}_i - \Delta\tilde{N}_p)\,\hbar\omega \tag{VIII.5.14}$$

(compare also McClung and Hellwarth, l.c.[3] on p. 268)

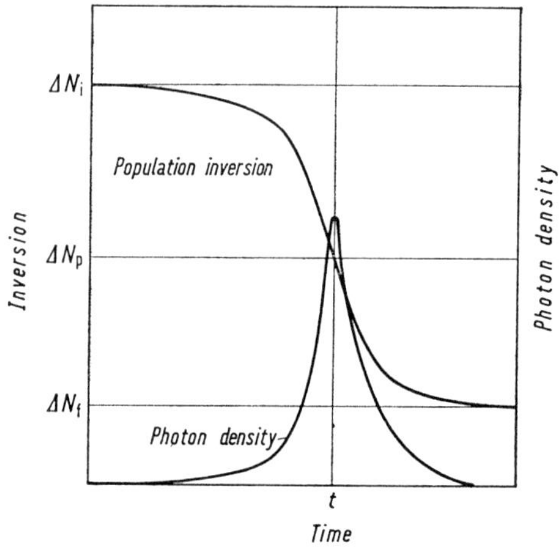

Fig. 71. Inversion and photon density in the giant pulse. [From Wagner and Lengyel: J. Appl. Phys. **34**, 2040 (1963), Fig. 1.]

dn/dt changes its sign according to (VIII.5.1) when at a time t_0

$$\Delta\tilde{N}(t_0) = \Delta\tilde{N}_p. \tag{VIII.5.15}$$

At this time, t_0, the peak power is reached. Inserting $\frac{2\varkappa_1}{W_{21}^0} = \Delta\tilde{N}_p$ into (VIII.5.10) yields for the peak photon number (neglecting n_i):

$$n_p = \frac{1}{2}\left(\Delta\tilde{N}_p \ln\left(\frac{\Delta\tilde{N}_p}{\Delta\tilde{N}_i}\right) - (\Delta\tilde{N}_p - \Delta\tilde{N}_i)\right). \tag{VIII.5.16}$$

In order to obtain the *peak power radiated*, one has to multiply as usual (VIII.5.16) by $2\varkappa\cdot\hbar\omega$.

An example of a computer solution of $n(t)$ is represented in Fig. 72.

For comparison with experimental data it should be borne in mind that in practical applications multimode laser action often occurs, and that this is not covered by the above theoretical treatment. In particular cases it also seems necessary to go beyond the rate equation treatment: Buley and Cummings[6] have shown by a computer solution of the semiclassical Eqs. (VII.1.32)—(VII.1.34), that an afterpulse of the light field may occur, even if the inversion $\Delta\tilde{N}$ is negative. This phenomenon can be explained by the super radiance effect: A certain time after the principal pulse has occured, the atoms start to oscillate in phase and give rise to super radiant emission.

[6] E. R. Buley and F. W. Cummings: Phys. Rev. **134**, A 1454 (1964).

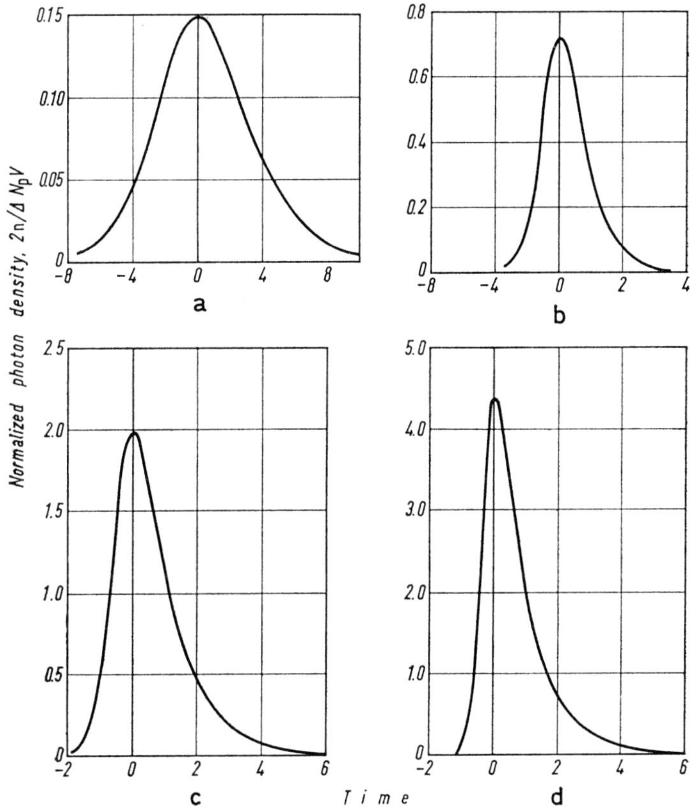

Fig. 72a—d. Photon density *vs* time in the central region of a giant pulse. (Time is measured in units of photon lifetime T in the Fabry-Perot interferometer. Origin at peak.)

a) $\log \Delta N_i/\Delta N_p = 0.5$, $\Delta N_i/\Delta N_p = 1.649$;
b) $\log \Delta N_i/\Delta N_p = 1.0$, $\Delta N_i/\Delta N_p = 2.718$;
c) $\log \Delta N_i/\Delta N_p = 1.5$, $\Delta N_i/\Delta N_p = 4.482$;
d) $\log \Delta N_i/\Delta N_p = 2.0$, $\Delta N_i/\Delta N_p = 7.389$.

[From WAGNER and LENGYEL: J. Appl. Phys. **34**, 2040 (1963), Fig. 3.]

IX. Further methods for dealing with quantum systems far from thermal equilibrium.

In this chapter we present several methods which were mostly derived very recently and which appear to be applicable not only to the laser but also to non-linear optics, spinwave theory and other problems.

Our treatment is fully quantum mechanical. We first discuss the *general form of the density matrix equation* which is not confined to the Markoffian assumption, so that it is an integral equation. We then show how one can establish an *exact generalized Fokker-Planck equation* for a c-number distribution function, which belongs to the general density matrix equation. We are further concerned with *elimination techniques* for atomic operators in coupled equations for field and atoms. We give exact elimination methods for inhomogeneously broadened lasers on the example of 2-level atoms using quantum mechanical Langevin equations.

Finally, we show how the atomic operators can be eliminated exactly within the density matrix equation or the Fokker-Planck equation.

IX.1. The general form of the density matrix equation[1]. We adopt the method of projection operators. We assume that a system can be described by a set of quantum states which are distinguished by a quantum number l. In the following we always assume that l stands for the complete set which characterizes the quantum states of the system. For instance, in the many-electron system, l could stand for all quantum numbers which describe the individual electrons.

In the following we shall use only the property that the projection operators P_{ik} form an algebraic ring defined by the relations

$$P_{ik} P_{lm} = P_{im} \delta_{kl}. \tag{IX.1.1}$$

By means of these projection operators it is possible to decompose every operator Ω into

$$\Omega = \sum_{lm} \langle l|\Omega|m\rangle P_{lm}. \tag{IX.1.2}$$

We now consider the most general form of a density matrix equation. We assume that the system under consideration is subject to external forces, for instance external electric and magnetic fields, and that it is coupled to external reservoirs (heat baths). Furthermore, we admit that the system itself has an internal interaction, e.g. in the case of spins, the spin-spin interaction or in the case of electrons the electron-electron interaction. These interactions cause transitions between the different quantum states. We assume, as usual, that the equation for $\varrho(t)$ is linear in ϱ. Then the equation for ϱ must be of the following form

$$\left.\begin{aligned} \varrho(t) = \varrho(t_0) + \int_{t_0}^{t} K^{(0)}(t,\tau)\,\varrho(\tau)\,d\tau + \int_{t_0}^{t} d\tau \Big\{ \sum_{iki'k'} K_{ikk'i'}(t,\tau) \times \\ \times P_{ik}\,\varrho(\tau) P_{k'i'} + \sum_{ik} K_{ik}^{(1)}(t,\tau) P_{ik}\,\varrho(\tau) + \sum_{ik} K_{ik}^{(2)}(t,\tau)\,\varrho(\tau) P_{ik}\Big\}, \end{aligned}\right\} \tag{IX.1.3}$$

t_0 is the initial time. As a consequence of causality the integral runs from t_0 to only t. The most general form in which the projection operators can occur is given by (IX.1.3), because higher powers in the projection operator can always be reduced according to (IX.1.1). Provided the kernels K depend only on τ, i.e. $K = K(\tau)$, Eq. (IX.1.3) reduces to an ordinary density matrix equation, which had been introduced e.g. in the theory of magnetic resonances[2] or in the laser[3]:

$$\left.\begin{aligned} \frac{d\varrho}{dt} = \sum_{iki'k'} K_{ikk'i'}(t)\, P_{ik}\,\varrho(t) P_{k'i'} + \sum_{ik} K_{ik}^{(1)}(t) P_{ik}\,\varrho(t) + \\ + \sum_{ik} K_{ik}^{(2)}(t)\,\varrho(t) P_{ik} + K^{(0)}(t)\,\varrho(t). \end{aligned}\right\} \tag{IX.1.4}$$

We now discuss some of the properties of this density matrix equation. As usual we have to require that ϱ is hermitian

$$\varrho = \varrho^{+} \tag{IX.1.5}$$

---

[1] H. HAKEN: Z. Physik **219**, 411 (1969). A rather broad class of density matrix equations has been derived by P.N. ARGYRES and P.L. KELLEY: Phys. Rev. **134**, A 98 (1964).

[2] R.K. WANGSNESS and F. BLOCH: Phys. Rev. **89**, 728 (1953). — P.N. ARGYRES, in: Magnetic and electric resonance and relaxation, p. 555, ed. by J. SMIDT. Amsterdam: North Holland Publ. Co. Inc. 1963.

[3] W. WEIDLICH and F. HAAKE: Z. Physik **186**, 203 (1965). — C.R. WILLIS: Phys. Rev. **147**, 406 (1960).

and that the total probability of finding the system in any state is $=1$, i.e.

$$\operatorname{tr}\varrho = 1. \tag{IX.1.6}$$

Taking the hermitian conjugate of (IX.1.3) we find

$$\left.\begin{aligned}\varrho^+(t) = \varrho^+(t_0) + \int_{t_0}^{t} d\tau \Big\{ &\varrho^+(\tau) K^{(0)*}(t,\tau) + \\ &+ \sum_{ik,i'k'} K^*_{ikk'i'}(t,\tau) P_{i'k'}\, \varrho^+(\tau) P_{ki} + \\ &+ \sum_{ik} K^{(1)*}_{ik}(t,\tau)\, \varrho^+(\tau) P_{ki} + \sum_{ik} K^{(2)*}_{ik}(t,\tau) P_{ki}\, \varrho^+(\tau) \Big\}\end{aligned}\right\} \tag{IX.1.7}$$

from which we immediately deduce by use of (IX.1.5)

$$K^{(0)*} = K^{(0)}, \tag{IX.1.8}$$

$$K^*_{ikk'i'} = K_{i'k'ki} \tag{IX.1.9}$$

and

$$K^{(1)*}_{ik} = K^{(2)}_{ki}. \tag{IX.1.10}$$

From the condition (IX.1.6) we find

$$\left.\begin{aligned}0 = \int_{t_0}^{t} d\tau \Big\{ &K^{(0)}(t,\tau) + \sum_{ik,i'k'} K_{ikk'i'}(t,\tau)\operatorname{tr}\big(P_{ik}\,\varrho(\tau) P_{k',i'}\big) + \\ &+ \sum_{ik} K^{(1)}_{ik}(t,\tau)\operatorname{tr}\big(P_{ik}\,\varrho(\tau)\big) + \sum_{ik} K^{(2)}_{ik}(t,\tau)\operatorname{tr}\big(\varrho(\tau) P_{ik}\big)\Big\}\end{aligned}\right\} \tag{IX.1.11}$$

(IX.1.11) is certainly fulfilled if*

$$\left.\begin{aligned}K^{(0)}(t,\tau) &+ \sum_{iki'k'} K_{ikk'i'}\operatorname{tr}\big(P_{k'k}\,\varrho(\tau)\big)\,\delta_{ii'} + \sum_{ik} K^{(1)}_{ik}(t,\tau)\operatorname{tr}\big(P_{ik}\,\varrho(\tau)\big) + \\ &+ \sum_{ik} K^{(2)}_{ik}(t,\tau)\operatorname{tr}\big(P_{ik}\,\varrho(\tau)\big) = 0.\end{aligned}\right\} \tag{IX.1.12}$$

If we require that condition (IX.1.12) is satisfied for arbitrary values of $\operatorname{tr}\big(P_{ik}\,\varrho(\tau)\big)$, we immediately obtain the conditions

$$\left.\begin{aligned}K^{(0)} &= 0, \\ \sum_{l} K_{lkil} + K^{(1)}_{ik} + K^{(2)}_{ik} &= 0.\end{aligned}\right\} \tag{IX.1.13}$$

From the density matrix equation we can easily derive equations for the expectation values of the projection operators. To this end we multiply the density operator equation with P_{lm} and take the trace. Using the cyclic property of the trace and the relation (IX.1.1), we find the equation

$$\left.\begin{aligned}\frac{d}{dt}\langle P_{lm}(t)\rangle = \langle P_{lm}(t_0)\rangle &+ \int_{t_0}^{t} K^{(0)}(t,\tau)\langle P_{lm}(\tau)\rangle\, d\tau + \\ &+ \int_{t_0}^{t} d\tau \Big\{\sum_{kk'} K_{mkk'l}(t,\tau)\langle P_{k'k}(\tau)\rangle + \\ &+ \sum_{k} K^{(1)}_{mk}(t,\tau)\langle P_{lk}(\tau)\rangle + \\ &+ \sum_{i} K^{(2)}_{il}(t,\tau)\langle P_{im}(\tau)\rangle\Big\}\end{aligned}\right\} \tag{IX.1.14}$$

* If the K's only depend on τ but not on t, (IX.1.12) is not only sufficient, but also necessary for (IX.1.11).

where the bracket indicates the average. Evidently the averaged Langevin equations of Chap. IV.5 are contained as special cases (with $P_{lm} = a_l^+ a_m$).

IX.2. Exact generalized Fokker-Planck equation: definition of the distribution function[1].

We consider a general quantum system described by the density matrix (IX.1.3) and the projection operators (IX.1.1).

We define the distribution function by*

$$f(v) = N \int \exp\left(-\sum_{ik} v_{ik} x_{ik}\right) \operatorname{tr}\left(\prod_{ik} \exp(x_{ik} P_{ik})\,\varrho\right) d\{x\}. \tag{IX.2.1}$$

N is the normalization constant, x_{ik} are classical complex quantities with

$$x_{ik} = -x_{ki}^*.$$

The v_{ik}'s are classical variables which are associated with each projection operator:

$$v_{ik} \leftrightarrow P_{ik}; \qquad v_{ik} = v_{ki}^*.$$

Most important for what follows is the sequence of the exponential operators which occur in the trace. We order these operators in the following way

$$O = O_L O_M O_R \equiv \prod_{ik} \exp(x_{ik} P_{ik}) \tag{IX.2.2}$$

where

$$\left.\begin{aligned} O_L = {} & (1,2)(1,3)(1,4)\ldots(1,n)(2,3)(2,4)\ldots \\ & \ldots(2,n)\ldots(n-2,n-1)(n-2,n)(n-1,n), \end{aligned}\right\} \tag{IX.2.3}$$

$$O_M = (1,1)(2,2)\ldots(n,n), \tag{IX.2.4}$$

$$\left.\begin{aligned} O_R = {} & (n,n-1)(n,n-2)(n-1,n-2)\ldots \\ & \ldots(n,2)(n-1,2)\ldots(3,2)(n,1)\ldots(3,1)(2,1). \end{aligned}\right\} \tag{IX.2.5}$$

We have used the following definitions

$$\left.\begin{aligned} & i < k, \\ & x_{ik} = \xi_{ik}, \\ & (i,k) = e^{\xi_{ik} P_{ik}} = (1 + \xi_{ik} P_{ik}), \end{aligned}\right\} \tag{IX.2.6}$$

$$\left.\begin{aligned} & i = k, \\ & x_{ii} = \xi_{ii} = \eta_i, \\ & (i,k) = e^{\xi_{ii} P_{ii}} \equiv e^{\eta_i P_{ii}} = 1 + (e^{\eta_i} - 1) P_{ii}, \end{aligned}\right\} \tag{IX.2.7}$$

$$\left.\begin{aligned} & i > k, \\ & x_{ik} = \tilde{\xi}_{ik}, \\ & (i,k) = e^{\tilde{\xi}_{ik} P_{ik}} = (1 + \tilde{\xi}_{ik} P_{ik}). \end{aligned}\right\} \tag{IX.2.8}$$

---

* This type of distribution function had been introduced by HAKEN, RISKEN and WEIDLICH: Z. Physik **206**, 355 (1967), where instead of P_{ik} we had used $a_i^+ a_k$. a_i^+ and a_k are the usual creation and annihilation operators of Fermions. In the one-particle subspace, $a_i^+ a_k$ just obeys the relation (IX.1.1). The definition (IX.2.1) secures that f is a real function (which is not always positive, however). (IX.2.1) allows for a number of modifications and extensions. Compare also Chap. IV.9.b.

[1] Here and in Chaps. IX.3, IX.4 and IX.5 we follow the paper of H. HAKEN: Z. Physik **219**, 411 (1969).

The definition (IX.2.1) allows us to calculate quantum mechanical expectation values by ordinary c-number procedures. For instance the expectation value of P_{lm} is given by

$$\operatorname{tr}(P_{lm}\,\varrho) = \int \dots \int v_{lm} f(v)\, d\{v\} \tag{IX.2.9}$$

where the integration runs over all variables $v_{ik} (= v_{ki}^*)$ over the whole complex plane.

If $\varrho(t)$ obeys the usual density matrix equation (IX.1.4) (which is of first order in its time derivative, the heat baths being Markoffian) it is even possible to calculate multitime averages of the form

$$\operatorname{tr}\left(P_{i_1k_1}(t_1) P_{i_2k_2}(t_2) \dots P_{i_nk_n}(t_n)\, \varrho(t_0)\right) \tag{IX.2.10}$$

with

$$t_1 > t_2 > \dots > t_n$$

by means of single-time distribution function. Since the treatment is identical to that of Chap. IV.11 we don't repeat it here.

IX.3. The exact generalized Fokker-Planck equation[1].

Theorem. The exact generalized Fokker-Planck equation belonging to (IX.1.3) and (IX.2.1) is given by

$$\left.\begin{aligned} f(t) &= f(t_0) + \int_{t_0}^{t} K^{(0)}(t,\tau) f(\tau)\, d(\tau) + \\ &+ \int_{t_0}^{t} d\tau \Big\{ \sum_{iki'k'} K_{ikk'i'}(t,\tau) \sum_{\lambda\mu} M_{k'i',ik;\lambda\mu}(x) v_{\lambda\mu} + \\ &+ \sum_{ik} K_{ik}^{(1)}(t,\tau) \sum_{\lambda\mu} N_{ik\lambda\mu}^{(1)}(x) v_{\lambda\mu} + \\ &+ \sum_{ik} K_{ik}^{(2)}(t,\tau) \sum_{\lambda\mu} N_{ik\lambda\mu}^{(2)}(x) v_{\lambda\mu} \Big\} f(\tau). \end{aligned}\right\} \tag{IX.3.1}$$

This equation is to be supplemented by the constraint $\sum v_{ii} = 1$.

Using the rule

$$x_{ik} \to -\frac{\partial}{\partial v_{ik}} \tag{IX.3.2}$$

we define the quantities M, $N^{(1)}$, $N^{(2)}$ as follows

$$M_{ik,k'i';\lambda\mu} = \sum_{mnl} A_{ikmn} B_{k'i'nl} \widetilde{C}_{ml\lambda\mu} e^{x_{nn} - x_{ll}} \tag{IX.3.3}$$

where

$$A_{ikmn} = \delta_{im} O_L^{k\to n} - x_{mi}\, \delta_{m<i} O_L^{k\to n}, \tag{IX.3.4}$$

$$B_{ikmn} = \delta_{kn} O_R^{m\leftarrow i} - x_{kn}\, \delta_{n<k} O_R^{m\leftarrow i}. \tag{IX.3.5}$$

$\delta_{m<i} = 1$ for $m < i$ and $= 0$ otherwise.

$$O_L^{m\to k} = \delta_{mk} + \sum x_{ml_1} x_{l_1l_2} \dots x_{l_jk}.$$

The summation runs over all ways from m to k with

$$m < l_1 < l_2 \dots < k \qquad \text{for } k \geqq m$$

$$O_L^{m\to k} = 0 \qquad \text{for } m > k \tag{IX.3.6}$$

$$O_R^{m\leftarrow l} = \delta_{ml} + \sum x_{ml_j} \dots x_{l_2l_1} x_{l_1l}.$$

The sum runs over all ways from l to m with

$$m > l_j > l_{j-1} \dots > l_2 > l_1 > l \quad \text{for } m \geqq l$$

$$O_R^{m\leftarrow l} = 0 \qquad \text{for } m < l \tag{IX.3.7}$$

[1] H. HAKEN: l.c.[1] on p. 274.

$$\widetilde{C}_{ik,lm} = \begin{cases} C_{iklm} = \delta_{il}(\delta_{km} - \delta_{k<m} x_{km}) & i<k,\ l<m \\ D_{ii,lm} = \delta_{il}\delta_{im} & i=k,\ l\leqq m \\ E_{ik,lm} = \delta_{km}(\delta_{il} - x_{li}\delta_{i<l})\, e^{-x_{ii}+x_{kk}} & i>k,\ l>m \end{cases} \tag{IX.3.8}$$
$$\widetilde{C}_{ik,lm} = 0 \text{ otherwise}$$

$$\left.\begin{aligned} N^{(1)}_{ikmn} &= \begin{cases} \left.\begin{aligned} H_{ik,mm} &= \delta_{im}(\delta_{km} - x_{km}\delta_{m<k}) \\ H_{ik,mn} &= \delta_{im}(\delta_{kn} - x_{kn}\delta_{n<k}) + \\ &\quad + x_{mn}\delta_{n<m}\delta_{in}(\delta_{kn} - x_{kn}\delta_{n<k}) \\ &\qquad m>n \end{aligned}\right\} i\geqq k \\ \sum_{m'n'} B_{ikm'n'} e^{x_{m'm'} - x_{n'n'}} \widetilde{C}_{m'n'mn} \qquad i<k \end{cases} \\ N^{(1)}_{ikmn} &= 0 \text{ for } i\geqq k,\ m<n \end{aligned}\right\} \tag{IX.3.9}$$

$$\left.\begin{aligned} N^{(2)}_{iklm} &= \begin{cases} \left.\begin{aligned} F_{ik,ll} &= \delta_{il}\delta_{kl} - x_{li}\delta_{l<i}\delta_{kl} \\ F_{ik,lm} &= \delta_{km}(\delta_{li} - x_{li}\delta_{l<i}) + \\ &\quad + \delta_{kl} x_{lm}\delta_{l<m}(\delta_{il} - x_{li}\delta_{l<i}) \\ &\qquad l<m \end{aligned}\right\} i\leqq k \\ \sum_{m'n'} A_{ikm'n'} \widetilde{C}_{m'n'lm} \qquad i>k. \end{cases} \\ N^{(2)}_{ikmn} &= 0 \text{ for } i\leqq k,\ l>m \end{aligned}\right\} \tag{IX.3.10}$$

Because the operator acting on f on the right-hand side of Eq. (IX.3.1) is *linear* in $v_{\lambda\mu}$, the Fouriertransform of f obeys a partial differential equation which is of only first order in its derivatives. Thus the solution of this equation may offer certain advantages over the direct solution of (IX.3.1).

IX.4. Derivation of the exact generalized Fokker-Planck equation. We first multiply the density matrix equation (IX.1.3) with O [see (IX.2.2)] from the left and then we take the trace over the total system. We then have to calculate the following expressions:

$$\mathrm{tr}(O P_{ik} \varrho P_{k'i'}), \tag{IX.4.1}$$
$$\mathrm{tr}(O P_{ik} \varrho), \tag{IX.4.2}$$
$$\mathrm{tr}(O \varrho P_{ik}). \tag{IX.4.3}$$

Using the cyclic properties of the trace we can rearrange (IX.4.1), (IX.4.2) and (IX.4.3) in the form

$$\mathrm{tr}(P_{k'i'} O P_{ik} \varrho), \tag{IX.4.4}$$
$$\mathrm{tr}(O P_{ik} \varrho), \tag{IX.4.5}$$
$$\mathrm{tr}(P_{ik} O \varrho), \tag{IX.4.6}$$

respectively.

Lemma 1:

$$O_L^{-1} = \left(1 - \sum_{i<k} \xi_{ik} P_{ik}\right). \tag{IX.4.7}$$

Proof: We denote the inverse of (i, k) by $\overline{(i, k)}$, so that

$$\left.\begin{aligned} O_L^{-1} = \overline{(n-1, n)}\,\overline{(n-2, n)} \ldots \overline{(l+1, l+2)} \ldots \\ \ldots \overline{(l, m+2)}\,\overline{(l, m+1)} \,|\, \overline{(l, m)}\,\overline{(l, m-1)} \ldots \overline{(1, 2)} \end{aligned}\right\} \tag{IX.4.8}$$

for $m+1 \leqq n$.

For $m = n$ we have instead of (IX.4.8)

$$O_L^{-1} = \overline{(n-1, n)}\,\overline{(n-2, n)} \ldots \overline{(l+1, l+3)}\,\overline{(l+1, l+2)} \,|\, \overline{(l, n)}\,\overline{(l, n-1)} \ldots \tag{IX.4.8'}$$

We assume that the expression (IX.4.7) is correct up to the indices (l, m), i.e. up to the bar, so that we may write

$$\left.\begin{aligned} O_L^{-1} = (1 - \xi_{n-1,n} P_{n-1,n} \cdots - \xi_{ik} P_{ik} \cdots \\ - \xi_{l,m+1} P_{l,m+1}) \,|\, (1 - \xi_{l,m} P_{lm}) \cdots \end{aligned}\right\} \tag{IX.4.9}$$

if $m+1 \leqq n$.

If $m = n$ we have instead of (IX.4.9):

$$\left.\begin{aligned} O_L^{-1} = (1 - \xi_{n-1,n} P_{n-1,n} \cdots - \xi_{ik} P_{ik} \cdots \\ - \xi_{l+1,l+2} P_{l+1,l+2}) \,|\, (1 - \xi_{l,n} P_{l,n}) \cdots \end{aligned}\right\} \tag{IX.4.9'}$$

We investigate the left neighbour of l, m. According to (IX.4.8), (IX.4.8') the left neighbours have the following indices

$$l, m+1; l, m+2; \quad \ldots, \tag{IX.4.10}$$

or

$$(l+1, l+2), (l+1, l+3), \quad \ldots. \tag{IX.4.11}$$

At any rate we have always

$$m+1, \ldots n > l \quad (\text{because } m > l) \tag{IX.4.12}$$

so that the following inequalities for an arbitrary pair (i, k) occurring on the left of the bar result

$$\left.\begin{aligned} i \geqq l \\ k > i \end{aligned}\right\} \rightarrow k > l. \tag{IX.4.13}$$

From the definition of O_L we have further $m > l$.

From these inequalities it is evident that k is unequal l and thus always

$$P_{ik} P_{lm} = 0. \tag{IX.4.14}$$

This proves our assertion up to $(l, m-1)$.

Lemma 2:

$$O_R^{-1} = (1 - \sum_{i>k} \tilde{\xi}_{ik} P_{ik}). \tag{IX.4.15}$$

Proof: Analogous to that of Lemma 1.

Definitions:

We introduce the decomposition:

$$O_L = O_{LL}^{l,m}(l, m) O_{LR}^{l,m} \tag{IX.4.16}$$

where

$$O_{LL}^{lm} = (1, 2)(1, 3) \ldots (1, n)(2, 3)(2, 4) \ldots (2, n) \ldots (l, m-1) \text{ if } m > l+1 \tag{IX.4.17}$$

and

$$\left.\begin{aligned}O_{LL}^{lm} &= (1, 2)(1, 3)\ldots(1, n)(2, 3)(2, 4)\ldots(2, n)\ldots(l-1, n-1)(l-1, n)\\ &\text{if } m = l+1,\end{aligned}\right\} \quad \text{(IX.4.18)}$$

$$\left.\begin{aligned}O_{LR}^{lm} &= (l, m+1)(l, m+2)\ldots(n-2, n-1)(n-2, n)(n-1, n)\\ &\text{if } m+1 \leqq n,\end{aligned}\right\} \quad \text{(IX.4.19)}$$

$$\left.\begin{aligned}&= (l+1, l+2)(l+1, l+3)\ldots(n-2, n-1)(n-2, n)(n-1, n)\\ &\text{if } m = n.\end{aligned}\right\} \quad \text{(IX.4.20)}$$

In a similar way we define

$$O_R = O_{RL}^{l,m}(l, m)\, O_{RR}^{l,m}, \quad \text{(IX.4.21)}$$

$$\left.\begin{aligned}O_{RL}^{lm} &= (n, n-1)(n, n-2)(n-1, n-2)(n, n-3)(n-1, n-3)\\ &\cdot (n-2, n-3)\ldots(l+1, m)\\ &\text{if } l+1 \leqq n,\end{aligned}\right\} \quad \text{(IX.4.22)}$$

$$\left.\begin{aligned}O_{RL}^{lm} &= (n, n-1)(n, n-2)(n-1, n-2)(n, n-3)(n-1, n-3)\\ &\cdot (n-2, n-3)\ldots(m+3, m+1)(m+2, m+1)\\ &\text{if } l = n,\end{aligned}\right\} \quad \text{(IX.4.23)}$$

$$\left.\begin{aligned}O_{RR}^{lm} &= (l-1, m)(l-2, m)\ldots(m+1, m)(n, m-1)(n-1, m-1)\ldots\\ &\ldots(m, m-1)\ldots(n, 1)(n-1, 1)\ldots(3, 1)(2, 1)\\ &\text{if } l-1 > m,\end{aligned}\right\} \quad \text{(IX.4.24)}$$

$$\left.\begin{aligned}O_{RR}^{lm} &= (n, m-1)(n-1, m-1)\ldots\\ &\ldots(m, m-1)(n, m-2)(n-1, m-2)\ldots\\ &\ldots(4, 2)(3, 2)(n, 1)(n-1, 1)\ldots(3, 1)(2, 1)\\ &\text{if } l = m+1.\end{aligned}\right\} \quad \text{(IX.4.25)}$$

Simple relations for derivatives

Evidently we have for $l < m$

$$\frac{\partial O}{\partial \xi_{lm}} = \frac{\partial O_L}{\partial \xi_{lm}} O_M O_R \quad \text{(IX.4.26)}$$

with

$$\frac{\partial O_L}{\partial \xi_{lm}} = O_{LL}^{lm} P_{lm} O_{LR}^{lm}, \quad \text{(IX.4.27)}$$

for $l = m$:

$$\frac{\partial O}{\partial \xi_{ll}} = O_L \frac{\partial O_M}{\partial \xi_{ll}} O_R, \quad \text{(IX.4.28)}$$

where

$$\frac{\partial O_M}{\partial \xi_{ll}} = P_{ll} O_M = O_M P_{ll}, \quad \text{(IX.4.29)}$$

for $l > m$

$$\frac{\partial O}{\partial \tilde{\xi}_{lm}} = O_L O_M \frac{\partial O_R}{\partial \tilde{\xi}_{lm}} \quad \text{(IX.4.30)}$$

with

$$\frac{\partial O_R}{\partial \tilde{\xi}_{lm}} = O_{RL}^{lm} P_{lm} O_{RR}^{lm}. \quad \text{(IX.4.31)}$$

We now come to the task of expressing (IX.4.4), (IX.4.5) and (IX.4.6) by derivatives of O with respect to x_{ik} (or $\xi_{ik}, \tilde{\xi}_{ik}$). We do this in several steps.

Rules for shifting projection operators and using derivatives.

Lemma 3:

$$P_{ik}O = \sum_{m,n} A_{ikmn} O_L P_{mn} O_M O_R . \tag{IX.4.32}$$

where A_{ikmn} is defined in (IX.3.4), (IX.3.6).

Proof:

This formula follows from the fact that 1) the P_{lm}'s span the basis of an algebraic ring, so that we can always express a function of the P_{ik}'s by a linear combination of the P_{lm}'s and that 2) O_L, O_M, O_R possess inverse operators. We multiply Eq. (IX.4.32) by O_L^{-1} from the left and by $O_R^{-1}\, O_M^{-1}$ from the right. This leaves us with the equation

$$O_L^{-1} P_{ik} O_L = \sum_{m,n} A_{ikmn} P_{mn} . \tag{IX.4.33}$$

When we take the matrix element with respect to the states ϕ_m and ϕ_n on both sides we find immediately

$$A_{ikmn} = \langle \phi_m (P_{ik} - \sum_{\nu < i} \xi_{\nu i} P_{\nu k}) O_L \phi_n \rangle \tag{IX.4.34}$$

where we have inserted the explicit form of O_L^{-1} according to Eq. (IX.4.7). We further introduce the abbreviation

$$O_L^{m\to k} = \langle \phi_m O_L \phi_k \rangle . \tag{IX.4.35}$$

Expanding O_L into the powers of P_{ik} and using the property (IX.1.1) of the projection operators we immediately find the formulas (IX.3.6). Introducing (IX.4.35), (IX.3.6) into (IX.4.34) we obtain (IX.3.4).

Lemma 4:

$$O\, P_{ik} = \sum_{mn} B_{ikmn} O_L O_M P_{ik} O_R \tag{IX.4.36}$$

with

$$B_{ikmn} \quad \text{given by (IX.3.5), (IX.3.7)}$$

Proof: Analogous to that of Lemma 3.

We now express $O_L\, P_{mn}\, O_M\, O_R$ by a linear combination of derivatives of O with respect to x_{ik}.

Lemma 5:

For $i < k$

$$O_L P_{ik} = \sum_{l<m} C_{ik\,lm} \frac{\partial}{\partial \xi_{lm}} O_L . \tag{IX.4.37}$$

with $C_{ik\,lm}$ given in (IX.3.8)

Proof:

Using (IX.4.27) we write (IX.4.37) in the form

$$O_L P_{ik} = \sum_{l<m} C_{ik\,lm} O_{LL}^{lm} P_{lm} O_{LR}^{lm} . \tag{IX.4.38}$$

By multiplying Eq. (IX.4.38) with O_L^{-1} from the left and from the right we obtain

$$P_{ik} O_L^{-1} = \sum_{l<m} C_{ik\,lm} (O_{LR}^{lm})^{-1} P_{lm} (O_{LL}^{lm})^{-1} . \tag{IX.4.39}$$

The right-hand side of Eq. (IX.4.39) follows from

$$O_L^{-1} O_{LL}^{lm} = (O_{LR}^{lm})^{-1} (\overline{l, m}), \tag{IX.4.40}$$

$$O_{LR}^{lm} O_L^{-1} = (\overline{l, m}) (O_{LL}^{lm})^{-1} \tag{IX.4.41}$$

[compare (IX.4.16)] and

$$(\overline{l, m}) P_{lm} (\overline{l, m}) = P_{lm} \quad (\text{for } l \neq m). \tag{IX.4.42}$$

For the further evaluation of the operator product on the right-hand side of Eq. (IX.4.39) we note that

$$(O_{LR}^{lm})^{-1} P_{lm} (O_{LL}^{lm})^{-1} = -\frac{d}{d\xi_{lm}} (O_{LR}^{lm})^{-1} (\overline{l,m}) (O_{LL}^{lm})^{-1} = -\frac{d}{d\xi_{lm}} O_L^{-1}. \tag{IX.4.43}$$

Due to (IX.4.7), (IX.4.43) reduces to

$$(O_{LR}^{lm})^{-1} P_{lm} (O_{LL}^{lm})^{-1} = P_{lm}. \tag{IX.4.44}$$

Thus Eq. (IX.4.39) takes the simple form

$$P_{ik} O_L^{-1} = \sum_{l<m} C_{ik\,lm} P_{lm}. \tag{IX.4.45}$$

Taking the expectation value with respect to the states l and m on the left- and right-hand side we find immediately

$$C_{ik\,lm} = \langle \phi_l P_{ik} (1 - \sum_{\mu<\nu} \xi_{\mu\nu} P_{\mu\nu}) \phi_m \rangle, \tag{IX.4.46}$$

$$C_{ik\,lm} = \delta_{il} (\delta_{km} - \xi_{km} \delta_{k<m}). \tag{IX.4.47}$$

Lemma 6: for $i = k$

$$O_L P_{ii} O_M = \sum_{l \leqq m} D_{ii,\,lm} \frac{\partial}{\partial \xi_{lm}} (O_L O_M), \tag{IX.4.48}$$

with D given by (IX.3.8).

Proof:

Use of (IX.4.26), (IX.4.28) yields

$$O_L P_{ii} O_M = \sum_{l<m} D_{ii,\,lm} O_{LL}^{lm} P_{lm} O_{LR}^{lm} O_M + \sum_l D_{ii,\,ll} O_L P_{ll} O_M. \tag{IX.4.49}$$

We multiply (IX.4.49) from the left with O_L^{-1} and from the right with $O_M^{-1} O_L^{-1}$. This leaves us immediately with

$$P_{ii} O_L^{-1} = \sum_{l<m} D_{ii,\,lm} P_{lm} + \sum_l D_{ii,\,ll} P_{ll} O_L^{-1} \tag{IX.4.50}$$

which can be transformed by use of (IX.4.7) into

$$P_{ii} O_L^{-1} = \sum_{l<m} P_{lm} (D_{ii,\,lm} - D_{ii,\,ll} \xi_{lm}) + \sum_l D_{ii,\,ll} P_{ll} \tag{IX.4.51}$$

after the indices had been rearranged. We first take the expectation value with respect to the states l, l which yields

$$\left.\begin{aligned} D_{ii,\,ll} &= \left\langle \phi_l P_{ii} \left(1 - \sum_{\mu<\nu} \xi_{\mu\nu} P_{\mu\nu}\right) \phi_l \right\rangle \\ &= \delta_{il} - \delta_{il} \sum_{l<\nu} \xi_{l\nu} \delta_{\nu l} \\ &= \delta_{il}. \end{aligned}\right\} \tag{IX.4.52}$$

We now investigate $l \neq m$. Taking the matrix element on the left- and right-hand side with respect to the states l, m, we find

$$\left\langle \phi_l P_{ii}\left(1-\sum_{\mu<\nu}\xi_{\mu\nu}P_{\mu\nu}\right)\phi_m\right\rangle = (D_{ii,lm}-D_{ii,ll}\xi_{lm})\,\delta_{l<m} \qquad (l\neq m) \qquad \text{(IX.4.53)}$$

and by use of (IX.4.52)

$$D_{ii,lm}=0 \qquad \text{(IX.4.54)}$$

Lemma 7: $i>k$

$$\left.\begin{aligned} O_L P_{ik} O_M O_R &= \sum_{l>m} E_{ik,lm}\frac{\partial}{\partial\tilde{\xi}_{l,m}}O \\ &= \sum_{l>m} E_{ik,lm} O_L O_M \frac{\partial}{\partial\tilde{\xi}_{lm}} O_R . \end{aligned}\right\} \qquad \text{(IX.4.55)}$$

with $E_{ik\,lm}$ defined in (IX.3.8).

Proof: Analogous to that of Lemma 5.

A second way of shifting projection operators:

In the foregoing analysis we have first shifted the projection operators from the left or the right to the middle of the product of $O_L O_M O_R$. Then in the next step we have shifted the P_{ik}'s to their places which would correspond to their order in the decomposition (IX.4.16) and (IX.4.21). Under certain circumstances it leads to simpler formulas if we immediately shift the projection operators to their correct places.

Lemma 8: $i\leqq k$

$$P_{ik}O_L O_M = \sum_{l<m} F_{ik,lm}\left(\frac{\partial}{\partial\xi_{lm}}O_L\right)O_M + \sum_l F_{ik,ll}O_L\frac{\partial O_M}{\partial\xi_{ll}} . \qquad \text{(IX.4.56)}$$

where $F_{ik\,lm}$ is given by (IX.3.10).

Proof:

Multiplying this expression from the left by O_L^{-1} and from the right by $O_M^{-1}O_L^{-1}$ we obtain in the same manner as above

$$O_L^{-1}P_{ik} = \sum_{l<m} F_{ik,lm}P_{lm} + \sum_l F_{ik,ll}P_{ll}O_L^{-1} \qquad \text{(IX.4.57)}$$

and using (IX.4.7)

$$(O_L^{-1}P_{ik}) = \sum_{l<m} F_{ik,lm}P_{lm} + \sum_l F_{ik,ll}P_{ll}\left(1-\sum_{\mu<\nu}\xi_{\mu\nu}P_{\mu\nu}\right). \qquad \text{(IX.4.58)}$$

(IX.4.58) can be rearranged in the form

$$O_L^{-1}P_{ik} = \sum_{l<m} P_{lm}(F_{ik,lm}-F_{ik,ll}\xi_{lm}\delta_{l<m}) + \sum_l F_{ik,ll}P_{ll} \qquad \text{(IX.4.59)}$$

where use of (IX.1.1) had been made.

Taking expectation values on both sides of Eq. (IX.4.59) with respect to states l, m we obtain immediately (IX.3.10).

Lemma 9: $i\geqq k$

$$O_M O_R P_{ik} = \sum_{m\geqq n} H_{ikmn}\frac{\partial(O_M O_R)}{\partial\tilde{\xi}_{mn}} \qquad \text{(IX.4.60)}$$

here H_{ikmn} is defined in (IX.3.9).

Proof: Analogous to that of Lemma 8.

Evaluation of (IX.4.4), (IX.4.5), (IX.4.6).

We are now in a position to put our previous results together in order to calculate the expressions (IX.4.4), (IX.4.5) and (IX.4.6).

1. Evaluation of (IX.4.4).

$$P_{ik} O P_{k'i'} = \sum_{mn} A_{ikmn} O_L P_{mn} O_M O_R P_{k'i'} \tag{IX.4.61}$$

[compare (IX.4.32)].

Using

$$O_R P_{k'i'} = \sum_{m'n'} B_{k'i'm'n'} P_{m'n'} O_R \tag{IX.4.62}$$

we obtain

$$P_{ik} O P_{k'i'} = \sum_{m'n'} \sum_{mn} A_{ikmn} B_{k'i'm'n'} O_L P_{mn} O_M P_{m'n'} O_R. \tag{IX.4.63}$$

Using

$$O_M P_{m'n'} = P_{m'n'} O_M e^{\eta_{m'} - \eta_{n'}} \tag{IX.4.64}$$

we transform (IX.4.63) into

$$P_{ik} O P_{k'i'} = \sum_{m'n'} \sum_{mn} A_{ikmn} B_{k'i'm'n'} O_L P_{mn} P_{m'n'} O_M e^{\eta_{m'} - \eta_{n'}} O_R \tag{IX.4.65}$$

which by use of (IX.1.1) reduces to

$$\sum_{mnn'} A_{ikmn} B_{k'i'nn'} O_L P_{mn'} O_M e^{\eta_n - \eta_{n'}} O_R. \tag{IX.4.66}$$

According to (IX.4.37), (IX.4.48), (IX.4.55) we have quite generally

$$O_L P_{mn'} O_M O_R = \sum_{\lambda\mu} \tilde{C}_{mn'\lambda\mu} \frac{\partial}{\partial x_{\lambda\mu}} O \tag{IX.4.67}$$

where

$$\tilde{C}_{mn'\lambda\mu} = \left.\begin{cases} C_{mn'\lambda\mu} & \text{[Eq. (IX.4.47)]} \quad \text{for } m < n' \quad \lambda < \mu \\ D_{mn',\lambda\mu} & \text{[Eq. (IX.4.52), (IX.4.54)]} \quad \text{for } m = n' \quad \lambda \leqq \mu \\ E_{mn',\lambda\mu} & \text{[Eq. (IX.3.8)]} \quad \text{for } m > n'. \quad \lambda > \mu \end{cases}\right\} \tag{IX.4.68}$$

$= 0$ otherwise.

Inserting (IX.4.67) into (IX.4.66) yields

$$P_{ik} O P_{k'i'} = \sum_{mnn'\lambda\mu} A_{ikmn} B_{k'i'nn'} \tilde{C}_{mn'\lambda\mu} e^{\eta_n - \eta_{n'}} \frac{\partial}{\partial x_{\lambda\mu}} O. \tag{IX.4.69}$$

Thus we obtain finally

$$P_{ik} O P_{k'i'} = \sum_{\lambda\mu} M_{ik,k'i',\lambda\mu}(x) \frac{\partial}{\partial x_{\lambda\mu}} O \tag{IX.4.70}$$

with

$$M_{ik,k'i',\lambda\mu} = \sum_{mnl} A_{ikmn} B_{k'i'nl} \tilde{C}_{ml\lambda\mu} e^{\eta_n - \eta_l}. \tag{IX.4.71}$$

The coefficients A, B, $\tilde{C}$, are given by the expressions (IX.3.4), (IX.3.5) and (IX.4.68), respectively.

2. Evaluation of (IX.4.6).

We distinguish between the cases $i \leqq k$ and $i > k$. If $i \leqq k$ we have immediately according to (IX.4.56)

$$P_{ik} O = \sum_{l \leqq m} F_{ik,lm} \left(\frac{\partial}{\partial \xi_{lm}} O \right)$$

with $F_{ik,lm}$ given by (IX.3.10).

If $i>k$, we first use (IX.4.32) which yields

$$P_{ik}O=\sum_{mn} A_{ikmn}O_L P_{mn} O_M O_R. \tag{IX.4.72}$$

Inserting

$$O_L P_{mn} O_M O_R=\sum_{m',l'} \tilde{C}_{mn,m'l'} \frac{\partial}{\partial x_{m'l'}} O \tag{IX.4.73}$$

into (IX.4.72) we find

$$P_{ik}O=\sum_{l,m} \frac{\partial O}{\partial x_{lm}} J_{lm}, \quad \text{where} \quad J_{lm}=\sum_{m'n'} A_{ikm'n'}\tilde{C}_{m'n'lm}. \tag{IX.4.74}$$

3. Evaluation of (IX.4.5).

In a similar way we treat OP_{ik}. For $i>k$ we apply (IX.4.60). For $i<k$ we have

$$OP_{ik}=O_L O_M \sum_{mn} B_{ikmn} P_{mn} O_R. \tag{IX.4.75}$$

We find

$$OP_{ik}=\sum_{mn} B_{ikmn} O_L P_{mn} e^{\eta_m-\eta_n} O_M O_R. \tag{IX.4.76}$$

Inserting (IX.4.67), i.e.

$$O_L P_{mn} O_M O_R=\sum_{l',m'} \tilde{C}_{mn\,l'm'} \frac{\partial}{\partial x_{l'm'}} O \tag{IX.4.77}$$

into (IX.4.75) we obtain

$$OP_{ik}=\sum_{l'm'} N_{ik\,l'm'} \frac{\partial}{\partial x_{l'm'}} O \tag{IX.4.78}$$

with

$$N_{ik\,l'm'}=\sum_{mn} B_{ikmn} e^{\eta_m-\eta_n} \tilde{C}_{mn\,l'm'}. \tag{IX.4.79}$$

Final step: We insert the explicit expressions for (IX.4.4), (IX.4.5), (IX.4.6) into the equation

$$\left.\begin{aligned} \operatorname{tr}(O\varrho(t)) &= \operatorname{tr}(O\varrho(t_0)) \\ &+\int_{t_0}^{t} d\tau\Big\{\sum_{iki'k'} K_{ikk'i'}(t,\tau)\operatorname{tr}\left(P_{k'i'}OP_{ik}\varrho(\tau)\right) \\ &+\sum_{ik} K_{ik}^{(1)}(t,\tau)\operatorname{tr}\left(OP_{ik}\varrho(\tau)\right)+\sum_{ik} K_{ik}^{(2)}(t,\tau)\left(P_{ik}O\varrho(\tau)\right)\Big\}. \end{aligned}\right\} \tag{IX.4.80}$$

In order to obtain an equation for the distribution function f (IX.2.1), we multiply (IX.4.80) by

$$\exp\{-\sum_{ik} v_{ik} x_{ik}\}$$

and integrate over the space (x_{ik}). Thus the left-hand side of Eq. (IX.4.80) becomes identical with $f(t)$. The traces on the right-hand side of Eq. (IX.4.80) have the form of derivatives of O with respect to x_{ik}, which are multiplied by functions of x_{ik}. These operations can be immediately translated into ones with respect to the variables v_{ik}, if the following well-known rules are observed:

$$\left.\begin{aligned} \frac{\partial}{\partial x_{ik}} &\to \text{multiplication by } v_{ik} \\ \text{multiplication by } x_{ik} &\to -\frac{\partial}{\partial v_{ik}}. \end{aligned}\right\} \tag{IX.4.81}$$

Having these rules in mind we immediately find Eq. (IX.3.1).

In conclusion we discuss a constraint which stems from the normalization of any quantum system, which is expressed by

$$\sum_i P_{ii} = 1 . \tag{IX.4.82}$$

We include (IX.4.82) explicitly in f (IX.2.1) by the following new definition of f:

$$\left.\begin{aligned} f &= N \int d\{x\} \exp\Big\{-\sum_{ik} v_{ik} x_{ik}\Big\} \times \\ &\times \frac{1}{2\pi} \int_{-\infty}^{+\infty} d\xi \, e^{-i\xi} \operatorname{tr}\big(O_L e^{i\xi \sum_j P_{jj}} O_M O_R \varrho\big) . \end{aligned}\right\} \tag{IX.4.83}$$

Having in mind that $O_M = e^{\sum x_{jj} P_{jj}}$, we introduce new variables

$$\tilde{x}_{jj} = x_{jj} + i\xi$$

which leaves us with

$$\left.\begin{aligned} f &= N \int d\{x\} \exp\Big\{-\sum_{i \neq k} v_{ik} x_{ik}\Big\} \exp\Big\{-\sum_j v_{jj} \tilde{x}_{jj}\Big\} \times \\ &\times \operatorname{tr}(O \varrho) \cdot \frac{1}{2\pi} \int e^{i\xi(\sum v_{jj} - 1)} d\xi . \end{aligned}\right\} \tag{IX.4.84}$$

Evidently the definition (IX.2.1) of f agrees with that of (IX.4.84), if the constraint $\sum_j v_{jj} = 1$ is added to the definition (IX.2.1).

IX.5. Projection onto macroscopic variables. Our Eq. (IX.3.1) applies to a single atom with two energy levels, or a spin $\frac{1}{2}$, to multi-level atoms, as well as to a complicated many-body problem. For a discussion of its solutions, it is thus certainly necessary to treat specific examples*.

On the other hand, under certain circumstances it is possible to perform strong simplifications even in rather general cases. Indeed, in many cases of practical importance, the system under consideration consists of many equivalent subsystems. Examples are spins in external fields, or laser-atoms interacting with the lightfield. In these examples the individual subsystems interact with the external fields in a similar way, i.e. by their total (macroscopic) electric or magnetic moments (after certain phase-factors have been transformed away, see e.g. Chap. VI.4.

In these cases, it is advantageous, first to introduce projection operators for each subsystem μ:** $P_{ik,\mu}$, and to formulate the density matrix equation by means of these $P_{ik,\mu}$. It reads

$$\varrho(t) = L \varrho(\tau) \tag{IX.5.1}$$

where L is a linear operator [see Eq. (IX.1.3)]. In our present case, L consists of a sum over the subsystems:

$$L = \sum_\mu \tilde{L}_\mu(P_{ik,\mu}) \equiv \sum_\mu \tilde{L}(P_{ik,\mu}) \tag{IX.5.2}$$

where the latter equation expresses the fact that the subsystems are equivalent.

---

* e.g. in the case of a two-level atom, f is built up of δ-functions (W. Weidlich: Private communication).

** For the explicit example of two-level atoms see Chap. IV.10.

We now introduce macroscopic variables:

$$V_{ik} = \sum_{\mu} v_{ik\mu} \tag{IX.5.3}$$

and define

$$f(t) = \int d\{x\} \exp\left\{-\sum_{ik} x_{ik} V_{ik}\right\} \operatorname{tr}\left(\left(\prod_{\nu} O_\nu\right) \varrho\right) \tag{IX.5.4}$$

where

$$O_\nu = \prod_{ik} e^{x_{ik} P_{ik,\nu}}. \tag{IX.5.5}$$

Because the O_ν's commute, we may decompose

$$\prod_{\nu} O_\nu = \prod_{\nu}{}^{(\mu)} O_\nu \cdot O_\mu$$

where (μ) indicates, that the factor $\nu = \mu$ is to be omitted. We multiply Eq. (IX.5.1) from the left by O and take the trace with respect to the whole system:

$$\left.\begin{aligned} \operatorname{tr}(O\varrho) &= \operatorname{tr}(OL\varrho(\tau)) = \sum_{\mu} \operatorname{tr}(O\tilde{L}(P_{ik,\mu})\varrho) \\ &= \sum_{\mu} \operatorname{tr}^{(\mu)}\left(\left(\prod_{\nu}{}^{(\mu)} O_\nu\right) \operatorname{tr}_\mu (O_\mu \tilde{L}(P_{ik,\mu})\varrho)\right). \end{aligned}\right\} \tag{IX.5.6}$$

In IX.4 it had been shown that

$$\operatorname{tr}_\mu (O_\mu \tilde{L}(P_{ik,\mu})\varrho) = h\left(x_{ik}, \frac{\partial}{\partial x_{ik}}\right) \operatorname{tr}_\mu (O_\mu \varrho). \tag{IX.5.7}$$

Note that O_μ is a function of x_{ik} due to (IX.5.5).

The right-hand side of Eq. (IX.5.6) can thus be written in the form

$$\text{IX.5.6} = \sum_{\mu} \operatorname{tr}^{(\mu)}\left(\left(\prod_{\nu}{}^{(\mu)} O_\nu\right) h\left(x_{ik}, \frac{\partial}{\partial x_{ik}}\right) \operatorname{tr}_\mu (O_\mu \varrho)\right) \tag{IX.5.8}$$

$$= \sum_{\mu} \operatorname{tr}\left(\prod_{\nu}{}^{(\mu)} O_\nu \cdot h\left(x_{ik}, \frac{\partial}{\partial x_{ik}}\right) O_\mu \varrho\right). \tag{IX.5.9}$$

Note that we can always use the cyclic properties of the traces, if needed.

Now let us make use of the fact, that h is of first order in the derivatives $\partial/\partial x_{ik}$ [compare Eq. (IX.3.1) and (IX.4.81)]. According to Leibniz's rule of differentiating products, we immediately find:

$$\operatorname{tr}(O\varrho) = h\left(x_{ik}, \frac{\partial}{\partial x_{ik}}\right) \operatorname{tr}\left(\prod_{\nu} O_\nu \varrho\right). \tag{IX.5.10}$$

We multiply Eq. (IX.5.10) with

$$\exp\left(-\sum_{ik} x_{ik} V_{ik}\right)$$

and integrate over x_{ik}.

This leaves us with

$$f(V_{ik}, t) = \chi\left(V_{ik}, \frac{\partial}{\partial V_{ik}}\right) f(V_{ik}, \tau) \tag{IX.5.11}$$

where χ is a linear operator defined on the right-hand side of Eq. (IX.3.1), if the small v_{ik}'s are replaced by the big ones, V_{ik}'s. A simple analysis shows that the condition $\sum_i v_{ii} = 1$ is now to be replaced by $\sum V_{ii} = N$, where N is the number of subsystems. The importance of the result (IX.5.11) is the following: while $f(v_{ik}, t)$ of a single subsystem may show wild fluctuations (e.g. it consists of δ-functions), the macroscopic variables V_{ik} possess a smooth distribution, which changes but little. Thus in many cases of interest it can be shown that we may neglect derivatives, e.g. higher than second order, so that (IX.5.11) then actually reduces to an ordinary Fokker-Planck equation.

IX.6. Exact elimination of the atomic operators within quantum mechanical Langevin equations[1]. We treat a system of two-level atoms. For the atomic system we use flip operators

$$\alpha_\mu^+ = (a_2^+ a_1)_\mu \quad \text{and} \quad \alpha_\mu = (a_1^+ a_2)_\mu ,$$

where $a_{j,\mu}^+$ is the creation operator of an electron in the j-th state of the μ-th atom, while $a_{j,\mu}$ is the corresponding annihilation operator. The equations of motion then read [compare (IV.6.1), (IV.6.2)]

field equations

$$\left(\frac{d}{dt} - i\omega_\lambda + \varkappa_\lambda\right) b_\lambda^+ = i \sum_\mu g_{\mu\lambda}^* \alpha_\mu^+ + F_\lambda^+ , \tag{IX.6.1}$$

matter equations

$$\left(\frac{d}{dt} - i\nu_\mu + \gamma\right) \alpha_\mu^+ = -i \sum_{\lambda'} g_{\mu\lambda'} b_{\lambda'}^+ \sigma_\mu + \Gamma_{21,\mu} , \tag{IX.6.2}$$

$$\left(\frac{d}{dt} + \frac{1}{T_p}\right)\sigma_\mu = \frac{d_0}{T_p} + 2i\alpha_\mu \sum_{\lambda''} g_{\mu\lambda''} b_{\lambda''}^+ - 2i\alpha_\mu^+ \sum_{\lambda''} g_{\mu\lambda''}^* b_{\lambda''} + \Delta\Gamma_\mu . \tag{IX.6.3}$$

The quantities occurring in these equations are explained in Chap. VI.3.

From Eq. (IX.6.2) we eliminate the atomic operators α_μ^+ in the following way

$$\alpha_\mu^+ = \frac{1}{\hat\gamma}\left\{-i \sum_{\lambda'} g_{\mu\lambda'} b_{\lambda'}^+ \sigma_\mu + \Gamma_{21,\mu}\right\} \tag{IX.6.4}$$

where

$$\hat\gamma = \gamma - i\nu_\mu + \frac{d}{dt} . \tag{IX.6.5}$$

Inserting (IX.6.4) into (IX.6.3) we find

$$\left.\begin{aligned} \left(\frac{d}{dt} + \frac{1}{T_p}\right)\sigma_\mu = \frac{d_0}{T_p} + \Delta\Gamma_\mu + 2i \sum_{\lambda''} \left\{ g_{\mu\lambda''} b_{\lambda''}^+ \frac{1}{\hat\gamma^*} \Gamma_{21,\mu}^+ - g_{\mu\lambda''}^* b_{\lambda''} \frac{1}{\hat\gamma} \Gamma_{21,\mu}\right\} + \\ + \left\{ -2 \sum_{\lambda'',\lambda'} g_{\mu\lambda''} b_{\lambda''}^+ \frac{1}{\hat\gamma^*} g_{\mu\lambda'}^* b_{\lambda'} - 2 \sum_{\lambda''\lambda'} g_{\mu\lambda''}^* b_{\lambda''} \frac{1}{\hat\gamma} g_{\mu\lambda'} b_{\lambda'}^+ \right\}\sigma_\mu . \end{aligned}\right\} \tag{IX.6.6}$$

From this equation we determine σ_μ explicitly which yields

$$\left.\begin{aligned} \sigma_\mu = \left[\hat\gamma_p + 2\left\{\sum_{\lambda',\lambda''}\left(g_{\mu\lambda''} b_{\lambda''}^+ \frac{1}{\hat\gamma^*} g_{\mu\lambda'}^* b_{\lambda'} + g_{\mu\lambda''}^* b_{\lambda''} \frac{1}{\hat\gamma} g_{\mu\lambda'} b_{\lambda'}^+\right)\right\}\right]^{-1} \times \\ \times \left[\frac{d_0}{T_p} + \Delta\Gamma_\mu + 2i \sum_{\lambda''}\left\{ g_{\mu\lambda''} b_{\lambda''}^+ \frac{1}{\hat\gamma^*} \Gamma_{21,\mu}^+ - g_{\mu\lambda''}^* b_{\lambda''} \frac{1}{\hat\gamma} \Gamma_{21,\mu}\right\}\right] \end{aligned}\right\} \tag{IX.6.7}$$

where

$$\hat\gamma_p = \frac{1}{T_p} + \frac{d}{dt} . \tag{IX.6.8}$$

Inserting on the other hand (IX.6.4) into (IX.6.1), we find the equation

$$\left(\frac{d}{dt} - i\omega_\lambda + \varkappa_\lambda\right) b_\lambda^+ = \sum_{\mu,\lambda'} g_{\mu\lambda}^* \frac{1}{\hat\gamma} g_{\mu\lambda'} b_{\lambda'}^+ \sigma_\mu + i \sum_\mu g_{\mu\lambda}^* \frac{1}{\hat\gamma} \Gamma_{21,\mu} + F_\lambda^+ \tag{IX.6.9}$$

---

[1] V. ARZT, H. HAKEN, H. RISKEN, H. SAUERMANN, CH. SCHMID, and W. WEIDLICH: Z. Physik **197**, 207 (1966).

for the light field operators b_λ^+. In these equations the σ_μ's are determined by the explicit expressions (IX.6.7), so that (IX.6.9) is an equation for the light field operators alone. Due to the occurrence of the light field operators in the denominator of σ_μ, these equations are highly nonlinear and of a very complicated structure.

XI.7. Rate equations in quantized form[1]. To make contact with the rate equation treatment, we specialize the foregoing treatment to single-mode operation. If we refrain from explicity expressing σ_μ by b^+, b via (IX.6.7), we may retain (IX.6.6) as a generalized rate equation for σ_μ. By putting $b^+ = e^{i\Omega t}\,\tilde{b}^+$ and neglecting frequencies in $\tilde{b}^+$ which are small compared to γ, (IX.6.6) is simplified to a usual rate equation for σ_μ (including noise, however):

$$\dot{\sigma}_\mu = \frac{1}{T_p}(d_0 - \sigma_\mu) - 4|g_\mu|^2 \frac{\gamma}{(\Omega - \nu_\mu)^2 + \gamma^2}\, b^+ b\, \sigma_\mu + \Gamma_{\mu,tot} \tag{IX.7.1}$$

where

$$\left.\begin{aligned} \Gamma_{\mu,tot} &= \Delta\Gamma_\mu + 2ig_\mu b^+ \frac{1}{\frac{d}{dt} + i\nu_\mu + \gamma}\,\Gamma_{12,\mu} - \\ &\quad - 2ig_\mu^* \left(\frac{1}{\frac{d}{dt} - i\nu_\mu + \gamma}\,\Gamma_{21,\mu}\right) b. \end{aligned}\right\} \tag{IX.7.2}$$

Spezializing (IX.6.9) to single-mode operation and making the same approximations yields

$$\dot{b}^+ = (i\omega - \varkappa)\, b^+ + \sum_\mu |g_\mu|^2 \frac{1}{i(\Omega - \nu_\mu) + \gamma}\, b^+ \sigma_\mu + F_{tot}^+ \tag{IX.7.3}$$

where

$$F_{tot}^+ = F^+ + i\sum_\mu g_\mu^* \frac{1}{\frac{d}{dt} - i\nu_\mu + \gamma}\,\Gamma_{21,\mu}. \tag{IX.7.4}$$

When we multiply (IX.6.3) by b from the right and add the conjugate complex equation, we find

$$\frac{d}{dt}(b^+ b) = -2\varkappa b^+ b + \sum_\mu |g_\mu|^2 \sigma_\mu (b^+ b) \frac{2\gamma}{(\Omega - \nu_\mu)^2 + \gamma^2} + G_{tot} \tag{IX.7.5}$$

where

$$G_{tot} = F_{tot}^+ b + b^+ F_{tot}. \tag{IX.7.6}$$

Apart from some minor differences, the Eqs. (IX.7.1) and (IX.7.5) are those used by McCumber[2] in order to treat intensity fluctuations. When the noise sources (IX.7.1) and (IX.7.6) are split into

$$\Gamma_{\mu,tot} = \langle \Gamma_{\mu,tot} \rangle + \{\Gamma_{\mu,tot} - \langle \Gamma_{\mu,tot} \rangle\} \tag{IX.7.7}$$

and

$$G_{tot} = \langle G_{tot} \rangle + \{G_{tot} - \langle G_{tot} \rangle\} \tag{IX.7.8}$$

the reduced noise sources in the curly brackets may be identified with those of McCumber, as has been shown by Lax[3]. (The remaining averages describe spontaneous emission).

---

[1] See footnote [1] on p. 286.

[2] D. E. McCumber: Phys. Rev. **141**, 306 (1966).

[3] M. Lax: IEEE. J. Quantum Electronics ***QE*-3**, 37 (1967). These noise sources have a shot noise character. The important contribution stems from the Schrot-effect [compare W. Schottky, Ann. Phys. **57**, 541 (1918)] *of photons*. Because the existence of single electrons comes from the quantization of the Schrödinger wavefield in the same way as the existence of photons arises from the quantization of the electromagnetic field, Schottky's treatment represented in fact a *quantum theory of noise* connected with currents.

IX.8. Exact elimination of the atomic operators from the density matrix equation[1]. In order to eliminate the atomic variables from the density matrix equation (IV.7.68), we adopt a projection operator formalism based on the work of ZWANZIG[2] and, more specifically, of ARGYRES and KELLEY[3].

We consider the following decomposition of the density operator $W(t)$. [In this chapter we write $W(t)$ for the density operator of the coupled system active atoms + field and $\varrho(t)$ for the reduced density operator of the field.]

$$W(t) = \mathfrak{K}\, W(t) + (1 - \mathfrak{K})\, W(t). \tag{IX.8.1}$$

Here $\mathfrak{K}$ is a projection operator

$$\mathfrak{K}\,(1 - \mathfrak{K}) = 0 \tag{IX.8.2}$$

acting on the density operator. Its precise definition will be given below.

Further, we write (IV.7.68) as

$$W(t) = -i\,\mathfrak{L}\, W(t) \tag{IX.8.3}$$

where the Liouville operator $\mathfrak{L}$ is defined according to the right-hand side of (IV.7.68):

$$\mathfrak{L} = \mathfrak{L}_A + \mathfrak{L}_L + \mathfrak{L}_{AL} + i\Lambda_A + i\Lambda_L. \tag{IX.8.4}$$

Here $\mathfrak{L}_A, \mathfrak{L}_L, \mathfrak{L}_{AL}$ are the commutator operations with the Hamiltonians (IV.7.69) and refer to the free motion of the active atoms, the field, and the atom-field interaction, respectively. Λ_A and Λ_L represent the incoherent part of the motion of atoms and field according to pump and loss mechanisms [see (IV.7.70) and (IV.7.71)]. For later use we give here an alternative, equivalent definition of Λ_A:

$$\left.\begin{aligned}
\Lambda_A &= \sum_{\mu=1}^{N} \Lambda_\mu, \\
e^{\Lambda_A t}(a_2^+ a_1)_\mu &= e^{-\gamma_\perp t}(a_2^+ a_1)_\mu, \\
e^{\Lambda_A t}(a_1^+ a_2)_\mu &= e^{-\gamma_\perp t}(a_1^+ a_2)_\mu, \\
e^{\Lambda_A t}(a_2^+ a_2 - a_1^+ a_1)_\mu &= e^{-\gamma_\parallel t}(a_2^+ a_2 - a_1^+ a_1)_\mu, \\
e^{\Lambda_A t} 1_\mu &= 1_\mu - d_0 (e^{-\gamma_\parallel t} - 1)(a_2^+ a_2 - a_1^+ a_1)_\mu.
\end{aligned}\right\} \tag{IX.8.5)*}$$

Inserting the decomposition (IX.8.1) into the right-hand side of (IX.8.3) and acting then from the left, alternatively with $\mathfrak{K}$ and $(1 - \mathfrak{K})$, we obtain the following set of equations for the two quantities $\mathfrak{K}\, W(t)$ and $(1 - \mathfrak{K})\, W(t)$

$$\mathfrak{K}\, \dot{W}(t) = -i\,\mathfrak{K}\,\mathfrak{L}\,\mathfrak{K}\, W(t) - i\,\mathfrak{K}\,\mathfrak{L}\,(1 - \mathfrak{K})\, W(t)\,, \tag{IX.8.6a}$$

$$(1 - \mathfrak{K})\, \dot{W}(t) = -i\,(1 - \mathfrak{K})\,\mathfrak{L}\,\mathfrak{K}\, W(t) - i\,(1 - \mathfrak{K})\,\mathfrak{L}\,(1 - \mathfrak{K})\, W(t)\,. \tag{IX.8.6b}$$

To eliminate one quantity, e.g. $(1 - \mathfrak{K})\, W(t)$, from the set, we integrate (IX.8.6b) and reinsert the formal solution

$$\left.\begin{aligned}
(1 - \mathfrak{K})\, W(t) &= \exp[-i(1 - \mathfrak{K})\,\mathfrak{L}\, t]\,(1 - \mathfrak{K})\, W(0) - \\
&\quad - i \int_0^t ds\, \exp[-i(1 - \mathfrak{K})\,\mathfrak{L}\, s]\,(1 - \mathfrak{K})\,\mathfrak{L}\,\mathfrak{K}\, W(t - s)
\end{aligned}\right\} \tag{IX.8.7}$$

* The notation is the same as in Chap. (VI.12).

[1] F. HAAKE: Z. Physik **227**, 179 (1969) whom we will follow.

[2] R. ZWANZIG: Physica **30**, 1109 (1964).

[3] P.N. ARGYRES and P.L. KELLEY: Phys. Rev. **134**, A 98 (1964).

into (IX.8.6a). This yields

$$\left.\begin{aligned} \mathfrak{K}\dot{W}(t) &= -i\mathfrak{K}\mathfrak{L}\mathfrak{K}W(t) \\ &\quad - \int_0^t ds\,\mathfrak{K}\mathfrak{L}\exp[-i(1-\mathfrak{K})\mathfrak{L}s](1-\mathfrak{K})\mathfrak{L}\mathfrak{K}W(t-s) \\ &\quad - i\mathfrak{K}\mathfrak{L}\exp[-i(1-\mathfrak{K})\mathfrak{L}t](1-\mathfrak{K})W(0). \end{aligned}\right\} \tag{IX.8.8}$$

This "generalized master equation" was first derived using the projection formalism by ZWANZIG[2]. He defined the projector $\mathfrak{K}$ such that $\mathfrak{K}W(t)$ is the diagonal part of the density operator $W(t)$ in some specific representation. Here, however, following the ideas of ARGYRES and KELLEY[3] and, with respect to the laser, HAAKE[1], we define $\mathfrak{K}$ as the trace operation in the Hilbert space of the active atoms:

$$\left.\begin{aligned} \mathfrak{K} &= A\,\mathrm{tr}_A \quad \text{with} \quad \mathrm{tr}_A A = 1 \\ \mathfrak{K}W(t) &= A\,\mathrm{tr}_A W(t) \equiv A\,\varrho(t). \end{aligned}\right\} \tag{IX.8.9}$$

The "parameter" A is an operator in the Hilbert space of the atoms. It can be chosen arbitrarily within the constraint of normalization ensuring $\mathfrak{K}$ to be a projector. For reasons of formal convenience, we require it to obey simultaneously the equations

$$\mathfrak{L}_A \cdot A = 0, \qquad \Lambda_A \cdot A = 0 \tag{IX.8.10}$$

Then A is the "unsaturated" atomic density operator

$$A = \prod_{\mu=1}^{N} A_\mu; \qquad A_\mu = \frac{1-d_0}{2}(a_1^+ a_1)_\mu + \frac{1+d_0}{2}(a_2^+ a_2)_\mu. \tag{IX.8.11}$$

With this definition of $\mathfrak{K}$, we can regard the generalized master Eq. (IX.8.8) as an equation of motion for the reduced density operator $\varrho(t)$ of the field. Thus the exact elimination of the atomic coordinates is achieved. Exploiting the algebraic properties of the Liouville operator $\mathfrak{L}$ and the projector $\mathfrak{K}$ following from their definitions*, we may write this equation as

$$\dot{\varrho}(t) = -i(\mathfrak{L}_L + i\Lambda_L)\,\varrho(t) + \int_0^t ds\,K(s)\,\varrho(t-s) + I(t). \tag{IX.8.12}$$

Here the inhomogeneity, $I(t)$, and the integral kernel, $K(s)$, are given by

$$\left.\begin{aligned} I(t) &= -i\,\mathrm{tr}_A\,\mathfrak{L}_{AL}\exp[-i(1-\mathfrak{K})\mathfrak{L}t](1-\mathfrak{K})W(0) \\ K(s) &= -\mathrm{tr}_A\,\mathfrak{L}_{AL}\exp[-i(1-\mathfrak{K})\mathfrak{L}s]\,\mathfrak{L}_{AL}A. \end{aligned}\right\} \tag{IX.8.13}$$

For further use it is convenient to represent $K(s)$ as an expansion in the interaction Liouville operator $\mathfrak{L}_{AL}$:

$$\left.\begin{aligned} K(s) &= \mathrm{tr}_A\,\mathfrak{L}_{AL}\sum_{n=0}^{\infty}(-1)^n\int_0^s ds_1\int_0^{s_1} ds_2\ldots\int_0^{s_{2n-1}} ds_{2n}\times \\ &\quad \times\{E(s-s_1)\,E(s_1-s_2)\ldots E(s_{2n-1}-s_{2n})\,E(s_{2n})\}\cdot A \end{aligned}\right\} \tag{IX.8.14}$$

with

$$E(s) = \exp[-i(\mathfrak{L}_A + \mathfrak{L}_L + i\Lambda_A + i\Lambda_L)s](1-\mathfrak{K})\,\mathfrak{L}_{AL}.$$

An analogous expansion holds for $I(t)$.

The inhomogeneous integrodifferential Eq. (IX.8.12) for the reduced density operator of the field describes $\dot{\varrho}(t)$ so as to depend on the whole "history" of $\varrho(t)$.

* Operators referring to different degrees of freedom commute; the trace over a commutator vanishes; for a complete list and a discussion of these properties see HAAKE's original paper.

This is, of course, a consequence of the general rule that if a coupled system "atom + field" is undergoing a Markoff process, the reduced system "field" alone does not. Note, however, that the free motion of the field and the pure field damping, as described by $\mathfrak{L}_L$ and Λ_L, respectively, act instantaneously. Only the photon-photon interaction which is brought about by the atom-photon interaction $\mathfrak{L}_{AL}$ is retarded. As we shall see in the next chapter, the known adiabatic elimination schemes can be characterized so as to take the indirect photon interaction as an instantaneous one.

IX.9. Solution of the generalized field master Eq. (IX.8.12)[1]. Fortunately it is not necessary to keep in consideration all terms of the expansion (IX.8.14) of the integral kernel $K(s)$. The lowest order term, $n=0$, to which usual applications of the projector formalism are always restricted, is found to describe an action of the atomic system on the field and to neglect any reaction in the inverse direction. The next term, $n=1$, is the first to account for saturation effects in the atomic system which are of fundamental importance for the laser. As we shall show, these two terms which give the kernel correctly up to fourth order in the interaction Louville operator $\mathfrak{L}_{AL}$ are usually sufficient for the theory of the laser.

When the expansion terms are evaluated explicitly up to this order, it is found that the time behaviour of the kernel is characterized by four different retardation effects.

1) Relaxation of the atomic dipole moment produces a retardation of the form $e^{-\gamma_\perp s}$.

2) Relaxation of the atomic inversion contributes a retardation term $e^{-\gamma_\| s}$.

3) The field damping occurs as $e^{-\varkappa s}$.

4) The inhomogeneous broadening of the atomic line gives rise to the factors

$$\sum_{\mu=1}^{N} g_\mu^2 e^{-i(\nu_\mu-\omega)s}, \qquad \sum_{\mu=1}^{N} g_\mu^4 e^{-i(\nu_\mu-\omega)s}.$$

This is the most complex situation to be dealt with in laser theory. There is no difficulty in principle in coping with this problem but, in view of the limited space, we here confine ourselves to the special case of a homogeneous line in resonance ($\nu_\mu=\omega$) and assume that the field damping is small compared to the atomic damping constants ($\varkappa\ll\gamma, \gamma_\|$). The resulting equation of motion for the field density operator, expressed in the interaction representation in terms of the diagonal representation

$$\varrho(t)=\int d^2\alpha\, P(\alpha,\alpha^*,t)\,|\alpha\rangle\langle\alpha| \tag{IX.9.1}$$

with respect to coherent states [see (IV.9.11), (IV.9.12)], then becomes*

$$\begin{aligned}\dot P(\alpha,a^*,t)&=\varkappa\left\{\frac{\partial}{\partial\alpha}\alpha+\frac{\partial}{\partial\alpha^*}\alpha^*+2n_{th}\frac{\partial^2}{\partial\alpha\,\partial\alpha^*}\right\}P(\alpha,a^*,t)+\\&+\int_0^t ds\left[\left(\frac{\partial}{\partial\alpha}\alpha+\frac{\partial}{\partial\alpha^*}\alpha^*\right)\left(-\frac{Ng^2d_0}{\gamma_\perp}\varphi_1(s)+\frac{4Ng^4d_0}{\gamma_\perp^2\gamma_\|}\varphi_2(s)\alpha^*\alpha\right)+\right.\\&\left.+4\frac{Ng^2(1+d_0)}{4\gamma_\perp}\varphi_1(s)\frac{\partial^2}{\partial\alpha\,\partial\alpha^*}\right]P(\alpha,\alpha^*,t-s).\end{aligned} \tag{IX.9.2}$$

* We have omitted the inhomogeneity; as $I(t)\to 0$ for $t\to\infty$, at least as $e^{-\gamma_\| t}$, $e^{-\gamma_\perp t}$, it does not play any role for the stationary operating laser; it must, however, be considered in a discussion of transient effects. We have further neglected some correction terms in the integral kernel containing especially derivatives with respect to α, α^* of fourth order; by normalizing the field variables as in (VI.12), one can see that this is valid; for a detailed discussion see HAAKE's paper[1].

[1] See footnote [1] on p. 228.

There is now a remarkable similarity to the semiclassical Fokker-Planck Eq. (VI.12.28): We find terms describing the field damping, the linear gain, the nonlinearity and the fluctuation. The retardation functions $\varphi_1(s)$, $\varphi_2(s)$ are given by

$$\left.\begin{aligned}\varphi_1(s) &= \gamma_\perp e^{-\gamma_\perp s}\\ \varphi_2(s) &= \frac{\gamma_\perp^2 \gamma_\parallel}{(\gamma_\perp - \gamma_\parallel)^2}\left[e^{-\gamma_\parallel s} - \{1 + s(\gamma_\perp - \gamma_\parallel)\, e^{-\gamma_\perp s}\}\right].\end{aligned}\right\} \qquad \text{(IX.9.3)}$$

In the following we solve the generalized Fokker-Planck equation (IX.9.2) in successive steps of approximation.

1) *Adiabatic or "Markoff" approximation.* A well-known argument first put forward by VAN HOVE[2] allows us to approximate the integro-differential Eq. (IX.9.2) by a first order differential equation with respect to time: if the kernel relaxes much faster than $P(\alpha, \alpha^*, t)$ the retardation may be neglected completely. Then (IX.9.2) becomes

$$\dot{P}(\alpha,\alpha^*,t) = \Lambda P(\alpha,\alpha^*,t) \qquad \text{(IX.9.4)}$$

with*

$$\left.\begin{aligned}\Lambda &= \Lambda_L + \int_0^\infty K(s)\, ds =\\ &= \beta\left(\frac{\partial}{\partial\alpha}\alpha + \frac{\partial}{\partial\alpha^*}\alpha^*\right)(C - \alpha^*\alpha) + 4Q\frac{\partial^2}{\partial\alpha^*\partial\alpha}.\end{aligned}\right\} \qquad \text{(IX.9.5)}$$

In (IX.9.4) we recognize the semiclassical Fokker-Planck equation discussed in (VI.12) and (VI.13). The region of validity of the approximation now follows from VAN HOVE's argument and the solution of the Fokker-Planck equation. The relaxation of $P(\alpha, \alpha^* t)$ according to (IX.9.4) is governed by the eigenvalues** γ_{nm} of the differential operator Λ. The validity criterion therefore reads

$$\gamma_{nm} \ll \gamma_\perp, \gamma_\parallel \qquad \text{(IX.9.6)}$$

By inspection of RISKEN and VOLLMER's[3] results, we see that (IX.9.6) is very well fulfilled for the experimentally important γ_{nm} near threshold, whereas difficulties may occur far away from threshold.

2) *Unified retardation or first non-Markoffian approximation.* In this step we do not neglect retardation any longer but disregard the differences in the retardation of linear gain, nonlinearity, and fluctuation. We rather put in a somewhat ad hoc manner

$$\varphi_1(s) = \varphi_2(s) = \gamma_\perp e^{-\gamma_\perp s}. \qquad \text{(IX.9.7)}$$

Consequently, $K(s)$ obeys

$$\frac{\partial}{\partial s} K(s) = -\gamma_\perp K(s). \qquad \text{(IX.9.8)}$$

---

* The notation occurring here is the same as in (VI.12).

** The eigenvalues γ_{nm} are the λ_{nm}'s of (VI.13) multiplied by $\sqrt{\beta \cdot Q}$.

[2] L. VAN HOVE, Physica **23**, 441 (1957).

[3] H. RISKEN and H.D. VOLLMER: Z. Physik **201**, 323 (1967); **204**, 240 (1967).

Differentiating the integrodifferential equation for $P(t)$ with respect to time and combining linearly the equations for $\dot{P}(t)$ and $\ddot{P}(t)$ according to (IX.9.8), we get

$$\left.\begin{aligned}&\ddot{P}(\alpha,\alpha^*,t)+\gamma_\perp\left(1-\frac{1}{\gamma_\perp}\Lambda_L\right)\dot{P}(\alpha,\alpha^*,t)\\&=\gamma_\perp\left(\Lambda_L+\frac{1}{\gamma_\perp}K(0)\right)P(\alpha,\alpha^*,t)+\int_0^t ds\,\underbrace{(\dot{K}(s)+\gamma_\perp K(s))}_{=0}P(\alpha,\alpha^*,t-s)\\&=\left(\gamma_\perp\Lambda_L+\int_0^\infty ds\,K(s)\right)P(\alpha,\alpha^*,t)=\gamma_\perp\Lambda P(\alpha,\alpha^*,t)\,.\end{aligned}\right\}\quad\text{(IX.9.9)}$$

Here Λ is again the Fokker-Planck differential operator as given by (IX.9.5). As Λ_L is of order of $\varkappa$ and as we are already restricted to $\varkappa\ll\gamma_\perp,\gamma_\parallel$ we may equivalently state

$$\ddot{P}(\alpha,\alpha^*,t)+\gamma_\perp\dot{P}(\alpha,\alpha^*,t)=\gamma_\perp\Lambda P(\alpha,\alpha^*,t)\,.\tag{IX.9.10}$$

This second order partial differential equation, which approximates the integrodifferential Eq. (IX.9.2) in the sense mentioned above, has been derived and solved recently by GNUTZMANN and WEIDLICH[4] on the basis of the ansatz (VI.11.9), (VI.11.10).

3) Exact treatment.

The integrodifferential Eq. (IX.9.2) is equivalent to a fourth order differential equation with respect to time. To show this we observe that $K(s)$ obeys

$$\dddot{K}(s)+(2\gamma_\perp+\gamma_\parallel)\ddot{K}(s)+\gamma_\perp(\gamma_\perp+2\gamma_\parallel)\dot{K}(s)+\gamma_\perp^2\gamma_\parallel K(s)=0\,.\tag{IX.9.11}$$

Evaluating $\ddot{P}$, $\dddot{P}$, $\ddddot{P}$ by differentiating (IX.9.2) and again combining linearly in view of (IX.9.11), we get

$$\left.\begin{aligned}&\ddddot{P}(t)+(2\gamma_\perp+\gamma_\parallel)\left(1-\frac{1}{2\gamma_\perp+\gamma_\parallel}\Lambda_L\right)\dddot{P}(t)+\\&\quad+\gamma_\perp(\gamma_\perp+2\gamma_\parallel)\left(1-\frac{K(0)+(2\gamma_\perp+\gamma_\parallel)\Lambda_L}{\gamma_\perp(\gamma_\perp+2\gamma_\parallel)}\right)\ddot{P}(t)+\\&\quad+\gamma_\perp^2\gamma_\parallel\left(1-\frac{\dot{K}(0)+(2\gamma_\perp+\gamma_\parallel)K(0)+\gamma_\perp(\gamma_\perp+2\gamma_\parallel)\Lambda_L}{\gamma_\perp^2\gamma_\parallel}\right)\dot{P}(t)\\&=\gamma_\perp^2\gamma_\parallel\left(\Lambda_L+\frac{\ddot{K}(0)+(2\gamma_\perp+\gamma_\parallel)\dot{K}(0)+\gamma_\perp(\gamma_\perp+2\gamma_\parallel)K(0)}{\gamma_\perp^2\gamma_\parallel}\right)P(t)\\&=\gamma_\perp^2\gamma_\parallel\left(\Lambda_L+\int_0^\infty ds\,K(s)\right)P(t)=\gamma_\perp^2\gamma_\parallel\Lambda P(t)\,.\end{aligned}\right\}\quad\text{(IX.9.12)}$$

Again Λ is the Fokker-Planck operator (IX.9.5).

---

[4] U. GNUTZMANN and W. WEIDLICH: Phys. Letters **27** A, 179 (1968). — U. GNUTZMANN: Z. Physik **222**, 283 (1969); **225**, 416 (1969).

In order to gain an insight into the character of the solutions of this equation, we present a somewhat simplified version of it arising from the neglect of*

$$\left.\begin{aligned} \frac{\Lambda_L}{2\gamma_\perp+\gamma_\parallel} &\ll 1\,, \\ \frac{K(0)+(2\gamma_\perp+2\gamma_\parallel)\Lambda_L}{\gamma_\perp(\gamma_\perp+2\gamma_\parallel)} &\ll 1\,, \\ \frac{\dot K(0)+(2\gamma_\perp+\gamma_\parallel)K(0)+\gamma_\perp(\gamma_\perp+2\gamma_\parallel)\Lambda_L}{\gamma_\perp^2\gamma_\parallel} &\ll 1\,. \end{aligned}\right\} \qquad \text{(IX.9.13)}$$

Then the following equation results:

$$\ddddot{P}(t)+(2\gamma_\perp+\gamma_\parallel)\dddot{P}(t)+\gamma_\perp(\gamma_\perp+2\gamma_\parallel)\ddot{P}(t)+\gamma_\perp^2\gamma_\parallel\dot{P}(t)=\gamma_\perp^2\gamma_\parallel\Lambda P(t)\,. \qquad \text{(IX.9.14)}$$

4) *Remarks on the solutions.*

Let us first state the important fact that the stationary solutions of the integro-differential Eq. (IX.9.2) and of all the corresponding differential Eqs. (IX.9.4), (IX.9.9), (IX.9.10), (IX.9.12), (IX.9.14) coincide. In all cases we find the well-known stationary distribution function (VI.13.3) of the Fokker-Planck theory.

The three differential Eqs. (IX.9.4), (IX.9.10), (IX.9.14) can all be treated in the same manner. They can all be written in the form

$$P(t)=\sum_{m=0}^{\infty}\sum_{n=-\infty}^{+\infty}A_{nm}T_{nm}(t)F_{nm}(\alpha,\alpha^*)\,. \qquad \text{(IX.9.15)}$$

Here the F_{nm}'s are the eigenfunctions of the Fokker-Planck differential operator [see (VI.13)]:

$$\left.\begin{aligned} \Lambda F_{nm} &= -\gamma_{nm}F_{nm}; \quad m=0,1,2,\ldots \\ \gamma_{nm} &\geqq 0 \qquad\qquad\quad n=0,\pm1,\pm2,\ldots \end{aligned}\right\} \qquad \text{(IX.9.16)}$$

The time functions $T_{nm}(t)$ have to be calculated from the respective differential equations:

$$\dot T_{nm}(t)+\gamma_{nm}T_{nm}(t)=0\,, \qquad \text{(IX.9.4a)}$$

$$\ddot T_{nm}(t)+\gamma_\perp\dot T_{nm}(t)+\gamma_\perp\gamma_{nm}T_{nm}(t)=0\,, \qquad \text{(IX.9.10a)}$$

$$\left.\begin{aligned} &\ddddot{T}_{nm}(t)+(2\gamma_\perp+\gamma_\parallel)\dddot{T}_{nm}(t)+\gamma_\perp(\gamma_\perp+2\gamma_\parallel)\ddot T_{nm}(t)+ \\ &+\gamma_\perp^2\gamma_\parallel\dot T_{nm}(t)+\gamma_\perp^2\gamma_\parallel\gamma_{nm}T_{nm}(t)=0\,. \end{aligned}\right\} \qquad \text{(IX.9.14a)}$$

Applying the Hurwitz criterion to the corresponding secular equations, we find that no instabilities can occur in the adiabatic or Markoff approximation (IX.9.4) and the first non-Markoff approximation (IX.9.10), whereas for (IX.9.14) we have the stability condition

$$\gamma_{nm}<\frac{\gamma_\perp(\gamma_\perp+2\gamma_\parallel)}{2\gamma_\perp+\gamma_\parallel}-\frac{\gamma_\perp^2\gamma_\parallel}{(2\gamma_\perp+\gamma_\parallel)^2}\,. \qquad \text{(IX.9.17)}$$

Formally (IX.9.17) can certainly be violated extremely far above threshold. It should be noted, however, that for high field intensities the total expansion (IX.8.14) must be used and summed up.

---

* These approximations may be estimated by the following somewhat oversimplified argument: the operators occurring in the 2nd and 3rd relation have the same structure as Λ, their eigenvalues $\tilde\gamma_{nm}$ are therefore principally known; the neglect is justified if $\tilde\gamma_{nm}\ll\gamma_\perp,\gamma_\parallel$.

5) *Non-Markoffian effects.* All non-Markoffian effects now follow from the secular equations corresponding to (IX.9.10a) and (IX.9.14a). Whereas the "Markoffian" $T_{nm}(t)$ is characterized by a single real relaxation constant (γ_{nm}), the non-Markoffian $T_{nm}(t)$ according to (IX.9.10a) and (IX.9.14a) has two and four relative terms, respectively, within general complex damping constants. Therefore the physically important observables, such as amplitude and intensity correlation functions, do not decay monotonically in time but rather show damped oscillations (compare also Chap. VI.2). This behavior is observable in some semi-conductor lasers.

X. Appendix. Useful operator techniques.

In the Chap. X.1 we briefly remind the reader how to describe the quantum mechanical harmonic oscillator by means of annihilation and creation operators of photons and how to write expectation values or matrix elements by means of Dirac's notation. In the second chapter we derive some useful operator relations. In the third chapter the formal solution of the Schrödinger equation is written down, which gives a natural access to time ordering operators. In Chap. X.4 we describe the disentangling technique of FEYNMAN, which allows us to deal in particular with exponential functions of operators. In the last chapter we apply this theorem to the transformation of exponential operators which contain Bose operators.

X.1. The harmonic oscillator. The Schrödinger equation reads

$$H\Phi(x)=E\Phi(x) \tag{X.1.1}$$

where the Hamiltonian is given by

$$H=-\frac{\hbar^2}{2m}\frac{d^2}{dx^2}+\frac{m}{2}\omega^2x^2. \tag{X.1.2}$$

We use as usual the momentum operator

$$p=\frac{\hbar}{i}\frac{d}{dx}. \tag{X.1.3}$$

It obeys the commutation relation

$$[p,x]=\frac{\hbar}{i}. \tag{X.1.4}$$

We introduce linear combinations of x and p in the form

$$b=\frac{1}{\sqrt{2\hbar}}\left(\sqrt{m\omega}\;x+\frac{i}{\sqrt{m\omega}}\,p\right), \tag{X.1.5}$$

$$b^+=\frac{1}{\sqrt{2\hbar}}\left(\sqrt{m\omega}\;x-\frac{i}{\sqrt{m\omega}}\,p\right). \tag{X.1.6}$$

We will show that (X.1.5) acts as an annihilation operator whereas (X.1.6) acts as a creation operator of a single oscillator quantum. Using (X.1.5) and (X.1.6) and the commutation relation (X.1.4) the Hamiltonian (X.1.2) transforms into

$$H=\hbar\omega(b^+b+\tfrac{1}{2}) \tag{X.1.7}$$

whereas the commutation relation between (X.1.5) and (X.1.6) reads

$$bb^+-b^+b=1. \tag{X.1.8}$$

We write the energy in the form

$$E = \hbar\omega(n + \tfrac{1}{2}). \tag{X.1.9}$$

We will prove that n is an integer number $n = 0, 1, 2, \ldots$. We first show that E is positive. We form the expectation value

$$E = \int \Phi^*(x)\left(-\frac{\hbar^2}{2m}\frac{d^2}{dx^2} + \frac{m}{2}\omega^2 x^2\right)\Phi(x)\,dx \tag{X.1.10}$$

which can be transformed into

$$= \int \left(\frac{\hbar^2}{2m}\left|\frac{d\Phi}{dx}\right|^2 + \frac{m}{2}\omega^2 x^2 |\Phi|^2\right) dx > 0 \tag{X.1.11}$$

by partial integration. Introducing

$$H' = b^+ b \tag{X.1.12}$$

as a new Hamiltonian and using (X.1.9) we write the Schrödinger Eq. (X.1.1) in the form

$$b^+ b\,\Phi = n\,\Phi. \tag{X.1.13}$$

Because n has a lower bound we may assume that we have found the solution Φ_0 of (X.1.13) with the lowest eigenvalue n_0:

$$b^+ b\,\Phi_0 = n_0\,\Phi_0. \tag{X.1.14}$$

We multiply Eq. (X.1.14) from the left with the annihilation operator b

$$b b^+ b\,\Phi_0 = n_0 b\,\Phi_0 \tag{X.1.15}$$

and find with aid of the commutation relation (X.1.8)

$$b^+ b\,(b\,\Phi_0) = (n_0 - 1)\,(b\,\Phi_0). \tag{X.1.16}$$

Thus it appears as if we had found a new wave function with a still lower eigenvalue $n_0 - 1$ in contradiction to the assumption. This contradiction is solved if

$$b\,\Phi_0 = 0 \tag{X.1.17}$$

holds, from which it follows immediately that

$$n_0 = 0. \tag{X.1.18}$$

Using (X.1.17) and the explicit form of b (X.1.5), we obtain the equation

$$\left(x + \frac{\hbar}{m\omega}\frac{d}{dx}\right)\Phi_0 = 0 \tag{X.1.19}$$

with the well-known solution of the oscillator ground state

$$\Phi_0 = N \cdot e^{-\frac{m\omega}{2\hbar}x^2} \tag{X.1.20}$$

where N is the normalization constant.

We now demonstrate how to build up all the higher wave functions. We start from (X.1.13) which we multiply by b^+ from the left.

$$b^+ b^+ b\,\Phi_n = n b^+\,\Phi_n \tag{X.1.21}$$

Using the commutation relation (X.1.8) we immediately find

$$b^+ b\,(b^+\,\Phi_n) = (n + 1)\,b^+\,\Phi_n \tag{X.1.22}$$

which means that

$$b^+ \Phi_n = \Phi_{n+1} \cdot C_{n+1} \tag{X.1.23}$$

is a new solution of the Schrödinger Eq. (X.1.13) with a new eigenvalue $n+1$. In (X.1.23) C_{n+1} is a c-number factor which serves for the normalization of Φ_{n+1} if Φ_n is normalized and which we will determine below. Starting from the lowest eigenfunction Φ_0 the function belonging to the n'th level is given by

$$\Phi_n = \frac{1}{\sqrt{n!}} (b^+)^n \Phi_0, \tag{X.1.24}$$

as can be shown by using the explicit form of C_n [cf. (X.1.35)] and the process of induction.

We now introduce the bracket notation as used in this book*. Usually we have to deal in quantum mechanics with integrals of the form

$$\int \Phi^*(x) \chi(x)\, dx \tag{X.1.25}$$

which contain the product of two functions. We now write (X.1.25) in the form

$$\langle \Phi^* \chi \rangle \qquad \text{(X.1.26)*}$$

and show that

$$\langle (b^+ \Phi)^* \chi \rangle = \langle \Phi^* b \chi \rangle \tag{X.1.27}$$

holds. The proof follows directly by using the explicit form (X.1.25) with (X.1.6) and performing a partial integration with respect to the derivative which appears in p in (X.1.6). In the same way one proves

$$\langle (b\Phi)^* \chi \rangle = \langle \Phi^* b^+ \chi \rangle . \tag{X.1.28}$$

We now claim that

$$\langle (b^+ \Phi_n)^* b^+ \Phi_n \rangle = (n+1) \langle \Phi_n^* \Phi_n \rangle \tag{X.1.29}$$

holds. For the proof we use (X.1.27) which yields for the left-hand side of (X.1.29)

$$\langle \Phi_n^* b b^+ \Phi_n \rangle \tag{X.1.30}$$

Using the commutation relation (X.1.8) we find

$$\text{(X.1.30)} = \langle \Phi_n^* (b^+ b + 1) \Phi_n \rangle \tag{X.1.31}$$

We now use the fact that Φ_n is an eigenfunction of the Hamiltonian, so that

$$\text{(X.1.31)} = \langle \Phi_n^* \Phi_n \rangle (n+1) \tag{X.1.32}$$

results, which completes our proof.

It is a simple matter to show explicitly that the eigenfunctions are *orthogonal* to each other by using the explicit form (X.1.24), the relation (X.1.27), and the property (X.1.17). The orthogonality is, of course, a consequence of the hermiticity of the Hamiltonian and the non-degeneracy.

We now determine the factor C_{n+1} in (X.1.23).

Using (X.1.23) we write the left-hand side of (X.1.29) in the form

$$C_{n+1}^2 \langle \Phi_{n+1}^* \Phi_{n+1} \rangle, \tag{X.1.33}$$

---

* This notation is connected with Dirac's original one by

$$\underset{\text{present book}}{\langle \phi^* \chi \rangle} = \underset{\text{Dirac}}{\langle \phi | \chi \rangle}$$

which according to (X.1.29) must be equal to

$$(n+1)\langle \Phi_n^* \Phi_n \rangle. \tag{X.1.34}$$

When we assume Φ_n and Φ_{n+1} to be normalized, a comparison between (X.1.33) and (X.1.34) shows

$$C_{n+1} = \sqrt{n+1}. \tag{X.1.35}$$

Thus (X.1.23) may be written more explicitly in the form

$$b^+ \Phi_n = \sqrt{n+1}\, \Phi_{n+1}. \tag{X.1.36}$$

In a similar way one proves

$$b \Phi_n = \sqrt{n}\, \Phi_{n-1}. \tag{X.1.37}$$

X.2. Operator relations for Bose operators[1]. In connection with the quantum mechanical treatment of the laser, often commutations between b^+ or b and a function of b or b^+ must be performed. We claim that

$$\left.\begin{aligned} &b(b^+)^n - (b^+)^n b = n(b^+)^{n-1} \\ \text{and}\qquad &b^+(b)^n - (b)^n b^+ = -n(b)^{n-1} \end{aligned}\right\} \tag{X.2.1}$$

hold and that more generally

$$\left.\begin{aligned} &b f(b^+, b) - f(b^+, b)\, b = \frac{\partial f}{\partial b^+} \\ \text{and}\qquad &b^+ f(b^+, b) - f(b^+, b)\, b^+ = -\frac{\partial f}{\partial b} \end{aligned}\right\} \tag{X.2.2}$$

are valid. The proof of (X.2.1) can be done by complete induction using the commutation relation (X.1.8) as follows:

For $n=1$ Eq. (X.2.1) is true, because it reduces to the commutation relation (X.1.8). Let us assume now that Eq. (X.2.1) has been proved for all n up to $n=m$. In order to show that it is also valid for $n=m+1$ we rearrange

$$b(b^+)^{m+1} - (b^+)^{m+1} b = b^+\big(b(b^+)^m - (b^+)^m b\big) + (b^+)^m \tag{X.2.3}$$

using the commutation relation (X.1.8). Using the relation (X.2.1) for $n=m$ we obtain immediately:

$$b(b^+)^{m+1} - (b^+)^{m+1} b = (m+1)(b^+)^m. \tag{X.2.4}$$

From the validity of (X.2.1) for $n=m$ thus follows the validity for $n=m+1$ and hence for all positive integer n.

In order to prove (X.2.2) we assume that f can be represented by a power series in b^+ and b, where any given sequence of operators b^+, b is allowed. Because b commutes with all powers of b, the proof of the first relation (X.2.2) immediately reduces to that of (X.2.1). Note, however, that one must not change the given order of the operators on the right-hand side, i.e. the differentiation with respect to b^+ must be done according to the product rule occurring in usual differentiation, i.e.

$$\frac{\partial(g_1 g_2)}{\partial b^+} = \frac{\partial g_1}{\partial b^+} g_2 + g_1 \frac{\partial g_2}{\partial b^+}. \tag{X.2.5}$$

[1] These relations are scattered over the literature on quantum field theory. In our representation we follow the exercises to our lectures on quantum theory, 1960.

We now claim that

$$e^{\alpha b} b^{+} e^{-\alpha b} = b^{+} + \alpha, \tag{X.2.6}$$

$$e^{\beta b^{+}} b\, e^{-\beta b^{+}} = b - \beta. \tag{X.2.7}$$

For the proof of (X.2.6) we put

$$f(\alpha) = e^{\alpha b} b^{+} e^{-\alpha b}. \tag{X.2.8}$$

Differentiation of (X.2.8) with respect to α and using the commutation relation (X.1.8) yields

$$f'(\alpha) = e^{\alpha b}(b b^{+} - b^{+} b) e^{-\alpha b} = 1, \tag{X.2.9}$$

which has the solution

$$f(\alpha) = \alpha + \text{const}. \tag{X.2.10}$$

The constant is determined from the initial condition that $f(0) = b^{+}$. Thus (X.2.6) follows. The proof of (X.2.7) is completely analogous. We now claim that

$$e^{\alpha b^{+} b} b^{+} e^{-\alpha b^{+} b} = b^{+} e^{\alpha}, \tag{X.2.11}$$

$$e^{\alpha b^{+} b} b\, e^{-\alpha b^{+} b} = b\, e^{-\alpha} \tag{X.2.12}$$

holds. For the proof we introduce the operator function

$$g(\alpha) = e^{\alpha b^{+} b} b^{+} e^{-\alpha b^{+} b}. \tag{X.2.13}$$

Differentiating $g(\alpha)$ with respect to α and using the commutation relation (X.2.8) yields

$$g' = e^{\alpha b^{+} b}(b^{+} b b^{+} - b^{+} b^{+} b) e^{-\alpha b^{+} b} = e^{\alpha b^{+} b} b^{+} e^{-\alpha b^{+} b} = g, \tag{X.2.14}$$

which possesses the solution

$$g = \text{const}\, e^{\alpha}. \tag{X.2.15}$$

The initial condition for $\alpha = 0$ fixes the constant, which must be equal to the operator b^{+}. The proof of (X.2.12) is analogous.

X.3. Formal solution of the Schrödinger equation. We consider a general quantum system, which is not necessarily the harmonic oscillator, and which is described by the usual Schrödinger equation

$$\dot{\psi}(t) = \frac{1}{i\hbar} H(t)\, \psi(t). \tag{X.3.1}$$

We solve this equation by an iteration procedure for a small time interval dt starting from $t = t_0$:

$$\psi(t_0 + dt) = \psi(t_0) + \frac{1}{i\hbar} H(t_0) \psi(t_0)\, dt. \tag{X.3.2}$$

On account of the smallness of dt higher powers can be neglected, so that we can write instead of (X.3.2)

$$\psi(t_0 + dt) = e^{\frac{1}{i\hbar} H(t_0)\, dt}\, \psi(t_0). \tag{X.3.3}$$

Repeating the same procedure for the next time-interval we find

$$\psi(t_0 + 2dt) = e^{\frac{1}{i\hbar} H(t_0 + dt)\, dt}\, e^{\frac{1}{i\hbar} H(t_0)\, dt}\, \psi(t_0) \tag{X.3.4}$$

and after N steps

$$\psi(t) = \prod_{\nu=1}^{N} e^{\frac{1}{i\hbar} H(t_0+(\nu-1)\,dt)\,dt}\,\psi(t_0). \qquad \text{(X.3.5)}$$

Note that H does not in general commute for different times. Therefore the exponential functions must be applied in such a manner that the operators with a lower time are applied prior to operators belonging to a later time*. When we adopt this prescription for the operator H, we can write (X.3.5) equally well in the form

$$\psi(t) = e^{\frac{1}{i\hbar}\sum_{\nu=1}^{N} H(t_0+(\nu-1)\,dt)\,dt}\,\psi(t_0). \qquad \text{(X.3.6)}$$

As we make the time interval dt smaller and smaller, the sum in (X.3.6) can be finally replaced by an integral so that the formal solution of (X.3.1) reads

$$\psi(t) = \exp\left\{\frac{1}{i\hbar}\int_{t_0}^{t} H(\tau)\,d\tau\right\}\psi(t_0). \qquad \text{(X.3.7)}$$

From (X.3.7) it is quite simple to proceed to a representation which is usually used in time-dependent perturbation theory. The formal expansion of the exponential function in (X.3.7) yields

$$\psi(t) = \left\{\sum_{\mu=0}^{\infty}\left(\frac{1}{i\hbar}\right)^{\mu}\frac{1}{\mu!}\int_{t_0}^{t} H(\tau_1)\,d\tau_1\int_{t_0}^{t} H(\tau_2)\,d\tau_2\ldots\int_{t_0}^{t} H(\tau_\mu)\,d\tau_\mu\right\}\psi(t_0). \qquad \text{(X.3.8)}$$

This expression can be rearranged when use is made of the time-ordering prescription: the operators must always be applied in such a way that operators with the smaller index are standing to the right of operators with the higher index. This yields**

$$\psi(t) = \left\{\sum_{\mu=1}^{\infty}\left(\frac{1}{i\hbar}\right)^{\mu}\int_{t_0}^{t} d\tau_\mu\int_{t_0}^{\tau_\mu} d\tau_{\mu-1}\ldots\int_{t_0}^{\tau_2} d\tau_1\,H(\tau_\mu)\,H(\tau_{\mu-1})\,\ldots\,H(\tau_1)\right\}\psi(t_0). \qquad \text{(X.3.9)}$$

X.4. Disentangling theorem[1]. As we have seen in the preceding chapter, the formal solution of the Schrödinger equation can be written in the form of an exponential function. Usually the Hamiltonian consists of several parts which do not commute with each other. Therefore one is not allowed to split the exponential function of a sum of these contributions into a product of exponential functions of the single contributions. The theorem we are going to introduce now shows us which corrections are necessary if such splitting into exponential factors is done. We adopt the notation that s is an ordering parameter in the sense in which time

---

* For references see X.4.

** A discussion of this expansion and its connection with the S-matrix approach of Heisenberg and other formalisms as well is given by F. J. DYSON: Phys. Rev. **75**, 486, 1736 (1947).

[1] In our representation we follow more or less R. P. FEYNMAN: Phys. Rev. **84**, 108 (1951). — The classical solution of the equation $e^z = e^x e^y$ for z, where x and y are noncommuting operators, given by HAUSDORF, involves an expansion of z into an infinite series of terms homogeneous in y. The successive terms are found from a recursion formula, which is known as the BAKER-CAMPBELL-HAUSDORF formula: F. HAUSDORF: Ber. Verhandl. Sächs. Akad. Wiss. Leipzig, Math.-Naturw. Kl. **58**, 19 (1906). For further discussions and treatments of this and related problems see: W. MAGNUS: Commun. Pure Appl. Math. **7**, 649 (1954). — J. WEI: J. Math. Phys. **4**, 1337 (1963). — R. KARPLUS and J. SCHWINGER: Phys. Rev. **73**, 1020 (1948). — M. LUTZKY: J. Math. Phys. **9**, 1125 (1968).

was used above. s need not necessarily be the time, but it may coincide with it in certain cases. The theorem now states that for large N ($N \to \infty$)

$$\exp\left(\frac{1}{N}\sum_{s=1}^{N}(A_s+B_s)\right)=\exp\left(\frac{1}{N}\sum_{s=1}^{N}A_s\right)\cdot\exp\left(\frac{1}{N}\sum_{s=1}^{N}\tilde{B}_s\right) \tag{X.4.1}$$

with

$$\tilde{B}_s=U_s^{-1}B_sU_s \quad\text{and}\quad U_s=\exp\left(\frac{1}{N}\sum_{t=1}^{s-1}A_t\right). \tag{X.4.2}$$

Note that by forming U_s^{-1} from U_s we obtain an operator expression where the operators with *larger* ordering indices act first. We indicate this reversed ordering by a prime behind the operator. Thus

$$U_s^{-1}=\exp\left\{-\frac{1}{N}\sum_{t=1}^{s-1}A_t'\right\}$$

where the primed operators with larger ordering indices act first.

If we put $\frac{1}{N}=ds$ and take the limit of very large N the sums may be replaced by integrals so that the theorem acquires the form

$$\exp\left(\int_0^1 ds\,(A_s+B_s)\right)=\exp\left(\int_0^1 ds\,A_s\right)\exp\left(\int_0^1 ds\,\tilde{B}_s\right) \tag{X.4.3}$$

with

$$\tilde{B}_s=U_s^{-1}B_sU_s \quad\text{and}\quad U_s=\exp\left(\int_0^s dt\,A_t\right). \tag{X.4.4}$$

If the operators A_s and B_s do not explicity depend on the ordering parameter, a special case of (X.4.3) is given by

$$\exp(A+B)=\exp A\,\exp\left(\int_0^1 ds\,e^{-sA}Be^{sA}\right). \tag{X.4.5}$$

The theorem (X.4.1) or (X.4.5) can be extended to arbitrary functionals, which are multiplied by an exponential factor. An alternative form to (X.4.1) and its corresponding form (X.4.3) and (X.4.5) reads

$$\exp\left(\frac{1}{N}\sum_{s=1}^{N}(A_s+B_s)\right)=\exp\left(\frac{1}{N}\sum_{s=1}^{N}\bar{B}_s\right)\exp\left(\frac{1}{N}\sum_{s=1}^{N}A_s\right) \tag{X.4.6}$$

with

$$\bar{B}_s=\exp\left(\frac{1}{N}\sum_{s'=s+1}^{N}A_{s'}\right)B_s\exp\left(-\frac{1}{N}\sum_{s'=s+1}^{N}A_{s'}'\right). \tag{X.4.7}$$

We prove (X.4.1). The proof of the other theorems is then obvious. The proof can be done in the following steps: We introduce

$$G\equiv\exp\left(\frac{1}{N}\sum_s U_s^{-1}B_sU_s\right) \tag{X.4.8}$$

which we write according to the ordering convention in the form

$$\prod_{s=1}^{N}\exp\left(\frac{1}{N}U_s^{-1}B_sU_s\right). \tag{X.4.9}$$

We now use the relation

$$e^{U^{-1}CU}=U^{-1}e^{C}U \tag{X.4.10}$$

which holds for the arbitrary operators U and C (provided U^{-1} exists) and which can be proven by a power series expansion of the exponential functions on both sides.

This yields

$$G = \prod_{s=1}^{N} U_s^{-1} \exp\left(\frac{1}{N} B_s\right) U_s \tag{X.4.11}$$

or, written explicitly

$$U_N^{-1} e^{\frac{1}{N} B_n} U_N U_{N-1}^{-1} e^{\frac{1}{N} B_{n-1}} U_{N-1} \dots U_2 U_1^{-1} e^{\frac{1}{N} B_1} U_1 . \tag{X.4.12}$$

Because of

$$U_{j+1} U_j^{-1} = e^{\frac{1}{N} \sum_{s=1}^{j} A_s} e^{-\frac{1}{N} \sum_{s=1}^{j-1} A'_s} = e^{\frac{1}{N} A_j} \tag{X.4.13}$$

we obtain

$$\left.\begin{aligned} G &= \underbrace{e^{-\frac{1}{N}\sum_{s=1}^{N-1} A'_s} e^{\frac{1}{N} B_N}} \quad \underbrace{e^{\frac{1}{N} A_{N-1}} e^{\frac{1}{N} B_{N-1}}} \dots \underbrace{e^{\frac{1}{N} A_1} e^{\frac{1}{N} B_1}} \\ &= e^{-\frac{1}{N}\sum_{s=1}^{N} A'_s} e^{\frac{1}{N}(A_N + B_N)} e^{\frac{1}{N}(A_{N-1}+B_{N-1})} \dots e^{\frac{1}{N}(A_1+B_1)} \\ &= \exp\left(-\frac{1}{N}\sum_{s=1}^{N} A'_s\right) \exp\left(\frac{1}{N}\sum_{s=1}^{N} (A_s + B_s)\right). \end{aligned}\right\} \tag{X.4.14}$$

In passing from the first row to the second we have put

$$e^{\frac{1}{N} A_j} e^{\frac{1}{N} B_j} = e^{\frac{1}{N}(A_j + B_j)} \tag{X.4.15}$$

Since A_j and B_j do not commute [otherwise the theorem (X.4.1) becomes trivial], (X.4.15) does not hold exactly, but implies the limit $N \to \infty$. More exactly, the term linear in $1/N$ is kept, whereas the higher order terms are neglected.

The proof shows clearly in which sense the theorem (X.4.1) must be applied: The operators $\widetilde{B}_s$ and A_s are ordered within the individual exponential operators, but the operator $\exp\left(\frac{1}{N}\sum_{s=1}^{N} \widetilde{B}_s\right)$ must be applied prior to the total operator $\exp\left(\frac{1}{N}\sum_{s=1}^{N} A_s\right)$. In the next chapter we will meet a case in which the ordering prescription is only partly lifted. In that case, A and B should bear two indices s, s' and the disentangling theorem is applied to s, but the ordering convention is kept for s' for the *full* product on the right-hand side of Eq. (X.4.1).

X.5. Disentangling theorem for Bose operators[1]. We consider a problem which we are often confronted with in laser theory, namely the case when the field described by b^+, b is coupled to another system of particles, e.g. electrons. Thus the total Hamiltonian has the form

$$H = H_p + H'_{osc} + \hbar g^* b^+ + \hbar g b . \tag{X.5.1}$$

H_p is the particle Hamiltonian, H'_{osc} is given by $H'_{osc} = \hbar\omega b^+ b$. The coupling coefficients g^* and g may still be functions of the particle variables or operators. If we have the field coupled to classical sources, we drop H_p and treat the coupling constants g as given classical time-dependent functions.

[1] See R.P. Feynman, l.c.[1] on p. 299.

According to (X.3.7) the formal solution of the Schrödinger equation with the Hamiltonian (X.5.1) is given by

$$\psi(t)=\exp\left(\frac{1}{i\hbar}\int_0^t H_\tau d\tau\right)\psi(0). \tag{X.5.2}$$

Note that H need not depend explicitly on τ. Nevertheless, we use the (time-) ordering index τ in order to apply the disentangling theorem (see below). According to the operator-ordering convention, the exponential operator can be written in the manner

$$\exp\left(\frac{1}{i\hbar}\int_0 H_{p,\tau} d\tau\right)\underbrace{\exp\left(\frac{1}{i\hbar}\int_0 (H'_{osc}+\hbar g^* b+\hbar g b)_\tau d\tau\right)}_{U}. \tag{X.5.3}$$

We note that in all the following considerations the time ordering with respect to the particle operators, which occur in H_p and g, g^*, is kept throughout all the calculations*. We will show, however, that we can get rid of the time ordering of the Bose operators using the above derived disentangling theorem*. Using (X.4.3) and passing over from the dimensionless quantities s to time, we can write U occurring in (X.5.3) in the form

$$U=e^{-i\omega b^+ b t}\cdot\exp\left(-i\int_0^t (g_\tau^*\tilde{b}_\tau^+ + g_\tau\tilde{b}_\tau)\,d\tau\right), \tag{X.5.4}$$

where according to (X.2.12) and (X.2.11)

$$\tilde{b}_\tau=e^{i\omega b^+ b\tau}b_\tau e^{-i\omega b^+ b\tau}=b_\tau e^{-i\omega\tau}, \tag{X.5.5}$$

$$\tilde{b}_\tau^+=e^{i\omega b^+ b\tau}b_\tau^+ e^{-i\omega b^+ b\tau}=b_\tau^+ e^{i\omega\tau}. \tag{X.5.6}$$

Note that now the exponential function containing $i\omega b^+ bt$ acts after the other exponential function in (X.5.4). We now apply the disentangling theorem a second time, namely to the second exponential function in (X.5.4). We thus investigate

$$V=\exp\left(-i\int_0^t (g_\tau^* b_\tau^+ e^{i\omega\tau}+g_\tau b_\tau e^{-i\omega\tau}\right)d\tau. \tag{X.5.7}$$

According to the disentangling theorem (X.4.3) we find

$$V=\exp\left(-ib^+\int_0^t g_\tau^* e^{i\omega\tau}d\tau\right)\cdot\exp\left(-i\int_0 g_\tau\hat{b}_\tau e^{-i\omega\tau}d\tau\right) \tag{X.5.8}$$

with

$$\hat{b}_\tau=U_\tau^{-1}b_\tau U_\tau, \tag{X.5.9}$$

$$U_\tau=\exp\left(-ib^+\int_0^\tau g_\sigma^* e^{i\omega\sigma}d\sigma\right). \tag{X.5.10}$$

The application of the rule (X.2.7) yields

$$\hat{b}_\tau=b_\tau-i\int_0 g_\sigma^* e^{i\omega\sigma}d\sigma. \tag{X.5.11}$$

---

* Compare the remarks at the end of Chap. X.4. In the present context we do not explicitly distinguish between s and s'.

Putting our intermediate results (X.5.4), (X.5.8), (X.5.10) (X.5.11) together, we obtain the following result: The original exponential operator

$$\exp\left(\frac{1}{i\hbar}\int_0^t H_\tau d\tau\right) \equiv \exp\left(-i\int_0^t\left(\frac{1}{\hbar}H_{p\tau}+\omega(b^+b)_\tau+g_\tau^* b_\tau^+ +g_\tau b_\tau\right)d\tau\right) \tag{X.5.12}$$

can be written in the form

$$\left.\begin{aligned}&\exp\left(\frac{1}{i\hbar}\int_0^t H_{p,\tau}d\tau\right)\exp(-i\omega t b^+b)\times\\ &\quad\times\exp\left(-ib^+\int_0^t g_\tau^* e^{i\omega\tau}d\tau\right)\exp\left(-ib\int_0^t g_\tau e^{-i\omega\tau}d\tau\right)\times\\ &\quad\times\exp\left(-\int_0^t\int_0^\tau g_\tau g_\sigma^* e^{-i\omega(\tau-\sigma)}d\tau\,d\sigma\right).\end{aligned}\right\} \tag{X.5.13}$$

It is most remarkable here that b need no longer bear the time-ordering index, so that the exponential functions are quite simple with respect to the annihilation and creation operators b, b^+. If H_p and g, g^* depend on particle operators, the time ordering must still be observed with respect to these operators. If H_p does not appear in the original Hamiltonian and the g's are c-number functions, the time-ordering index t may be dropped, and only the usual time dependence of these functions need be kept.

We finally show what the formulas look like when the exponential function depending on the oscillator energy is not pulled out to the left-hand side but to the right-hand side. We start with the expression

$$U=\exp\left\{-i\int_0^t\left(\omega(b^+b)_\tau+g_\tau^* b_\tau^+ +g_\tau b_\tau\right)d\tau\right\}, \tag{X.5.14}$$

where g_τ, g_τ^* may still depend on further operators. The application of the disentangling theorem (X.4.6) and its extension to continuous order parameters yields

$$U=\exp\left\{-i\int_0^t\left(g_\tau^*\bar{b}_\tau^+ +g_\tau\bar{b}_\tau\right)d\tau\right\}\exp(-i\omega t b^+b) \tag{X.5.15}$$

with

$$\left.\begin{aligned}\bar{b}_\tau^+ &= \exp\left(-i\omega(t-\tau)b^+b\right)b^+\exp\left(i\omega(t-\tau)b^+b\right)\big|_\tau,\\ \bar{b}_\tau &= \exp\left(-i\omega(t-\tau)b^+b\right)b\exp\left(i\omega(t-\tau)b^+b\right)\big|_\tau.\end{aligned}\right\} \tag{X.5.16}$$

According to (X.2.11), (X.2.12) the expressions (X.5.16) reduce to

$$\left.\begin{aligned}\bar{b}_\tau^+ &= b_\tau^+ e^{-i\omega(t-\tau)},\\ \bar{b}_\tau &= b_\tau e^{i\omega(t-\tau)}\end{aligned}\right\} \tag{X.5.17}$$

so that we find

$$U=\exp\left\{-i\int_0^t\left(g_\tau^* b_\tau^+ e^{-i\omega(t-\tau)}+g_\tau b_\tau e^{i\omega(t-\tau)}\right)d\tau\right\}\exp(-i\omega t b^+b). \tag{X.5.18}$$

A comparison with formula (X.5.7) shows that (X.5.18) can be obtained from (X.5.7) by the following replacements

$$\left.\begin{aligned}g_\tau^* e^{i\omega\tau} &\to g_\tau^* e^{-i\omega(t-\tau)},\\ g_\tau e^{-i\omega\tau} &\to g_\tau e^{i\omega(t-\tau)}.\end{aligned}\right\} \tag{X.5.19}$$

This allows us to transcribe the result (X.5.13) to our present case. We then obtain

$$\left.\begin{aligned}&\exp\left\{-i\int_0^t\left(\frac{1}{\hbar}H_{p\tau}+\omega(b^+b)_\tau+g_\tau^*b_\tau^++g_\tau b_\tau\right)d\tau\right\}\\&\quad=\exp\left\{-i\int_0^t\frac{d\tau}{\hbar}H_{p\tau}\right\}\exp\left\{-ib^+\int_0^t g_\tau^*e^{-i\omega(t-\tau)}d\tau\right\}\times\\&\times\exp\left\{-ib\int_0^t g_\tau e^{i\omega(t-\tau)}d\tau\right\}\exp\left\{-\int_0^t\int_0^\tau g_\tau g_\sigma^*e^{-i\omega(\tau-\sigma)}d\tau\,d\sigma\right\}\times\\&\quad\times\exp\{-i\omega t b^+b\}.\end{aligned}\right\}\tag{X.5.20}$$

In practical applications often the expectation value with respect to the vacuum state must be calculated:

$$\left\langle\Phi_0^*\exp\left\{-\frac{i}{\hbar}\int_0^t H_\tau d\tau\right\}\Phi_0\right\rangle,\tag{X.5.21}$$

Because the application of the annihilation operator b to Φ_0 to the right-hand side and of b^+ to Φ_0^* on the left-hand side yields zero, we obtain

$$(\text{X.5.21})=\exp\left\{-\frac{i}{\hbar}\int_0^t H_{p\tau}d\tau-\int_0^t\int_0^\tau g_\tau g_\sigma^*e^{i\omega(\tau-\sigma)}d\tau\,d\sigma\right\},\tag{X.5.22}$$

Thus the particle coordinates (or operators) are now coupled directly to each other via the second term in the exponential. Due to the double integral over τ and σ, this interaction is retarded.

In other applications the statistical average must be evaluated in the following form

$$Z^{-1}\,\text{trace}\left\{\exp\left(-\frac{i}{\hbar}\int_0^t H_\tau d\tau\right)\exp(i\omega t b^+b)\exp\left\{-\frac{\hbar\omega b^+b}{kT}\right\}\right\}\tag{X.5.23}$$

with

$$Z=\text{trace}\left\{\exp\left\{-\frac{\hbar\omega b^+b}{kT}\right\}\right\}.\tag{X.5.24}$$

The additional operator $\exp(i\omega t b^+b)$ is chosen as usual so that it compensates $\exp(-i\omega t b^+b)$ which occurs during the disentangling process. We do not repeat the individual steps here but merely quote the final result [2]

$$\left.\begin{aligned}(\text{X.5.23})=\exp&\left\{-\frac{i}{\hbar}\int_0^t H_{p\tau}d\tau-\int_0^t\int_0^\tau g_\tau g_\sigma^*e^{-i\omega(\tau-\sigma)}d\tau\,d\sigma(n_{th}+1)-\right.\\&\left.-\int_0^t\int_\tau^t g_\tau g_\sigma^*e^{i\omega(\tau-\sigma)}d\tau\,d\sigma\,n_{th}\right\}\end{aligned}\right\}\tag{X.5.25}$$

where $n_{th}=\dfrac{1}{e^{\hbar\omega/kT}-1}$.

[2] H. Haken: Z. Physik **155**, 223 (1959) (We have corrected here for a misprint.)

Sachverzeichnis.

(Deutsch-Englisch.)

Bei gleicher Schreibweise in beiden Sprachen sind die Stichwörter nur einmal aufgeführt.

Subject Index.

(English-German.)

Where English and German spelling of a word is identical the German version is omitted.

Satz, Druck und Bindearbeiten: Universitätsdruckerei H. Stürtz AG, Würzburg